Applications in Plant Biotechnology
Focus on Plant Secondary Metabolism and Plant Molecular Pharming

Editors:

Abdullah Makhzoum

Department of Biological Sciences and Biotechnology
Botswana International University of Science and Technology
Palapye, Botswana

Kathleen Hefferon

Department of Microbiology
Cornell University, Ithaca, NY, USA

CRC Press
Taylor & Francis Group
Boca Raton London New York

CRC Press is an imprint of the
Taylor & Francis Group, an **informa** business
A SCIENCE PUBLISHERS BOOK

First edition published 2023
by CRC Press
6000 Broken Sound Parkway NW, Suite 300, Boca Raton, FL 33487-2742

and by CRC Press
4 Park Square, Milton Park, Abingdon, Oxon, OX14 4RN

CRC Press is an imprint of Taylor & Francis Group, LLC

Library of Congress Cataloging-in-Publication Data (applied for)

ISBN: 978-0-367-34446-7 (hbk)
ISBN: 978-1-032-39824-2 (pbk)
ISBN: 978-1-003-00886-6 (ebk)

DOI: 10.1201/9781003008866

Typeset in Times New Roman
by Radiant Productions

Preface

This book is the fruition of three years of effort made by all co-authors and editors. It summarizes recent innovative applications in plant biotechnology in two important areas: plant secondary metabolism and plant molecular pharming.

In the section on plant secondary metabolism, the first chapter focuses on recent innovations and discoveries using the medicinal plant species *Withania somnifera* as a pharma factory and its tremendous potential in the medical and pharmaceutical industries. The second chapter highlights several very promising biotechnological approaches and applications used to produce a number of important tropane alkaloids. The third chapter illustrates the applications and mechanisms of specific plant secondary metabolites as anticancer drugs and bioactive compounds, and their mode of action. The fourth chapter describes the *in silico* and computational analysis of African medicinal plants and reports the content of some databases and libraries as CamMedNP, ConMedNP, NANPDB, p-ANAPL, AfroDb, AfroCancer, AfroMalariaDb, Afrotryp, Ethiopia (ETM-DB), and South Africa (SANCDB). The fifth chapter discusses the ethno botanical uses of the *Strychnos* species found in the Southern African region and explores a variety of chemical compounds extracted from different *Strychnos* species, while addressing the antimicrobial activity and toxicity of these compounds. The sixth chapter explains the role of *Agrobacterium rhizogenes* as a vector in transforming medicinal plants and for inducing hairy roots. Hairy roots are a profound tool used in plant metabolic engineering to enhance secondary metabolite production and accumulation.

The second section of the book deals with plant molecular pharming and its applications. This section initiates from the seventh chapter, and explains different methods and tools used in plant molecular pharming as well the challenges ahead for the production of recombinant proteins, and potential solutions. The eighth chapter highlights some important livestock diseases and the potential of plant-based vaccines to protect against them. The go-to-market potential in addition to cost-effectiveness and other advantages of plant-based vaccines are presented in relation to a number of United Nations Sustainable Development Goals (SDGs). The ninth chapter considers recent advancements in engineering of chloroplasts, which can serve as a target organelle for the production of biomolecules or the improvement of agronomic traits. The chloroplast has been extensively explored and engineered to produce valued molecules such as enzymes, secondary metabolites, protein drugs, vaccines, anti-microbial peptides, and interfering RNAs for over 30 years. Chapter ten addresses the applications of plant viruses for the expression and synthesis of vaccines against several infectious human diseases and cancers as well as other therapeutic proteins. This chapter also discusses various investigative reports using well-known and important plant viruses such as the Tobacco mosaic virus, Cowpea mosaic virus and Potato virus X, which have been employed for biotechnological, prophylactic and therapeutic purposes. The eleventh chapter discusses the use of plant molecular pharming to manufacture biologics against HIV. It addresses progress in the development of plant-derived HIV vaccines, the risks and challenges associated with vaccine production in plants, as well as the benefits of vaccine production in plants compared to other expression systems.

In summary, this book covers two increasingly growing fields of plant biotechnology, by drawing on some of the most prominent research groups in their respective fields as co-authors. The research described in this book should be helpful for any who are interested in both current and future aspects of plant biotechnology for many years to come.

Contents

Acknowledgment of Reviewers

Reviewers

Dr. Marina Clemente
Instituto Tecnológico de Chascomús (INTECH)
CONICET-UNSAM
Intendente Marino Km 8,2; CC 164 (B7130IWA)
Chascomús; Provincia de Buenos Aires; Argentina
Tel:+54-2241-430323; FAX:+54-2241-424048
E-mail: marinaclemente@hotmail.com
https://intech.conicet.gov.ar/
www.iib.unsam.edu.ar/iib-unsam/investigacion.html

Professor Sumita Jha
Department of Botany, University of Calcutta'35 Ballygunge Circular Road, Kolkata 700019, India

Professor Sergio Rosales Mendoza
Faculty of Chemical Sciences
Center for Research in Biomedicine and Health
Office Address: Manuel Nava 6 CP. 78210 San Luis Potosi
Telephone: (+52) 4448262440
Email: rosales.s@uaslp.mx
Website: https://www.researchgate.net/profile/Sergio_Rosales-Mendoza
ORCID: 00-0003-2569-7329

Dr. Mohammad Tahir Waheed
Assistant Professor
Course Coordinator
Department of Biochemistry
Quaid-i-Azam University
45320, Islamabad, Pakistan
Member: International Society of Plant Molecular Farming (ISPMF)
Member: European Plant Science Organization (EPSO)
Member: Asian Council of Science Editors (ACSE)
Associate Editor:
1. BMC Plant Biology
2. BMC Developmental Biology
3. American Journal of Biochemistry and Biotechnology
Tel: 0092-51-90643207

Cell: 0092-333-9964114
https://scholar.google.com.pk/citations?user=5cYbyykAAAAJ&hl=en&oi=ao
https://www.researchgate.net/profile/Mohammad_Waheed2
https://bmcplantbiol.biomedcentral.com/about/editorial-board
https://bmcdevbiol.biomedcentral.com/about/editorial-board
https://thescipub.com/journals/ajbb/editors

Dr. Mohamed Elhiti
Associate Professor
Department of Botany, Faculty of Science,
Tanta University, Tanta, 31527, Egypt

Professor Ed Rybicki
Director, Biopharming Research Unit
Molecular & Cell Biology Department
University of Cape Town

Dr. Sarah Bushra Nasir
Abdus Salam School of Sciences,
Department of Life Sciences,
Nusrat Jahan College,
Chenab Nagar, Pakistan.

Abdulbaset Azizi
Department of Plant Protection
University of Kurdistan,
Sanandaj, Iran

Dr. Emmanuel Margolin
Postdoctoral Scientist
Bioharming research Unit (Department of Molecular and Cell Biology), University of Cape Town
And Viral Vaccine Development Group and Human Papillomavirus Research Group (Department of Pathology), Institute of Infectious Diseases and Molecular Medicine, Faculty of Health Scienc-es, University of Cape Town

Dr. Mohammed Kamil Sherif
Central Medical Centre
Palapye, Botswana
E-mail: centralmedicentre@gmail.com

I-Plant Secondary Metabolism

CHAPTER 1

Withania somnifera: A Future Pharma Factory

Tarun Halder, Subrata Kundu and *Biswajit Ghosh**

Introduction

Humans have used plants accessible within their territory for diet and curative purposes since their civilization began. This traditional system of medicine based on belief and practices has been used as a therapy for most human ailments for over hundreds of centuries (Winters 2006, Kulkarni and Dhir 2008, Mirjalili et al. 2009a, Rayees et al. 2012, 2013). Irrespective of the progression in drug development methods, several species of plants are unparalleled as major ingredients of current medicine. Due to our existing industrialized lifestyle, we are continuously exposed to an over abundance of chemicals and different environmental pollutants that ultimately intensified the incidence of complicated neurodegenerative, cardiovascular diseases, and cancers. Perceptible antagonistic effects of modern drugs used in the management of complicated ailments ultimately resulted in renewed attention towards herbs and medicinal plants as an alternative source.

Withania somnifera (L.) Dun, is one of the important medicinal plants in India belonging to the Solanaceae family. It is commonly known as 'Ashwagandha', Winter Cherry, Indian ginseng. In the monographs of the World Health Organization (WHO), Ashwagandha has been considered as one of the important medicinal plants and also has been included in the list of top thirty two prime concerned medicinal plants by the National Medicinal Plant Board of India (http://www.nmpb.nic.in) owing to its huge demand in both domestic and international markets (Mirjalili et al. 2009a, Singh et al. 2015). In Ayurveda and other traditional systems of medicine, it is one of the most valued medicinal plants and it has been in use for more than 3000 years. It is widely used in traditional Indian medicine systems for curing a variety of ailments. It

Plant Biotechnology Laboratory, Post Graduate Department of Botany, Ramakrishna Mission Vivekananda Centenary College, Rahara, Kolkata-700118, India.
Emails: tarunhalder13@gmail.com; subratakundu83@gmail.com
* Corresponding author: ghosh_b2000@yahoo.co.in

possesses adaptogenic, tonic analgesic, antipyretic, anti-inflammatory, and abortifacient properties and is one of the most extensively used plants in various systems of medicine (Chopra et al. 1958). There are several reports along with numerous clinical trials that support the use of *W. somnifera* for hepatotoxicity (Bhattacharya et al. 2000a), anxiety (Bhattacharya et al. 2000b), cognitive (Bhattacharya et al. 1995), neurological disorders (Kuboyam et al. 2005, Pandey et al. 2018), inflammation (Al-Hindawi et al. 1992, Noh et al. 2016), hyperlipidemia (Visavadiya and Narasimhacharya 2007) and Parkinson's disease (Ahmad et al. 2005). The leaves are also known to act as an insect repellent (Schmelze et al. 2008). The steroidal lactones known as withanolides (a group of biologically active oxygenated ergostane type steroidal lactones) are found both in leaves and roots parts of the plant (Kaushik et al. 2017). These compounds have been intensely investigated because of their pronounced anti-tumor properties and novel steroidal structure. Alkaloids constitute another major group of components that have been isolated from *W. somnifera.* Some alkaloids have also been isolated from the roots of *W. somnifera*, among them, withanine is the main alkaloid comprising 38% of the total alkaloid material (Atal et al. 1975). The major withanolides withaferin A, withanolide A show antitumor and cytotoxic activities (Mondal et al. 2010, Siddique et al. 2014, Kuboyama et al. 2014). *W. somnifera* is predominantly propagated through seeds (Rao et al. 2012). Its traditional cultivation has been limited due to a low percentage of seed viability, poor germination and seedling survival, and low yield of withanolides from natural sources (Vakeswaran and Krishnasamy 2003). In addition, infestation with pathogens and pests poses a serious challenge in its commercial cultivation and improvement (Sharma et al. 2011). In recent years, there is enormous demand for the enhanced synthesis of pharmacologically important metabolites through biotechnological interventions (Singh et al. 2015, Ray et al. 2019). Unfortunately, the concentration of pharmacologically important withanolides in *W. somnifera* is quite low, ranging from 0.001 to 0.5% of dry weight (DW) (Mirjalili et al. 2009b). Nevertheless, the chemical synthesis of withanolides is unwieldy and there is limited information on biosynthesis and regulation of secondary metabolites. In this regard, cell or organ cultures are found to be of immense potential for the mass production of secondary metabolites. Cell suspensions usually synthesize metabolites at a faster rate owing to their uniform and active growth. Apart from these techniques, genetic transformation (both *Agrobacterium tumefaciens* and *A. rhizogenes* mediated transformation) has emerged as a powerful tool for engineering the plants for overexpressing desired metabolites and to decipher molecular functions of selected genes. The production of withanolides in the plant could be monitored through seasonal changes or growth periods. In this chapter, we have analyzed detailed exploration on this plant carried out by different researchers from various locations globally and their remarkable influences on research from some of the available literature to illuminate the pharmacological importance of *W. somnifera.*

Historical Background

W. somnifera has been widely used as a remedy in the Ayurvedic system of medicine in India that can be traced back to the years' BC (Singh et al. 2011). In Sanskrit, ashwagandha means "horse's smell", probably originating from the odor of its root that resembles the sweat of horses. The species name '*somnifera*' means "sleep bearing" in Latin. Traditional uses of Ashwagandha among tribal peoples in Africa include fevers and inflammatory conditions. In the Indian Ayurvedic system of medicine, this plant is described as a powerful rejuvenating herb. This plant was labeled in ayurvedic text, i.e., the "Charaka" and the "Sushruta Samhitas". Nevertheless, the root of Ashwagandha has been regarded as a tonic, aphrodisiac, narcotic, diuretic, anthelmintic, astringent, thermogenic, and stimulant. Due to its Ayurvedic history, it has been used as a "Rasayana", a treatment that releases emotional tension and physical discomfort, and creates a foundation of wellness and stability in the body. Ashwagandha has been used for thousands of years to target everything from the negative effects of stress in the human body to strengthening the immune system throughout the seasons. In Africa, traditional uses of Ashwagandha among tribal peoples include fevers and inflammatory diseases. In Yemen, the dried leaves are ground into a paste that is used for treating burns and wounds, as well as for sunscreen. For external healing, the berries and leaves have been applied to tumors, tubercular glands, carbuncles, and ulcers. Indeed, Ashwagandha has become the backbone of many multi ingredient Ayurvedic and other systems of medicine that have been used for a wide assortment of ailments.

Different Plant Parts Use for Medicinal Purpose

The leaves of *W. somnifera* are a little bitter and are recommended to treat fever and painful swellings. The crude preparation of the plant has also been found to be very active against several pathogenic bacteria. Sore eyes, ulcers, and swellings can be cured using a fomentation of the leaves (Table 1). The leaves are also used as a compelling and an anthelmintic, can be crushed of the tissue and applied to tumors and ulcers, it is also consumed as a vegetable and used as livestock fodder (Kirtikar and Basu 1991). Other uses of the leaves are to heal open as well as septic and inflamed wounds and abscesses and to treat inflammation, hemorrhoids, rheumatism, and syphilis. Major active compounds of withanolides like withaferin A are found mostly in the leaves of this plant, first of all, withaferin A under the withanolides group have been elucidated from the leaves by Lavie's group (Lavie et al. 1965). Israel chemotypes of Ashwagandha, production of withaferin A from the intact plants are about 0.2–0.3% based on DW of leaves (Abraham et al. 1968) compared to Indian chemotypes and were compared with withaferin A based on DW and it is absent in roots, stems and fruits (Gupta et al. 1996). The common effects of *W. somnifera* root extracts have cumulative properties on humoral immune responses. In the animal blood system, the roots are accomplished to increase the number of white blood cells. It helps to relieve insomnia by the processing of nervous cells and possesses a slight tranquilizing activity that promotes sound sleep. It also helps regulate long term blood sugar levels and is also a beneficial treatment for weight loss. The roots of *W. somnifera* are a natural alterative, aphrodisiac, removing obstructions, diuretic, narcotic, sedative, and restorative. According to different levels of experiments based on research reports, Ashwagandha roots powder can be an effective herbal supplement from the traditional as well as a modern system of medicine for the treatment of various types of diseases. Ashwagandha root is a popular male sex tonic in India. It helps to cure erectile dysfunction and improve male sperm count. It also helps to reduce bad cholesterol levels which are responsible for hypertension and cardiovascular problems. Recent studies revealed that the ashwagandha root contains steroidal properties which can be effective in treating inflammation. It is also used to treat low back pain and sciatica. It is generally safe to use, as it is purely natural and free from side effects. The

Table 1: Extracts of different plant organs of *W. somnifera* extracted by various solvents and their different pharmacological activities.

Plant parts	Extract types	Action on target cells	References
Root	Chloroform	Anti-cancer (Liver, breast, colon), Prostate cancer	Siddique et al. 2014
	Chloroform	Against chemically induced Fatigue	Biswal et al. 2013
	Alcoholic	Anti-tumor	Khazal et al. 2013
	Alcoholic	Antimicrobial	Khanchandani et al. 2019
	Aqueous	Antimicrobial	Kumari et al. 2020
	Ethanolic	Anti-cancer (Breast cancer)	Maliyakkal et al. 2013
	Ethanolic	Anti-cancer (Prostate cancer)	Kim et al. 2020
	Ethanolic of N-118	Neuroprotectant (Against cerebral stroke)	Ahmad et al. 2015
	Ethanolic of 101R, 118R and 128R	Immunomodulatory	Kushwaha et al. 2012a, Kushwaha et al. 2012b
	Ethanolic	Anti-cancer (Cervical)	Jha et al. 2014
	Aqueous	Anti-cancer (Cervical)	Nile et al. 2019
	Aqueous	Anti-stress (reduced in T-cell population and up-regulated Th1 Cytokines in chronically stressed mouse)	Khan et al. 2006
	Aqueous	Anti-stress	Jain and Saxena 2009
	Aqueous	Alzheimer's disease	Sehgal et al. 2012
	Aqueous	Neuroprotective (Enhanced memory and attenuated hippocampal neurodegeneration through inducing glutathione biosynthesis)	Baitharu et al. 2013, 2014
	Methanolic	Hepatoprotective role in acetaminophen-intoxicated rats	Devkar et al. 2016
	Methanolic	Parkinson's disease	De Rose et al. 2016
	Methanolic	Neuroprotective effects	Bhattarai and Han 2014
Leaf	Methanolic	Anticancer (TIG1, U2OS, and HT1080) by activating p53, apoptosis pathway & cell cycle arrest)	Widodo et al. 2008
	Methanolic	Hypoglycemic (Increased uptake of glucose in myotubes and adipocytes)	Gorelick et al. 2015
	Methanolic	Neuroprotection through activation of neuronal proteins, oxidative stress and DNA damage	Konar et al. 2011

Table 1 contd. ...

...Table 1 contd.

Plant parts	**Extract types**	**Action on target cells**	**References**
	Methanolic	Anticancer (Neuroblastoma)	Kuboyama et al. 2014
	Methanolic	Antimicrobial (against methicillin resistant *Stephylococcus aureus* and *Enterococcus* sps.)	Bisht and Rawat 2014
	Methanolic	Antiproliferative (against MCF-7, HCT116 and HepH2 cell lines)	Alfaifi et al. 2016
	Methanolic	Antimicrobial activity	Dhiman et al. 2016
	Methanolic	Antibacterial (triagainst *Salmonella typhi*) and Antioxidant	Alam et al. 2012
	Ethanolic	Antimicrobial activity	Dhiman et al. 2016
	Hydroalcoholic	Anti-cancer (Breast Cancer)	Nema et al. 2013
	Aqueous	Anti-cancer	Wadhwa et al. 2013
Stem	Methanolic	Cytotoxic activity against cancer cell lines MDA-MB-231 (Human breast cancer cell lines)	Srivastava et al. 2015
	Ethanolic	Cytotoxic activity against Cancer cell lines MDA-MB-231 (Human breast cancer cell lines)	Srivastava et al. 2015
Fruit/ berry	Ethanolic	Alzheimer's disease	Jayaprakasam et al. 2010
	Methanolic	Antioxidant and antibacterial activities	Alam et al. 2012

root contains an alkaloid somniferine. Roots are also used to prepare the herbal remedy of ashwagandha, which has been traditionally used to treat various symptoms and disorders. The roots of the plant are categorized as Ramayana's, which are reputed to promote health and longevity by augmenting defense against disease, arresting the aging process, revitalizing the body in debilitated conditions, increasing the capability of the individual to resist adverse environmental factors and creating a sense of mental wellbeing. The medicinal importance of bioactive constituents from different parts of *W. somnifera* are represented in Table 1.

Pharmacological Importance

Pharmacology is the science of drugs, including the study of those substances that interact with living systems through chemical processes, by binding to regulatory molecules and activating or inhibiting normal body processes. Although all the development in synthetic chemistry and drug development progresses, plants are still a significant source of modern medicinal preparations. Many species of plants are known for their medicinal values and are used to treat several human ailments and diseases. Medicinal properties of these plants are linked with their roots, seeds, flowers, leaves, fruits, or the whole plant itself. It has been mentioned that the medicines from plants are better in bioactivity and are found to possess curative properties against various infectious and non-infectious diseases when compared to chemical and synthetic medicines (Itokawa et al. 2008, Ezzat et al. 2019). Therefore the bioactive compound is the fundamental molecule that acts like a drug. These bioactive compounds are mostly derived from plant secondary metabolites, and many naturally occurring pure compounds have become medicines. Active lead compounds can also be further modified to enhance the biological profiles and developed as clinical trial candidates (Itokawa et al. 2008). It is fact that natural plant product research and development (R&D) potentially plays a vital innovative role in drug discovery.

The plentiful therapeutic applications of *W. somnifera* are linked to the occurrence of natural phytochemicals primarily withanolides and also some other secondary metabolites, which can be found at different levels in various plant parts like roots, stems, and leaves and are together responsible for the different pharmacological activity. Several withanolides have been obtained and characterized over the past five decades. This type of steroid has attracted significant attention from numerous researchers, not only because of its complex structural features but because of its multiple bioactivities and potential in drug research and development (Ichikawa et al. 2006, Misra et al. 2008, Chen et al. 2011, Dar et al. 2015, Ahmad and Dar 2017). More than 12 alkaloids, 40 withanolides, and quite a few sitoindosides have been reported from this plant species (Mishra et al. 2000, Mirjalili et al. 2009a). Withanolides are a group of steroidal lactones with atoms C-22 and C-26 bridged by D-lactone functionality and are considered potential active pharmacological compounds specific to this species (Abou-Douh 2012). The chemical diversity among different withanolides is created due to glycosylation, hydroxylation or the creation of additional rings along with the addition of side chains in the steroid backbone (Fig. 1). Ichikawa et al. (2006) and Misra et al. (2008) reported that the therapeutic potential of the species *W. somnifera* is mainly ascribed towards the presence of several withanolides, especially the noted compound withaferin A and withanolide A. These withanolides have demonstrated significant pharmacological activities (Fig. 2). *Withania somnifera* contains a spectrum of varied bioactive secondary metabolites which allows it to have a wide range of biological importance. In preclinical studies, it has revealed antimicrobial, anti-inflammatory, anti-tumor, anti-stress, neuroprotective, cardio-protective, and anti-diabetic properties. Moreover, it has confirmed the ability to reduce reactive oxygen species, modulate mitochondrial function, regulate apoptosis, reduce inflammation and improve endothelial function (Dar et al. 2015).

W. somnifera contains a spectrum of diverse phytochemicals enabling it to have a broad range of biological implications. In pre-clinical studies, it has shown antimicrobial, anti-inflammatory, antitumor, anti-stress, neuroprotective, cardio-protective, and anti-diabetic properties. Additionally, it has demonstrated the ability to reduce reactive oxygen species, modulate mitochondrial function, regulate apoptosis, and reduce inflammation and enhance endothelial function. In view of these pharmacologic properties, *W. somnifera* is a potential drug candidate to treat various clinical conditions, particularly related to the nervous system. In the field of plant science, *W. somnifera* has always been the center of attraction within the research community for various pharmacological activities such as anti-diabetic, cardio-protective, antimicrobial, anti-inflammatory, neuroprotective, etc. (Tripathi and Verma 2014, Dhar et al. 2015). Furthermore, it has the potentiality to cure a wide range of cancer cells such as breast, colon, cervical, lung, prostrate, etc. (Fig. 3). The utilization of active compounds on cancer cells acts as multiple modes of action on target cells in different ways such as induced cell cycle arrest, apoptosis, cytotoxicity, inhibition of

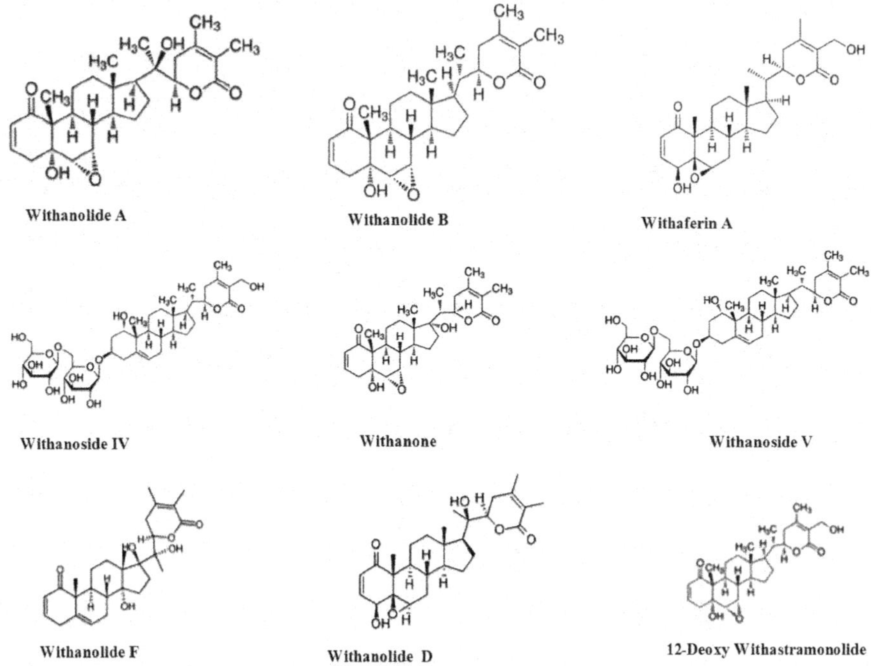

Figure 1: Chemical structure of various withanolides from *W. somnifera*.

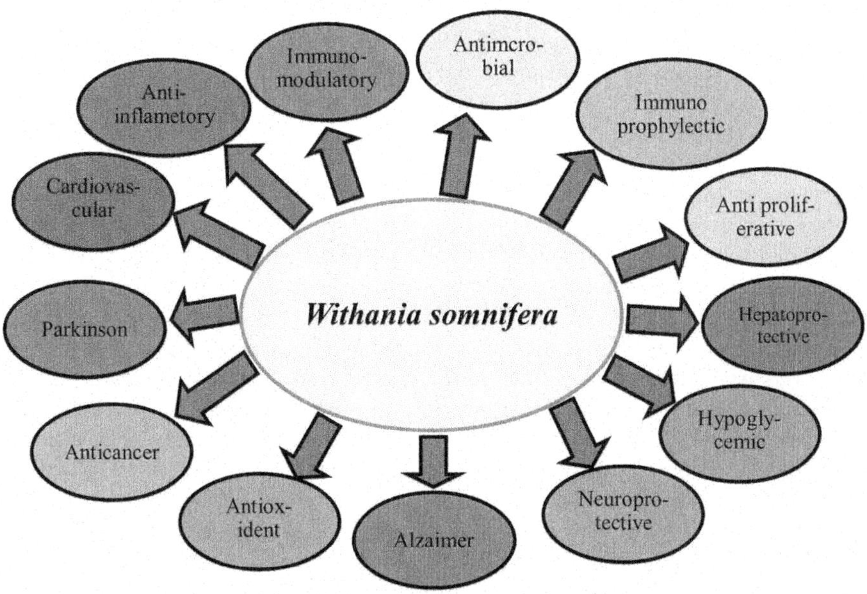

Figure 2: Different pharmacological activities of *W. somnifera*.

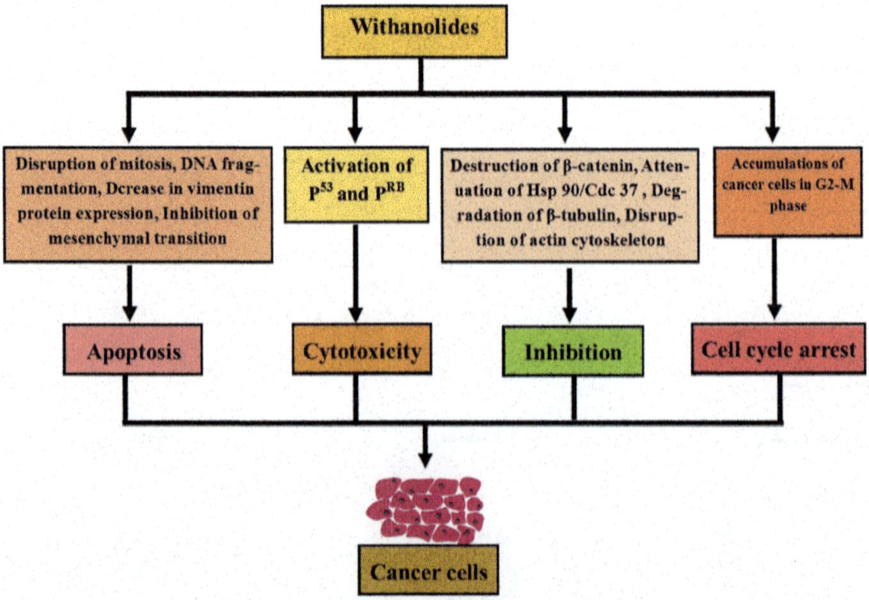

Figure 3: Mechanisms of action of withanolides on cancer cells.

angiogenesis and also overwhelming various oncogenic pathways (Palliyaguru et al. 2016). Adaptogenic or the stress response regulating effects of *W. somnifera* has been reported as a pharmacological description in the 1980s (Kumar et al. 2017). Since then, several observations of this plant have revealed ginseng like effects on various animal models, now it is often referred to as Indian ginseng (Kulkarni and Dhir 2008).

Apart from the medicinal properties and various preclinical studies performed on *W. somnifera* or its ingredients, for different pharmacological actions, several clinical studies have also been done by various authors (Table 2). *W. somnifera* was found to recover the seminal plasma levels of antioxidant enzymes, vitamins A, C, and E and corrected fructose in infertile men. Moreover, the treatment also significantly increased serum T and LH and reduced the treatment of male sexual dysfunction and infertility levels of FSH and PRL, which is useful to evaluate the spermatogenic activity indication of healthy semen oligospermic patients (Ahmad et al. 2010, Ambiye et al. 2013). It was studied by different researchers on infertile men where the rate of sperm apoptosis and cytoplasmic ROS levels were significantly higher on the different age groups in comparison with control subjects, likewise, the seminal plasma levels which contain the essential metal ions like Zn^{2+}, Fe^{2+}, Cu^{2+} and Au^{2+} were low. In another case, a double blind, placebo controlled study performed on 50 patients with ICD-10 anxiety disorders, it was observed that by the end of the first month of treatment with an ethanol extract of *Withania* 250 mg twice a day, 72% of patients showed moderate to excellent improvement and in about half of these cases, benefits were observed within the first fortnight (Andrade et al. 2000). The group of patients who received naturopathic therapy involving deep breathing relaxation technique, a standard multi vitamin and *W. somnifera* (300 mg b.i.d.), were found to recover effectively (Cooley et al. 2009). A study conducted on breast cancer patients undergoing combination chemotherapy with oral *W. somnifera* (2 g every 8 h)

Table 2: Pharmacological importance of different withanolides.

Sl. No.	Activity	Bioactive compounds	References
1	Anticancer	Withanolide A, Withanolide D	Jayaprakasam et al. 2003, Mandal et al. 2008, Aalinkeel et al. 2010
		Withalongolide A	Subramanian et al. 2014
		Withalongolide A-4,19,27-triacetate	Mondal et al. 2010, kuai et al. 2017
		Withanolide A, Withanoside IV, Withanoside VI	Kuboyama et al. 2014, Grin et al. 2012
		5,6-de-Epoxy-5-en-7-one-17-hydroxy withaferin A	Siddique et al. 2014
		Withaferin A, carnosol	Aliebrahimi et al. 2018, Zhang et al. 2012, Stan et al. 2008, Fong et al. 2012, Panjamurthy et al. 2009, Samadi et al. 2012, Thaiparambil et al. 2011
2	Hypoglycemic	4 β-hydroxywithanolide E	Takimoto et al. 2014
		Withaferin A	Gorelick et al. 2015
3	Anti microbial	17βhydroxywithanolide K	Choudhary et al. 2010
		Withaferin A	Subramanian and Sethi 1969
		14,15β-epoxywithanolide I	Choudhary et al. 2010
		Withanolide F	Choudhary et al. 2010
		Withanolide D	Choudhary et al. 2010
		Withanoside-IV, Withanoside-V, Withaferin-A, 12-deoxy Withastramonolide, Withanolide-A, and Withanolide-B	Caputi et al. 2018
4	Anti-inflammatory	Withaferin A	Noh et al. 2016
		3β-hydroxy-2,3-dihydrowithanolide F	Budhiraja et al. 1984
		Withanone	Pandey et al. 2018
		Sitoindoside VII and VIII	Mishra et al. 2000
		Withanolide D	Pawar et al. 2011
5	Antioxidant	Withaferin A	Bhattacharya et al. 1997, Bhattacharya and Muruganandam 2003
6	Neuroprotection	Withanolide A	Kataria et al. 2012, Kurapati et al. 2013, Baitharu et al. 2014
		Withanolide A, Withanoside IV, and Withanoside VI	Zhao et al. 2002

Table 2 contd. ...

...Table 2 contd.

Sl. No.	Activity	Bioactive compounds	References
		Withanoside IV	Kuboyama et al. 2006
		Withanone	Pandey et al. 2018
7	Immunosuppressive	Withaferin A	Shohat et al. 1978
		Withanolide E	Shohat et al. 1978
		5,20α(R)-dihydroxy 6α,7α-epoxy-1-oxo-(5 α)-with a 2,24 dienolide	Bahr and Hansel 1982
8	CNS related	3β-hydroxy-2,3-dihydrowithanolide F	Budhiraja et al. 1984
9	Alzheimer's disease	Withanolide A	Sehgal et al. 2012
		Sitoindosides VII–X	Bhattacharya et al. 1995
		Withanoside I	Kuboyama et al. 2006
		Withanoside	Kuboyama et al. 2005, 2006
		Withanolide sulfoxide	Mulabagal et al. 2009
10	Cardiovascular	3β-hydroxy-2,3-dihydrowithanolide F	Budhiraja et al. 1984
		Withaferin A	Ravindran et al. 2015
11	Hepatoprotective	3β-hydroxy-2,3-dihydrowithanolide F	Budhiraja et al. 1986
12	Amnesia	Sitoindosides VII–X, and withaferin-A	Schliebs et al. 1997
13	Poor immunity	Withanolide A	Bani et al. 2006
14	Stress relief	Sitoindosides VII and VIII	Bhattacharya and Muruganandam 2003

showed a notable and remarkable anticancer activity. Another study with the inspirations of neural health parameters after the *in vitro* culture of mild TBI, the culture model treated traumatically injured cells like SH-SY5Y human neuroblastoma cells, which were used as a useful model for analysis of neuroprotective treatments by *Withania* extract. The pharmacological importance is given below in Table 2. Ashwagandha is supported by many poly herbal preparations against various diseases, some available drugs found from the market which totally based on withanolides (Table 3).

Antimicrobial Activity

Several studies have been conducted in different biological models and it was revealed that the leaves and roots of *W. somnifera* have potent antimicrobial activity. It has been reported that leaf extract (6.25 mg/ml and 12.5 mg/ml) inhibited the growth of five Gram-negative pathogenic bacteria (*Escherichia coli, Salmonella typhi, Citrobacter freundii, Pseudomonas aeruginosa,* and *Klebsiella pneumonia*) (Alam et al. 2012) and some Gram-positive clinical isolates of methicillin-resistant *Staphylococcus aureus* and *Enterococcus* spp. (Bisht and Rawat 2014). In another study, the crude leaf

Table 3: List of some available commercial pharmaceutical products from *W. somnifera* for therapeutic uses.

Sl. No	Product name	Manufacturer	Use of the products
1	100% Natural *W. somnifera* extract Alkaloids, Withanolides Ashwagandha extract	Xi'an Saina Biological Technology Co., Ltd.	Antiallergy, antihistamine, antipyretic, pain-relieving, local anesthetic, antibacterial
2	Amaybion	Aimil Pharmaceuticals Pvt. Ltd	Anti-ageing
3	Amrutha Kasthuri	Pankajakasthuri Herbals India Ltd.	Convalescence, recuperation from general debility, neurasthenia, and better disease resistance
4	Amry-gel	Aimil Pharmaceuticals Pvt. Ltd	Health-Tonic
5	Arshadi pills	Dehlvi Remedies	Stress, depression, cardiac tonic
6	Articulin-F	Eisen Pharmaceuticals Pvt. Ltd	Useful in rheumatoid Arthritis
7	Ashvagandha	Morpheme Remedies	Combating stress
8	Ashvagandha	The Himalaya Drug Co.	Relieves Chronic stress related physiological abnormalities
9	Ashwagandh tablets	Himalyan Drugs Co.	Anti-stress
10	Ashwagandha	Ayurceutics	Stress reliever
11	Ashwagandha extract	Nanjing Zelang Medical Technology Co., Ltd.	used in sedative and pain relief, lowering blood pressure, treat chronic inflammation, improve sexual function, etc.
12	Ashwagandha pills	Herbal Hills	Stress and revitalizer
13	Ashwagandha/*W. somnifera* extract	Wuxi Gorunjie Natural-Pharma Co., Ltd.	Anti-inflammatory, Anti-arthritic, Stress reliever
14	Ashwagandhahills	Herbal Hills	Stress and revitalizer
15	Ashwagandharista	Baidynath Ayurved Bhawan	Nerve tonic, memory and cognition improvement, better power of concentration, relieves mental tension, natural sleep induction, and recovery from nervous and general debility
16	Aswal Plus	Gufic Ltd.	Ageing, Anti-stress, Adaptogenic, Rejuvenating agent
17	Brento	Zandu Pharmaceutical Works Ltd.	Nerve tonic, Combats stress
18	Dabur Ashwagandha cooleyrna	Dabur	Combating stress
19	Geriforte (tablet and syrup)	The Himalaya Drug Co.	Stress relief and relief from insomnia

Table 3 contd. ...

...Table 3 contd.

Sl. No	Product name	Manufacturer	Use of the products
20	Gestone	Zandu Pharmaceuticals works Ltd	Preeclampsia, Placental insufficiency, Threatened abortion, and pregnant anemia
21	Himalaya Ashwagandha	The Himalaya Drug Co.	Stress management
22	Himalaya massage oil	The Himalaya Drug Co.	Body relaxation, stress relief, and relief from insomnia, backache
23	Imunocin	Nukem Remedies Ltd.	Immunomodulator
24	Keshari Kalpa	Baidyanath Ayurved Bhavan Ltd.	Aphrodisiac, Health tonic
25	Lactare	TTK Pharma Pvt. Ltd.	Improved lactation
26	Lacton	Baidyanath Ayurved Bhavan Ltd.	Improved lactation
27	Lovemax	BACFO Pharmaceuticals Ltd	Vigour and vitality promotion
28	Medispermina	Mesi Products (p) Pvt. Ltd.	Azoospermia, Oligospermia, Impotence, Infertility
29	Mentat and Mentat DS	The Himalaya Drug Co.	Memory and learning disorders, behavioural disorders, ADHD, anxiety and stress related disorders, mental fatigue and an adjuvant in AD and PD
30	MindCare and MindCare Jr	The Himalaya Drug Co.	Mental alertness, mental fatigue and occasional irritability
31	Mustong	TTK Pharma Pvt. Ltd.	Revitaliser
32	Natural 80 mesh American ginseng root extract	Qingdao Fraken International Trading Co., Ltd.	Withanolides possess remarkable antibacterial, antitumor, antiarthritic, anti-inflammatory, and immunosuppressive properties
33	Nutramax *W. somnifera* extract (10:1)	Hunan Nutramax Inc.	Anti-stress, hypotensive, antispasmodic, bradicardic, respiratory stimulant activities
34	Nutramax-AE (100%)	Hunan Nutramax Inc.	Antiallergy, Antihistamine, Antibacterial
35	One Be	Lupin Lab Ltd.	Stress, Ageing, Adaptogen, Rejuvenator, Immuno modulator
36	Sioton	Albert David Ltd.	Improved libido, Sexual fatigue
37	Stir	Targof Pure Drug Ltd.	Stress, Male subfertility, Oligospermia, Adaptogen, Neurostimulator
38	StressCare	The Himalaya Drug Co.	Supports cortisol levels

Table 3 contd. ...

...Table 3 contd.

Sl. No	Product name	Manufacturer	Use of the products
39	Stresscom	Dabur India Ltd.	Relieves anxiety neurosis, physical and mental stress, and relieves general debility and depression
40	Stresswin	Baidynath Ayurved Bhawan	Combating exertion, reduction in anxiety, strain, and stress, improvement of stamina, relief from disturbed sleep, mental alertness
41	Vigomax	Charak Pharmaceuticals Pvt., Ltd.	Vigour and vitality enhancement
42	Vigorex SF	Zandu Pharmaceuticals works Ltd	Stress reliever, Improves strength
43	Vital plus	Mukthi Pharma	Recovery from impotence, oligospermia, and recovery from general weakness, fatigue
44	Amul Memory Milk	Amul	Reduces anxiety, stress

* Sources from Sangwan et al. 2004, Bharti et al. 2016 and Mandlik and Namdeo 2020

extracts of *W. somnifera* were tested against some pathogenic bacteria and it was found that 100 µl of extracts (100 mg/ml) efficiently inhibits the growth of all pathogenic bacteria (Pandit et al. 2013). It is also reported that zinc oxide nanoparticle synthesis from the leaf extract, 100 µg/ml nanoparticle showed a significant antibacterial effect on *Enterococcus faecalis* and *Staphylococcus aureus* (Malaikozhundan et al. 2020). Methanolic leaf extract of *W. somnifera* has shown anti-bacterial activity against Gram-positive clinical isolates of methicillin-resistant *Staphylococcus aureus* and *Enterococcus* spp. The mechanism of antimicrobial activity of plant extracts acknowledged cytotoxicity, gene silencing, and immunopotentiation, it is also reported that the dichloromethane and ethyl acetate extracts of *W. somnifera* exhibits a good bacteriocidal, and fungicidal activity against *Staphylococcus aureus*, methicillin resistant *Staphylococcus aureus* and *Trichophyton mentagrophytes* (Mwitari et al. 2013). In a mice model salmonellosis was found within various vital organs, it was reduced by the treatment of *W. somnifera* (Owais et al. 2005). *W. somnifera* treatments on validated experimental models like Guinea pigs and mice reduced the infection against several pathogenic bacteria like *E. coli*, *Listeria monocytogenes*, *Bordeteria pertrusis*, etc. (Teixeira et al. 2006). In another study, it was found that aqueous extracts of *W. somnifera* root and leaves have potential antimicrobial activity against a wide range of bacterial strains (Singariya et al. 2012). The antibacterial property of *W. somnifera* plant extract against the Urinary Tract Infections (UTI) and liver infection bacterial pathogens like *Escherichia coli* and *Klebsiella pneumonia* treated as a multidrug constituent (Mishra and Patnaik 2020). The induction of apoptosis of *Leishmania donovani* by the different doses of withanolides through DNA nicks, cell cycle arrest at the sub G0/G1 phase with concomitant increased in oxygen species (ROS), decreased mitochondrial potential and blocking of protein kinase-C pathways has also been reported (Grover et al. 2012, Chandrasekaran et al. 2013).

Anticancerous Activity

According to World Health Organization (WHO), cancer is a large group of diseases that can start in almost any organ or tissue of the body when abnormal cells grow uncontrollably, go beyond their usual boundaries to invade adjoining parts of the body or spread to other organs. The latter process is called metastasis and is a major cause of death from cancer. It is a major public health problem worldwide and in most countries cancer, ranks as the second most common cause of death following cardiovascular diseases. The global cancer burden is estimated to have risen to 18.1 million new cases and 9.6 million deaths in 2018. It has been reported that one in five men and one in six women worldwide develop cancer during their lifetime, and one in eight men and one in eleven women die from the disease (WHO 2018). There are two hundred types of cancer of which the common types in men are lung, prostate, colorectal, stomach and liver cancer whereas in women breast, colorectal, lung, cervical and thyroid cancer are the most common.

The success of cancer treatment is quite insufficient despite enormous developments in surgery, irradiation and chemotherapy. Nowadays, the main approach to treat cancer is the administration of anticancer compounds that inhibit cancer growth in several ways such as impairing mitosis; targeting the cancer cells' energy source, enzymes, and hormones and triggering apoptosis activity (Croce 2008). Though these chemotherapeutic drugs proficiently prevent cancerous cells, they also induce toxicity along with adverse side effects in the patients (Kroschinsky et al. 2017).

Natural products derived from different biological organisms play a crucial role in anticancer therapy. In this context, anticancer drugs from natural compounds of plants and their derivatives have proven to be more effective with negligible side effects against various cancers. Numerous research published over the last few decades indicates the unique characters of *W. somnifera* to suppress various types of cancer, *W. somnifera* has been used as an Ayurvedic remedy for the treatment of various types of cancer for several thousands of years. Various types of cancer cells treated by this plant's extracts are prostate, colon, lung, breast, leukemia, pancreatic, renal, head and neck cancer of humans, etc. (Yadav et al. 2010, Patel et al. 2013, Nema et al. 2013). The anticancerous potential of bioactive withanolides from *Withania* has been studied by numerous research groups and various mechanisms of action such as cell differentiation induction, cytotoxicity, cancer chemoprevention and cyclooxygenase-2 (COX-2) inhibition and a potential to inhibit the enzyme quinine reductase has been explored. Crude leaf extract of *W. somnifera* kills cancer cells with five different pathways such as p53 signaling, granulocyte macrophage colony stimulating factor (GM-CFS) signaling, apoptosis signaling, death receptor signaling and G2-M DNA related various damage regulation pathway (Widodo et al. 2008). Extract of *Withania somnifera* decreased the activities of some key enzymes of Kreb's cycle, such as alpha-ketoglutarate dehydrogenase, succinate dehydrogenase, isocitrate dehydrogenase and malate dehydrogenase, which in turn starve to death of colon cancer cells (Muralikrishnan et al. 2010). Studies on various cell lines of different tissue such as HCT-15 (colon), PC-3, DU-145 (prostrate), A-549 (lung) and IMR-32 (neuroblastoma) disrupt by 50% ethanolic extracts of root, stem and leaves (Yadav et al. 2010). Analysis of molecular docking to recognize the use of withaferin-A and withanone for the development of cancer drugs (Vaishnavi et al. 2012). ROS-induced apoptosis by the effect of withaferin A to function on anticancerous activity on melanoma cells and breast cancer cells by deafening of Bcl-2/Bax and Bcl-2/Bim ratios, the cascade of this apoptosis effect on down regulation of Bcl-2 in mitochondria, translocation of Bax to the mitochondrial membrane, abrogation of transmembrane potential, the release of cytochrome into the cytosol and activation of caspases-9 and 3 (Mayola et al. 2011, Dar et al. 2019). Extract of *W. somnifera* with Withanoloid A exhibits potent inhibitory activity against IKKβ kinase activity in nano molar concentration, which in turn inhibits TNF induced NF-kB activation (Kaileh et al. 2007). A therapeutic compound like Withaferin A was isolated from *W. somnifera*, is more effective in inhibiting colon and breast cancer cell growth (Thaiparambil et al. 2011). In another improvement report, Withaferin A intensified on radiation tempted apoptosis in human renal cancer cells by excessive generation of ROS, inhibition of Bcl-2 and dephosphorylation of AKT (Choi et al. 2011). Withaferin A plays several important roles such as the induction of mitochondrial dysfunction, and oxidative stress in cancer cell lines like human leukemia HL-60 cells by stimulating many possible events accountable simultaneously for mitochondrial independent and dependent apoptosis pathways (Malik et al. 2007, Sehrawat et al. 2019). Another report illustrated that the dose and time dependent action of withanolide D on myeloid (K-562) and lymphoid (MOLT- 4) cells without affecting normal lymphocytes, where withanolide D activates N-SMase 2 with modulate phosphorylation of JNK and p38MP K and induces apoptosis in both myeloid and lymphoid cells in leukemia patients (Malik et al. 2007). Withaferin A treatment to inhibit the growth of mesothelioma by preventing proteasome and by inducing apoptosis (Yang et al. 2012). Inhibition of mammary tumor growth via vimentin expression by the treatment with withaferin A active compound to snooping with cytoskeletal architecture protein β-tubulin (Antony et al. 2014, Lee et al. 2015). Significantly introverted results of withaferin A to cure breast cancer cells with decreasing the size of cells, tumor area and numbers of cells in a transgenic mouse. Application of the same treatments on human breast cancer cells to induction of apoptosis and reduction of complex-III activity by dose dependent treatments of withaferin A (Hahm et al. 2011, 2013, Kim and Singh 2014). Molecular docking application and the structural changes between mutant and wild type p53 proteins, explored the therapeutic potential of Withaferin A and Withanone for rebuilding of wild type p53 protein function in cancer cells (Sundar et al. 2019). *In silico* experiment of withaferin A in arresting the development of breast cancer cells via targeting estrogen receptor protein (Ali et al. 2020).

Neuroprotective Activity

The preparations of *W. somnifera* have been extensively used in different central nervous system related disorders. It has been reported that the extract of the plant has plenty of potentials to overcome the excitotoxicity and oxidative damage in experimental animal models, due to the presence of diverse phytochemical constituents (Parihar and Hemnani 2003). The different tissue extracts of the plants have the capability to control various neurotransmitters. It was reported that corticosterone was suppressed by the extracts with a gradual increase in serotonin level in the hippocampus (Bhatnagar et al. 2009). *W. somnifera* roots contain active compounds like Withanolide A and withanoside IV that encourages neurite extension in cultured neurons and rodents injected with Aβ 25–35 (Kuboyama et al. 2002). The crude extract containing withanone protect scopolamine-induced toxicity in both neuronal and glial cells, neuronal cell markers such as NF-H, MAP-2, PSD-95, GAP-43, and glial cell marker glial fibrillary acidic protein (GFAP) has been made inactive by the induction of scopolamine and DNA damage and oxidative stress markers were markedly attenuated by *W. somnifera* (Konar et al. 2011). Attenuation of toxic glial cells by the extracts of *W. somnifera* to balance the expression of GFAP and heat shock protein (HSP70), neural cell adhesion molecule (NCAM) and mortalin (Kumar et al. 2014). Root powder extract of *W. somnifera* markedly rescued the numerous degenerating cells in CA2 and CA3 sub-areas of rats' hippocampus subjected to immobilization stress (Jain et al. 2001). Semi-purified extract of the roots or its derivatives of *W. somnifera* effect on plaque pathology, behavioral deficits, accumulation of β-amyloid peptides (Aβ) and oligomers in the brains of middle-aged Alzheimer's disease of transgenic mice by improving low density lipoprotein receptor-related protein in brain micro vessels and liver (Sehgal et al. 2012). The root extract stimulated neurite outgrowth extensions in human neuroblastoma cell lines (Zhao et al. 2002), and the leaf extract saved retinoic acid discriminated C6 and IMR-32 cells from glutamate toxicity (Kataria et al. 2012). The axons are neural cells mainly protracted by withanolide-A, and dendrites by withanolides IV and VI while withanoside IV induced together axonal and dendritic rejuvenation and synaptic connection in rat cortical neurons damaged by amyloid-b (Ab) (Thaiparambil et al. 2011, Dahikar et al. 2012). In a few cases, *Withania* extract acts as an antioxidant and cholinergic modulator and has beneficial effects in age-dependent neurodegenerative condition such as Canine cognitive dysfunction (CCD) and Alzheimer's disease (AD) therapy (Singh and Ramassamy 2017). Scopolamine (SC)-induced mice were treated with an alcoholic leaf extract of *Withania*, with the effect on muscarinic subtype acetylcholine receptors on mouse cerebral cortex and hippocampus to reduce loss of memory symptoms (Konar et al. 2019).

Anti-inflammatory Activity

Anti-inflammatory effects of *W. somnifera* extracts have been reported in a variety of rheumatological disorders, reducing complete adjuvant-induced inflammation in rats and decreasing inflamed rat serum containing α2-glycoprotein (Anbalagan and Sadique 1981). Root powder of this plant was found to have a powerful inhibitory effect on nephritis, proteinuria, and other inflammatory indicators such as nitric oxide (NO), interleukin (IL)-6 and tumor necrosis factor (TNF)-a and ROS in a mouse model of lupus (Minhas et al. 2011). It is also reported that the root extract effects to exhibited anti-inflammatory and restorative activity by determining edema, necrosis, neutrophil infiltration in trinitro-benzyl-sulfonic acid (TNBS), induced inflammatory bowel disease (Pawar et al. 2011). Anti-inflammatory activities are due to the presence of biologically active steroids of this plant withaferin-A. It has been reported that doses consisting of 600 and 800 mg/kg body weight of root powder significantly decreased the severity of arthritis by suppressing inflammatory mediators and increased the function of motor activity in the animal's system (Gupta and Singh 2014). Inflammation in cystic fibrosis, irritable bowel syndrome and arthritic inflammation have been treated by the effective function of *W. somniferous* containing withanolides on inhibiting through the antioxidant effect of COX-2 generation, NF-κB activation and inhibition of endothelial cell protein C receptor with release of cytokines, thereby inducing depletion of inflammatory mediators (Mulabagal et al. 2009, Ku et al. 2014). Active compound withaferin A decreases the synthesis of inflammatory mediators like histamine, prostaglandins, interleukins and cytokines (Gupta and Singh 2014). It also affects human umbilical vein endothelial cells (HUVECs) to reduce phorbol-12-myristate-13-acetate (PMA)-induced shedding of endothelial cell protein-C-receptor (EPCR) by inhibiting TNF-a and interleukin (IL)-1β, as well as the effect of withaferin A attenuated PMA-stimulated phosphorylation of p^{38}, extracellular regulated kinases (ERK)-1/2, and c-Jun N-terminal kinase (JNK) (Ku et al. 2014). Withaferin A prevents Iκ-β degradation and phosphorylation, which subsequently blocks NFκ-β/DNA binding, NFκ-β translocation, gene transcription in Murine fibro sarcoma L929sA cells and human embryonic kidney 293T cells (Kaileh et al. 2007). In several rodent models, *W. somnifera* has analgesic activity on various pain management therapies (Sabina et al. 2009).

Immunomodulatory Activity

The authenticated *W. somnifera* extracts, coded as WST and WS2, were studied by a research group to investigate their immunomodulatory activities in authenticated experiments on Swiss albino mice models. Their study revealed crude extracts as an important anti-allergic agent, the anti-allergic activity of this plant is more prominent with WS2 as compared to WST. WS2 extract has effectively reduced ovalbumin induced paw edema (Aggarwal et al. 1999). In another study, root powder extract of *W. somniferous* repressed the function of mitogen induced lymphocyte proliferation and DTH reaction in rats (Rasool and Varalakshmi 2006). The root extract also enhanced total white blood cell count as well as inhibited delayed type hypersensitivity reactions and enhanced phagocytic activity of macrophages (Davis and Kuttan 2002). Oral administration of 75% methanolic extract of ashwagandha significantly increases total leucocyte count in normal BALB/c mice and in mice with leucopenia induced by a sub-lethal dose of γ (*gama*) radiation (Kuttan 1996). *W. somnifera* root extract can upregulate Th1 dominant polarization in BALB/C mice and enhance supports such as the humoral and cell mediated immune responses (Bani et al. 2006). Withanolide E and withaferin A have specific effects on T lymphocytes and both B and T lymphocytes (Gautam et al. 2004). *W. somnifera* extract has been stimulated to increase the production of nitric oxide through the induction of nitric oxide synthase, which in turn affects the macrophage's cytotoxicity against microorganisms (Iuvone et al. 2003). It is also been reported that ashwagandha extract has a beneficial effect on T-cells and may provide immunity against bacterial intracellular propagation (Siddiqui et al. 2012). *W. somnifera* extracts also regulate the intestinal cells of IL-7 in IEC-6, which initiate immunoprotection in the body (Mwitari et al. 2013). Recently network pharmacology of *W. somnifera* was studied, where five bioactive compounds are capable to regulate immunomodulators by 15 immune system pathways on six different (F2, GSK3B, IL2, NFKBIA, RIPK2 and HSPA6) immune targets to reach effective immunomodulation (Chandran and Patwardhan 2017).

Nephroprotective Activity

In vitro studies on *W. somnifera* offers an option to improve a reproducible practice for nephroprotective activities (Kushwaha et al. 2016). It has been reported that an injection of gentamicin (40 mg/kg) was given intraperitoneally daily once for five days and in the *W. Somnifera* treated WST group (received *W. somnifera* orally (500 mg/kg/day) it as a single dose in the morning for the test duration) increased the nephroprotective activity (Kushwaha et al. 2016). The root extract of

Hemidesmus indicus and *W. Somnifera* was used orally on Wistar albino rats for 10 days concurrently with gentamicin. Subsequent nephrotoxicity was evaluated from a single dose injection of gentamicin. The nephroprotective effect was measured after 10 days of administering and show high levels of nephroprotective activity (Padhya et al. 2018). *In vivo* nephroprotective and nephrocurative function of *W. somnifera* in gentamicin induced nephrotoxic rats with different experimental treatments has also been reported (Govindappa et al. 2019). The serum and kidney tissue were analyzed to examine renal function after following a combination of treatment such as saline control for 21 days, gentamicin nephrotoxic control for 8 days, alcoholic extract of *W. somnifera* for 13 days + simultaneous administration of gentamicin and *W. somnifera*, from day 14 to 21, gentamicin for 8 days + alcoholic extract of *W. somnifera* from day 9 to 21. It was observed that *W. somnifera* effectively functions as a nephroprotective and nephrocurative activity with the significant restoration of renal function with meaningfully reduced blood urea nitrogen, creatine, alkaline phosphatase, gamma glutamyl transferase, albumin, total protein, calcium, potassium and kidney malon dialdehyde concentrations (Govindappa et al. 2019). In another experiment with three different oral doses of *W. somnifera* root extract (250, 500, and 750 mg/kg) for 14 days before gentamicin treatment and thereafter concurrently with gentamicin (100 mg/kg) for 8 days showed nephrotoxicity of gentamicin treated rats with the significant increase in kidney weight, creatinine, urea, urinary protein, and glucose, and a significant decrease in body weights and potassium. In contrast, *W. somnifera* extract (500 mg/kg) significantly reversed these changes as evidenced microscopically when compared to the other two doses of *W. somnifera* (250 and 750 mg/kg), and there were no significant changes in the levels of sodium in the experimental animals compared to control (Jeyanthi and Subramanian 2009).

Cardio-protective Activity

Myocardial infarction and myocardial ischemia reperfusion injury, in an extensive range of patients which takes place in patients, ranging from survivors who have been hospitalized for cardiac arrest to myocardial infarction victims and patients suffering cardiac surgery and significant major public health afflictions. The treatment with the extract of *W. somnifera* was found to be helpful in the proliferation of the heart rate, contractility, relaxation and decrease preload along with improved antioxidant enzymes and inhibition of lipid peroxidation comparable to vitamin E, and improved cardio-protective antioxidant (Mohanty et al. 2004). Increased antioxidant activity contributes to cardio-protection by upregulated Bcl-2, an anti-apoptotic protein and decreased Bax, a pro-apoptotic protein as well as reduction of terminal deoxynucleotidyl transferase biotin dUTP nick end labeling (TUNEL) positivity, a hallmark of apoptosis (Mohanty et al. 2008). Another experiment studied by Thirunavukkarasu et al. (2006) establishs the energy enhancing properties formulation containing *Withania* which conceded heart and mentioned its use as a dietary supplement for cardio-protection. Positively reformed the formulation of the myocardial energy substrate, improved cardiac function and reduced restriction size (Thirunavukkarasu et al. 2006). In another study, "Marutham", a polyherbal formulation containing *W. somnifera* was found cardio-protective and antioxidant activity on isoproterenol induced ischemic rats (Prince et al. 2008).

Anti-diabetic Activity

The anti-diabetic activity of aqueous extract of *W. somnifera* has also been reported with a significantly improved insulin sensitivity index in diabetic rats (Anwer et al. 2008). Root and leaf extracts of *W. somnifera* for the treatment against diabetes mellitus showed a significant reduction of blood sugar with hypoglycaemic and hypolipidaemic properties on streptozotocin-induced diabetic rats (Udayakumar et al. 2009, Sarangi et al. 2013). The reduction of blood sugar by the Ayurvedic polyherbal formulations such as Dianix, Trasina has also been reported in humans (Mutalik et al. 2005, Gauttam and Kalia 2013). In a dose dependent way of anti-diabetic activity of *W. somnifera* root and leaf extracts showed to enlighten the glucose uptake in skeletal myotubes and adipocytes, leaf extract showed very fast prominent results than the root extract of *W. somnifera* (Gorelick et al. 2015). Withaferin A moderates multiple low doses of Streptozotocin in Type 1 diabetes mellitus in Swiss albino mice. Withaferin A enriched the multiple low doses of Streptozotocin induced oxidative and nitrosative stress along with those capable of anti-inflammatory effect by reducing in the levels of IL-6 and TNF-α cytokines compared to diabetic mice, withaferin A can effectively combat Streptozotocin to induced Type 1 diabetes mellitus via modulation of Nrf2/NFκB signaling (Tekula et al. 2018). Nanoparticles synthesis from the leaf extract with the treatment of zinc oxide can affect the treatment of inhibition activity of α-amylase and α-glucosidase due to demonstrating its anti-diabetic potential by the treatment with 100 μg/ml zinc oxide nanoparticle (Malaikozhundan et al. 2020). Recently it has been studied that *W. somnifera* root extract effects on improvements in insulin resistance activity and β-cell dysfunction by improvement of Dipeptidyl peptidase-4 (DPP-4) inhibitors potential in type 2 diabetic rats (Ram and Krishna 2021).

Anti-stress/Adaptogenic Activity

The effect of *W. somnifera* as potent anti-stress agents in the animal system has also been reported (Singh et al. 2001, Tiwari et al. 2014). Anti-stress activity can increase plasma corticosterone, phagocytic index, physical endurance, cardiac activity, augmentation level of Th-1 cytokines, the proliferation of T lymphocytes, prevention of stress induced ulcers and induced hepatotoxicity by carbon tetrachloride (Tomi et al. 2005, Khan et al. 2006, Al-Qirim et al. 2007). Experimental studies of some animal systems have been studied to show some effective results by the treatment of Ashwagandha. In mice T cell population and upregulated Th1 cytokines by the treatment of aqueous roots extract fraction (Khan et al. 2006). Roots extract of Ashwagandha has been used in human subjects specifically serum cortisol levels reduction with except any side effect (Sharma et al. 2018). A poly herbal formulation called EuMil, markedly available compounds to enhance the level of cerebral monoamine (nor-adrenaline, dopamine, and 5-hydroxytryptamine) to induce chronic electroshock stress (Bhattacharya and Muruganandam 2003). It is also reported that free aqueous fraction isolated from the roots exhibited anti-stress action in a dose dependent manner in mice (Khare 2007). "Perment" is another polyherbal formulation showed anti-depressant and anxiolytic activity in rats, which was partially due to activation of adrenergic and serotonergic systems (Ramanathan et al. 2011). Another report has been exhibited to active products from *W. somnifera* is glycowithanolides can affect on anti-anxiety medication against pentylenetetrazole induced in rats, which was comparable which is analogous of anti-depressants molecules, it is also the addition of tribulin, an endocoid to effect of clinical anexity of rat brain (Bhattacharya et al. 2000b).

Withanolides and their Biosynthesis

The biosynthetic pathway of withanolides are not well explored (Kirson et al. 1977, Nittala and Lavie 1981, Ray and Gupta 1994). The pathway starts with the activation of acetate to form acetyl coenzyme A then fuses with another acetyl-CoA to form mevalonic acid (Fig. 4). The living system synthesizes metabolites using only R-mevalonic acid, which later loses one carbon moiety to form IPP. Geranyl pyrophosphate (GPP) is formed by head-to-tail condensation of 3, 3-dimethyl allyl pyrophosphate (DMAPP) with IPP. A second IPP molecule then condenses with trans GPP to form farnesyl pyrophosphate (FPP), a substrate of squalene synthase (SQS). SQS catalyzes squalene production by a head-to-head condensation of two FPPs in the presence of NADPH as a coenzyme, which is then quickly oxidized to squalene 2, 3-epoxide. The latter undergoes ring closure to form lanosterol which gives rise to several steroidal triterpenoid backbones via 24-methylene cholesterol as an intermediate. Based on *W. somnifera* feeding experiments by Lockley et al. (1976), 24-methylene cholesterol was proposed as a withanolide precursor, where

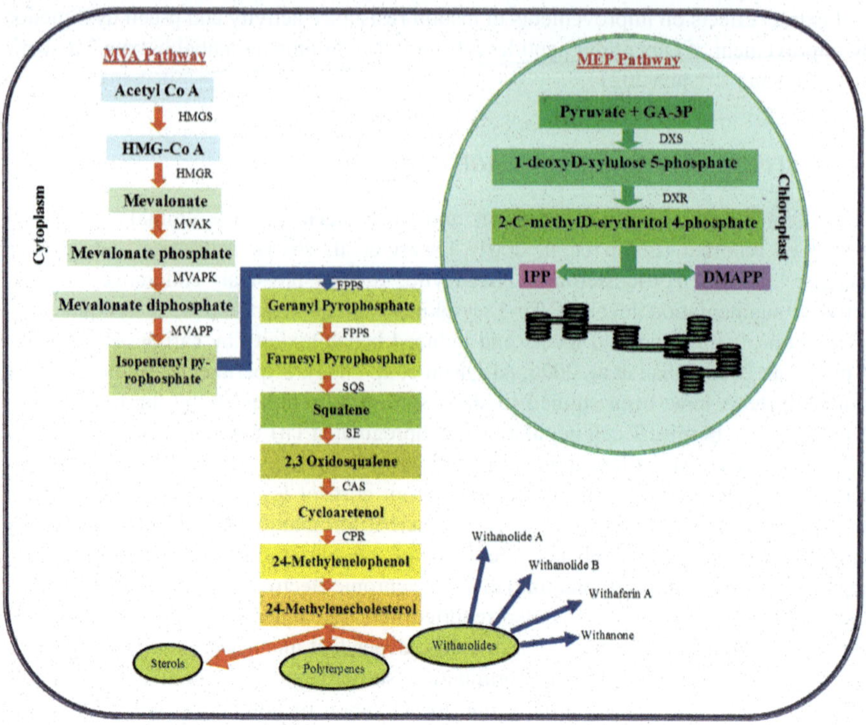

Figure 4: Withanolides biosynthetic pathway.

Withaferin A and Withanolide D incorporated the radioactive precursors, while labeled 24-(R, S)-methyl-cholesterol failed to do so. Withanolides are formed by hydroxylation of C 22 and δ-lactonization between C 22-C 26 of 24-methylene cholesterol (Mirjalili et al. 2009b). The formation of squalene from two molecules of FPP is a very critical step to regulate the metabolic flow toward withanolide biosynthesis rather than the usual isoprenoid pathway. This is catalyzed by important key enzyme squalene synthase, a 47 kDa transmembrane dual function enzyme (Gupta et al. 2012). FPP can be directed by the overexpression of SQS to the production of squalene, a precursor molecule of triterpenoids. It has been suggested that active biosynthesis of withanolides occurs when sterol formation is sacrificed (Kamisako et al. 1984, Flores-Sánchez et al. 2002).

In Vitro Culture for Enhanced Synthesis of Withanolides

The genetic diversity of important medicinal plants are decreasing sharply due to the indiscriminate harvesting of wild populations. To overcome this problem organized cultivation or shifting cultivation techniques has been widely adopted. Nevertheless, the quantities of phytochemicals in cultivated medicinal plants compared to wild populations vary greatly due to several factors such as harsh environmental conditions, diseases, etc. In such circumstances, the implication of *in vitro* tissue culture techniques was found to be an alternate system for the even production of important secondary metabolites. The *in vitro* system also possesses the capacity for higher production and accumulation of metabolites owing to the active growth and higher rate of metabolism within a short time (Fig. 5). *In vitro* cultures have been effectively used for enhanced production of some major secondary metabolites such as Withanolide A, Withanolide B and Withaferin A in *W. somnifera* (Table 4). It has been reported that 1.14-fold and 1.20-fold increase in the content of withaferin A and withanone in leaves of *in vitro* derived plants as compared to field plants, whereas roots of the *in vitro* culture exhibited 1.10-fold higher accumulation of withanolide A compared with the field grown parent plants (Sivanandhan et al. 2011). In another study, *in vitro* regenerated plants showed two-fold higher withaferin A and withanolide A and ten-fold increase in withanone contents as compared to field grown plants (Sharada et al. 2007). Similarly, a marked increase in withanolide A production was observed in *in vitro* shoot cultures (Sabir et al. 2008). Through modulation of sucrose concentration, media formulations and type and concentration of Plant growth regulators (PGR's) the production of secondary metabolites can also be modified (Nagella and Murthy 2010). It has been reported that culture of shoot tips of *W. somnifera* proliferating on B5 medium accumulated maximum withaferin A (0.09%), whereas withanolide D accumulation was maximal (0.065%) in hormone free medium (Ray and Jha 2001). In a recent report, different types and concentrations of nitrogen sources (adenine sulfate, L-glutamine, potassium nitrate ammonium nitrate and sodium nitrate) and various carbon sources (sucrose, glucose, maltose and fructose) were modified to enhance shoot multiplication and withanolides accumulation (Sivanandhan et al. 2015b). It was reported that MS (Murashige and Skoog 1962) medium with 6-Benzylaminopurine (BAP) (6.66 μM) and Indole-3-acetic acid (IAA) (1.71 μM) with 20 mg/l L-glutamine and 4% sucrose improved the multiple shoot formation while the change of sucrose from 4% to 6%, induced maximum withanolides content (Sivanandhan et al. 2015b).

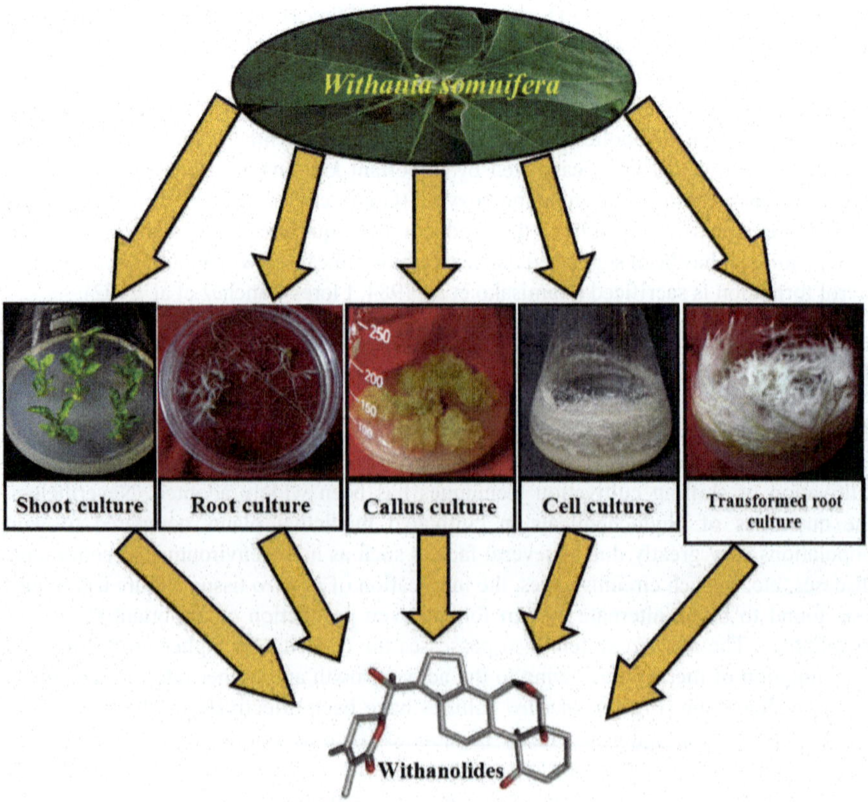

Figure 5: Different modes of *in vitro* cultures of *W. somnifera*.

Shoot Culture

The first shoot culture for the production of withanolides was established by Roja et al. (1991). Production of Withaferin A and Withanolide D has been reported from the shoot tip cultures in MS medium supplemented with 1 mg/l BAP (Ray and Jha 2001). The presence of 4% sucrose resulted in increased accumulation of withanolides production (Withanolide D 0.08% and Withaferin A 0.16%). The shoot culture was established using nodal segments as explants using experimental cell lines of *W. somnifera* plants (RS Selection-1 and RS Selection-2) (Sangwan et al. 2007). Two hormones combinations, i.e., BAP and kinetin (Kn) were influenced by the morphogenetic response and changed the pattern of withanolide A biosynthesis. There was a considerable increase of Withanolide A in the *in vitro* shoot cultures as compared to the aerial parts of field grown plants. Displayed in the shoot culture of RS-Selection-1 raised at 1.00 ppm of BAP and 0.50 ppm of Kn produced the highest (0.238% based on DW) concentration of withanolide. A significant variation in the levels of withanolide A was observed (ten-fold, 0.014–0.14 mg/g) based on fresh weight (FW) upon changing the hormone composition (Sangwan et al. 2007). In *in vitro* system, the concentration of withaferin A ranged between 0.27 and 7.64 mg/g DW, in case of *in situ* system range of withaferin A was between 8.06 and 36.31 mg/g

Table 4: *In vitro* culture and production of withanolides.

Culture types	Media	Active compounds	References
Shoot tips	BAP (1.0 mg/l) and Coconut milk (10%)	Withaferin A	Ray and Jha 2001
Root	½ MS with IAA (2.85 μM) and IBA (9.85 μM)	Withanolides	Wadegaonkar et al. 2006
Shoot	BAP (0.4 μM) and IAA (0.4 μM)	Withanolide A	Sharada et al. 2007
Callus culture	TDZ (0.5 mg/l) and α-Naphthalene acetic acid (NAA) (0.5 mg/l)	Chlorogenic acid, Withaferin A	Adil et al. 2019
Shoot	BAP (1.0 mg/l) and Kn (1.0 mg/l)	Withanolide A	Sabir et al. 2008
Adventitious root	IBA (0.5 mg/l), ½ MS with IBA (0.5 mg/l)	Withanolide A	Praveen and Murthy 2010
Adventitious root	2,4-D (2.0 mg/l) and Kn (0.2 mg/l), ½ MS with IBA (0.5 mg/l) and IAA (0.1 mg/l), 1/2 MS with IBA (0.5 mg/l) and IAA (0.1 mg/l)	Withanolide A, Withanone, Withanolide B, Withaferin A, 12-deoxywithastramonolide, Withanoside V, Withanoside IV	Sivanandhan et al. 2012
Adventitious root	2,4-D (2.0 mg/l) Kn (0.2 mg/l), ½ MS with IBA (0.5 mg/l) and IAA (0.1 mg/l), ½ MS with IBA (0.5 mg/l) and IAA (0.1 mg/l)	Withanolide A, Withanolide B, Withaferin A, Withanoside V, Withanoside IV	Sivanandhan et al. 2012
Shoot	BAP (0.6 mg/l) and spermidine (20 mg/l)	Withaferin A, Withanone, Withanolide B, Withanolide A	Sivanandhan et al. 2013
Shoot	BAP (1.5 mg/l), IAA (0.3 mg/l) and L-glutamine (20 mg/l)	Withaferin A, Withanone, Withanolide B, Withanolide A	Sivanandhan et al. 2014b
Shoot	MSO	Withaferin A, Withanone, Withanolide B, Withanolide A	Sivanandhan et al. 2014b
Adventitious root	IBA (2.0 mg/l) and IAA (0.5 mg/l)	Withanolide A	Senthil et al. 2015
Adventitious root	IBA (1.0 mg/l) and IAA (0.25 mg/l)	Withanolide A	Senthil et al. 2015
Shoot	BAP (1.0 mg /l)	Withaferin A	Senthil et al. 2015
Shoot and Adventitious root	BAP (4.44 μM)	Withaferin A, Withanolide A	Parameswari et al. 2017
Adventitious root	IAA (0.025 mg/l) and IBA (0.01 mg/l)	Withaferin A, Withanolide A, Withanolide B and 12-deoxywithastramonolide, Withanoside IV, Withanoside V	Rangaraju et al. 2018

DW. The highest amount of withaferin A found in *in vitro* system is 7.37 mg/g DW and *in situ* system is 41.42 mg/g DW respectively (Johny et al. 2015). Old nodal explants of *W. somnifera* derived shoots were culture on half strength of MS liquid medium for five weeks, after five weeks later it has given highest accumulation of biomass and withaferin A content. Shoots cultured in half and full strength MS liquid media showed Withaferin A was produced in relatively high amounts (1.30% and 1.10% DW) in respectively as compared to field-grown plants under full sunlight (0.85% DW) (Mir et al. 2014). Elicitation of shoot inoculum mass (2 g/l FW) with 100 μM of salicylic acid (SA) treatment with 4 h exposure time, then the inoculum was cultured on 0.6 mg/l BAP and 20 mg/l spermidine, After 4 weeks of culture in 20 ml, liquid medium recorded higher production of withanolides A (8.48 mg/g DW), withanolides B (15.47 mg/g DW, withaferin A (29.55 mg/g DW) and withanone 23.44 mg/g DW) that were 1.14 to 1.18-fold higher than elicitation with methyl jasmonate (MeJ) elicited at 100 μM after 5 weeks of culture (Sivanandhan et al. 2013).

Adventitious root Culture

The plant roots are a rich source of pharmacologically active secondary metabolites. However, harvesting the natural roots from field conditions is destructive for the plants. Therefore, *in vitro* adventitious root culture technology is considered as an alternative source for the production of valuable secondary metabolites. The technology of this culture has been explored for the last several decades to produce secondary metabolites due to the rapid development of roots. *In vitro* production of withanolides has been reported from adventitious root culture from different types of explants like leaves, internodes, cotyledons, etc. (Sivanandhan et al. 2012). High biomass accumulations and contents enhancements of withanolides in the adventitious root culture depend on the maturity of culture systems and the treatments of elicitors in cultures can produce large quantities of metabolites (Smetanska 2008). In suspension culture, the adventitious roots production of withanolide A gradually increased with the biomass. Effect of withanolides production and growth rate of roots depends on pH range, carbon source concentrations and some influence of PGR' (Murthy and Praveen 2013, Rangaraju et al. 2018). The maximum biomass (7.48 ± 0.25 g/l) accumulation was observed in medium containing with 4.93 μM Indole-3-butyric acid (IBA) and production of withanolide A (204.98 ± 0.87 μg/l DW) and withaferin A (227.15 ± 0.57 μg/l DW) accumulation was also recorded in IBA contained half-strength Murashiga-Skoog media (Murugesan and Senthil 2017). Basal medium supplemented with 2.85 μM IAA acid and 9.85 μM IBA achieved maximum number of roots with 100% response and find 47 g FW of roots from 1 g roots within 6 weeks culture and maximum concentration of withanolides (10 mg/g DW) was obtained from the culture in the bioreactor (Wadegaonkar et al. 2006). Callus-derived roots were cultured on elicitor treated medium which was induced to increase the biosynthesis of secondary metabolites in diverse plant organ cultures (El-Ghaouth et al. 1994).

Hairy root Culture

The non transformed cultures need an exogenous phytohormone supply in basal medium and the growth rate is very slow, resulting in poor or negligible secondary metabolite synthesis. Therefore, alternative methods like hairy root cultures were organized for a higher growth rate along with synthesis of root derived secondary metabolites (Verpoorte et al. 1993, Lodhi et al. 1996, Giri and Narasu 2000). Hairy roots are genetically unique and have biosynthetic stability to introduce T-DNA along with additional genes for alteration of metabolic pathways and production of useful metabolites or compounds of interest (Sivanandhan et al. 2014a). The first hairy root culture for the production of withanolides was established by Hamil et al. (1987). There are several reports on the production of hairy root cultures of *W. somnifera* using *A. rhizogenes* (Table 5). The transformation efficiency to explants for transformed root culture is influenced by various parameters likes *Agrobacterium* stain, co-cultivation medium, duration of co-cultivation

Table 5: Synthesis of different withanolides from hairy root culture of *W. somnifera*.

Explants	*A. rhizogenes* strains	Secondary metabolites	References
Leaf	LBA 9402, A4 and LBA 9360	Withaferin A	Banerjee et al. 1994
Leaf and stem	LBA 9402	Withanolide D	Ray et al. 1996
Leaf discs	MTCC 2364, MTCC 532	-	Pawar et al. 2004
Leaf	LBA 9402, A4	Withaferin A, Withanolide D	Bandyopadhyay et al. 2007
Leaf segments	R1601	Withanolide A	Murthy et al. 2008
Leaf	ATCC 15834	Withaferin A	Doma et al. 2012
Leaf	ATCC 15834	Withanolide A	Doma et al. 2012
Leaf segments	R1601	Withanolide A	Praveen and Murthy 2012
Petiole	R1000	Withaferin A	Saravanakumar et al. 2012
Cotyledon and leaf	R1601	Withanolide A	Praveen and Murthy 2013
Leaf	R1000	Withanolide A, Withanone, Withaferin A	Sivanandhan et al. 2013
Shoot tips	ATCC15834		Ara et al. 2014
Leaf	R1000		Sivanandhan et al. 2014a
Leaf segments	ATCC15834	Withaferin A	Varghese et al. 2014
Leaf	R1000	Withaferin A	Sivanandhan et al. 2015a
Leaf segments	R1000	Withaferin A, Withanone, Withanolide A	Thilip et al. 2015
Leaf	A4	Withanolide A,Withanolide A	Pandey et al. 2016
Shoot tips	ATCC15834	-	Dehdashti et al. 2017
Leaf	A4, 1600 and 8196	Withaferin A, Withanolide A, Withanolide B and 12-deoxywithastramonolide	Johny et al. 2018

and concentration of acetosyringone (Murthy et al. 2008). Among various explants, cotyledons and leaves show a positive response towards infection with *A. rhizogenes* (Praveen and Murthy 2012). Carbon source also plays an important role in differential accumulation of withaferin A when grown in a medium containing 4% (w/v) sucrose A (Praveen and Murthy 2013). The role of macro elements and nitrogen sources has also been considered to monitor their effects on hairy root production and accumulation of withanolides (Kumar et al. 2005). Application of elicitors to hairy root cultures could also improve the secondary metabolite production in plant cell/organ culture (Sivanandhan et al. 2013). A positive correlation between elicitor treatment and withanolide production could be established in *W. somnifera* hairy roots (Thilip et al. 2019). The various concentrations of elicitors like MeJ and SA which leads to increased withaferin A, withanolide A and withanone production were found by different authors (Ketchum et al. 1999, Doma et al. 2012). Though elicitation has resulted in enhancement of secondary metabolites, there is limited information of the molecular mechanisms of elicitation involved in triggering the production of bioactive molecules.

Cell suspension Culture

Medicinal plants are considered the most fashionable source of life saving medicines for the majority of the world's population. The natural habitats of medicinal plants are fast declining by continuous use. Scientists encourage the productions of active compounds possibilities of investigation, first cell culture for withanolides was established by Yu et al. (1974), cell cultures technique it is an alternative method for the production of plant secondary metabolites (Table 6). It is a flexible system that is easily manipulated to increase secondary metabolites production that may contribute to the scale up and optimization in our daily life or industries. In cell suspension culture generally the production of withanolides specifically withanolide A and withaferin A are produces a very smaller amount (Ciddi 2006, Nagella and Murthy 2011). Synthesis of active compounds under the cell suspension culture depends on some effective factors like carbon source, PGR's, salt concentrations, macro and micronutrient compositions, etc. (Sivanandhan et al. 2013). In high concentration of ammonium becomes toxic if it is not metabolized immediately, these inhibit the effect of cellular metabolism. NO_3^- is first reduced to NH_4^+ before being incorporated into amino acids (Scheible et al. 2004). It was observed that the accumulation of higher NO_3^- and lower NH_4^+ concentrations favor both cell growth and withanolide production (Sivanandhan et al. 2013). As a carbon source in cell culture, sucrose (2–4%) was found to be best for the biomass accumulation for the withanolide A production in comparison to glucose, fructose and maltose because upon hydrolysis of sucrose produces glucose, which is readily utilized for cell growth (Lancien et al. 1999).

Bioreactor on Withanolides Production

Bioreactor technology to modern biotechnological science is more important for the enhancing of products during a short time within small places. The major challenge for the production of phytochemicals from cell, tissue and organ cultures of this plant has been faced by different researchers in recent years. Adventitious roots, transformed roots, shoot

Table 6: Synthesis of different withanolides from cell suspension culture of *W. somnifera*.

Culture explants	Culture medium	Secondary metabolite(s)	References
Hypocotyl	2,4-D (2.0 mg/l), Kn (1.0 mg/l) and Glutamine (292 mg/l)	Withaferine A	Ciddi 2006
Hypocotyl	Infection with *Agrobacterium tumefaciens* MTCC-2250	Withaferine A	Baldi et al. 2008
Leaf segments	2,4-D (2.0 mg/l) and Kn (0.5 mg/l)	Withanolide A	Nagella and Murthy 2010
Leaf	2,4-D (3.0 mg/l) and Kn (0.5 mg/l)	Withanolide A, Withanone	Sabir et al. 2011
Leaf segments	2,4-D (2.0 mg/l) and Kn (0.5 mg/l)	Withanolide A	Nagella and Murthy 2011
Roots	Picloram (1.0 mg/l), Kn (0.5 mg/l) and L-glutamine (200 mg/l)	Withanolide A, Withanone, Withanolide B, Withaferin A	Sivanandhan et al. 2013
Roots	Picloram (1.0 mg/l), Kn (0.5 mg/l) and L-glutamine (200 mg/l)	Withanolide A, Withanone, Withanolide B, Withaferin A, 12 deoxywithanstramonolide, Withanoside IV, Withanoside V	Sivanandhan et al. 2014c
Leaf	NAA (0.5 mg/l) and BAP (1.0 mg/l)	Withaferine A	Jain et al. 2014
Leaf	2,4-D (1.5 mg/l) and Kn (0.2 mg/l)	Withaferin A	Ahlawat et al. 2016
Leaf	2,4-D (1.5 mg/l) and Kn (0.2 mg/l)	Withanolide A, Withanone, Withaferin A	Ahlawat et al. 2017

and cell culture have been established in various configurations of bioreactor for industrial and research purposes such as stirred tank, bubble column, balloon type bubble bioreactors, etc. Different volumes of bioreactors have been used for the cultivation of various plant tissues. Bubble column bioreactor (5 lit.) for the cell culture of *W. somnifera* has been used by Ahlawat et al. (2017) to improve plant material production used cell suspension culture was inoculated through inoculation port into the bubble column bioreactor with containing MS liquid medium supplemented with 30 g/l sucrose, 1.5 mg/l 2,4-dichlorophenoxyacetic acid (2,4-D) and 0.5 mg/l Kn under the dark conditions after 22 days of culture contain 13.9 ± 0.17 g/l biomass and active compounds like withanolide A (4.19 ± 0.07 mg/l DW), withaferin A (0.22 ± 0.04 mg/l DW), and withanone (3.88 ± 0.05 mg/l DW) they have been recorded. Sivanandhan et al. (2014c) reported enhanced biomass production in Industrial bioreactor (7 lit.) for cell culture with elicitor treatments, after 28 days of culture biomass and active compounds increased significantly. In the case of roots culture in bubble column bioreactor (2.5 lit.) with containing ½ MS liquid medium supplemented with 15 g/l sucrose, 2.85 μM IAA and 9.85 μM IBA and isolate 47 g of root after 42 days from the 1 g of fresh root inoculum (Wadegaonkar et al. 2006). Another study on adventitious roots culture on bubble column bioreactor has been performed by Senthil et al. (2015), after 45 days of culture they got a higher accumulation of biomass and active compounds.

Metabolic Engineering

Metabolic engineering is renovating the metabolic pathways through variation of enzymatic activities either channelizing metabolic flux toward targeted metabolite or diverting flux from the undesirable compound and the synthesis of pharmaceutically or clinically important compound (Verpoorte et al. 2002, Lessard et al. 2002). The target of one of the important enzyme synthesizing genes is SQS for potential target as a branch point of withanolide biosynthesis of secondary metabolite pathway in *W. somnifera* (Patel et al. 2016). The functional role of SQS in plants for triterpene sterol biosynthesis has been reported to numerous plants (Devarenne et al. 2002, Wentzinger et al. 2002). Various reports are revealed that enzymatic genes expression of isoprenoid pathway has been modified by different expression to enhance the downstream carbon flow for the production of specific active compounds (Chen et al. 2000, Hey et al. 2006). In few cases, precursors feeding of isoprenoid of IPP and DMAPP is to boost the production of downstream metabolites (Kumar et al. 2012). The biosynthetic pathway of Withanolides has been comprehensively studied due to its pharmacological and medicinal importance. Withanolides are synthesized by distracting the metabolite flux away from the isoprenoid pathway through the reductive condensation of farnesyl diphosphate to squalene by the enzymatic activity of the SQS. The SQS gene was over-expressed using *Agrobacterium tumefaciens* gene transformation and cell suspension cultures were established to identify its effect on withanolide synthesis, the study confirmed that a significant 4 fold enhancement in SQS activity and 2.5 fold improvement in withanolide A content were detected in the suspension cultures compared to the non-transformed culture (Grover et al. 2012). In another report, SQS genes were also overexpressed in transgenic plants with a concomitant increment of withanolide A and withaferin A in the root and leaves up to 2–2.25 and 4–4.5 fold respectively (Patel et al. 2015). Cycloartenol synthase (CAS) is an important enzyme in the withanolide biosynthetic pathway, catalyzing cyclization of 2, 3 oxidosqualene into cycloartenol. The overexpression of CAS in transgenic *W. somnifera* increased the withanolide content to the extent of 1.06 to 1.66 fold compared to non-transformed controls (Mishra et al. 2016). These results validate metabolic engineering of isoprenoid pathway in *W. somnifera* with subsequently improved synthesis of withanolides.

Conclusions

Withania somnifera, is a miracle shrub conventionally used as traditional medicine for several ailments in different parts of the world. Numerous scientific reports have established the prospective therapeutic application of several classes of secondary metabolites including withanolides, sitoindosides and other alkaloids extracted from different parts of the plant against diverse diseases. Studies conducted in cellular models have established that root and leaf extracts have potent anticancer, immunomodulatory and anti-inflammatory activities. Due to its immense potential, there is an increasing demand for crude extract as well as purified secondary metabolites within pharmacological industries. Therefore, alternate ways to improve the production of bioactive withanolides is very much required considering the rising market demand. The incorporation of high throughput molecular biology techniques has resulted in identification of superior genotypes with a higher level of therapeutically important secondary metabolites. The progress in bioinformatics research has provided diverse tools that are valuable in the screening of rate limiting enzymes associated with the biosynthesis pathway of important secondary metabolites. Due to the low abundance of these bioactive compounds, large biomass is exploited for medicinal purposes in pharmaceutical industries to achieve upward commercial demand. The application of these high throughput technologies will ultimately produce higher amounts of secondary metabolites and meet the demand of pharmaceutical industries.

Acknowledgments

The authors are thankful to the Department of Biotechnology (DBT), Government of India (DBT-GI) for financial support. Also acknowledged, DST-FIST and RUSA program for infrastructural facilities. The authors are thankful to Swami Kamalasthananda, Principal, Ramakrishna Mission Vivekananda Centenary College for providing facilities.

References

Aalinkeel, R., Z. Hu, B.B. Nair, D.E. Sykes, J.L. Reynolds, S.D. Mahajan, and S.A. Schwartz. 2010. Genomicanalysis highlights the role of the JAK STAT signaling in the antiproliferative effects of dietary flavonoid 'Ashwagandha' in prostate cancer cells. Evid. Based Complement. Alternat. Med. 7: 177–187.

Abou-Douh, A.M. 2002. New withanolides and other constituents from the fruit of *Withania somnifera*. Archiv der Pharmazie: JPMC. 335: 267–276.

Abraham, A., I. Kirson, E. Glotter, and D.A. Lavie. 1968. Chemotaxonomic study of *Withania somnifera* (L.) dunal. Phytochem. 7: 957–962.

Adil, M., B.H. Abbasi, and I. ul Haq. 2019. Red light controlled callus morphogenetic patterns and secondary metabolites production in *Withania somnifera* L. Biotechnol. Rep. 24: e00380.

Aggarwal, R., S. Diwanay, and P. Patki. 1999. Studies on immunomodulatory activity of *Withania somnifera* (Ashwagandha) extracts in experimental immune inflammation. J. Ethnopharmacol. 97: 27–35.

Ahlawat, S., P. Saxena, A. Ali, and M.Z. Abdin. 2016. *Piriformospora indica* elicitation of withaferin A biosynthesis and biomass accumulation in cell suspension cultures of *Withania somnifera*. Symbiosis. 69: 37–46.

Ahlawat, S., P. Saxena, A. Ali, S. Khan, and M.Z. Abdin. 2017. Comparative study of withanolide production and the related transcriptional responses of biosynthetic genes in fungi elicited cell suspension culture of *Withania somnifera* in shake flask and bioreactor. Plant Physiol. Biochem. 114: 19–28.

Ahmad, H., K. Khandelwal, S.S. Samuel, S. Tripathi, K. Mitra, R.S. Sangwan, R. Shukla, and A.K. Dwivedi. 2015. Neuro protective potential of a vesicular system of a standardized of a new chemotype of *Withania somnifera* Dunal (NMITLI118RT+) against cerebral stroke in rats. Drug Deliv. 23: 2630–2641.

Ahmad, M., S. Saleem, A.S. Ahmad, M.A. Ansari, S. Yousuf, M.N. Hoda, and F. Islam. 2005. Neuroprotective effects of *Withania somnifera* on 6-hydroxydopamine induced Parkinsonism in rats. Hum. Exp. Toxicol. 24: 137–147.

Ahmad, M.K., A.A. Mahdi, K.K. Shukla, N. Islam, S. Rajender, D. Madhukar, S.N. Shankhwar, and S. Ahmad. 2010. *Withania somnifera* improves semen quality by regulating reproductive hormone levels and oxidative stress in seminal plasma of infertile males. Fertil. Steril. 94: 989–996.

Ahmad, M., and N.J. Dar. 2017. *Withania somnifera*: Ethnobotany, pharmacology, and therapeutic functions. In sustained energy for enhanced human functions and activity (pp. 137–154). Academic Press.

Alam, N., M. Hossain, M.A. Mottalib, S.A. Sulaiman, S.H. Gan, and M.I. Khalil. 2012. Methanolic extracts of *Withania somnifera* leaves, fruits and roots possess antioxidant properties and antibacterial activities. BMC Complement. Altern. Med. 12: 175.

Alfaifi, M.Y., K.A. Saleh, M.A. El-Boushnak, S.E.I. Elbehairi, M.A. Alshehri, and A.A. Shati. 2016. Antiproliferative activity of the methanolic extract of *Withania somnifera* leaves from Faifa Mountains, Southwest Saudi Arabia, against several human cancer cell lines. Asian Pac. J. Cancer Prev. 17: 2723–2726.

Al-Hindawi, M.K., S.H. Al-Khafaji, and M.H. Abdul-Nabi. 1992. Anti-granuloma activity of Iraqi *Withania somnifera*. J. Ethnopharmacol. 37: 113–116.

Ali, M.A., M.A. Farah, K.M. Al-Anazi, S.H. Basha, F. Bai, J. Lee, F. Al-Hemaid, A.H. Mahmoud, and W.A. Hailan. 2020. *In Silico* Elucidation of the plausible inhibitory potential of withaferin A of *Withania Somnifera* medicinal herb against breast cancer targeting estrogen receptor. Curr. Pharm. Biotechno. DOI: https://doi.org/10.2174/1389201021666200129121843.

Aliebrahimi, S., S.M. Kouhsari, S.S. Arab, A. Shadboorestan, and S.N. Ostad. 2018. Phytochemicals, withaferin A and carnosol, overcome pancreatic cancer stem cells as c-Met inhibitors. Biomed. Pharmacother. 106: 1527–1536.

Al-Qirim, T.M., A. Zafir, and N. Banu. 2007. Comparative antioxidant potential of *Rauwolfia serpentina* and *Withania somnifera* on cardiac tissues. FASEB J. 21: 271–271.

Ambiye, V.R., D. Langade, S. Dongre, P. Aptikar, M. Kulkarni, and A. Dongre. 2013. Clinical evaluation of the spermatogenic activity of the root extract of Ashwagandha (*Withania somnifera*) in oligospermic males: a pilot study. Evid. Based Complementary Altern. Med. https://doi.org/10.1155/2013/571420.

Anbalagan, K., and J. Sadique. 1981. Influence of an Indian medicine (Ashwagandha) on acute phase reactants in inflammation. Indian J. Exp. Biol. 19: 245–249.

Andrade, C., A. Aswath, S.K. Chaturvedi, M. Srinivasa, and R. Raguram, R. 2000. A double-blind, placebo-controlled evaluation of the anxiolytic efficacy of an ethanolic extract of *Withania somnifera*. Indian J. Psychiatry 42: 295–301.

Antony, M.L., J. Lee, E.R. Hahm, S.H. Kim, A.I. Marcus, V. Kumari, X. Ji, Z. Yang, C.L. Vowell, P. Wipf, G.T. Uechi, N.A. Yates, G. Romero, S.N. Sarkar, and S.V. Singh. 2014. Growth arrest by the antitumor steroidal lactone withaferin A in human breast cancer cells is associated with downregulation and covalent binding at cysteine 303 of beta tubulin. J. Biol. Chem. 289: 1852e1865.

Anwer, T., M. Sharma, K.K. Pillai, and M. Iqbal. 2008. Effect of *Withania somnifera* on insulin sensitivity in non-insulin-dependent diabetes mellitus rats. Basic Clin. Pharmacol. 102: 498–503.

Ara, T.A.L.A.T., and A.K. Choudhary. 2014. Study on efficacy of two strains (ATCC 15834 and MTCC 532) of *Agrobacterium rhizogenes* on hairy root induction of *Withania somnifera*. Int. J. Biotechnol. Res. 4: 1–8.

Atal, C.K., O.P. Gupta, K. Ranghunathan, and K.L. Dhar. 1975. Pharmacognosy and phytochemistry of *Withania somnifera* (Linn.) Dunal (Ashwagandha). Central Council for Research in Indian Medicine and Homeopathy, New Delhi, India. 11: 6–18.

Bahr, V., and R. Hansel. 1982. Immunomodulating Properties of 5, 20α (R)-dihydroxy-6α, 7α-epoxy-1-oxo-(5α)-witha-2, 24-dienolide and solasodine. Planta. Med. 44: 32–33.

Baitharu. I., V. Jain, S.N. Deep, K.B. Hota, S.K. Hota, D. Prasad, and G. Ilavazhagan. 2013. *Withania somnifera* root extract ameliorates hypobaric hypoxia induced memory impairment in rats. J. Ethnopharmacol. 145: 431–441.

Baitharu, I., V. Jain, S.N. Deep, S. Shroff, J.K. Sahu, P.K. Naik, and G. Ilavazhagan. 2014. Withanolide A pre vents neurodegeneration by modulating hippocampal glutathione biosynthesis during hypoxia. PLoS One 9: e105311.

Baldi, A., D. Singh, and V.K. Dixit. 2008. Dual elicitation for improved production of withaferin A by cell suspension cultures of *Withania somnifera*. Appl. Biochem. Biotechnol. 151: 556.

Bandyopadhyay, M., S. Jha, and D. Tepfer. 2007. Changes in morphological phenotypes and withanolide composition of Ri-transformed roots of *Withania somnifera*. Plant Cell Rep. 26: 599–609.

Banerjee, S., A.A. Naqvi, S. Mandal, and P.S. Ahuja. 1994. Transformation of *Withania somnifera* (L.) Dunal by *Agrobacterium rhizogenes*: infectivity and phytochemical studies. Phytother. Res. 8: 452–455.

Bani, S., M. Gautam, F.A. Sheikh, B. Khan, N.K. Satti, K.A. Suri, G.N. Qazi, and B. Patwardhan. 2006. Selective Th1 upregulating activity of *Withania somnifera* aqueous extract in an experimental system using flow cytometry. J. Ethnopharmacol. 107: 107–115.

Bharti, V.K., J.K. Malik, and R.C. Gupta. 2016. Ashwagandha: multiple health benefits. In Nutraceuticals (pp. 717–733). Academic Press.

Bhatnagar, M., D. Sharma, and M. Salvi. 2009. Neuroprotective effects of *Withania somnifera* Dunal: A possible mechanism. Neurochem. Res. 34: 1975–1983.

Bhattacharya, S.K., A. Kumar, and S. Ghosal. 1995. Effects of glycowithanolides from *Withania somnifera* on animal model of Alzheimer's disease and perturbed central cholinergic markers of cognition in rats. Phytother. Res. 9: 110–113.

Bhattacharya, S.K., K.S. Satyan, and S. Ghosal. 1997. Antioxidant activity of glycowithanolides from *Withania somnifera*. Indian J. Exp. Biol. 35: 236–239.

Bhattacharya, S.K., A. Bhattacharya, and A. Chakrabarti. 2000a. Adaptogenic activity of Siotone, a polyherbal formulation of Ayurvedic rasayanas. Indian J. Exp. Biol. 38: 119–128.

Bhattacharya, S.K., A. Bhattacharya, K. Sairam, and S. Ghosal. 2000b. Anxiolytic-antidepressant activity of *Withania somnifera* glycowithanolides: an experimental study. Phytomedicine 7: 463–469.

Bhattacharya, S.K., and A.V. Muruganandam. 2003. Adaptogenic activity of *Withania somnifera*: an experimental study using a rat model of chronic stress. Pharmacol. Biochem. Behav. 75: 547–555.

Bhattarai, J.P., and S.K. Han. 2014. Phasic and tonic type A γ-aminobutryic acid receptor mediated effect of *Withania somnifera* on mice hippocampal CA1 pyramidal neurons. J. Ayurveda and Integr. Med. 5: 216–222.

Bisht, P., and V. Rawat. 2014. Antibacterial activity of *Withania somnifera* against Gram positive isolates from pus samples. Ayu. 35: 330–332.

Biswal, B.M., S.A. Sulaiman, H.C. Ismail, H. Zakaria, and K.I. Musa. 2013. Effect of *Withania somnifera* (Ashwagandha) on the development of chemotherapy induced fatigue and quality of life in breast cancer patients. Integr. Cancer Ther. 12: 312–322.

Budhiraja, R.D., S. Sudhir, and K.N. Garg. 1984. Anti-inflammatory activity of 3 beta-hydroxy-2,3-dihydron withanolide F. Planta Med. 50: 134–136.

Budhiraja, R.D., K.N. Garg, S. Sudhir, and B. Arora. 1986. Protective effect of 3-ss-hydroxy-2, 3-dihydrow ithanolide F against CCl_4 induced hepatotoxicity. Planta Med. 52: 28–29.

Caputi, F.F., E. Acquas, S. Kasture, S. Ruiu, S. Candeletti, and P. Romualdi. 2018. The standardized *Withania somnifera* Dunal root extract alters basal and morphine induced opioid receptor gene expression changes in neuroblastoma cells. BMC Complement Altern. Med. 18: 9.

Chandran, U., and B. Patwardhan. 2017. Network ethnopharmacological evaluation of the immunomodulatory activity of *Withania somnifera*. J. Ethnopharmacol. 197: 250–256.

Chandrasekaran, S., A. Dayakar, J. Veronica, S. Sundar, and R. Maurya. 2013. An *in vitro* study of apoptotic like death in *Leishmania donovani* promastigotes by withanolides. Parasitol Int. 62: 253–261.

Chen, D., H. Ye, and G. Li. 2000. Expression of a chimeric farnesyl diphosphate synthase gene in *Artemisia annua* L. transgenic plants via *Agrobacterium tumefaciens* mediated transformation. Plant. Sci. 155: 179–185.

Chen, L.X., H. He, and F. Qiu. 2011. Natural withanolides: an overview. Nat. Prod. Rep. 28: 705–740.

Choi, M.J., E.J. Park, K.J. Min, J.W. Park, and T.K. Kwon. 2011. Endoplasmic reticulum stress mediates withaferin A induced apoptosis in human renal carcinoma cells. Toxicol. *In Vitro* 25: 692–698.

Chopra, R.N., I.C. Chopra, K.L. Honda, and L.D. Kapur. 1958. In: Indigenous Drugs of India, Vol. 12, U.M. Dhu and Sons Pvt. Ltd. Calcutta: 436.

Choudhary, M.I., S. Hussain, S. Yousuf, and A. Dar. 2010. Chlorinated and diepoxy with anolides from *Withania somnifera* and their cytotoxic effects against human lung cancer cell line. Phytochem. 71: 2205–2209.

Ciddi, V. 2006. Withaferin A from cell cultures of *Withania somnifera*. Indian J. Pharm. Sci. 68.

Cooley, K., O. Szczurko, D. Perri, E.J. Mills, B. Bernhardt, Q. Zhou, and D. Seely. 2009. Naturopathic care for anxiety: a randomized controlled trial ISRCTN78958974. PLoS One. 4: e6628.

Croce, C.M. 2008. Oncogenes and cancer. N. Engl. J. Med. 358: 502–511.

Dahikar, P.R., N. Kumar, and Y. Sahni. 2012. Pharmacokinetics of *Withania somnifera* (ashwagandha) in healthy buffalo calves. Buffalo Bull. 31: 219.

Dar, N.J., A. Hamid, and M. Ahmad. 2015. Pharmacologic overview of *Withania somnifera*, the Indian ginseng. Cell. Mol. Life Sci. 72: 4445–4460.

Dar, P.A., S.A. Mir, J.A. Bhat, A. Hamid, L.R. Singh, F. Malik and T.A. Dar. 2019. An anticancerous protein fraction from *Withania somnifera* induces ROS dependent mitochondria mediated apoptosis in human MDA-MB-231 breast cancer cells. Int. J. Biol. Macromol. 135: 77–87.

Davis, L., and G. Kuttan. 2002. Effect of *Withania somnifera* on cell mediated immune responses in mice. J. Exp. Clin. Cancer Res. 21: 585–590.

De Rose, F., R. Marotta, S. Poddighe, G. Talani, T. Catelani, M.D. Setzu, and E. Acquas. 2016. Functional and morphological correlates in the drosophila LRRK2 loss of function model of Parkinson's disease: drug effects of *Withania somnifera* (Dunal) administration. PLoS One 11: e0146140.

Dehdashti, S.M., S. Acharjee, S. Kianamiri, and M. Deka. 2017. An efficient *Agrobacterium rhizogenes* mediated transformation protocol of *Withania somnifera*. Plant Cell Tissue Organ Cult. 128: 55–65.

Devarenne, T.P., A. Ghosh, and J. Chappell. 2002. Regulation of squalene synthase, a key enzyme of sterol biosynthesis, in tobacco. Plant. Physiol. 129: 1095–1106.

Devkar, S.T., A.D. Kandhare, A.A. Zanwar, S.D. Jagtap, S.S. Katyare, S.L. Bodhankar, and M.V. Hegde. 2016. Hepatoprotective effect of withanolide rich fraction in acetaminophen-intoxicated rat: decisive role of TNF-α, IL-1β , COX-II and iNOS. Pharm. Biol. 54: 2394–2403.

Dhar, N., S. Razdan, S. Rana, W.W. Bhat, R. Vishwakarma, and S.K. Lattoo. 2015. A decade of molecular understanding of withanolide biosynthesis and *in vitro* studies in *Withania somnifera* (L.) Dunal: prospects and perspectives for pathway engineering. Front. Plant Sci. 6: 1031.

Dhiman, R., N. Aggarwal, K.R. Aneja, and M. Kaur. 2016. *In vitro* antimicrobial activity of spices and medicinal herbs against selected microbes associated with juices. Int. J. microbial. 2016: 9015802.

Doma, M., G. Abhayankar, V.D. Reddy, and P.B. Kishor. 2012. Carbohydrate and elicitor enhanced withanolide (withaferin A and withanolide A) accumulation in hairy root cultures of *Withania somnifera* (L.). Ind. J. Expt. Biol. 50: 484–490.

El-Ghaouth, A.H.M.E.D., J. Arul, J. Grenier, N. Benhamou, A. Asselin, and R. Belanger. 1994. Effect of chitosan on cucumber plants: suppression of *Pythium aphanidermatum* and induction of defense reactions. Phytopathology 84: 313–320.

Ezzat, A., M. Wu, X.L. Li, and C.K. Kwoh. 2019. Computational prediction of drug–target interactions using chemogenomic approaches: an empirical survey. Brief. Bioinformatics. 20: 1337–1357.

Flores-Sánchez, I.J., J. Ortega-López, M.D.C. Montes-Horcasitas, and A.C. Ramos-Valdivia. 2002. Biosynthesis of sterols and triterpenes in cell suspension cultures of *Uncaria tomentosa*. Plant Cell Physiol. 43: 1502–1509.

Fong, M.Y., S. Jin, M. Rane, R.K. Singh, R. Gupta, and S.S. Kakar. 2012. Withaferin A synergizes the therapeutic effect of doxorubicin through ROS mediated autophagy in ovarian cancer. PLoS One 7: e42265.

Gautam, M., S.S. Diwanay, S. Gairola, Y.S. Shinde, S.S. Jadhav, and B.K. Patwardhan. 2004. Immune response modulation to DPT vaccine by aqueous extract of *Withania somnifera* in experimental system. Int. Immunopharmacol. 4: 841–849.

Gauttam, V.K., and A.N. Kalia. 2013. Development of polyherbal antidiabetic formulation encapsulated in the phospholipids vesicle system. J. Adv. Pharm. Technol. Res. 4: 108–117.

Giri, A., and M.L. Narasu. 2000. Transgenic hairy roots recent trends and applications. Biotechnol. Adv. 18: 1–22.

Gorelick, J., R. Rosenberg, A. Smotrich, L. Hanus, and N. Bernstein. 2015. Hypoglycemic activity of withanolides and elicitated *Withania somnifera*. Phytochem. 116: 283–289.

Govindappa, P.K., V. Gautam, S.M. Tripathi, Y.P. Sahni, and H.L.S. Raghavendra. 2019. Effect of *Withania somnifera* on gentamicin induced renal lesions in rats. Rev. Bras. Farmacogn. 29: 234–240.

Grin, B., S. Mahammad, T. Wedig, M.M. Cleland, L. Tsai, H. Herrmann, and R.D. Goldman. 2012. Withaferin A alters intermediate filament organization, cell shape and behavior. PLoS One 7: e39065.

Grover, A., S.P. Katiyar, J. Jeyakanthan, V.K. Dubey, and D. Sundar. 2012. Blocking Protein kinase C signaling pathway: mechanistic insights into the anti-leishmanial activity of prospective herbal drugs from *Withania somnifera*. BMC Genom. 13: S20.

Grover, A., G. Samuel, V.S. Bisaria, and D. Sundar. 2013. Enhanced withanolide production by overexpression of squalene synthase in *Withania somnifera*. J. Biosci. Bioeng. 115: 680–685.

Gupta, A., and S. Singh. 2014. Evaluation of anti-inflammatory effect of *Withania somnifera* root on collagen induced arthritis in rats. Pharm. Biol. 52: 308–320.

Gupta, A.P. 1996. Quantitative determination of withanferin A in different plant parts of *Withania somnifera* by TLC densitometry. J. Medi. Aro. Plant Sci. 18: 788–790.

Gupta, N., P. Sharma, R.J. Santosh Kumar, R.K. Vishwakarma, and B.M. Khan. 2012. Functional characterization and differential expression studies of squalene synthase from *Withania somnifera*. Mol. Biol. Rep. 39: 8803–8812.

Hahm, E.R., M.B. Moura, E.E. Kelley, B. Van Houten, S. Shiva, and S.V. Singh. 2011. Withaferin A induced apoptosis in human breast cancer cells is mediated by reactive oxygen species. PloS one 6: e23354.

Hahm, E.R., and S.V. Singh. 2013. Autophagy fails to alter withaferin A mediated lethality in human breast cancer cells. Curr. Cancer Drug Targets 13: 640–650.

Hamill, J.D., A.J. Parr, M.J. Rhodes, R.J. Robins, and N.J. Walton. 1987. New routes to plant secondary products. Bio/technology 5: 800–804.

Hey, S.J., S.J. Powers, M.H. Beale, N.D. Hawkins, J.L. Ward, and N.G. Halford. 2006. Enhanced seed phytosterol accumulation through expression of a modified HMG-CoA reductase. Plant Biotechnol. J. 4: 219–229.

Ichikawa, H., Y. Takada, S. Shishodia, B. Jayaprakasam, M.G. Nair, and B.B. Aggarwal. 2006. Withanolides potentiate apoptosis, inhibit invasion, and abolish osteoclastogenesis through suppression of nuclear factor-κB (NF-κB) activation and NF-κB–regulated gene expression. Mol. Cancer Ther. 5: 1434–1445.

Itokawa, H., S.L. Morris-Natschke, T. Akiyama, and K.H. Lee. 2008. Plant-derived natural product research aimed at new drug discovery. J. Nat. Med. 62: 263–280.

Iuvone, T., G. Esposito, F. Capasso, and A.A. Izzo. 2003. Induction of nitric oxide synthase expression by *Withania somnifera* in macrophages. Life Sci. 72: 1617–25.

Jain, S., S.D. Shukla, K. Sharma, and M Bhatnagar. 2001. Neuroprotective effects of *Withania somnifera* Dunn in hippocampal sub-regions of female albino rat. Phytother. Res. 15: 544–548.

Jain, S., and M.K.J.S. Banerjee. 2014. Increasing biomass accumulation and withaferin a production in cell suspension culture of *Withania somnifera*. Glob. J. Multidiscip. Stud. 3.

Jain, S.M., and P. Saxena. 2009. Protocols for *in vitro* cultures and secondary metabolite analysis of aromatic and medicinal plants. pp. 303–315. *In*: S. Mohan Jain (ed.). Methods in Molecular Biology, vol. 1391, 2nd edn. Humana Press, New York.

Jayaprakasam, B., K. Padmanabhan, and M.G. Nair. 2010. Withanamides in *Withania somnifera* fruit protect PC-12 cells from β-amyloid responsible for Alzheimer's disease. Phytother. Res. 24: 859–863.

Jayaprakasam, B., Y. Zhang, N.P. Seeram, and M.G. Nair. 2003. Growth inhibition of human tumor cell lines by withanolides from *Withania somnifera* leaves. Life Sci. 74: 125–132.

Jeyanthi, T., and P. Subramanian. 2009. Nephroprotective effect of *Withania somnifera*: a dose dependent study. Ren. fail. 31: 814–821.

Jha, A.K., M. Nikbakht, N. Capalash, and J. Kaur. 2014. Demethylation of RARβ 2 gene promoter by *Withania somnifera* in HeLa cell line. European J. Med. Plants 4: 503–510.

Johny, L., X. Conlan, D. Cahill, and A. Adholeya. 2015. *In vitro* and *in situ* screening systems for morphological and phytochemical analysis of *Withania somnifera* germplasms. Plant Cell Tissue Organ Cult. 120: 1191–1202.

Johny, L., X.A. Conlan, A. Adholeya, and D.M. Cahill. 2018. Growth kinetics and withanolide production in novel transformed roots of *Withania somnifera* and measurement of their antioxidant potential using chemiluminescence. Plant Cell Tissue Organ Cult. 132: 479–495.

Kaileh, M., W.V., A. Berghe, J. Heyerick, J. Horion, C. Piette, C. Libert, D. De Keukeleire, T. Essawi, and G. Haegeman. 2007. Withaferin a strongly elicits IkappaB kinase beta hyperphosphorylation concomitant with potent inhibition of its kinase activity. J. Biol. Chem. 282: 4253–4264.

Kamisako, W., K. Morimoto, I. Makino, and K. Isoi. 1984. Changes in triterpenoid content during the growth cycle of cultured plant cells. Plant Cell Physiol. 25: 1571–1574.

Kataria, H., R. Wadhwa, S.C. Kaul, and G. Kaur. 2012. Water extract from the leaves of *Withania somnifera* protect RA differentiated C6 and IMR-32 cells against glutamate induced excitotoxicity. PLoS One 7: e37080.

Kaushik, M.K., S.C. Kaul, R. Wadhwa, M. Yanagisawa, and Y. Urade. 2017. Triethylene glycol, an active component of Ashwagandha (*Withania somnifera*) leaves, is responsible for sleep induction. PloS one 12: e0172508.

Ketchum, R.E., D.M. Gibson, R.B. Croteau, and M.L. Shuler. 1999. The kinetics of taxoid accumulation in cell suspension cultures of *Taxus* following elicitation with methyl jasmonate. Biotechnol. Bioeng. 62: 97–105.

Khan, B., S.F. Ahmad, S. Bani, A. Kaul, K.A. Suri, N.K. Satti, and G.N. Qazi. 2006. Augmentation and proliferation of T lymphocytes and Th–1 cytokines by *Withania somnifera* in stressed mice. Int. Immunopharmacol. 6: 1394–1403.

Khanchandani, N., P. Shah, T. Kalwani, A. Ardeshna, and D. Dharajiya. 2019. Antibacterial and antifungal ctivity of Ashwagandha (*Withania somnifera* L.): a review. J. Drug Deliv. Ther. 9: 154–161.

Khare, C.P. 2007. Indian medicinal plants–an illustrated dictionary. First Indian Reprint, Springer (India) Pvt. Ltd., New Delhi. pp. 717–718.

Khazal, K.F., T. Samuel, D.L. Hill, and C.J. Grubbs. 2013. Effect of an extract of *Withania somnifera* root on estrogen receptor-positive mammary carcinomas. Anticancer Res. 33: 1519–1523.

Kim, S.H., and S.V. Singh. 2014. Mammary cancer chemoprevention by withaferin A is accompanied by *in vivo* suppression of self-renewal of cancer stem cells. Cancer Prev. Res. 7: 738–747.

Kim, S.H., K.B. Singh, E.R. Hahm, B.L. Lokeshwar, and S.V. Singh. 2020. *Withania somnifera* root extract inhibits fatty acid synthesis in prostate cancer cells. J. Tradit. Complement. Med. https://doi.org/10.1016/j.jtcme.2020.02.002.

Kirson, I., A. Abraham, and D. Lavie. 1977. Chemical analysis of hybrids of *Withania somnifera* L. (Dun.). 1. Chemotypes III (Israel) by Indian I (Delhi). Isr. J. Chem. 16: 20–24.

Kirtikar, K.R., and B.D. Basu. 1991. Indian Medicinal Plants. Vol. 2. Dehradun. International book distributors, p. 1182.

Konar, A., N. Shah, R. Singh, N. Saxena, S.C. Kaul, R. Wadhwa, and M.K. Thakur. 2011. Protective role of Ashwagandha leaf extract and its component withanone on scopolamine induced changes in the brain and brain derived cells. PLoS One 6: e27265.

Konar, A., R. Gupta, R.K. Shukla, B. Maloney, V.K. Khanna, R. Wadhwa, D.K. Lahiri, and M.K. Thakur. 2019. M1 muscarinic receptor is a key target of neuroprotection, neuroregeneration and memory recovery by i-Extract from *Withania somnifera*. Sci. Rep. 9: 1–15.

Kroschinsky, F., F. Stolzel, S. von Bonin, G. Beutel, M. Kochanek, M. Kiehl, and P. Schellongowski. 2017. New drugs, new toxicities: severe side effects of modern targeted and immunotherapy of cancer and their management. Crit. Care. 21: 89.

Ku, S.K., M.S. Han, and J.S. Bae. 2014. Withaferin A is an inhibitor of endothelial protein C receptor shedding *in vitro* and *in vivo*. Food Chem. Toxicol. 68: 23–29.

Kuai, R., C. Subramanian, P.T. White, B.N. Timmermann, J.J. Moon, M.S. Cohen, and A. Schwendeman. 2017. Synthetic high density lipoprotein nanodisks for targeted withalongolide delivery to adrenocortical carcinoma. Int. J. Nanomedicine 12: 6581.

Kuboyama, T., C. Tohda, J. Zhao, N. Nakamura, M. Hattori, and K Komatsu. 2002. Axon or dendrite predominant outgrowth induced by constituents from Ashwagandha. Neuroreport 13: 1715–1720.

Kuboyama, T., C. Tohda, and K. Komatsu. 2005. Neuritic regeneration and synaptic reconstruction induced by withanolide A. Br. J. Pharmacol. 144: 961–971.

Kuboyama, T., C. Tohda, and K. Komatsu. 2006. Withanoside IV and its active metabolite, sominone, attenuate Aβ (25–35) induced neurodegeneration. Eur. J. Neurosci. 23: 1417–1426.

Kuboyama, T., C. Tohda, and K. Komatsu. 2014. Effects of Ashwagandha (roots of *Withania somnifera*) on neurodegenerative diseases. Biol. Pharm. Bull. 37: 892–897.

Kulkarni, S.K., and A. Dhir. 2008. *Withania somnifera*: an Indian ginseng. Prog. Neuro-Psychopharmacol. Biol. Psychiatry 32: 1093–1105.

Kumar, P., R. Singh, A. Nazmi, D. Lakhanpal, H. Kataria , and G. Kaur. 2014. Glioprotective effects of Ashwagandha leaf extract against lead induced toxicity. Biomed. Res. Int. 2014: 182029.

Kumar, S., F.M. Hahn, E. Baidoo, T.S. Kahlon, D.F. Wood, C.M. McMahan, K. Cornish, J.D. Keasling, H. Daniell, and M.C. Whalen. 2012. Remodeling the isoprenoid pathway in tobacco by expressing the cytoplasmic mevalonate pathway in chloroplasts. Metab. Eng. 14: 19–28.

Kumar, V., K.N.C. Murthy, S. Bhamid, C.G. Sudha, and G.A. Ravishankar. 2005. Genetically modified hairy roots of *Withania somnifera* Dunal: a potent source of rejuvenating principles. Rejuv. Res. 8: 37–45.

Kumar, V., A. Dey, and S.S. Chatterjee. 2017. Phytopharmacology of Ashwagandha as an anti-diabetic herb. In Science of Ashwagandha: Preventive and Therapeutic Potentials (pp. 37–68). Springer, Cham.

Kumari, M., R.P. Gupta, D. Lather, and P. Bagri. 2020. Ameliorating effect of *Withania somnifera* root extract in *Escherichia coli* infected broilers. Poultry Science. https://doi.org/10.1016/j.psj.2019.11.022.

Kurapati, K.R., V.S. Atluri, T. Samikkannu, and N.P. Nai. 2013. Ashwagandha (*Withania somnifera*) reverses beta-amyloid1-42 induced toxicity in human neuronal cells: implications in HIV associated neurocognitive disorders (HAND). PLoS One 8: e77624.

Kushwaha, S., S. Roy, R. Maity, A. Mallick, V.K. Soni, P.K. Singh, and C. Mandal. 2012a. Chemotypical variations in *Withania somnifera* lead to differentially modulated immune response in BALB/c mice. Vaccine. 30: 1083–1093.

Kushwaha, S., V.K. Soni, P.K. Singh, N. Bano, A. Kumar, R.S. Sangwan, and S. Misra Bhattacharya. 2012b. *Withania somnifera* chemotypes NMITLI 101R, NMITLI 118R, NMITLI 128R and withaferin A protect *Mastomys coucha* from *Brugia malayi* infection. Parasite Immunol. 34: 199–209.

Kushwaha, V., M. Sharma, P. Vishwakarma, M. Saini, and K. Saxena. 2016. Biochemical assessment of nephroprotective and nephrocurative activity of *Withania somnifera* on gentamicin induced nephrotoxicity in experimental rats. Int. J. Res. Med. Sci. 4: 298–302.

Kuttan, G. 1996. Use of *Withania somnifera* Dunal as an adjuvant during radiation therapy. Indian J. Exp. Biol. 34: 854–856.

Lancien, M., S. Ferrario-Méry, Y. Roux, E. Bismuth, C. Masclaux, B. Hirel, P. Gadal, and M. Hodges. 1999. Simultaneous expression of NAD dependent isocitrate dehydrogenase and other Krebs cycle genes after nitrate resupply to short term nitrogen starved tobacco. Plant Physiol. 120: 717–726.

Lavie, D., E. Glotter, and Y. Shvo. 1965. Constituents of *Withania somnifera* Dun. Part IV. The structure of withaferin A. J. Chem. Soc. (Resumed). 1371: 7517–7531.

Lee, J.H., J.E. Kim, Y.J. Jang, C.C. Lee, T.G. Lim, S.K. Jung, E. Lee, S.S. Lim, Y.S. Heo, S.G. Seo, and J.E. Son. 2015. Dehydroglyasperin C suppresses TPA-induced cell transformation through direct inhibition of MKK4 and PI3K. Mol. Carcinogen. 55: 552–562.

Lessard, P.A., H. Kulaveerasingam, G.M. York, A. Strong, and A.J. Sinskey. 2002. Manipulating gene expression for the metabolic engineering of plants. Metab. Eng. 4: 67–79.

Lockley, W.J., H.H. Rees, and T.W. Goodwin. 1976. Biosynthesis of steroidal withanolides in *Withania somnifera*. Phytochem. 15: 937–939.

Lodhi, A.H., R.J.M. Bongaerts, R. Verpoorte, S.A. Coomber, and B.V. Charlwood. 1996. Expression of bacterial isochorismate synthase (E C 5.4.99.6) in transgenic root cultures of *Rubia peregrina*. Plant Cell Rep. 16: 54–7.

Malaikozhundan, B., J. Vinodhini, M.A.R. Kalanjiam, V. Vinotha, S. Palanisamy, S. Vijayakumar, B. Vaseeharan, and A. Mariyappan. 2020. High synergistic antibacterial, antibiofilm, antidiabetic and antimetabolic activity of *Withania somnifera* leaf extract-assisted zinc oxide nanoparticle. Bioproc. Biosyst. Eng. p: 1–15. https://doi.org/10.1007/s00449-020-02346-0.

Malik, F., A. Kumar, S. Bhushan, S. Khan, A. Bhatia, K.A. Suri, G.N. Qazi, and J. Singh. 2007. Reactive oxygen species generation and mitochondrial dysfunction in the apoptotic cell death of human myeloid leukemia HL-60 cells by a dietary compound withaferin A with concomitant protection by N-acetyl cysteine. Apoptosis 12: 2115–33.

Maliyakkal, N., N. Udupa, K.S. Pai, and A. Rangarajan. 2013. Cytotoxic and apoptotic activities of extracts of *Withania somnifera* and *Tinospora cordifolia* in human breast cancer cells. Inter. J. Appl. Res. in Nat. Prod. 6: 1–10.

Mamidi, P., and A.B. Thakar. 2011. Efficacy of Ashwagandha (*Withania somnifera* Dunal. Linn.) in the management of psychogenic erectile dysfunction. Ayu. 32: 322.

Mandal, C., A. Dutta, A. Mallick, S. Chandra, L. Misra, R.S. Sangwan, and C. Mandal. 2008. Withaferin A induces apoptosis by activating p38 mitogen activated protein kinase signaling cascade in leukemic cells of lymphoid and myeloid origin through mitochondrial death cascade. Apoptosis. 13: 1450–1464.

Mandlik, D.S., and A.G. Namdeo. 2020. Pharmacological evaluation of Ashwagandha highlighting its healthcare claims, safety, and toxicity aspects. J. Diet. Suppl. 1–44.

Mayola, E., C. Gallerne, D.D. Esposti, C. Martel, S. Pervaiz, L. Larue, B. Debuire, A. Lemoine, C. Brenner, and C. Lemaire. 2011. Withaferin A induces apoptosis in human melanoma cells through generation of reactive oxygen species and down-regulation of Bcl-2. Apoptosis 16: 1014–1027.

Minhas, U., R. Minz, and A Bhatnagar. 2011. Prophylactic effect of *Withania somnifera* on inflammation in a non-autoimmune prone murine model of lupus. Drug Discov. Ther. 5: 195–201.

Mir, B.A., S.A. Mir, and S. Koul. 2014. *In vitro* propagation and withaferin A production in *Withania ashwagandha*, a rare medicinal plant of India. Physiol. Mol. Biol. Pla. 20: 357–364.

Mirjalili, M.H., E. Moyano, M. Bonfill, R.M. Cusido, and J. Palazón. 2009a. Steroidal lactones from *Withania somnifera*, an ancient plant for novel medicine. Molecules 14: 2373–2393.

Mirjalili, M.H., S.M. Fakhr-Tabatabaei, H. Alizadeh, A. Ghassempour, and F. Mirzajani. 2009b. Genetic and withaferin A analysis of Iranian natural populations of *Withania somnifera* and *W. coagulans* by RAPD and HPTLC. Nat. Prod. Commu. 4: 337–346.

Mishra, L.C, B.B. Singh, and S. Dagenais. 2000. Scientific basis for the therapeutic use of *Withania somnifera* (ashwagandha): a review. Altern. Med. Rev. 5: 334–346.

Mishra, L., P. Mishra, A. Pandey, R.S. Sangwan, N.S. Sangwan, and R. Tuli. 2008. Withanolides from *Withania somnifera* roots. Phytochem. 69: 1000–1004.

Mishra, D., and S. Patnaik. 2020. GCMS Analysed phytochemicals and antibacterial activity of *Withania Somnifera* (L.) Dunal extract in the context of treatment to liver cirrhosis. Biomed. Pharmacol. http://dx.doi.org/10.13005/bpj/1862.

Mishra, S., S. Bansal, B. Mishra, R.S. Sangwan, J.S. Jadaun, and N.S. Sangwan. 2016. RNAi and homologous over expression based functional approaches reveal triterpenoid synthase gene cycloartenol synthase is involved in downstream withanolide biosynthesis in *Withania somnifera*. PLoS One 11: e0149691.

Mohanty, I., D.S. Arya, A. Dinda, K.K. Talwar, S. Joshi, and S.K. Gupta. 2004. Mechanisms of cardioprotective effect of *Withania somnifera* in experimentally induced myocardial infarction. Basic Clin. Pharmacol. Toxicol. 94: 184–190.

Mohanty, I.R., D.S. Arya, and S.K. Gupta. 2008. *Withania somnifera* provides cardioprotection and attenuates ischemia–reperfusion induced apoptosis. Clin. Nutr. 27: 635–642.

Mondal, S., C. Mandal, R. Sangwan, and S. Chandra. 2010. Withanolide D induces apoptosis in leukemiaby targeting the activation of neutral sphingomyelinase ceramide cascade mediated by synergistic activation of c-Jun N-terminal kinase and p38 mitogen activated protein kinase. Mol. Cancer. 9: 239.

Mulabagal, V., G.V. Subbaraju, C.V. Rao, C. Sivaramakrishna, D.L. Dewitt, D. Holmes, B. Sung, B.B. Aggarwal, H.S. Tsay, and M.G. Nair. 2009. Withanolide sulfoxide from Aswagandha roots inhibits nuclear transcription factor kappa B, cyclooxygenase and tumor cell proliferation. Phytother. Res. 23: 987–992.

Muralikrishnan, G., S. Amanullah, M.I. Basha, A.K. Dinda, and F. Shakeel. 2010. Modulating effect of *Withania somnifera* on TCA cycle enzymes and electron transport chain in azoxymethane induced colon cancer in mice. Immunopharmacol. Immunotoxicol. 32: 523–7.

Murashige, T., and F. Skoog. 1962. A revised medium for rapid growth and bioassays with tobacco tissue cultures. Physiol. Plant. 15: 473–497.

Murthy, H.N., C. Dijkstra, P. Anthony, D.A. White, M.R. Davey, J.B. Power, E.J. Hahn, and K.Y. Paek. 2008. Establishment of *Withania somnifera* hairy root cultures for the production of withanolide A. J. Integr. Plant Biol. 50: 975–981.

Murthy, H.N., and N. Praveen. 2013. Carbon sources and medium pH affects the growth of *Withania somnifera* (L.) Dunal adventitious roots and withanolide A production. Nat. Prod. Res. 27: 185–189.

Murugesan, P., and K. Senthil. 2017. Influence of auxin an biomass production and withanolide accumulation in adventitious root culture of indian rennet: *Withania Coagulans*. Asian J. Pharm. Clin. Res. 10: 131–134.

Mutalik, S., M. Chetana, B. Sulochana, P.U. Devi, and N. Udupa. 2005. Effect of Dianex, a herbal formulation on experimentally induced diabetes mellitus. Phytother. Res. 19: 409–415.

Mwitari, P.G., P.A. Ayeka, J. Ondicho, E.N. Matu, and G.C.C. Bii. 2013. Antimicrobial activity and probable mechanisms of action of medicinal plants of Kenya: *Withania somnifera, Warbugia ugandensis, Prunus africana* and *Plectrunthus barbatus*. PLoS One 8: 1–9.

Nagella, P., and H.N. Murthy. 2010. Establishment of cell suspension cultures of *Withania somnifera* for the production of withanolide A. Bioresour. Technol. 101: 6735–6739.

Nagella, P., and H.N. Murthy. 2011. Effects of macroelements and nitrogen source on biomass accumulation and withanolide A production from cell suspension cultures of *Withania somnifera* (L.) Dunal. Plant Cell Tissue Organ Cult. 104: 119–124.

Nema, R., S. Khare, P. Jain, and A. Pradhan. 2013. Anticancer activity of *Withania somnifera* (leaves) flavonoids compound. Int. J. Pharm. Sci. Rev. Res. 19: 103–106.

Nile, S.H., A. Nile, E. Gansukh, V. Baskar, and G. Kai. 2019. Subcritical water extraction of withanosides and withanolides from ashwagandha (*Withania somnifera* L.) and their biological activities. Food Chem. Toxicol. 132: 110659.

Nittala, S.S., and D. Lavie. 1981. Chemistry and genetics of withanolides in *Withania somnifera* hybrids. Phytochem 20: 2741–2748.

Noh, E.J., M.J. Kang, Y.J. Jeong, J.Y. Lee, J.H. Park, H.J. Choi, S.M. Oh, K.B. Lee, D.J. Kim, J. Shin, and S.D. Cho. 2016. Withaferin A inhibits inflammatory responses induced by *Fusobacterium nucleatum* and *Aggregatibacter actinomycetem* comitans in macrophages. Mol. Med. Rep. 14: 983–988..

Owais, M., K. Sharad, A. Shehbaz, and M. Saleemuddin. 2005. Antibacterial efficacy of *Withania somnifera* (ashwagandha) an indigenous medicinal plant against experimental murine salmonellosis. Phytomed. 12: 229–235.

Padhya, M.R., S.D. Jogdand, and J. Bhattacharjee. 2018. Evaluation & comparison of nephroprotective effect of *Hemidesmus indicus* Linn. & *Withania somnifera* Linn. on gentamicin induced nephrotoxicity in rats. Int. J. Basic Clin. Pharmacol. 7: 691.

Palliyaguru, D.L., S.V. Singh, and T.W. Kensler. 2016. *Withania somnifera*: from prevention to treatment of cancer. Mol. Nutr. Food Res. 60: 1342–1353.

Pandey, A., S. Bani, P. Dutt, N.K. Satti, K.A. Suri, and G.N. Qazi. 2018. Multifunctional neuroprotective effect of Withanone, a compound from *Withania somnifera* roots in alleviating cognitive dysfunction. Cytokine. 102: 211–221.

Pandey, V., R. Srivastava, N. Akhtar, J. Mishra, P. Mishra, and P.C. Verma. 2016. Expression of *Withania somnifera* steroidal glucosyl transferase gene enhances withanolide content in hairy roots. Plant Mol. Biol. Rep. 34: 681–689.

Pandit, S., K.W. Chang, and J.G. Jeon. 2013. Effects of *Withania somnifera* on the growth and virulence properties of *Streptococcus mutans* and *Streptococcus sobrinus* at sub MIC levels. Anaerobe. 19: 1–8.

Panjamurthy, K., S. Manoharan, M.R. Nirmal, and L. Vellaichamy. 2009. Protective role of Withaferin A on immunoexpression of p53 and bcl-2 in 7,12-dimethylbenz (a) anthracene induced experimental oral carcinogenesis. Investig. New Drugs 27: 447–452.

Parameswari, M., E. Laras, K. Joni, M. Valizadeh, and S. Kalaiselvi. 2017. A study on the influence of plant growth regulator on shoot multiplication and evaluation of major withanolides in *Withania somnifera*. World J. Pharm. Pharm Sci. 6: 842–853.

Parihar, M.S., and T. Hemnani. 2003. Phenolic antioxidants attenuate hippocampal neuronal cell damage against kainic acid induced excitotoxicity. J. Biosci. 28: 121–128.

Patel, K., R.B. Singh, and D.K. Patel. 2013. Pharmacological and analytical aspects of withaferin A: A concise report of current scientific literature. Asian Pac. J. Reprod. 2: 238–243.

Patel, N., P. Patel, S.V. Kendurkar, and B.M. Khan. 2015. Overexpression of squalene synthase in *Withania somnifera* leads to enhanced withanolide biosynthesis. Plant Cell Tissue Organ Cult. 122: 409–420.

Patel, N., P. Patel, and B.M. Khan. 2016. Metabolic engineering: Achieving new insights to ameliorate metabolic profiles in *Withania somnifera*. In Medicinal plants-recent advances in research and development (pp. 191–214). Springer, Singapore.

Pawar, P., S. Gilda, S. Sharma, S. Jagtap, A. Paradkar, K. Mahadik, P. Ranjekar, and A. Harsulkar. 2011. Rectal gel application of *Withania somnifera* root extract expounds anti-inflammatory and muco restorative activity in TNBS induced inflammatory bowel disease. BMC Complement. Altern. Med. 11: 34.

Pawar, P.K., and V.L. Maheshwari. 2004. *Agrobacterium rhizogenes* mediated hairy root induction in two medicinally important members of family Solanaceae. Indian J. Biotechnol. 3: 414–417.

Praveen, N., and H.N. Murthy. 2010. Production of withanolide A from adventitious root cultures of *Withania somnifera*. Acta. Physiol. Plant. 32: 1017–1022.

Praveen, N., and H.N. Murthy. 2012. Synthesis of withanolide A depends on carbon source and medium pH in hairy root cultures of *Withania somnifera*. Ind. Crop. Prod. 35: 241–243.

Praveen, N., and H.N. Murthy. 2013. Withanolide A production from *Withania somnifera* hairy root cultures with improved growth by altering the concentrations of macro elements and nitrogen source in the medium. Acta. Physiol. Plant. 35: 811–816.

Prince, P.S.M., S. Suman, P.T. Devika, and M. Vaithianathan. 2008. Cardioprotective effect of 'Marutham' a polyherbal formulation on isoproterenol induced myocardial infarction in Wistar rats. Fitoterapia. 79: 433–438.

Ramanathan, M., B. Balaji, A. Justin, N. Gopinath, M. Vasanthi, and R.V. Ramesh. 2011. Behavioural and neurochemical evaluation of Perment® an herbal formulation in chronic unpredictable mild stress induced depressive model. Indian J. Exp. Biol. 49: 269–275.

Rangaraju, S., A.N. Lokesha, and C.R. Aswath. 2018. Standardization of various factors for production of adventitious roots in selected varieties of *Withania somnifera* and estimation of total withanolides content by high performance liquid chromatography. Acta. Sci. Agricul. 2: 100–109.

Ram, H., and A. Krishna. 2021. Improvements in Insulin Resistance and β-Cells Dysfunction by DDP-4 Inhibition Potential of *Withania somnifera* (L.) Dunal Root Extract in Type 2 Diabetic Rat. Biointerface Res. Appl. Chem. 11: 8141–8155.

Rao, S., V.K. Teesta, A. Bhattrai, K. Khushi, and S. Bhat. 2012. *In vitro* propagation of *Withania somnifera* and estimation of withanolides for neurological disorders. J. Pharmacogn. 3: 85–87.

Rasool, M., and P. Varalakshmi. 2006. Immunomodulatory role of *Withania somnifera* root powder on experimental induced inflammation: an *in vivo* and *in vitro* study. Vasc. Pharmacol. 44: 406–410.

Ravindran, R., N. Sharma, S. Roy, A.R. Thakur, S. Ganesh, S. Kumar, J. Devi, and J. Rajkumar. 2015. Interaction studies of *Withania somnifera's* key metabolite withaferin A with different receptors associated with cardiovascular disease. Curr. Comput Aid Drug 11: 212–221.

Ray, A., S. Jena, T. Halder, A. Sahoo, B. Kar, J. Patnaik, B. Ghosh, P.C. Panda, N. Mahapatra, and S. Nayak. 2019. Population genetic structure and diversity analysis in *Hedychium coronarium* populations using morphological, phytochemical and molecular markers. Ind. Crops Prod. 132: 118–133.

Ray, A.B., and M. Gupta. 1994. Withasteroides: A growing group of naturally occurring steroidal lactones. Prog. Chem. Org. Nat. Prod. 63: 1–106.

Ray, S., B. Ghosh, S. Sen, and S. Jha. 1996. Withanolide production by root cultures of *Withania somnifera* transformed with *Agrobacterium rhizogenes*. Planta Med. 62: 571–573.

Ray, S., and S. Jha. 2001. Production of withaferin A in shoot cultures of *Withania somnifera*. Planta Med. 67: 432–436.

Rayees, S., R.K. Johri, G. Singh, M.K. Tikoo, S. Singh, S.C. Sharma, N.K. Satti, V.K. Gupta, and Y.S. Bedi. 2012. Sitopaladi, a polyherbal formulation inhibits IgE mediated allergic reactions. Spatula DD. 2: 75–82.

Rayees, S., A. Kumar, S. Rasool, P. Kaiser, N.K. Satti, P.L. Sangwan, S. Singh, R.K. Johri, and G. Singh. 2013. Ethanolic extract of Alternanthera sessilis (AS-1) inhibits IgE-mediated allergic response in RBL-2H3 cells. Immunol. Investig. 42: 470–480.

Roja, G., M.R. Heble, and A.T. Sipahimalani. 1991. Tissue cultures of *Withania somnifera*: morphogenesis and withanolide synthesis. Phytother. Res. 5: 185–187.

Sabina, E., S. Chandel, and M.K. Rasool. 2009. Evaluation of analgesic, antipyretic and ulcerogenic effect of withaferin A. Int. J. Integr. Biol. 6: 52–56.

Sabir, F., N.S. Sangwan, N.D. Chaurasiya, L.N. Misra, and R.S. Sangwan. 2008. *In vitro* withanolide production by *Withania somnifera* L. cultures. Zeitschrift für Naturforschung C. 63: 409–412.

Sabir, F., R.S. Sangwan, J. Singh, L.N. Misra, N. Pathak, and N.S. Sangwan. 2011. Biotransformation of withanolides by cell suspension cultures of *Withania somnifera* (Dunal). Plant Biotechnol. Rep. 5: 127–134.

Samadi, A.K., S.M. Cohen, R. Mukerji, V. Chaguturu, X. Zhang, B.N. Timmermann, M.S. Cohen, and E.A. Person. 2012. Natural withanolide withaferin A induces apoptosis in uveal melanoma cells by suppression of Akt and c-MET activation. Tumour Biol. 33: 1179–1189.

Sangwan, R.S., N.D. Chaurasiya, L.N. Misra, P. Lal, G.C. Uniyal, R. Sharma, N.S. Sangwan, K.A. Suri, G.N. Qazi, and R. Tuli. 2004. Phytochemical variability in commercial herbal products and preparations of *Withania somnifera* (Ashwagandha). Curr. Sci. 86: 461–465.

Sangwan, R.S., N.D. Chaurasiya, P. Lal, L. Misra, G.C. Uniyal, R. Tuli, and N.S. Sangwan. 2007. Withanolide A biogeneration in *in vitro* shoot cultures of Ashwagandha (*Withania somnifera* Dunal), a Main Medicinal Plant in Ayurveda. Chem. Pharm. Bull. 55: 1371–1375.

Sarangi, A., S. Jena, A.K. Sarangi, and B. Swain. 2013. Anti-diabetic effect of *Withania somnifera* root and leaf extracts on streptozotocin induced diabetic rats. J. Cell & Tissue Res. 13: 3597.

Saravanakumar, A., A. Aslam, and A. Shajahan. 2012. Development and optimization of hairy root culture systems in *Withania somnifera* (L.) Dunal for withaferin A production. Afr. J. Biotechnol. 11: 37–68.

Scheible, W.R., R. Morcuende, T. Czechowski, D. Osuna, C. Fritz, N. Palacios-Rojas, O. Thimm, M.K. Udvardi, and M. Stitt. 2004. Genome-wide reprogramming of primary and secondary metabolism, protein synthesis, cellular growth processes and signaling infrastructure by nitrogen. Plant Physiol. 136: 2483–2499.

Schliebs, R., A. Liebmann, S.K. Bhattacharya, A. Kumar, S. Ghosal, and V. Bigl. 1997. Systemic administration of defined extracts from *Withania somnifera* (Indian Ginseng) and Shilajit differentially affects cholinergic but not glutamatergic and GABAergic markers in rat brain. Neurochem. Int. 30: 181–190.

Schmelzer, G.H., and A. Gurib-Fakim. 2008. Plants resources of tropical Africa. Backhuys Publishers. 11: 34–37.

Sehgal, N., A. Gupta, R.K. Valli, S.D. Joshi, J.T. Mills, E. Hamel, P. Khanna, S.C. Jain, S.S. Thakur, and V. Ravindranath. 2012. *Withania somnifera* reverses Alzheimer's disease pathology by enhancing low density lipoprotein receptor related protein in liver. Proc. Natl. Acad. Sci. 109: 3510–3515.

Sehrawat, A., S.K. Samanta, E.R. Hahm, C. St Croix, S. Watkins, and S.V. Singh. 2019. Withaferin A mediated apoptosis in breast cancer cells is associated with alterations in mitochondrial dynamics. Mitochondrion 47: 282–293.

Senthil, K., M. Jayakodi, P. Thirugnanasambantham, S.C. Lee, P. Duraisamy, P.M. Purushotham, K. Rajasekaran, S. Nancy Charles, I. Mariam Roy, A.K. Nagappan, G.S. Kim, Y.S. Lee, S. Natesan, T.S. Min, and T.J. Yang. 2015. Transcriptome analysis reveals *in vitro* cultured *Withania somnifera* leaf and root tissues as a promising source for targeted withanolide biosynthesis. BMC Genom. 16: 14.

Sharada, M., A. Ahuja, K.A. Suri, S.P. Vij, R.K. Khajuria, V. Verma, and A. Kumar. 2007. Withanolide production by *in vitro* cultures of *Withania somnifera* and its association with differentiation. Biol. Plantarum. 51: 161–164.

Sharma, A., I. Sharma, and P.K. Pati. 2011. Post-infectional changes associated with the progression of leaf spot disease in *Withania somnifera*. J. Plant Pathol. 397–405.

Sharma, A.K., I. Basu, and S. Singh. 2018. Efficacy and safety of ashwagandha root extract in subclinical hypothyroid patients: a double-blind, randomized placebo-controlled trial. J. Altern. Complement Med. 24: 243–248.

Shohat, B., I. Kirson, and D. Lavie. 1978. Immunosuppressive activity of two plant steroidal lactones withaferin A and withanolide. E. Biomedicine 28: 18–24.

Siddique, A.A., P. Joshi, L. Misra, N.S. Sangwan, and M.P. Darokar. 2014. 5 6-De-epoxy-5-en-7-one-17-hydroxy withaferin A a new cytotoxic steroid from *Withania somnifera* L. Dunal leaves. Nat. Prod. Res. 28: 392–398.

Siddiqui, N.A., S. Sing, M.M. Siddiqui, and T.H. Khan. 2012. Immunomodulatory effect of *Withania somnifera, Asparagus racemose* and *Picrorhiza kurroa* roots. Int. J. Pharmacol. 8: 108–14.

Singariya, P., K.K. Moury, and P. Kumar. 2012. Antimicrobial activity of the crude extracts of *Withania somnifera* and *Cenchrus setigerus*. Invitro. Phcog. J. 4: 60–65.

Singh, B., A.K. Saxena, B.K. Chandan, D.K. Gupta, K.K. Bhutani, and K.K. Anand. 2001. Adaptogenic activity of a novel, withanolide-free aqueous fraction from the roots of *Withania somnifera* Dun. Phytother. Res. 15: 311–318.

Singh, M., and C. Ramassamy. 2017. *In vitro* screening of neuroprotective activity of Indian medicinal plant *Withania somnifera*. J. Nutr. Sci. 6: e54.

Singh, N., M. Bhalla, P. de Jager, and M. Gilca. 2011. An overview on ashwagandha: a Rasayana (Rejuvenator) of Ayurveda. Afr. J. Tradit. Complement. Altern. Med. 8: 208–213.

Singh, P., R. Guleri, V. Singh, G. Kaur, H. Kataria, B. Singh, G. Kaur, S.C. kaul, R. Wadhwa, and P.K. Pati. 2015. Biotechnological interventions in *Withania somnifera* (L.) Dunal. Biotechnol. Genet. Eng. Rev. 19: 1–20.

Sivanandhan, G., T.S. Mariashibu, M. Arun, M. Rajesh, S. Kasthurirengan, N. Selvaraj, and A. Ganapathi. 2011. The effect of polyamines on the efficiency of multiplication and rooting of *Withania somnifera* (L.) Dunal and content of some withanolides in obtained plants. Acta Physiol. Plant. 33: 2279–2288.

Sivanandhan, G., M. Arun, S. Mayavan, M. Rajesh, T.S. Mariashibu, M. Manickavasagam, N. Selvaraj, and A. Ganapathi. 2012. Chitosan enhances withanolides production in adventitious root cultures of *Withania somnifera* (L.) Dunal. Ind. Crops Prod. 37: 124–129.

Sivanandhan, G., M. Rajesh, M. Arun, M. Jeyaraj, G.K. Dev, A. Arjunan, M. Manickavasagam, M. Muthuselvam, N. Selvaraj, and A. Ganapathi. 2013. Effect of culture conditions, cytokinins, methyl jasmonate and salicylic acid on the biomass accumulation and production of withanolides in multiple shoot culture of *Withania somnifera* (L.) Dunal using liquid culture. Acta Physiol. Plant. 35: 715–728.

Sivanandhan, G., N. Selvaraj, A. Ganapathi, and M. Manickavasagam. 2014a. An efficient hairy root culture system for *Withania somnifera* (L.) Dunal. Afr. J. Biotechnol. 13: 4141–4147.

Sivanandhan, G., N. Selvaraj, A. Ganapathi, and M. Manickavasagam. 2014b. Improved production of withanolides in shoot suspension culture of *Withania somnifera* (L.) Dunal by seaweed extracts. Plant Cell Tissue Organ Cult. 119: 221–225.

Sivanandhan, G., N. Selvaraj, A. Ganapathi, and M. Manickavasagam. 2014c. Enhanced biosynthesis of withanolides by elicitation and precursor feeding in cell suspension culture of *Withania somnifera* (L.) Dunal in shake flask culture and bioreactor. PLoS One 9: e104005.

Sivanandhan, G., C. Arunachalam, N. Selvaraj, A.A. Sulaiman, Y.P. Lim, and A. Ganapathi. 2015a. Expression of important pathway genes involved in withanolides biosynthesis in hairy root culture of *Withania somnifera* upon treatment with *Gracilaria edulis* and *Sargassum wightii*. Plant Physiol. Biochem. 91: 61–64.

Sivanandhan, G., N. Selvaraj, A. Ganapathi, and M. Manickavasagam. 2015b. Effect of nitrogen and carbon sources on *in vitro* shoot multiplication, root induction and withanolides content in *Withania somnifera* (L.) Dunal. Acta Physiol. Plant. 37: 12.

Smetanska, I. 2008. Production of secondary metabolites using plant cell cultures. In Food biotechnology (pp. 187–228). Springer, Berlin, Heidelberg.

Srivastava, S., R.S. Sangwan, S. Tripathi, B. Mishra, L.K. Narnoliya, L.N. Misra, and N.S. Sangwan. 2015. Light and auxin responsive cytochrome P450s from *Withania somnifera* Dunal: cloning, expression and molecular modelling of two pairs of homologue genes with differential regulation. Protoplasma 252: 1421–1437.

Stan, S.D., Y. Zeng, and S.V. Singh. 2008. Ayurvedic medicine constituent withaferin a causes G2 and M phase cell cycle arrest in human breast cancer cells. Nutr. Cancer. 60: 51–60.

Subramanian, C., H. Zhang, R. Gallagher, G. Hammer, B. Timmermann, and M. Cohen. 2014. Withanolides are potent novel targeted therapeutic agents against adrenocortical carcinomas. World J. Surg. 38: 1343.

Subramanian, S.S., and P.D. Sethi. 1969. Withaferin–A from *Withania coagulants* roots. Current Science (India) 38: 267–268.

Sundar, D., Y. Yu, S.P. Katiyar, J.F. Putri, J.K. Dhanjal, J. Wang, A.N. Sari, E. Kolettas, S.C Kaul, and R. Wadhwa. 2019. Wild type p53 function in p53 Y220C mutant harboring cells by treatment with Ashwagandha derived anticancer withanolides: bioinformatics and experimental evidence. J. Exp. Clin. Cancer Res. 38: 103.

Takimoto, T., Y. Kanbayashi, T. Toyoda, Y. Adachi, C. Furuta, K. Suzuki, T. Miwa, and M. Bannai. 2014. 4β-Hydroxywithanolide E isolated from *Physalis pruinosa* calyx decreases inflammatory responses by inhibiting the NF-κB signaling in diabetic mouse adipose tissue. Int. J. Obes. 38: 1432–1439.

Teixeira, S.T., M.C. Valadares, S.A. GonÃ§alves, A. De Melo, and M.L.S. Queiroz. 2006. Prophylactic administration of *Withania somnifera* extract increases host resistance in *Listeria monocytogenes* infected mice. Int. Immunopharmacol. 6: 1535–42.

Tekula, S., A. Khurana, P. Anchi, and C. Godugu. 2018. Withaferin A attenuates multiple low doses of Streptozotocin (MLD-STZ) induced type 1 diabetes. Biomed. Pharmacother. 106: 1428–1440.

Thaiparambil, J.T., L. Bender, T. Ganesh, E. Kline, P. Patel, Y. Liu, M. Tighiouart, P.M. Vertino, R.D. Harvey, A. Garcia, and A.I. Marcus. 2011. Withaferin A inhibits breast cancer invasion and metastasis at sub-cytotoxic doses by inducing vimentin disassembly and serine 56 phosphorylation. Int. J. Cancer. 129: 2744–2755.

Thilip, C., C.S. Raju, K. Varutharaju, A. Aslam, and A. Shajahan. 2015. Improved *Agrobacterium rhizogenes* mediated hairy root culture system of *Withania somnifera* (L.) Dunal using sonication and heat treatment. 3 Biotech. 5: 949–956.

Thilip, C., V.M. Mehaboob, k. Varutharaju, K. Faizal, P. Raja, A. Aslam, and A. Shajahan. 2019. Elicitation of withaferin A in hairy root culture of *Withania somnifera* (L.) Dunal using natural polysaccharides. Biologia. 74: 961–968.

Thirunavukkarasu, M., S. Penumathsa, B. Juhasz, L. Zhan, M. Bagchi, T. Yasmin, M.A. Shara, H.S. Thatte, D. Bagchi, and N. Maulik. 2006. Enhanced cardiovascular function and energy level by a novel chromium (III) supplement. Biofactors 27: 53–67.

Tiwari, R., S. Chakraborty, M. Saminathan, K. Dhama, and S.V. Singh. 2014. Ashwagandha (*Withania somnifera*): Role in safeguarding health, immunomodulatory effects, combating infections and therapeutic applications: A review. J. Biol. Sci. 14: 77–94.

Tomi, H., M. Yoshida, and K. Kishida. 2005. *Withania somnifera* Dunal extracts for increasing male sperm count. Nippon Shinyaku Co., Ltd., USA, Patent No. 6,866,872.

Tripathi, V., and J. Verma. 2014. Current updates of Indian antidiabetic medicinal plants. Int. J. Res. Pharm. Chem. 4: 114–118.

Vakeswaran, V., and V. Krishnasamy. 2003. Improvement in storability of Ashwagandha (*Withania somnifera* Dunal) seeds through pre storage treatments by triggering their physiological and biochemical properties. Seed Technol. 203–203.

Varghese, S., R. Keshavachandran, B. Baby, and P.A. Nazeem. 2014. Genetic transformation in ashwagandha *Withania somnifera* (L.) Dunal for hairy root induction and enhancement of secondary metabolites. J. Trop. Agri. 52: 39–46.

Udayakumar, R., and S. Kasthurirengan, T.S. Mariashibu, M. Rajesh, V.R. Anbazhagan, S.C. Kim, A. Ganapathi, and C.W. Choi. 2009. Hypoglycaemic and hypolipidaemic effects of *Withania somnifera* root and leaf extracts on alloxan induced diabetic rats. Int. J. Mol. Sci., 10: 2367–2382.

Verpoorte, R., R. Van der Heijden, J. Schripsema, J.H.C. Hoge, and H.J.G. Ten Hoopen. 1993. Plant cell biotechnology for the production of alkaloids: present status and prospects. J. Nat. Prod. 56: 186–207.

Visavadiya, N.P., and A.V.R.L. Narasimhacharya. 2007. Hypocholesteremic and antioxidant effects of *Withania somnifera* (Dunal) in hypercholesteremic rats. Phytomedicine 14: 136–142.

Vaishnavi, K., N. Saxena, N. Shah, R. Singh, K. Manjunath, M. Uthayakumar, S.P. Kanaujia, S.C. Kaul, K. Sekar, and R. Wadhwa. 2012. Differential activities of the two closely related withanolides, Withaferin A and Withanone: bioinformatics and experimental evidences. PLoS One 7: e44419.

Wadegaonkar, P.A., K.A. Bhagwat, and M.K. Rai. 2006. Direct rhizogenesis and establishment of fast growing normal root organ culture of *Withania somnifera* Dunal. Plant Cell Tissue Organ Cult. 84: 223–225.

Wadhwa, R., R. Singh, R. Gao, N. Shah, N. Widodo, T. Nakamoto, and S. Kaul. 2013. Water extract of Ashwagandha leaves has anticancer activity: identification of an active component and its mechanism of action. PLoS One 8: e77189.

Wentzinger, L.F., T.J. Bach, and M.A. Hartmann. 2002. Inhibition of squalene synthase and squalene epoxidase in tobacco cells triggers an upregulation of 3-hydroxy-3-methylglutaryl coenzyme a reductase. Plant Physiol. 130: 334–346.

Widodo, N., Y. Takagi, B.G. Shrestha, T. Ishii, S.C. Kaul, and R. Wadhwa. 2008. Selective killing of cancer cells by leaf extract of Ashwagandha: components, activity and pathway analyses. Cancer Lett. 262: 37–47.

Winters, M. 2006. Ancient medicine, modern use: *Withania somnifera* and its potential role in integrative oncology. Altern. Med. Rev. 11: 269–277.

Yadav, B., A. Bajaj, M. Saxena, and A.K. Saxena. 2010. *In vitro* anticancer activity of the root, stem and leaves of *Withania somnifera* against various human cancer cell lines. Indian J. Pharm. Sci. 72: 659.

Yang, H., Y. Wang, V.T. Cheryan, W. Wu, C.Q. Cui, L.A. Polin, H.I. Pass, Q.P. Dou, A.K. Rishi, and A. Wali. 2012. Withaferin A inhibits the proteasome activity in mesothelioma *in vitro* and *in vivo*. PLoS One 7: e41214.

Yu, P.L.C., M.M. El-Olemy, and S.J. Stohs. 1974. A phytochemical investigation of *Withania somnifera* tissue cultures. Lloydia. 37: 593–597.

Zhang, H., A.K. Samadi, M.S. Cohen, and B.N. Timmermann. 2012. Antiproliferative withanolides from the Solanaceae: a structure–activity study. Pure. Appl. Chem. 84: 1353–1367.

Zhao, J., N. Nakamura, M. Hattori, T. Kuboyama, C. Tohda, and K. Komatsu. 2002. Withanolide derivatives from the roots of *Withania somnifera* and their neurite outgrowth activities. Chem. Pharm. Bull. 50: 760–765.

Chapter 2

Biotechnological Approaches for Tropane Alkaloids Production

Ryad Amdoun,[1] *Boualem Harfi,*[2] *Asma Moussous,*[3] *Abdullah Makhzoum*[4] and *Lakhdar Khelifi*[3,*]

Introduction

Plants produce a wide range of secondary metabolites that are an important source of biomolecules used in the agri-food and pharmaceutical industries (Balandrin and Klocke 1988, Sasson 1991, Mulabagal and Tsay 2004). Secondary metabolites are distributed in three major families: phenolic compounds, isoprenoids, and nitrogen compounds. In this latter family, alkaloids are the most important in their number and diversity (Bruneton 1987). There are several groups of alkaloids known to date including the group of tropane alkaloids such as hyoscyamine ($C_{17}H_{23}NO_3$) and scopolamine ($C_{17}H_{21}NO_4$). These two alkaloids are widely used in human and veterinary medicine (Houmani and Cosson 2000). They are abundant in *Solanaceae* such as *Mandragora*, *Atropa*, *Hyoscyamus*, *Datura*, *Scopolia*, and *Duboisia* which are used as antispasmodic, sedative, anesthetic, and mydriatic (Yamada and Tabata 1997).

The search for profitable production, independent of the environmental hazards inducing fluctuations of yield, pushes for other more stable and efficient production routes for the production of these metabolites. Biotechnology can be a good alternative for this production (Guignard et al. 1985, Verpoorte and Memelink 2002, Vanisree et al. 2004). While biotechnological production of certain metabolites from microorganisms has been made on an industrial scale, the use of plant cells remains limited. However, extensive plant metabolism research has established bioreactor production processes that aim to increase metabolite yields (Kim et al. 2002, Georgiev et al. 2007).

[1] Institut National de la Recherche Forestière (INRF), Baïnem-BP 37, Cheraga, Algiers, Algeria.
[2] Centre de Recherche en Biotechnologies, Ali Mendjeli, Constantine, Algeria.
[3] Laboratoire des Ressources Génétiques et Biotechnologie, Ecole Nationale Supérieure Agronomique (ENSA)- Kasdi Merbah, El-Harrach, 16200 Algiers, Algeria.
[4] Department of Biological Sciences & Biotechnology, Botswana International University of Science & Technology, Botswana.
* Corresponding author: lakhdar.khelifi@edu.ensa.dz

Tropane alkaloids

Alkaloids are nitrogenous organic compounds generally of vegetable origin having a basic reaction and possessing pharmacological properties. Nitrogen gives them an alkaline property, hence the name alkaloids (Guignard et al. 1985, Bruneton 1987). Thetropanic denomination comes from their fundamental nucleus which is the tropane. These are esters of tropanol alcohol or scopanol with tropic acid. Alkaloids crystallize and dissolve in organic solvents and generally not in water. They are distinct from other nitrogen compounds such as amino acids, betalaines, peptides, amino sugars, nitrogen vitamins, pophyrins and low molecular weight alkylamines (Bruneton 1987).

Alkaloids are biosynthesized enzymatically (Fig. 1) in the endoplasmic reticulum (Guignard et al. 1985, Hashimoto and Yamada 1987). Two precursors of the biosynthetic

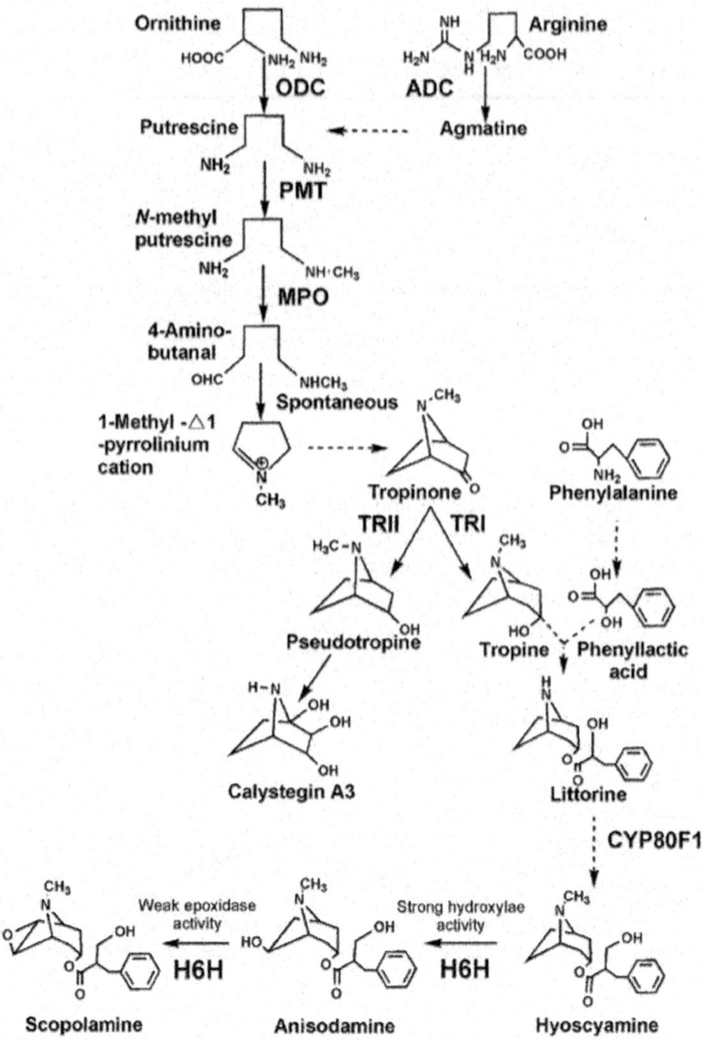

Figure 1: Tropan alkaloids biosynthetic pathway in *Solanaceae* (Xia et al. 2016).

chain of hyoscyamine and scopolamine are amino acids: ornithine and arginine (Nakjima et al. 1993). After a series of transformation initiated by *ornithine decarboxylase* (ODC) and *arginine decarboxylase* (ADC) as well as *putrecin N-methyl-transferase*, tropinone is formed. Due to the activity of *tropinone reductase I*, tropinone is reduced to tropin which is esterified by the tropic acid thereby giving hyoscyamine (Nakjima et al. 1993). The biosynthesis of hyoscyamine takes place at the root level (Kitamura et al. 1996), after which it migrates to the aerial parts where it is either accumulated (Kitamura et al. 1996) or biotransformed to scopolamine by hydroxylation followed by epoxidation which is very intense in young plants and less efficient in older plants (Cosson et al. 1976). The transformation of hyoscyamine into scopolamine takes place through two enzymes: *hyoscyamine 6 β hydroxylase* (H6H) (Hashimoto et al. 1986) and hyoscyamine 6 β epoxidase mainly located in the leaves (Hashimoto et al. 1986).

Production by Biotechnology

Plant tissue culture for the production of biomolecules has attracted growing interest since the late 1960s (Mulabagal and Tsay 2004). Advances in plant biotechnology in tissue culture have opened up a promising alternative pathway for traditional agriculture for the production of tropane alkaloids (Dicosmo and Misawa 1995, Mulabagal and Tsay 2004).

Various cell culture systems have been studied and optimized to improve production. These optimized systems could be exploited for the industrial production of biomolecules of interest. Additionally increased yields, biotechnology production offers the advantage of being independent of climate fluctuations, diseases and pests (Mulabagal and Tsay 2004). Several types of cultures have been tested. They can be classified into two categories, undifferentiated tissue cultures and differentiated tissue cultures.

Undifferentiated Tissue Culture

Early work on the production of alkaloids, and secondary metabolites in general, involves undifferentiated tissue culture such as callusand cell suspensions (Table 1).

However, several studies highlight the interests of cell differentiation and the performance of plant tissues for the biosynthesis of alkaloids (Lindsey and Yeoman 1983, Kitamura et al. 1985, Biondi et al. 2002). These are synthesized at the root level. It would therefore require a tissue specialized in the root. The hairy roots (HRs) induced by *Agrobacterium rhizogenes* bacteria is a performing tissue (Giri and Narasu 2000).

Hairy roots (HRs)Vs Undifferentiated Tissue Cultures

The HRs grown in hormone-free B5 (Gamborg et al. 1968) medium have the highest biomass than HRs cultured in the same medium with phytohormones (2 mg of ANA + 0.02 mg of BAP). There is a loss of 40% for biomass, 20% for the growth rate due to dedifferentiation (callus formation). The cell suspension, prepared from callus induced on *D. stramonium* hypocotyl by E_{17} strain of *A. Tumefaciens*, has much lower growth characteristics than HRs cultured in B5 medium supplemented with phytohormones (Amdoun et al. 2007).

Table 1: Tropane alkaloids from plant cell cultures.

Plant name	**Culture type**	**Reference**
Atropa belladonna	Cell suspension, callus	Bhandary et al. 1969
Scopolia parviflora	Callus	Tabata et al. 1972
Hyoscyamus niger	Callus	Yamada and Hashimoto 1982
Atropa belladonna	Callus	Lindsey and Yeoman 1983
Datura innoxia	Callus	Lindsey and Yeoman 1983
D. clorantha	Callus	Lindsey and Yeoman 1983
D. stramonium	Callus	Lindsey and Yeoman 1983
Hyoscyamus niger	Callus	Lindsey and Yeoman 1983
Solanum dulcamara	Callus	Lindsey and Yeoman 1983
S. nigrum	Callus	Lindsey and Yeoman 1983
Duboisia leichhardtii	Callus	Yamada and Endo 1984
Duboisia myoporoides	Callus	Kitamura et al. 1985
Atropa belladonna	Callus	Simola et al. 1988
Hyoscyamus muticus L.	Callus	Basu and Chand 1998
Hyoscyamus muticus	Callus	Biondi et al. 2002

HRs dedifferentiated under the influence of phythormones (calli formation), show a loss of 73% of hyoscyamine content than HRs grown in hormone-free B5 medium. The hyoscyamine content of the E_{17} cell suspension is higher than that of the HRs cultured in B5 medium with phytohormones (Amdoun et al. 2007). *A. Tumefaciens* strain would cause physiological deregulation (Cosson and Kuntzmann-Cougoul 1979). The latter would induce an important call for metabolites necessary for the biosynthesis of alkaloids, or the tumor transformation would directly induce an exaltation of alkaloid biosynthesis (Goldmann 1977, Cosson and Kuntzmann-Cougoul 1979).

The production of tropan alkaloids therefore requires a root specialized tissue. The HR system is efficient. This type of morphogenetic structure exceeds the cell suspensions growing in hyoscyamine content. In case of callogenesis of HRs, its production performances are extremely reduced, hence the elimination of callogenic lines during the characterization and selection of HRs lines.

Factors controlling the induction of HRs

The infection mechanism with *A. rhizogenes* is three steps: bacterial-plant adhesion, activation of virulence *vir* genes and transfer of plasmid T-DNA to the host cell. The genes are activated by three chemical signals from the plant: the monosaccharides (glucose), the acidic pH and the phenolic compounds, including acetosyringone. The induction rate of HRs strongly depends on the interaction Genotype x Bacterial strain. Some genotypes are reactive to infection, others are recalcitrant (Ecran and Taskin 1999). The different strains of *A. rhizogenes* show variable virulence. Some do not induce any reaction, others only the formation of callus, and some the formation of HRs (Ecran and Taskin 1999, Park and Facchini 2000, Giri et al. 2001).

Virulence also depends on the bacterial concentration. Low concentration means low availability of bacteria for transformation of plant cells, while high concentration decreases the efficiency of transformation because of competitive inhibition between bacteria (Kumar et al. 1991). The best bacterial concentration is at $DO_{600} = 0.5$ (Dhakulkar et al. 2005).

Phenolic activators of virulence such as acetosyringone, added at concentrations of 30 to 50 μM, play a role in the induction efficiency of HRs and allow an earlier appearance of roots (Tao and Li 2006).

Some monosaccharides such as fructose, galactose, glucose and xylose activate *vir* genes. In some cases they are needed for processing (Ankenbauer and Nester 1990). This effect is accentuated in the presence of acetosyringone at adequate doses and in an acid medium (Cangelosi et al. 1990, Kumar et al. 1991, Wise et al. 2005).

Acid pH is also a very important factor for HRs induction (Mantis and Winans 1992, Li et al. 2002, Gao and Lynn 2005). Transformation-related genes, such as *virD2*, code better in an acidic environment (Satchel et al. 1985).

Other factors such as the type, age and physiological state of the explant (stem, root, hypocotyl, etc.) play a role in the efficiency of induction (Diouf 1996, Karmarkar et al. 2001).

Induction protocols

Induction of HRs can be performed according to different types of protocols in three steps: Culture and activation of *A. rhizogenes*, infection and plant-bacterium co-culture and finally, isolation of induced root hair.

a. Culture and Activation of *A. rhizogenes*

The bacteria are maintained and cultured in Petri dishes containing an agar culture medium such as nutrient Agar (Giri et al. 2001, Dhakulkar et al. 2005) or YEB medium (Tao and Li 2006). The plates are incubated at 27° ± 1°C for 16 to 72 h (Kovalenko and Maliuta 2003, Dhakulkar et al. 2005, Amdoun et al. 2007, Tao and Li 2006, Giri et al. 2001). Overactivation of the *vir* genes is achieved by adding acetosyringone to the suspension medium of the bacterium (Kovalenko and Maliuta 2003) and/or the plant-bacterium co-culture medium (Giri et al. 2001 Tao and Li 2006).

b. Infection and Co-culture Plant-bacteria

Several infection techniques exist. The so-called immersion technique is used by different authors:

Park and Facchini (2000) adopt the following protocol: Seedlings and callus of *Papaver somniferum* and *California popy* were randomly wounded by a scalpel and immersed in the bacterial suspension for 10 to 15 minutes, then dried with sterile filter paper and grow on B5 medium in the dark. After 2 days of co-culture, the explants were transferred to B5 medium containing 20 g/l of sucrose and 50 mg/l of paramomycin. After 4 to 5 weeks of culture, roots appear at the sites of infection.

Dhakulkar et al. (2005) use different types of explants isolated from vitrosemis. They are precultured for 2 days on MS (Murashige and Skoog 1962) medium (Explants 1 cm in length), were immersed for 30 minutes in conical containers containing the bacterial suspension. Subsequently, the explants are dried and transferred to MS½ medium. Three days later, the explants are transferred to MS medium containing 400 mg/l of cefotaxime to eliminate the bacteria. The concentration of cefotaxime is then reduced by half each week to 50 mg/l. finally the bacteria-free culture is transferred to B5 medium. All cultures are kept in complete darkness at 25° ± 2°C until the roots appear.

Tao et al. (2006) pre-cultured the leaf explants in solid MS medium at different concentrations of acetosyringone (0, 10, 30 and 40 µM) for 2 days, the explants are then immersed in the suspension of *Agrobacterium* containing 20 µM acetosyringone and incubated for a certain time (0 to 30 minutes). Subsequently, the explants are dried with sterile filter paper and transferred to the same preculture medium for co-culture. After 4 days, the explants are transferred to fresh medium (MS ½) containing 500 mg/l of carbenicillin and 200 mg/l of canamycin. Every 4 days, the explants are transferred to MS ½ medium containing 250 mg/l of canamycin and 150 mg/l of carbenicillin until the formation of root hair.

Amdoun et al. (2006) used deposition infection: Stem explants are infected with a syringe containing the bacterial suspension. The inoculated explants are cultured on solid MS medium containing 250 mg/l of cefotaxime. Co-cultivation is done in the dark at 26° ± 1°C. Transplants are necessary to avoid the development of the bacteria in the medium until the appearance of the transformed roots.

c. Isolation and Culture

The induced roots are excised when they reach a certain length. Dhakulkar et al. (2005) indicated that neoformed roots are separated from the explant when they reach a length of 4 to 5 cm. Amdoun et al. (2006) report that transformed roots are severed when they reach about 2 cm. The excised roots are placed on MS medium. The culture is then maintained in complete darkness at 25° –26° ± 2°C (Dhakulkar et al. 2005, Amdoun et al. 2006). When the roots show their ability to grow by elongating and developing a branched structure, they are able to be transferred into a liquid medium. The culture media used varies according to the authors, but it is a determining criterion for the success of the cultivation of the root hairs.

Transformation of HRs is usually confirmed by the presence of *rol* genes (*rolB*) in the root genome, detected by PCR based on the protocol described by Dechaux and Boitel-Conti (2005). The oligonucleotide primer sequences used for PCR amplification of the *rolB* gene were as follows: 5′-ATGGATCCCAAATTGCTATTCCTTCCACGA-3′ (forward) and 5′-TTAGGCTTCTTTCTTCAGGTTTACTGCAGC-3′ (reverse).

Main Biotechnological Strategies for Improving Productivity

For production of tropic alkaloids to become economically viable, optimized processes must be developed to produce high yields (Berlin and Sasse 1985). Various biotechnological strategies for improving yields have been developed, including: selection of efficient lines of HRs, optimization of the composition of the culture medium, elicitation, precursor input, permeabilization of cells, trapping of molecules released in the liquid medium and optimization of environmental conditions.

Characterization and Selection of Highly Productive HRs Lines

Induced HRs lines have different growth properties and individual biochemical properties. They must therefore be characterized and selected. Unbranched and/or callogenic lines, a sign of chaotic growth, are systematically eliminated. The lines obtained by Amdoun (2010) noted AK1, AK2, AZK1 and AZK2 and those of Harfi et al. (2016) have different characters. The variation in growth rate and the morphological differences observed in the HRs lines can be attributed, on the one hand, to the different T-DNA insertion sites in the plant cell genome, and on the other hand the number of copies of T-DNA inserted (Furze et al. 1987).

Successful HRs are able to grow without any hormonal contribution to the culture medium (Giri and Narasu 2000). They are very dense, genetically more stable over time and endowed with rapid but variable growth depending on the species and varieties considered (Giri and Narasu 2000, Kovalenko and Maliuta 2003).

The selected lines thus prove to be the plant material of choice for the production of alkaloids. This type of tissue produces significant levels of secondary metabolites compared to other types of plant material cultured (cell suspensions, calli, normal root) (Bourgaud et al. 1997, Shanks and Morgan 1999, Raoa and Ravishankarb 2002, Souret et al. 2003). The productivity of certain lines of efficient HRs may even exceed those of whole plants (Kovalenko and Maliuta 2003, Ono and Tian 2011). In contrast to cell cultures, HRs culture is characterized by high genetic and synthetic stability (Giri and Narasu 2000, Guillon et al. 2006, Hu and Du 2006). Medina-Bolivar and Flores (1995) report that *Hyoscyamus muticus* HRs showed levels of hyoscyamine biosynthesis equal to or greater than the roots of the whole plant and maintained the same biosynthetic capacity for more than 12 years. The productivity of certain lines of efficient HRs may even exceed those of whole plants (Kovalenko and Maliuta 2003, Ono and Tian 2011). In addition, HRs can produce new compounds that cannot be synthesized by unprocessed roots (Hu and Du 2006).

Elicitation

Secondary metabolites are generally biosynthesized in abundance as a response to stress (Hutcheson 1998, He et al. 2002). Elicitation is a treatment that mimics stress. It consists of the use of biotic and/or abiotic substances, called elicitors, to stimulate the biosynthesis of specific secondary metabolites (Vasconsuelo and Boland 2007).

Efficiency of elicitors

The efficiency of the elicitors depends on the complex interaction between the elicitor and the plant. The same elicitor can stimulate secondary metabolism for different cell lines, and the same cell line can respond favorably to various elicitors. The concentration of the elicitor, the environmental conditions and the culture medium are all factors that significantly affect the intensity of the response to elicitation (Vasconsuelo and Boland 2007).

A model of the elicitation mechanism states that after the binding of the elicitor with a receptor located in the plasma membrane causing the activation of G-protein, the latter stimulates adenylate cyclase (AC) and phospholipase C (PLC). This increases the level of secondary messengers (cAMP, DAG, IP3) coupled with the activation of their target kinases (PKA, PKC). The flow of Ca^{2+} thus increases through the plasma membrane. A rapid and cascade activation of the protein kinase will take place thus inducing changes in the phosphorylation of MAPKs (Mitogen Activated Protein Kinases) and, in some cases, their translocation into the nucleus. This ultimately leads to the activation of transcription of enzymes in the biosynthetic pathways of secondary metabolites (Vasconsuelo and Boland 2007).

Examples of some elicitors used

a. The Jasmonates

Jasmonates (jasmonic acid: AJ, jasmonate methyl: MeJA) occupy a privileged place in elicitation in higher plants (Herbert 2001, Zabetakis et al. 1999). These are stress hormones that act as a defense regulator in plants (Gundlach et al. 1992, Wasternack and Parthier 1997, Reymond and Farmer 1998, Blechert et al. 1995, Doares et al. 1995, Dar et al. 2015). A transmembrane receptor is activated following an external signal (e.g., stress) leading to the release into the cytoplasm of linolenic acid present in the plasma membrane. Linolenic acid is then converted to jasmonic acid (Farmer and Ryan 1992). Disorders of alkaloid biosynthesis are among those regulated by jasmonates (Memelink et al. 2001). The addition of jasmonic acid to *Brugmansia candida* HRs culture medium induced a 12-fold increase in hyoscyamine content (Spollansky et al. 2000).

While jasmonates improve the production of tropane alkaloids, they are biomass inhibitors (Kang et al. 2004, Deng 2005). Taking into account simultaneously these two antagonistic responses (biomass and alkaloid content) during the AJ elicitation, Amdoun et al. (2018) were able to optimize the concentration of AJ for maximal response without affecting biomass using the desirability function defined by Derringer and Suich (1980).

b. Salicylic and Acetylsalicylic Acids

Salicylic and acetylsalicylic acids are widely used as elicitors for the improvement of alkaloid production in different species (Zhao et al. 2000, Lee et al. 2001), hence their use as elicitors for the biosynthesis of Hyoscyamine and Scopolamine.

When a plant cell receives a signalling molecule on its plasma membrane, a signal transduction network is triggered leading to the activation of the target genes or the biosynthesis of transcription factors that regulate the genes encoding the enzymes involved in the pathways, known as biosynthesis of the desired metabolites (Zhao et al. 2005).

Salicylic acid is one of these signalling molecules. It is a known inducer of systemic acquired resistance (SAR) of the plant-pathogen interaction. It is known for its inductive properties of gene expression related to the biosynthesis and production of some classes of secondary metabolites (Taguchi et al. 2001), such as indolic and tropanic alkaloids (Zhao et al. 2000, Kang et al. 2004).

A GC/MS screening of 343 root hair lines from *Datura stramonium*, *D. tatula* and *D. innoxia* reveals the presence of 13 alkaloids. Hyoscyamine is the dominant alkaloid, to increase the production of this alkaloid in selected lines; elicitations of salicylic acid and acetylsalicylic acid have been used. After an exposure time of 24 h, the results show that the concentration of 0.1 mM is the best for the two elicitors. The highest observed hyoscyamine content of 17.94 ± 0.14 mg/gDW is obtained in the root hair of *D. tatula* (Harfi et al. 2018).

A combined ANN-RSM (artificial neural network-Response Surface Methodology) modelling approach consisting in applying a quadratic model of RSM to the responses predicted by the ANN made it possible to statistically explore the optimal regions of the response to elicitation. The results show that elicitation of *D. stramonium* root hair with high concentrations of salicylic acid (1.5 to 2.1 mM) is statistically positive provided that the exposure time is short (8 h). After this shock elicitation, the hyoscyamine could be harvested from the medium following a permeabilization of the roots. A fresh medium without an elicitor could be reinjected to avoid the inhibition of root growth (Amdoun et al. 2019).

c. Mineral Salts

Elicitation of *D. stramonium* HRs (line S9) with a concentration of 2 g of NaCl/l makes it possible to triple the hyoscyamine content compared to the control HRs (not elicited) (Khelifi et al. 2011).

Potassium (KCl) and calcium ($CaCl_2$) chlorides tested for different exposure times and concentrations significantly influence the biosynthesis of hyoscyamine in HRs from three *Datura* species (*D. stramonium, D. innoxia* and *D. tatula*) (Harfi et al. 2016). Optimal HRs elicitation combines a concentration of 2g/l KCl with a 10-hour exposure time for *D. tatula* and a 24-hour time for *D. stramonium* and *D. innoxia*. KCl elicitation resulted in improvements of 2.32, 1.99, and 1.85-fold over control treatments, respectively, for the HRs of *D. stramonium, D. tatula*, and *D. innoxia*. For $CaCl_2$, the best hyoscyamine content of 16.978 mg/gDW is observed in *D. tatula* HRs elicited with 2 g/l for 24 hours (Harfi et al. 2016).

In general, salt stress elicitation inhibits the growth of HRs. The inhibitory effect of KCl and $CaCl_2$ is all the more important when the concentration is high and the exposure time is long (Harfi et al. 2016).

d. *Pseudomonas* spp.

Pseudomonas are used as elicitors for the improvement of the root growth of plants and the strengthening of their defense mechanisms. The effect of the latter is associated with the production of secondary metabolites with an elicitor effect such as antibiotics, sideriophores and phytohormones (Latour and Lemanceau 1997, Van Loon et al. 2003).

The application of *Pseudomonas* spp. strains as elicitors can improve the biomass and hyoscyamine and scopolamine yields of the *Datura* spp. HRs (Moussous et al. 2018). Indeed, two species, *Pseudomonas fluorescens* (strains: P64, P66, C7R12) and *Pseudomonas putida* (strain PP01) were used in two contact times of 10 and 5 days for the elicitation of the HRs of *Datura stramonium, D. tatula* and *D. innoxia*. The biomass as well as the hyoscyamine content of the three species was improved after elicitation by *P. fluorescens* strains. A maximum increase of 431% in the hyoscyamine content of *D. tatula* HRs elicited for 5 days with strain C7R12 and a 583% increase for *D. stramonium* HRs elicited with the same strain for 10 days were obtained. The highest scopolamine content is obtained in the *D. tatula* HRs elicited with the C7R12 strain for 5 days with an increase of 471% compared to the non-elite control (Moussous et al. 2018).

After a 5-day exposure to *Pseudomonas* ssp., the authors observed a growth stimulation of the HRs biomass of the three *Datura* species. However, for a 10-day exposure time to *Pseudomonas* ssp., some treatments stimulate the growth of HRs and the best biomass result is obtained in the line DI elicited for 10 days by strain C7R12 with an improvement of 177% compared to non-elicited control.

Cell Permeabilization

In general, biomolecules biosynthesized by plant cells accumulate in the vacuoles. To liberate them, these molecules must cross two barriers: the plasma membrane and the tonoplast. Cell permeabilization is a function of the pore formation of plant cell membranes allowing the entry and exit of various molecules. Plant cell permeability can be measured by the enzymatic activity of *hexokinase*, *glucose 6-phosphate dehydrogenase*, *isocitrate dehydrogenase*, *malic acid* and *citrate synthetase* (Brodelius et al. 1988). There is a wide range of permeabilizer that enhances enzymatic activity or causes the release of intracellular biomolecules (Parr et al. 1984, Felix 1982). Among the permeabilizing agents used: Isopropanol, dimethylsulfoxide (DMSO), chitosan and Tween 20, ultrasound and high electric field pulses (Brodelius et al. 1988, Beaumont and Knorr 1987, Knorr and Teutonico 1986, Dornenburg and Knorr 1993, Khelifi et al. 2011).

The purpose of the permeabilization of the cells thus consists in treating the HRs with permeabilization agents to release the desired biomolecule in the liquid culture medium in the interest of harvesting it without destroying the plant material (HRs).

By simultaneously applying two optimization strategies, elicitation with different cell permeabilization agents, Khelifi et al. (2011) found that *D. stramonium* HRs treated with Tween 20 released at least 25% hyoscyamine into the culture medium. The best combination elicitation/permeabilization consists of combining 10 μM jasmonic acid with 1.5% Tween 20 for 24 hours.

Precursor Feeding

The exogenous contribution in the culture medium of precursors of the molecules of interest can improve the yield. The concept is based on the idea that any intermediate or intervening compound or one at the beginning of the biosynthetic pathway of a secondary metabolite is likely to increase the yield of the final product when it is provided (Mulabagal and Tsay 2004). Several studies show that this strategy is effective in improving various secondary metabolites (Silvestrini et al. 2002, Moreno et al. 1993, Whitmer et al. 1998, Whitmer et al. 2002). For example, the addition of phenylalanine stimulates the production of rosmarinic acid (Ellis and Towers 1970), capsaicin (Lindsey 1986) and taxol (Fett-Neto et al. 1994).

The effects of tropane alkaloids precursors in transformed root cultures of *Datura innoxia* Mill. have been studied by (Boitel-Conti et al. 2000). These are: L-ornithine, L-arginine, L-phenylalanine, DL-β-phenyllactic acid and tropinone. These authors report that the contribution of precursors (0.5 m mol/l) stimulates the rate of hyoscyamine when combined with a treatment with Tween 20. The improvement is over 40% for L-phenylalanine and over 60% for DL-phenyllactic acid.

Optimization of Culture Medium Composition

The composition of the culture medium is decisive for the improvement of the growth of HRs and the yield of tropic alkaloids (Amdoun et al. 2009, Amdoun et al. 2010).

Influence of nutrients

The biosynthesis of alkaloids depends on many factors including nutritional status. Mineral elements such as NO_3^-, Ca^{2+}, K^+, SO_4^{2-} and carbohydrates are known for their positive influences on the biomass and alkaloid content of HRs (Gontier et al. 1994, Saenz-Carbonell and Loyola-Vargas 1996, Sikuli and Demeyer 1997, Nussbaumer et al. 1998, Pinñol et al. 1999, Lanoue et al. 2004). The work of Amdoun et al. (2009) show the positive effects of NO_3^-, $H_2PO_4^-$, Ca^{2+} and the NO_3^-/Ca^{2+} interaction on the hyoscyamine content of the eluted HRs of *Datura stramonium*.

Nutritional limitation influence

The control of nutrient requirements such as NO_3^-, $H_2PO_4^-$, K^+, Ca^{2+} and carbohydrates is likely to improve the alkaloid content. However, when the plant is limited in nutrients, the defense mechanisms are induced (Harborne 1997). Solavetivone production in *Hyoscyamus muticus* HRs can be synergistically enhanced by limiting phosphate in the culture medium. This effect was more pronounced when elicitation with mushroom extracts is applied (Dunlop and Curtis 1991).

The study of the response to elicitation as a function of the composition of the culture medium was carried out by Amdoun et al. (2009). This is a jasmonic acid elicitation of *D. stramonium* HRs grown in B5 medium (Gamborg et al. 1968) with different concentrations of NO_3^-, $H_2PO_4^-$, K^+, Ca^{2+} and Mg^{2+}. The results show that NO_3^- is the most influential element. The first 12 days of culture are important to allow nitrate to be efficient on the production of hyoscyamine and the intensity of response to elicitation.

HRs limited in nitrate at the beginning of their culture do not respond significantly to elicitation. These results show that the combined strategy of a nutritional limitation followed by elicitation proposed by Dunlop and Curtis (1991) is relevant only after a stage of good mineral nutrition. In fact, the response to elicitation is greatly amplified when NO_3^- of culture medium B5 is completely consumed.

References

Amdoun, R., L. Khelifi, B. Zarouri, M. Slaoui, and S. Amroune. 2006. Production de chevelus racinaires chez deux espèces de *Datura* par transformation génétique *in vitro*. Revue Biotechnologies végétales 00: 7–9.

Amdoun, R., M. Khelifi–Slaoui, G. Hadjimi, S. Amroune, and L. Khelifi. 2007. Etude des propriétés de croissance et du contenu en hyoscyamine d'une culture de chevelus racinaires et de suspensions cellulaires de *Datura stramonium* L. Revue Biotechnologies végétales 1: 7–9.

Amdoun, R., L. Khelifi, M. Khelifi-Slaoui, S. Amroune, E.-H. Benyoussef, D.V. Thi, C. Assaf-Ducrocq, and E. Gontier. 2009. Influence of minerals and elicitation on *Datura stramonium* L. tropane alkaloid production: Modelization of the *in vitro* biochemical response. Plant Science 177: 81–87.

Amdoun, R. 2010. Optimisation de la production par voiebiotechnologique des alcaloïdestropaniques à partir de chevelus racinaires de *Datura stramonium* L. Doctoral dissertation, 77 p.

Amdoun, R., L. Khelifi, M. Khelifi-Slaoui, S. Amroune, M. Asch, C. Assaf-Ducrocq, and E. Gontier. 2010. Optimization of the culture medium composition to improve the production of hyoscyamine in elicited *Datura stramonium* L. hairy roots using the response surface methodology (RSM). International Journal of Molecular Sciences 11(11): 4726–4740.

Amdoun, R., L. Khelifi, M. Khelifi-Slaoui, S. Amroune, M. Asch, C. Assaf-ducrocq, and E. Gontier. 2018. The desirability optimization methodology (DOM) a tool to predict two antagonist responses in biotechnological systems: Case of biomass growth and hyoscyamine content in elicited *Datura starmonium* hairy roots. Iranian J. Biotech. 2018 January;16(1): 11–19.

Amdoun, R., E.H. Benyoussef, A. Benamghar, and L. Khelifi. 2019. Prediction of hyoscyamine content in *Datura stramonium* L. hairy roots using different modeling approaches: Response Surface Methodology (RSM), Artificial Neural Network (ANN) and Kriging. Biochemical Engineering Journal 144: 8–17.

Ankenbauer, R.G., and E.W. Nester. 1990. Sugar-mediated induction of *Agrbacterium tumefaciens* virulence genes: Structural specificity and activities of monosaccharides. Journal of Bacteriology 172(11): 6442–6446.

Balandrin, M.J., and J.A. Klocke. 1988. Medicinal, aromatic and industrial materials from plants. pp. 1–36. *In*: Bajaj, Y.P.S. (ed.). Biotechnology in Agriculture and Forestry. Medicinal and Aromatic Plant. 4, Springer-Verlag, Berlin, Heidelberg.

Basu, P., and S. Chand. 1998. Tropane alkaloids from callus cultures, differentiated and shoots of *Hyoscyamus muticus* L. Journal of Plant Biochemistry and Biotechnology 7(1): 39–42.

Beaumont, M.D., and D. Knorr. 1987. Effects of immobilizing agents and procedures on viability of cultured celery (*Apium graveolens*) cells. Biotechnology Letters 9(6): 377–382.

Berlin, J., and F. Sasse. 1985. Selection and screening techniques for plant cell cultures. Advanced Biochemistry and Engeneering 31: 99–132.

Bhandary, S.R., H.A. Collin, E. Thomas, and H.E. Street. 1969. Root, callus, and cell suspension cultures, from *Atropa belladonna* L. and *Atropa belladonna*, cultivar lutea Döll. Annals of Botany 33(4): 647–656.

Biondi, S., K.M. Scaramagli, K.M. Oksman-Caldentey, and F. Poli. 2002. Secondary metabolism in root and callus cultures of *Hyoscyamus muticus* L.: the relationship between morphological organisation and response to methyl jasmonate. Plant Science 163(3): 563–569.

Blechert, S., W. Brodschelm, S. Hölder, L. Kammerer, T.M. Kutchan, M.J. Mueller, Z.Q. Xia, and M.H. Zenk. 1995. The octadecanoid pathway: Signal molecules for the regulation of secondary pathways. Proc. Natl. Acad. Sci. U.S.A. 92: 4099–4105.

Boitel-Conti, M., J.C. Laberche, A. Lanoue, C. Ducrocq, and B.S. Sangwan-Norreel. 2000. Influence of feeding precursors on tropane alkaloid production during an abiotic stress in *Datura innoxia* transformed roots. Plant Cell, Tissue and Organ Culture 60(2): 131–137.

Bourgaud, F., V. Bouque, E. Gontier, and A. Guckert. 1997. Hairy root cultures for the production of secondary metabolites. AgBiotech News Inf. 9(9): 205–208.

Brodelius, P.E., C. Funk, and R.D. Shillito. 1988. Permeabilization of cultivated plant cells by electroporation for release of intracellularly stored secondary products. Plant Cell Reports 7(3): 186–188.

Bruneton, J. 1987. Elément de phytochimie et pharmacognosie, Paris : Lavoisier - Tech. & doc., 584 p.

Cangelosi, G.A., R.G. Ankebauer, and E.W. Nester. 1990. Sugars induce the *Agrobacterium* virulence genes through a periplasmic binding protein and a transmembrane signal protein. Genetic 87: 6708–6712.

Cosson, L., J.C. Vaillant, and E. Dequeant. 1976. Les alcaloïdes tropaniques des feuilles du *Duboisia myoporoides* neocaledonien. Phytochemistry 15(5): 818–820.

Cosson, L., and N. Kuntzmann-Cougoul. 1979. Tropan alkaloids (hyoscyamine and scopolamine), model for regulation studies of the production of *Duboisia myoporoides* and *Datura* sp. growing in controlled conditions. In II International Symposium on Spices and medicinal Plants 96: 135–142.

Dar, T.A., M. Uddin, M.M.A. Khan, K.R. Hakeem, and H. Jaleel. 2015. Jasmonates counter plant stress: A review. Environmental and Experimental Botany 115: 49–57.

Dechaux, C., and M. Boitel-Conti. 2005. A strategy for overaccumulation of scopolamine in *Datura innoxia* hairy root cultures. Acta Biol. Cracov. Bot. 47: 101–107.

Deng, F. 2005. Effects of glyphosate, chlorsulfuron, and methyl jasmonate on growth and alkaloid biosynthesis of jimsonweed (*Datura stramonium* L.). Pesticide Biochemistry and Physiology 82: 16–26.

Derringer, G., and R. Suich. 1980. Simultaneous optimization of several response variables. Journal of Quality Technology 12(4): 214–219.

Dhakulkar, S., T.R. Ganapathi, S. Bhargava, and V.A. Bapat. 2005. Induction of hairy roots in *Gmilina arborea* Roxb. and production of verbascoside in hairy root. Plant Science 69: 812–818.

Dicosmo, F., and M. Misawa. 1995. Plant cell and tissue culture: Alternatives for metabolite production. Biotechnology Advances 13(3): 425–453.

Diouf, D. 1996. La transformation génétique des Casuarinaceae: un outil pour l'étude moléculaire des symbioses actinorhiziennes. ThèseDoct. Univ. Paris VII-Denis Diderot. p.127.

Doares, S.H., T. Syrovets, E.W. Weiler, and C.A. Ryan. 1995. Oligogalacturonides and chitosan activate plant defensive genes through the octadecanoid pathway. Proc. Natl. Acad. Sci. U.S.A. 92: 4095–4098.

Dörnenburg, H., and D. Knorr. 1993. Cellular permeabilization of cultured plant tissues by high electric field pulses or ultra high pressure for the recovery of secondary metabolites. Food Biotechnology 7(1): 35–48.

Dunlop, D.S., and W.R. Curtis. 1991. Synergistic response of plant hairy-root cultures to phosphate limitation and fungal elicitation. Biotechnology Progress 7: 434–438.

Ellis, B.E., and G.H.N. Towers. 1970. Biogenesis of rosmarinic acid in Mentha. Biochemical Journal 118(2): 291–297.

Ercan, A.G., and M. Taşkin. 1999. *Agrobacterium rhizogenes*-mediated hairy root formation in some *Rubia tinctorum* L. population grown in Turkey. Tr. J. of Botany 23: 373–377.

Farmer, E.E., and C.A. Ryan. 1992. Octadecanoid precursors of jasmonic acid activate the synthesis of wound-inducible proteinase inhibitors. Plant Cell 4: 129–134.

Felix, H.1982. Permeabilized cells. Analytical Biochemistry 120(2): 211–234.

Fett-Neto, A.G., S.J. Melanson, S.A. Nicholson, J.J. Pennington, and F. DiCosmo. 1994. Improved taxol yield by aromatic carboxylic acid and amino acid feeding to cell cultures of *Taxus cuspidata*. Biotechnology and Bioengineering 44(8): 967–971.

Furze, J.M., J.D. Hamill, A.J. Parr, R.J. Robins, and M.J.C. Rhodes. 1987. Variations in morphology and nicotine alkaloid accumulation in protoplast-derived hairy root cultures of *Nicotiana rustica*. Journal of Plant Physiology 131(3): 237–246.

Gamborg, O.L., R.A. Miller, and K.Ojima. 1968. Nutrient requirements of suspension cultures of soybean root cells. Exp. Cell Res. 50: 151–158.

Gao, R., and D.G. Lynn. 2005. Environmental pH sensing: Resolving the virA/virG two component system inputs for *Agrobacterium pathogenesis*. Journal of Bacteriology (187)6: 2182–2189.

Georgiev, M.I., A.I. Pavlov, and T. Bley. 2007. Hairy root type plant *in vitro* systems as sources of bioactive substances. Applied Microbiology and Biotechnology 74(6): 1175–1185.

Giri, A., and M.L. Narasu. 2000. Transgenic hairy roots: Recent trends and applications. Biotechnol. Adv. 18: 1–22.

Giri, A., S.T. Ravindra, V. Dhingra, and M.L. Narasu. 2001. Influence of different strains of *Agrobacterium rhizogenes* on induction of hairy roots and artemisinin production in *Artemisia annua*. Curr. Sci. (81) 4: 378–382.

Goldmann, A. 1977. La culture des tissus et des cellules des végétaux. Masson Ed., p.202. 141

Gontier, E., B.S. Sangwan, and J.N. Barbotin. 1994. Effects of calcium, alginate, and calciumalginate immobilization on growth and tropane alkaloid levels of stable suspension cell line of *Datura innoxia* Mill. Plant Cell Rep. 9(13): 533–536.

Guignard, J.L., L. Cosson, and M. Henry. 1985. Abrégé de phytochimie. Ed. Masson. 224 p.

Guillon, S., J. Trémouillaux-Guiller, P.K. Pati, M. Rideau, and P. Gantet. 2006. Hairy root research: Recent scenario and exciting prospects. Curr. Opin. Plant Biol. 9: 341–346.

Gundlach, H., M.J. Müller, T.M. Kutchan, and M.H. Zenk. 1992. Jasmonic acid is a signal transducer in elicitor-induced plant cell cultures. Proc. Natl. Acad. Sci. U.S.A. 89: 2389–2393.

Harborne, J.B. 1997. Biochemical plant ecology. pp. 503–516. *In*: Dey, P.M., and J.B. Harborne (eds.). Plant Biochemistry, Academic Press.

Harfi, B., M. Khelifi-Slaoui, M. Bekhouche, R. Benyammi, K. Hefferon, A. Makhzoum, and L. Khelifi. 2016. Hyoscyamine production in hairy roots of three *Datura* species exposed to high-salt medium. *In Vitro* Cellular & Developmental Biology-Plant 52(1): 92–98.

Harfi, B., L. Khelifi, M. Khelifi-Slaoui, C. Assaf-Ducrocq, and E. Gontier. 2018. Tropane alkaloids GC/MS analysis and low dose elicitors' effects on hyoscyamine biosynthetic pathway in hairy roots of Algerian *Datura* species. Scientific Reports 8(1): 1–8.

Hashimoto, T., Y. Yukimune, and Y. Yamada. 1986. Tropane alkaloid production in *Hyoscyamus* root cultures. Journal of Plant Physiology 124(1-2): 61–75.

Hashimoto, T., and Y. Yamada. 1987. Purification and characterization of hyoscyamine 6β-hydroxylase from root cultures of *Hyoscyamus niger* L. Hydroxylase and epoxidase activities in the enzyme preparation. European Journal of Biochemistry 164(2): 277–285.

He, C.Y., T. Hsiang, and D.J. Wolyn. 2002. Induction of systemic disease resistance and pathogen defence responses in *Asparagus officinalis* inoculated with nonpathogenic strains of *Fusarium oxysporum*. Plant Pathol. 51: 225–230.

Herbert, R.B. 2001. The biosynthesis of plant alkaloids and nitrogenous microbial metabolites. Nat. Prod. Rep. 18: 50–65.

Houmani, Z., and L. Cosson. 2000. Quelques espèces algériennes a alcaloïdes tropnaiques. Ethnopharmacology. Edt. ERGA.: 205–219.

Hu, Z.B., and M. Du. 2006. Hairy root and its application in plant genetic engineering. Journal of Integrative Plant Biology 48(2): 121–127.

Hutcheson, S.W. 1998. Current concepts of active defense in plants. Annu. Rev. Phytopathol. 36: 59–90.

Kang, S.M., H.Y. Jung, Y.M. Kang, D.J. Yun, J.D. Bahk, J.K. Yang, and M.S. Choi. 2004. Effects of methyljasmonate and salicylic acid on the production of tropane alkaloids and the expression of PMT and H6H in adventitious root cultures of *Scopolia parviflora*. Plant Science 166: 745–751.

Karmarkar, S.H., R. Keshavachandran, P.A. Nazeem, and D. Girija. 2001. Hairy root induction in Adapathiyan (Holostemmaada-kodien k. SCHUM.). Journal of Tropical Agriculture 39: 102–107.

Khelifi, L., B. Zarouri, R. Amdoun, B. Harfi, A. Morsli, and M. Khelifi-Slaoui. 2011. Effects of elicitation and permeabilization on hyoscyamine content in *Datura stramonium* hairy roots. Adv. Environ. Biol. 5: 329–334.

Kim, Y., B. Wyslouzil, and P. Weathers. 2002. Secondary metabolism of hairy root cultures in bioreacteurs. *In Vitro* Cell. Dev. Biol. Plant 38: 1–10.

Kitamura, Y., H. Miura, and M. Sugii. 1985. Alkaloid composition and atropine esterase activity in callus and differentiated tissues of *Duboisia myoporoides* R. BR. Chemical and Pharmaceutical Bulletin 33(12): 5445–5448.

Kitamura, Y., M. Shigehiro, and T. Ikenaga. 1996. Tropic acid moiety of atropine may be recycled in Duboisia. Phytochemistry 42(5): 1331–1334.

Knorr, D., and R.A. Teutonico. 1986. Chitosan immobilization and permeabilization of *Amaranthus tricolor* cells. Journal of Agricultural and Food Chemistry 34(1): 96–97.

Kovalenko, P.G., and S.S. Maliuta. 2003. An effect of transformation by Ri-plasmids and elicitors on licorice cells and secondary metabolites production. Ukr. Bioorg. Acta 1(1): 50–60.

Kumar, V., B. Jones, and M.R. Davey. 1991. Transformation by *Agrobacterium rhizogenes* and regeneration of transgenic shoots of the wild soybean *Glycine argyrea*. Plant Cell Reports 10(3): 135–138.

Lanoue, A., M. Boitel-Conti, C. Dechaux, J.C. Laberche, P. Christen, and B. Sangwan-Norreel. 2004. Comparison of growth properties, alkaloid production and water uptake of tow selected *Datura* hairy root lines. Acta Biologica Cracoviensia Serie Botanica 46: 185–192.

Latour, X., and P. Lemanceau. 1997. Métabolisme carboné et énergétique des *Pseudomonas* spp. fluorescents saprophytes à oxydase positive. Agronomie 17(9-10): 427–443.

Lee, K.T., H. Hirano, T. Yamakawa, T. Kodama, Y. Igarashi, and K. Shimomura. 2001. Response of transformed root culture of *Atropa belladonna* to salicylic acid stress. Journal of Bioscience and Bioengineering 91(6): 586–589.

Li, L., Y. Jia, Q. How, T.C. Charl, E.W. Nester, and S.Q. Pan. 2002. A global Ph senor: *Agrobacterium* senor protein *chev*G regulates acid inducible genes on its two chromosomes and Ti plasmid. *Microbiology* (99)19: 12369–12374.

Lindsey, K., and M.M. Yeoman. 1983. The relationship between growth rate, differentiation and alkaloid accumulation in cell cultures. Journal of Experimental Botany 34(8): 1055–1065.

Lindsey, K. 1986. Incorporation of [14C] phenylalanine and [14C] cinnamic acid into capsaicin in cultured cells of *Capsicum frutescens*. Phytochemistry 25(12): 2793–2801.

Mantis, N.J., and S.C. Winas. 1992. The *Agrobacterium tumefaciens Vir* gene transcriptionally induced by acid pH and other stress stimuli. Journal of Bacteriology (174)4: 1189–1195.

Medina-Bolivar, F., and H.E. Flores. 1995. Selection for hyoscyamine and cinnamoyl putrescine overproduction in cell and root cultures of *Hyoscyamus muticus*. Plant Physiology 108(4): 1553–1560.

Memelink, J., R. Verpoorte, and J.W. Kijne. 2001. ORCAnization of jasmonate responsive gene expression in alkaloid metabolism. TRENDS in Plant Science 6(5): 212–219.

Moreno, P.R.H., R. Van der Heijden, and R. Verpoorte. 1993. Effect of terpenoid precursor feeding and elicitation on formation of indole alkaloids in cell suspension cultures of *Catharanthus roseus*. Plant Cell Reports 12(12): 702–705.

Moussous, A., C. Paris, M. Khelifi-Slaoui, M. Bekhouche, D. Zaoui, S.M. Rosloski, A. Makhzoum, S. Desobry, and L. Khelifi. 2018. *Pseudomonas* spp. increases root biomass and tropane alkaloid yields in transgenic hairy roots of *Datura* spp. *In Vitro* Cellular & Developmental Biology-Plant 54(1): 117–126.

Mulabagal, V., and H.S. Tsay. 2004. Plant cell cultures—an alternative and efficient source for the production of biologically important secondary metabolites. Int. J. Appl. Sci. Eng. 1(2): 29–48.

Murashige, T., and F. Skoog. 1962. A revised medium for rapid growth and bioassays with tobacco tissue culture. Physio. Plant. 15(3): 473–497.

Nakakjima, K., T. Hashimoto, and Y. Yamada. 1993. Two tropinone reductases with different stereospecificities are short-chaindeshydrogenases evolved from a common ancestor. Proc. Natl. acad. Sci. U.S.A. 9: 9591–9595.

Nussbaumer, P., I. Kapétanidis, and P. Christen. 1998. Hairy root of *Daturacandida* x *aurea*: Effect of culture medium composition on growth and biosynthesis. Plant Cell Rep. 17: 405–409.

Ono, N.N., and L. Tian. 2011. The multiplicity of hairy root cultures: Prolific possibilities. Plant Science 180(3): 439–446.

Park, S.U., and P.J. Facchini. 2000. *Agrobacterium rhizogenes*-mediated transformation of *Opiumpoppy*, *Papaver somniferum* L., and *California poppy*, *Eschscholzia californica* Cham. Root cultures. J. of Exper. Bot. (51)347: 1005–1016.

Parr, A.J., R.J. Robins, and M.J.C. Rhodes. 1984. Permeabilization of Cinchona ledgeriana cells by dimethyl sulfoxide. Effect on alkaloid release and long term membrane integrity. Plant Cell Rep. 3: 262–5.

Pinñol, M.T., J. Palazón, R.M. Cusidó and M. Ribó. 1999. Influence of calcium ion-concentration in the medium on tropane alkaloid accumulation in *Datura stramonium* hairy roots. Plant Science 141: 41–49.

Rao, S.R., and G.A. Ravishankar. 2002. Plant cell cultures: Chemical factories of secondary metabolites. Biotechnology Advances 20(2): 101–153.

Reymond, P., and E.E. Farmer. 1998. Jasmonate and salicylate as global signals for defense gene expression. Current Opinion in Plant Biology 1(5): 404–411.

Saenz-Carbonell, L., and V.M. Loyola-Vargas. 1996. *Datura stramonium* hairy roots tropane alkaloid content as a response to changes in Gamborg's B5 medium. Appl. Biochem. Biotechnol. 3(61): 321–337.

Sasson, A. 1991. Production of useful biochemicals by higher-plant cell cultures: Biotechnological economic aspects. Option méditerranées – série séminaire. 14: 59–74.

Shanks, J.V., and J. Morgan. 1999. Plant hairy root culture. Current Opinion in Biotechnology 10: 151–155.

Sikuli, N.N., and K. Demeyer. 1997. Influence of the ion-composition of the medium on alkaloid production by hairy roots of *Datura stramonium*. Plant Cell Tissue Organ Cult. 3(47): 261–267.

Silvestrini, A., G. Pasqua, B. Botta, B. Monacelli, R. van der Heijden, and R. Verpoorte. 2002. Effects of alkaloid precursor feeding on a *Camptotheca acuminata* cell line. Plant Physiology and Biochemistry 40(9): 749–753.

Simola, L.K., S. Nieminen, A. Huhtikangas, M. Ylinen, T. Naaranlahti, and Lounasmaa, M. 1988. Tropane alkaloids from *Atropa belladonna*, Part II. Interaction of origin, age, and environment in alkaloid production of callus cultures. Journal of Natural Products 51(2): 234–242.

Souret, F.F., Y. Kim, B.E. Wyslouzil, K.K. Wobbe, and P.J. Weathers. 2003. Scale-up of *Artemisia annua* L. hairy root culture produces complex patterns of terpenoid gene expression. Biotechnol. Bioeng. (83): 653–667.

Spollansky, T.C., S.I. Pitta-Alvarez, and A.M. Giulietti. 2000. Effect of jasmonic acid and aluminum on production of tropane alkaloids in hairy root cultures of *Brugmansia candida*. Electron. J. Biotechnol. 1(3): 72–75.

Stachel, S.E., E. Messens, M. Van Montagu, and P. Zambryski. 1985. Identification of the signal molecules produced by wounded plant cells that activate T-DNA transfer in *Agrobacterium tumefaciens*. Nature 318(6047): 624.

Tabata, M., H. Yamamoto, N. Hiraoka, and M. Konoshima. 1972. Organization and alkaloid production in tissue cultures of *Scopolia parviflora*. Phytochemistry 11(3): 949–955.

Taguchi, G., T. Yazawa, N. Hayashida, and M. Okazaki. 2001. Molecular cloning and heterologous expression of novel glucosyltransferases from tobacco cultured cells that have broad substrate specificity and are induced by salicylic acid and auxin. European Journal of Biochemistry 268(14): 4086–4094.

Tao, J., and L. Li. 2006. Genetic transformation of *Torenia fournieri* L. mediated by *Agrobacterium rhizogenes*. South Afri. J. of Botany 72: 211–216.

Van Loon, L., and P.A.H.M. Bakker. 2003. Signalling in rhizobacteria-plant interactions. pp. 297–330. *In*: Kroon, H., and E.J.W. Visser (eds.). Root Ecology. Springer, Berlin, Heidelberg.

Vanisree, M., C.Y. Lee, S.F. Lo, S.M. Nalawade, C.Y. Lin, and H.S. Tsay. 2004. Studies on the production of some important secondary metabolites from medicinal plants by plant tissue cultures. Bot. Bull. Acad. 45: 1–22.

Vasconsuelo, A., and R. Boland. 2007. Molecular aspects of the early stages of elicitation of secondary metabolites in plants. Plant Science 172: 861–875.

Verpoorte, R., and J. Memelink. 2002. Engineering secondary metabolite production in plants. Current Opinion in Biotechnology 13(2): 181–187.

Wasternack, C., and B. Parthier. 1997. Jasmonate-signalled plant gene expression. Trends Plant Sci. 2: 302–307.

Whitmer, S., C. Canel, D. Hallard, C. Gonçalves, and R. Verpoorte. 1998. Influence of precursor availability on alkaloid accumulation by transgenic cell line of *Catharanthus roseus*. Plant Physiology 116(2): 853–857.

Whitmer, S., R. van der Heijden, and R. Verpoorte. 2002. Effect of precursor feeding on alkaloid accumulation by a tryptophan decarboxylase over-expressing transgenic cell line T22 of *Catharanthus roseus*. Journal of Biotechnology 96(2): 193–203.

Wise, A.A., L. Voinov, and A.N. Binns. 2005. Inter subunit complementation of sugar signal transduction in VirA hereridimers and post translational regulation of VirA activity in *Agrobacterium tumefaciens*. Journal of Bacteriology 189(1).

Xia, K., X. Liu, Q. Zhang, W. Qiang, J. Guo, X. Lan, and Z. Liao. 2016. Promoting scopolamine biosynthesis in transgenic Atropa belladonna plants with pmt and h6h overexpression under field conditions. Plant Physiology and Biochemistry 106: 46–53.

Yamada, Y., and T. Hashimoto. 1982. Production of tropane alkaloids in cultured cells of *Hyoscyamus niger*. Plant Cell Reports 1(3): 101–103.

Yamada, Y., and T. Endo. 1984. Tropane alkaloid production in cultured cells of *Duboisia leichhardtii*. Plant Cell Reports 3(5): 186–188.

Yamada, Y., and M. Tabata. 1997. Plant biotechnology of tropane alkaloids. Plant Biotechnology 14(1): 1–10.

Zabetakis, L., R. Edwards, and D. O'Hagan. 1999. Elicitation of tropane alkaloid biosynthesis in transformed root cultures of *Datura stramonium*. Phytochemistry 50: 53–56.

Zhao, J., W.H. Zhu Q. Hu, and X.W. He. 2000. Improved alkaloid production in *Catharanthus roseus* suspension cell cultures by various chemicals. Biotechnology Letters 22(15): 1221–1226.

Zhao, J., L.C. Davis, and R. Verporte. 2005. Elicitor signal transduction leading to production of plant secondary metabolites. Biotechnology Advances 23(4): 283–333.

CHAPTER 3

Anticancer Mechanisms of Plant Secondary Metabolism

David O Nkwe

Introduction

Cancer is a complex physiological condition characterized by uncontrollable proliferation of cells. This abnormal cell behavior primarily results from genetic mutations that alter activities of key proteins regulating cellular processes. To combat cancer, different treatment options are available; and these include radiotherapy, surgery and chemotherapy. Cisplatin, a platinum-based drug, was the first chemotherapeutic agent described in the mid-1960s, and was approved for clinical use in 1978 (Makovec 2019). However, one drawback with platinum-based compounds is that, upon administration, they bind to thiol-containing plasma proteins in the bloodstream, which subsequently inactivates the compounds (Tanida et al. 2012). Other drawbacks include severe side effects such as nausea, bone marrow suppression, kidney toxicity, as well as acquired or intrinsic drug resistance (Lazarević et al. 2017). In addition, these drugs are effective on a limited number of cancer types, which limits their wide-clinical useage. Although the use of cisplatin was a milestone achievement in the late 1970s, it was historically preceeded by the use of organic compounds that are predominantly based on plant secondary metabolite such as the vinca alkaloid vincristine, derived from the leaves of *Catharanthus roseus*, and approved for clinical use in 1963 (Kuruppu et al. 2019, Seca and Pinto 2018). To date plant secondary metabolites continue to provide an attractive alternative to metal-based drugs due to their comparably low cellular toxicity (Fridlender et al. 2015, Seca and Pinto 2018).

The use of plants in the treatment of many diseases, including cancer, is as old as human history. Literature shows that approximately 80% of the world's population relies on traditional medicine, and about 60% of anticancer drugs are based on medicinal plants (Cragg et al. 2006). Various parts of plants (including the stem, roots, leaves and flowers) are used to prepare crude extracts that have anticancer properties (Iqbal et al. 2017).

Botswana International University of Science and Technology, Private Bag 16, Palapye, Botswana.
Email: nkwed@biust.ac.bw

Advances in chemical and biological techniques have made it possible to characterize the active compounds, as well as describe their mode of action on various cancer types. This has led to the development of new drug formulations derived from the active parent compounds. One of the best examples is the plant-derived anticancer compound paclitaxel, obtained from several *Taxus* species (Hao et al. 2017, Li et al. 2009, Witherup et al. 1990). Paclitaxel has been the basis of another anticancer drug called docetaxel that exhibits less severe side effects compared to paclitaxel (Eisenhauer and Vermorken 1998). Notably, plant metabolites are diverse and therefore have various molecular targets in cancer cells. This chapter focuses on selected modes of action of plant secondary metabolites, that include: (a) abolition of receptor tyrosine kinase activity; (b) topoisomerase inhibition; (c) inhibition of protein translation; and (d) cytoskeleton disruption.

Receptor Tyrosine Kinase Inhibitors

Receptor tyrosine kinases (RTKs) are transmembrane cell surface receptors that dimerise and become active upon ligand binding (Du and Lovly 2018). RTK ligands include several extracellular molecules such as growth factors, hormones, cytokines and neurotrophic factors. Specific tyrosine residues of active RTKs are autophosphorylated, thereby stimulating signal transduction through the RAS/MAPK and RAS/PI3K/AKT pathways, to promote cell proliferation, differentiation and migration as shown in Fig. 1 (Butti et al. 2018, Du and Lovly 2018, Regad 2015). One way to downregulate these pathways is through the clathrin-mediated or clathrin-independent endocytosis of the activated receptors, such as the epidermal growth factor receptor (EGFR) which has largely been used a model for the biology of RTKs (Goh and Sorkin 2013). The internalized ligand-receptor complexes are eventually degraded in the lysosomes or proteasomes (Mohapatra et al. 2013, Sangwan and Park 2006). As RTKs are important in many aspects of the biology of cells, genetic alteration of the RTKs or their ligands can result in cell transformation and the eventual development of a variety of cancers. For example, mutations leading to the overexpression of EGFRs have been observed in some cancers including breast cancer (10–30% of cases); glioblastomas (30–50%); colorectal cancer (25–82%); and non-small-cell lung cancer (5–20%) (Regad 2015). Remarkably, almost two decades ago, 30/58 (52%) of genes encoding RTKs were implicated in human cancers (Blume-Jensen and Hunter 2001); and these numbers may have increased over the years as more studies are conducted on the etiology of various cancer types.

Given the number of cancers linked to RTK signaling pathways, these have been the target for cancer therapy since the early 1980s, with about 37 drugs approved by the United States Food and Drug Administration (Bhullar et al. 2018). One of the latest RTK inhibitors described is leelamine (also called dehydroabietylamine), originally extracted from the bark of pine trees. Leelamine is thought to have multiple targets in different cancer cells [see review (Merarchi et al. 2019)], but here it is discussed in the context of RTKs. In a study conducted to investigate its mode of action on melanoma cell line, leelamine interfered with cholesterol transport, where cholesterol accumulated in late endosome/lysosomal compartments, resulting in cell death (Fig. 2) (Gowda et al. 2017, Kuzu et al. 2014). Under normal physiological conditions, cholesterol together with sphingolipids are enriched on special and highly organised plasma membrane microdomains called lipids rafts (Alonso and Millán 2001). These microdomains are scaffolds for proteins involved in signaling and intracellular trafficking. Additionally,

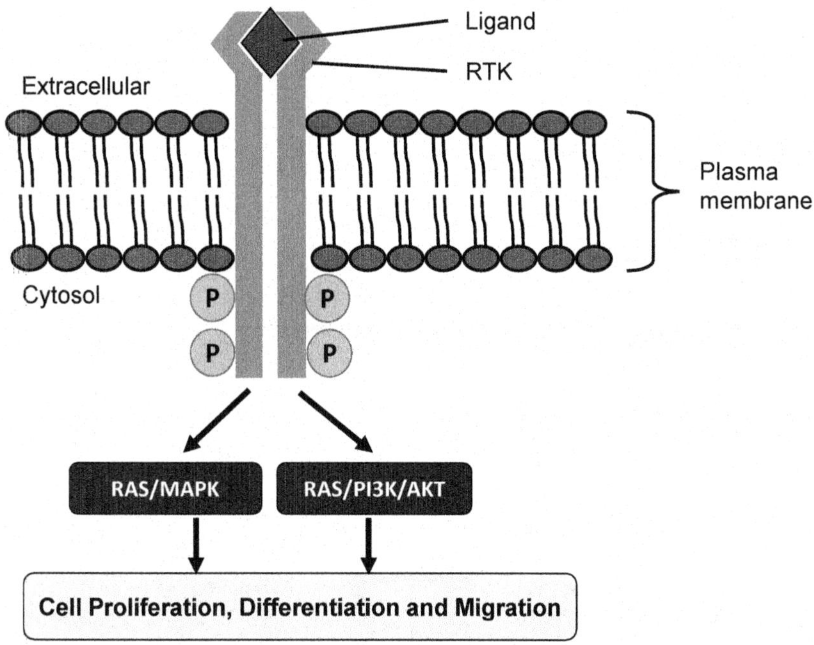

Figure 1: Receptor Tyrosine Kinase (RTK) activation. A ligand binds to the inactive RTK, resulting in dimerization of RTK monomers, which activates the receptor. In turn, the receptor stimulates signaling pathways such as the RAS/MAPK or RAS/PI3K/AKT, leading to events that promote cellular activities such as proliferation, differentiation and migration.

cholesterol regulates the fluidity of the phospholipid membrane bilayer. High cholesterol levels alter lipid rafts and may subsequently increase the distribution and signaling of RTKs. In cancer cells with elevated cholesterol levels, the use of compounds such as leelamine that interfere with cholesterol metabolism would disrupt the function of lipids rafts, thereby impeding RTK signaling pathways. Kuzu and co-researchers demonstrated that leelamine disrupted pathways involving mitogen-activated protein kinase (MAPK), serine/threonine Kinase 3 (AKT3) and signal transducer and activator of transcription 3 (STAT3). Therefore, leelamine offers an alternative to the treatment of cancers in which the above pathways are over-activated, such as in certain breast cancers (Sangai et al. 2012). Although Kuzu and co-workers argue that the primary mechanism by which cell death induced in melanoma cancer was through cholesterol accumulation, and was caspase-independent, a more recent report shows that in three different breast cancer cell lines (MCF-7, MDA-MB and SUM159), the mechanism by which leelamine exerts its effect was caspase-dependent (Sehrawat et al. 2017). Caspases are primary effectors on intrinsic and extrinsic apoptotic pathways. Therefore, it is likely that leelamine may be effective against several cancer types, employing different mechanisms to induce cell death, therefore making it a broad-spectrum anticancer agent. As leelamine and its derivatives do not have apparent systemic toxicity in mice with xenografted melanoma cells (Gowda et al. 2014), it is a promising anticancer agent that disrupts RTK signaling. Although not yet tested in human subjects, having tested the compound in an animal model has brought it a step closer to human clinical trials.

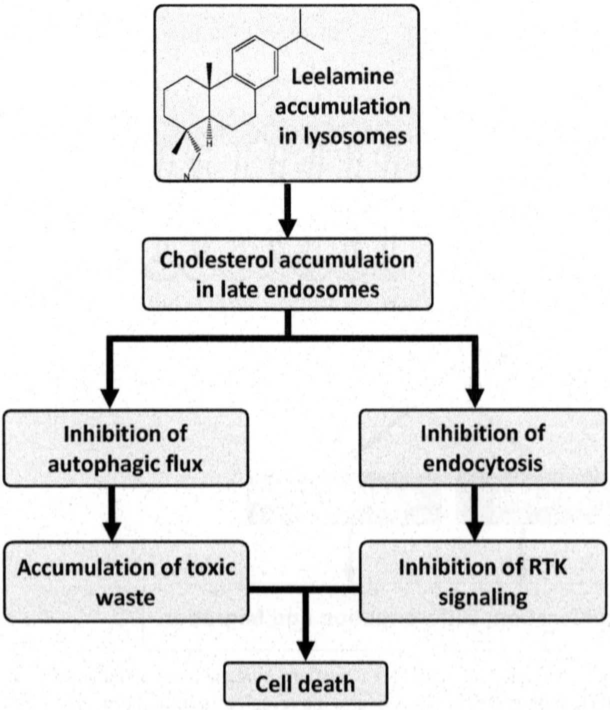

Figure 2: A summary of cell death induction by leelamine in melanoma cells. Leelamine accumulation in lysosomes result in cholesterol accumulation in late endosome. This results in, amongst others, inhibition of endocytosis and receptor tyrosine kinase signaling, and ultimately cell death ensues (Gowda et al. 2017, Kuzu et al. 2014).

Topoisomerase Inhibitors

DNA topology has gained interest as a viable target for cancer therapy. Regulating proteins that maintain DNA topology affects DNA metabolism in cells. During transcription in eukaryotic cells, there is formation of positive and negative DNA supercoils ahead of and behind the transcription machinery, respectively (Liu and Wang 1987). The negative supercoils may promote RNA:DNA hybrids, also known as R-loops. These loops compromise transcription and replication efficiencies, and cause genome instability. This supercoiling is alleviated by topoisomerases, which are highly conserved enzymes that bind the supercoiled DNA and produce transient breakage of one strand, thus enabling the DNA to untwist and relax (Champoux 2001).

Humans have six topoisomerases, namely TOP1, TOP1mt, TOP2α, TOP2β, TOP3α and TOP3β, with diverse and specialised functions (Pommier et al. 2016). TOP1 is a 90,726 Da protein, thought to be involved in the early stages of transcription and replication where it resolves R-loops (Aguilera and Garcia-Muse 2012, Stuckey et al. 2015). It interacts with transcription factor for RNA polymerase II, factor D (TFIID) which binds to promoters. In addition, TOP1 has been implicated in regulating gene expression by assisting in nucleosome disassembly at the promoters, before transcription

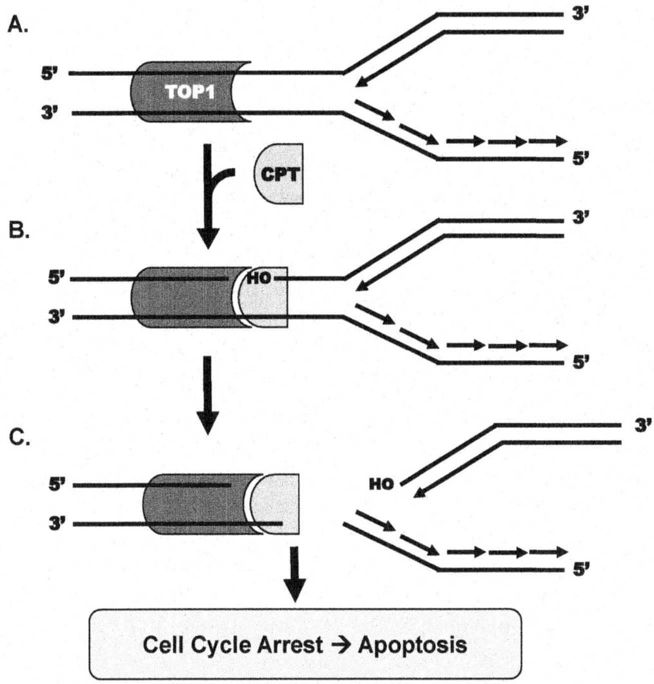

Figure 3: Camptothecin inhibits topoisomerase 1 (TOP1). (A) Normal physiologic activity of TOP1 where it interacts with double strand DNA during replication (or transcription), creating a temporary nick on one strand that allows the DNA to untwist, relax and religate. (B) CPT binds the DNA-bound TOP1, thereby inhibiting religation of cleaved DNA strand. (C) The DNA is eventually broken up and this leads to cell cycle arrest and apoptosis.

initiation (Pedersen et al. 2012). Following the initiation of transcription, there is a pause attributed to covalent modifications of the carboxyl-terminal-domain of the RNA polymerase II. This pause is important as it allows the following processes to occur: histone methylation, mRNA capping, polyadenylation, and mRNA splicing (Eick and Geyer 2013). To proceed with transcription, TOP1 is recruited to the pause site to alleviate torsional stress-associated pause, thereby regulating the process (Baranello et al. 2016).

Lately, analysis of nearly 25,000 patient samples with various cancers revealed an overexpression of TOP1 protein in 51% of the tumours, along with amplification of *TOP2A* gene in nearly 4% of the tumours (Heestand et al. 2017), which makes these plausible therapeutic targets. In 1966, camptothecin, a pentacyclic alkaloid obtained from *Camptotheca acuminate*, was observed to kill cancer cells (Oberlies and Kroll 2004). An investigation into camptothecin mechanism of action showed that the compound bound and stabilised the TOP1-DNA cleavage complex, thereby abolishing religation of the cleaved DNA strand (Svejstrup et al. 1991). This, therefore, induces a cascade of reactions leading to apoptosis (Fig. 3). As the compound has low solubility, several derivatives have been made, with at least three in clinical use (Liu et al. 2015). Evidently, targeting topoisomerase has proven viable, and more of such compounds need to be identified in plants to produce more potent drugs.

Translation Inhibitors

Translation, the synthesis of proteins, is a well-regulated process that is important for growth and development, and it has the following sequential steps: initiation, elongation, termination and ribosome recycling (Sonenberg and Hinnebusch 2009). The upstream events of the initiation pathway involve assembly of the translation machinery. In eukaryotic systems, this assembly starts with the small ribosomal subunit, 40S, binding a ternary complex (TC) to form the 43S pre-initiation complex (PIC). The TC is made up of initiator Methionyl-tRNA and the GTP-bound eukaryotic initiation factor 2 (eIF2). Interaction of the TC and the 40S is enhanced by several initiation factors that include eIF1, 1A, 5 and the eIF3 complex. Concomitantly, the mRNA is activated through binding eIF4B and a trimeric eIF4F complex made up of eIF4A (5'UTR unwinding RNA helicase), eIF4G (scaffold protein) and eIF4E (cap-binding protein) (Grzmil and Hemmings 2012, Jackson et al. 2010). The activation process also involves the binding of poly-A tail binding protein (PABP) to poly-A tail of the mRNA. The interaction of the bound-PABP with eIF4G circularizes the mRNA such that the 5' and 3' ends are juxta-positioned. Next, the activated multi-factorial mRNA attaches to the 43S PIC, ready for scanning of the 5'UTR until recognition of the initiation codon, thus triggering the irreversible hydrolysis of the GTP-bound eIF2. The initiation factors are then displaced, and the 60S large ribosomal subunit joins the 40S subunit, forming the 80S initiation complex required for polypeptide elongation. Upregulation of key players in translation may tip the physiological scale towards uncontrollable proliferation of host cells, and thus cause cancer. In recent years, translation effectors have emerged as promising anticancer molecular targets, such as the cap-binding protein eIF4E, which has a molecular weight of 25,097 Da (Avdulov et al. 2004, Fang et al. 2019, Graff et al. 2007, Pettersson et al. 2014). Overexpression of eIF4E, has been reported in several cancers including gall-bladder cancer (Fang et al. 2019), renal cell carcinoma (Ichiyanagi et al. 2019), acute myeloid leukemia (Hariri et al. 2013), malignant melanoma (Carter et al. 2016) and prostate cancer (D'Abronzo and Ghosh 2018, D'Abronzo and Ghosh 2018).

One of the hallmarks of cancer is hypoxia, a state of low oxygen levels in cells, which favours transition of tumour cells from the non-motile epithelial to mesenchymal type, thereby promoting cell mobility and metastasis (Muz et al. 2015). These cell activities are promoted by the activity of hypoxia-inducible factor 1 (HIF1), a heterodimeric transcription factor made up of HIF1α and HIF1β subunits that are oxygen-regulated and constitutively expressed, respectively (Meijer et al. 2012). Under normal oxygen levels, the Von Hippel Lindau gene mediates ubiquitination of HIF1α, thus targeting it for proteasomal degradation; under physiological stress of low oxygen level, HIF1α accumulates (Fig. 4) (Laplante and Sabatini 2013). The expression of HIF1α is regulated by amongst others the kinase complex mTORC1 [**m**echanistic (or **m**ammalian) **t**arget **o**f **r**apamycin] (Fig. 3). mTORC1, activated by the PI3K/AKT pathway, increases HIF1α levels through phosphorylating 4E (eIF4E) binding protein 1 (4E-BP1) complex, one of the key players during translation (Gingras et al. 1999). The phosphorylation releases eIF4E from 4E-BP1 (Carter et al. 2016, Sonenberg 2008, Sonenberg and Gingras 1998). In a study that was conducted to search for inhibitors of the HIF1 pathway, celastrol was identified as a potent inhibitor of HIF1α accumulation (Ma et al. 2014). Celastrol, used in traditional Chinese medicine, is a quinone methide triterpenoid extracted from the roots of *Tripterygium* species (Inabuy et al. 2017, Lv et al. 2019). HIF1α inhibition was associated with desphosphorylation of mTOR and eIF4E, effectors known to

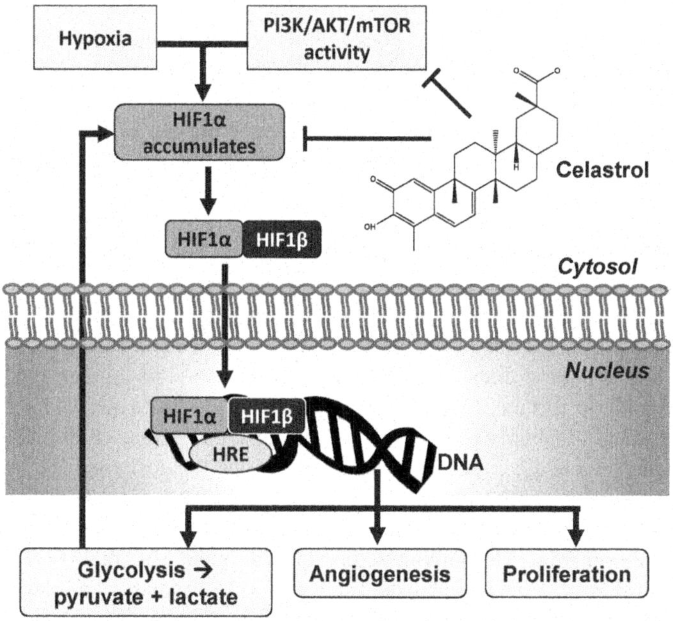

Figure 4: The HIF-1 pathway. Hypoxia (physiologic stress) and PI3K/AKT/mTOR activity contribute to HIF1α stabilization and accumulation. Stable HIF1α binds HIF1β to form a HIF1 complex which interacts with the hypoxia responsive element (HRE) of the DNA. Transcription of target genes is initiated culminating in glycolytic tumor metabolism, angiogenesis and cell proliferation. The end products of glycolysis, pyruvate and lactate, induce the accumulation of HIF1α and therefore increased HIF-1 activity. Celastrol can inhibit PI3K/AKT/mTOR signaling events or interfere with the HIF1α translation machinery, thus negatively impacting on HIF1α accumulation.

regulate its expression, therefore impairing the growth of human hepatoma tumour cells (Ma et al. 2014). Celastrol inhibits HIF-1α expression by disrupting formation of the initiation machinery for protein synthesis, without affecting the levels of HIF-1α mRNA (Han et al. 2014, Ma et al. 2014). Celastrol and its analog dihydrocelastrol have also been shown to cause apoptosis in human pancreatic tumour cells (Chakravarthy et al. 2013) and mantle cell melanoma (Xie et al. 2018). Celastrol's ability to modulate a number of molecular targets and inhibit proliferation cancer cells (Cascão et al. 2017) makes it a good broad-spectrum anticancer drug candidate, that needs to be developed further for human clinical trials.

Cytoskeleton Disruptors

The cytoskeleton is essential in the life of a cell because its elements are involved in various cellular activities. The main cytoskeleton elements are microtubules and microfilaments, made from tubulin and actin monomers, respectively. Due to the dynamic nature of the cytoskeleton, it performs the following functions: providing mechanical support to cells; forming tracks along which other biological molecules are ferried within a cell; cell division and migration. Mitotic cell division involves both the microtubules and microfilaments, with the former playing a role in the formation of mitotic spindles which facilitate the segregation of chromosomes following their duplication. Microfilaments, on the other hand, form a constriction band between the two yet-to-be separated daughter cells during cytokinesis (Sundquist and Ullman 2015). Given the role that the microtubules play in cell division, disrupting their function causes cell cycle arrest, and therefore apoptosis. This disruption has been associated with some secondary metabolites known to have antiproliferative capacities.

To understand why microtubules have become targets for cancer treatment, it is important to first appreciate their structural composition and morphology, in a broader sense. Microtubules are made up α- and β-tubulin subunits that associate to form heterodimers. The assembly of microtubules is nucleated by γ-tubulin (Moritz and Agard 2001). Then αβ-tubulin heterodimers are thread assembled head (β-tubulin) to tail (α-tubulin) into structural units called protofilaments. A total of thirteen protofilaments assemble laterally to form a 25-nm external diameter hollow structure. The positions of the α- and β-tubulin subunits dictate the polarity of the resultant cylindrical structure, where the exposed α and β ends constitute the minus and the positive ends, respectively. Heterodimers are added (i.e., polymerization) at the positive ends, whilst dissociation (i.e., depolymerization) occurs at the minus end. These processes are governed by guanosine triphosphate (GTP) binding to tubulin (Horio and Murata 2014, Hyman et al. 1992). Tubulin has an exchange and nonexchange GTP-binding sites, which are both essential for optimal assembly when GTP-bound. Hydrolysis of GTP at the exchange site leads to depolymerization of the tubulin, which is more pronounced at the minus end of the tubulin. As reviewed by Jordan and Wilson (Jordan and Wilson 2004), two kinds of dynamic situations have been observed *in vitro* and in cell models. The first one, called dynamic instability, involves non-equivalent changes in the length of the microtubule ends, with the plus end exhibiting rapid growth and shortening, compared to the minus end. In the second scenario, called treadmilling, the shortening and lengthening of the opposite ends are balanced. Although mechanisms directing these behaviors may not be entirely clear, regulatory proteins are particularly important (Akhmanova and Steinmetz 2019). Microtubule dynamics are key in the spatial and temporal organization of key subcellular processes such as mitosis. Therefore, chemicals that interfere with the tuning of these features, would affect fidelity of the cell cycle, which may result in apoptosis.

Mitosis is characterized by five phases: prophase, prometaphase, metaphase, anaphase and telophase. During the prophase, chromosomes condense and the mitotic spindle consisting of bipolar array of microtubules, separates the chromosomes. These spindle fibers radiate from the centrosome (the microtubule organizing center) and contact the chromosomes in the centromeres. Disrupting microtubule assembly or disassembly interferes with cell cycle progression, in turn leading to apoptosis. There are several plant-derived secondary metabolites that target the dynamics of microtubules. Two categories of microtubule inhibitors from plants are recognized: being microtubule stabilizers and destabilizers to promote polymerization and depolymerization, respectively. The best studied are taxanes (stabilizers) and vinca alkaloids (destabilizers). The two classical vinca alkaloids are vinblastine and vincristine, originally isolated from *Catharanthus roseus* (Jordan et al. 1991, Potier 1980). These compounds interact with the tubulin binding sites, thus preventing the aggregation of tubulin (Correia and Lobert 2001). Destabilizing microtubules compromises the integrity of mitotic spindle apparatus, hence arresting mitosis at metaphase. At lower concentrations that do not cause cell death, vinblastine inhibits microtubule dynamic instability and cell migration (Yang et al. 2010), consistent with the notion that when administered at different concentrations, some drugs may have varying effects on cells. Using new and advanced synthetic methods, the potency of vinblastine has previously been improved 100-fold (Carney et al. 2016). The ultrapotent vinblastine had higher affinity for tubulin and could disrupt dimer-dimer interface. This shows the potential for exploring synthetic chemistry methods to enhance the efficacy of existing anticancer drugs.

Notably, the role of some drugs can adversely be affected by proteins that interact with therapeutic targets. Such include microtubule associated proteins (MAPs) that are well documented in cancer research. For example, Tau, principally expressed in neurons, is aberrantly expressed in several cancers including prostate, breast, gastric and colorectal (He et al. 2014, Schroeder et al. 2019), where its role as a prognostic maker is variable. For instance, in gastric and breast cancers, Tau expression decreases sensitivity to paclitaxel (taxane) chemotherapy (He et al. 2014, Koo et al. 2015, Mimori et al. 2006), whilst in prostate cancer Tau is not a strong prognostic marker, given its heterogenous expression (Schroeder et al. 2019). These findings suggest that more systematic studies are required to understand how several proteins interact with key therapeutic targets in cancer cells, as these have the potential to affect the efficacy of plant-derived drugs.

Conclusion

Over the years, we have gained a lot of knowledge on the etiology of cancer and how different cancers can be treated. Advances in molecular cell biology have provided the tools to gain in-depth knowledge about the mechanisms of action, whilst synthetic chemistry has assisted in enhancing the potency of parent compounds, therefore widening the spectrum of available anticancer drugs. On one hand, gaps still exist in knowledge about the understanding of cellular components that regulate key therapeutic targets, while on the other hand there is need to explore a wide kingdom of plants for the much-needed natural compounds, to combat several medical conditions including cancer. There is a need to carry out more tests on animals, and human clinical trials are required for plant-based natural compounds so that these compounds can gain recognition as anticancer drugs. Plants, especially the marginally explored indigenous African fauna, offer a wealth of compounds whose anticancer mechanisms still need to be elucidated.

Acknowledgements

I am thankful to Dr. Nasir Mahmood for reviewing this chapter and for his useful comments. I also thank James Matshwele for drawing the structures of compounds in this chapter.

References

Aguilera, A., and T. Garcia-Muse. 2012. R loops: From transcription byproducts to threats to genome stability. Molecular Cell 46: 115–124.

Akhmanova, A., and M.O. Steinmetz. 2019. Microtubule minus-end regulation at a glance. Journal of Cell Science 132: jcs227850.

Alonso, M.A., and J. Millán. 2001. The role of lipid rafts in signalling and membrane trafficking in T lymphocytes. Journal of Cell Science 114: 3957.

Avdulov, S., S. Li, D. Van, D. Burrichter, M. Peterson, D.M. Perlman, J.C. Manivel, N. Sonenberg, D. Yee, P.B. Bitterman, and V.A. Polunovsky. 2004. Activation of translation complex eIF4F is essential for the genesis and maintenance of the malignant phenotype in human mammary epithelial cells. Cancer Cell 5: 553–563.

Baranello, L., D. Wojtowicz, K. Cui, B.N. Devaiah, H.J. Chung, K.Y. Chan-Salis, R. Guha, K. Wilson, X. Zhang, H. Zhang, J. Piotrowski, C.J. Thomas, D.S. Singer, B.F. Pugh, Y. Pommier, T.M. Przytycka, F. Kouzine, B.A. Lewis, K. Zhao, and D. Levens. 2016. RNA polymerase II regulates topoisomerase 1 activity to favor efficient transcription. Cell 165: 357–371.

Bhullar, K.S, N.O. Lagarón, E.M. McGowan, I. Parmar, A. Jha, B.P. Hubbard, and H.P.V. Rupasinghe. 2018. Kinase-targeted cancer therapies: Progress, challenges and future directions. Molecular Cancer 17: 48–48.

Blume-Jensen, P., and T. Hunter. 2001. Oncogenic kinase signalling. Nature 411: 355–365.

Butti, R., S. Das, V.P. Gunasekaran, A.S. Yadav, D. Kumar, and G.C. Kundu. 2018. Receptor tyrosine kinases (RTKs) in breast cancer: Signaling, therapeutic implications and challenges. Molecular Cancer 17: 34–34.

Carney, D.W., J.C. Lukesh, D.M. Brody, M.M. Brütsch, and D.L. Boger. 2016. Ultrapotent vinblastines in which added molecular complexity further disrupts the target tubulin dimer-dimer interface. Proceedings of the National Academy of Sciences of the United States of America 113: 9691–9698.

Carter, J.H., J.A. Deddens, N.R.I.V. Spaulding, D. Lucas, B.M. Colligan, T.G. Lewis, E. Hawkins, J. Jones, J.O. Pemberton, L.E. Douglass, and J.R. Graff. 2016. Phosphorylation of eIF4E serine 209 is associated with tumour progression and reduced survival in malignant melanoma. Br. J. Cancer 114: 444–453.

Cascão, R., J.E. Fonseca, and L.F. Moita.2017. Celastrol: A spectrum of treatment opportunities in chronic diseases. Front Med. (Lausanne) 4: 69–69.

Chakravarthy, R., M.J. Clemens, G. Pirianov, N. Perdios, S. Mudan, J.E. Cartwright, and A. Elia. 2013. Role of the eIF4E binding protein 4E-BP1 in regulation of the sensitivity of human pancreatic cancer cells to TRAIL and celastrol-induced apoptosis. Biology of the Cell 105: 414–429.

Champoux, J.J. 2001. DNA topoisomerases: Structure, function, and mechanism. Annual Review of Biochemistry 70: 369–413.

Correia, J.J., and S. Lobert. 2001. Physiochemical aspects of tubulin-interacting antimitotic drugs. Current Pharmaceutical Design 7: 1213–1228.

Cragg, G.M., D.J. Newman, and S.S. Yang. 2006. Natural product extracts of plant and marine origin having antileukemia potential. The NCI experience. Journal of Natural Products 69: 488–498.

D'Abronzo, L.S., and P.M. Ghosh. 2018. eIF4E Phosphorylation in Prostate Cancer. Neoplasia 20: 563–573.

Du, Z., and C.M. Lovly. 2018. Mechanisms of receptor tyrosine kinase activation in cancer. Molecular Cancer 17: 58.

Eick, D., and M. Geyer. 2013. The RNA Polymerase II carboxy-terminal domain (CTD) code. Chemical Reviews 113: 8456–8490.

Eisenhauer, E.A., and J.B. Vermorken. 1998. The taxoids. Drugs 55: 5–30.

Fang, D., J. Peng, G. Wang, D. Zhou, and X. Geng. 2019. Upregulation of eukaryotic translation initiation factor 4E associates with a poor prognosis in gallbladder cancer and promotes cell proliferation *in vitro* and *in vivo*. Int. J. Mol. Med. 44: 1325–1332.

Fridlender, M., Y. Kapulnik, and H. Koltai. 2015. Plant derived substances with anti-cancer activity: From folklore to practice. Frontiers in Plant Science 6.

Gingras, A.-C., S.P. Gygi, B. Raught, R.D. Polakiewicz, R.T. Abraham, M.F. Hoekstra, R. Aebersold, and N. Sonenberg. 1999. Regulation of 4E-BP1 phosphorylation: A novel two-step mechanism. Genes & Development 13: 1422–1437.

Goh, L.K., and A. Sorkin. 2013. Endocytosis of receptor tyrosine kinases. Cold Spring Harb Perspect. Biol. 5: a017459-a017459.

Gowda, R., S.V. Madhunapantula, A. Sharma, O.F. Kuzu, and G.P. Robertson. 2014. Nanolipolee-007, a novel nanoparticle-based drug containing leelamine for the treatment of melanoma. Mol. Cancer Ther. 13: 2328–2340.

Gowda, R., G.S. Inamdar, O. Kuzu, S.S. Dinavahi, J. Krzeminski, M.B. Battu, S.R. Voleti, S. Amin, and G.P. Robertson. 2017. Identifying the structure-activity relationship of leelamine necessary for inhibiting intracellular cholesterol transport. Oncotarget 8: 28260–28277.

Graff, J.R., B.W. Konicek, T.M. Vincent, R.L. Lynch, D. Monteith, S.N. Weir , P. Schwier, A. Capen, R.L. Goode, M.S. Dowless, Y. Chen, H. Zhang, S. Sissons, K. Cox, A.M. McNulty, S.H. Parsons, T. Wang, L. Sams, S. Geeganage, L.E. Douglass, B.L. Neubauer, N.M. Dean, K. Blanchard, J. Shou, L.F. Stancato, J.H. Carter, and E.G. Marcusson. 2007. Therapeutic suppression of translation initiation factor eIF4E expression reduces tumor growth without toxicity. J. Clin. Invest. 117: 2638–2648.

Grzmil, M., and B.A. Hemmings. 2012. Translation regulation as a therapeutic target in cancer. Cancer Research 72: 3891.

Han, X., S. Sun, M. Zhao, X. Cheng, G. Chen, S. Lin, Y. Guan, and X. Yu. 2014. Celastrol stimulates hypoxia-inducible factor-1 activity in tumor cells by initiating the ROS/Akt/p70S6K signaling pathway and enhancing hypoxia-inducible factor-1α protein synthesis. PLoS One 9: e112470-e112470.

Hao, J., H. Guo, X. Shi, Y. Wang, Q. Wan, Y-B. Song, L. Zhang, M. Dong, and C. Shen. 2017. Comparative proteomic analyses of two Taxus species (Taxus × media and Taxus mairei) reveals variations in the metabolisms associated with paclitaxel and other metabolites. Plant and Cell Physiology 58: 1878–1890.

Hariri, F., M. Arguello, L. Volpon, B. Culjkovic-Kraljacic, T.H. Nielsen, J. Hiscott, K.K. Mann, and K.L.B. Borden. 2013. The eukaryotic translation initiation factor eIF4E is a direct transcriptional target of NF-κB and is aberrantly regulated in acute myeloid leukemia. Leukemia 27: 2047–2055.

He, W., D. Zhang, J. Jiang, P. Liu, and C. Wu. 2014. The relationships between the chemosensitivity of human gastric cancer to paclitaxel and the expressions of class III β-tubulin, MAPT, and survivin. Medical Oncology 31: 950.

Heestand, G.M., M. Schwaederle, Z. Gatalica, D. Arguello, and R. Kurzrock. 2017. Topoisomerase expression and amplification in solid tumours: Analysis of 24,262 patients. Eur. J. Cancer 83: 80–87.

Horio, T., and T. Murata. 2014. The role of dynamic instability in microtubule organization. Frontiers in Plant Science 5.

Hyman, A.A., S. Salser, D.N. Drechsel, N. Unwin, and T.J. Mitchison. 1992. Role of GTP hydrolysis in microtubule dynamics: Information from a slowly hydrolyzable analogue, GMPCPP. Mol. Biol. Cell 3: 1155–1167.

Ichiyanagi, O., H. Ito, S. Naito, T. Kabasawa, H. Kanno, T. Narisawa, M. Ushijima, Y. Kurota, M. Ozawa, T. Sakurai, H. Nishida, T. Kato, M. Yamakawa, and N. Tsuchiya. 2019. Impact of eIF4E phosphorylation at Ser209 via MNK2a on tumour recurrence after curative surgery in localized clear cell renal cell carcinoma. Oncotarget 10: 4053–4068.

Inabuy, F.S., J.T. Fischedick, I. Lange, M. Hartmann, N. Srividya, A.N. Parrish, M. Xu, R.J. Peters, and B.M. Lange. 2017. Biosynthesis of diterpenoids in Tripterygium adventitious root cultures. Plant Physiol. 175: 92–103.

Iqbal, J., B.A. Abbasi, T. Mahmood, S. Kanwal, B. Ali, S.A. Shah, and A.T. Khalil. 2017. Plant-derived anticancer agents: A green anticancer approach. Asian Pacific Journal of Tropical Biomedicine 7: 1129–1150.

Jackson, R.J., C.U.T. Hellen, and T.V. Pestova. 2010. The mechanism of eukaryotic translation initiation and principles of its regulation. Nature reviews Molecular Cell Biology 11: 113–127.

Jordan, M.A., D. Thrower, and L. Wilson. 1991. Mechanism of inhibition of cell proliferation by vinca alkaloids. Cancer Research 51: 2212–2222.

Jordan, M.A., and L. Wilson. 2004. Microtubules as a target for anticancer drugs. Nature Reviews Cancer 4: 253–265.

Koo, D.-H., H.J. Lee, J-H. Ahn, D.H. Yoon, S.-B. Kim, G. Gong, B.H. Son, S.H. Ahn, and K.H. Jung. 2015. Tau and PTEN status as predictive markers for response to trastuzumab and paclitaxel in patients with HER2-positive breast cancer. Tumor Biology 36: 5865–5871.

Kuruppu, A.I., Paranagama, P., and C.L. Goonasekara. 2019. Medicinal plants commonly used against cancer in traditional medicine formulae in Sri Lanka. Saudi Pharmaceutical Journal 27: 565–573.

Kuzu, O.F., R. Gowda, A. Sharma, and G.P. Robertson. 2014. Leelamine mediates cancer cell death through inhibition of intracellular cholesterol transport. Mol. Cancer Ther. 13: 1690–1703.

Laplante, M., and D.M. Sabatini. 2013. Regulation of mTORC1 and its impact on gene expression at a glance. Journal of Cell Science 126: 1713.

Lazarević, T., A. Rilak, and Z.D. Bugarčić. 2017. Platinum, palladium, gold and ruthenium complexes as anticancer agents: Current clinical uses, cytotoxicity studies and future perspectives. European Journal of Medicinal Chemistry 142: 8–31.

Li, S., Y. Fu, Y. Zu, B. Zu, Y. Wang, and T. Efferth. 2009. Determination of paclitaxel and its analogues in the needles of Taxus species by using negative pressure cavitation extraction followed by HPLC-MS-MS. Journal of Separation Science 32: 3958–3966.

Liu, L.F., and J.C. Wang. 1987. Supercoiling of the DNA template during transcription. Proceedings of the National Academy of Sciences of the United States of America 84: 7024–7027.

Liu, Y.-Q., W.-Q. Li, S.L. Morris-Natschke, K. Qian, L. Yang, G.-X. Zhu, X.-B. Wu, A.-L. Chen, S.-Y. Zhang, X. Nan and K.-H. Lee. 2015. Perspectives on biologically active camptothecin derivatives. Med. Res. Rev. 35: 753–789.

Lv, H., L. Jiang, M. Zhu, Y. Li, M. Luo, P. Jiang, S. Tong, H. Zhang, and J. Yan. 2019. The genus Tripterygium: A phytochemistry and pharmacological review. Fitoterapia 137: 104190.

Ma, J., L.Z. Han, H. Liang, C. Mi, H. Shi, J.J. Lee, and X. Jin. 2014. Celastrol inhibits the HIF-1alpha pathway by inhibition of mTOR/p70S6K/eIF4E and ERK1/2 phosphorylation in human hepatoma cells. Oncology Reports 32: 235–242.

Makovec, T. 2019. Cisplatin and beyond: Molecular mechanisms of action and drug resistance development in cancer chemotherapy. Radiology and Oncology 53: 148–158.

Meijer, T.W.H., J.H.A.M. Kaanders, P.N. Span, and J. Bussink. 2012. Targeting hypoxia, HIF-1, and tumor glucose metabolism to improve radiotherapy efficacy. Clin. Cancer Res. 18: 5585.

Merarchi, M., Y.Y. Jung, L. Fan, G. Sethi, and K.S. Ahn. 2019. A brief overview of the antitumoral actions of leelamine. Biomedicines 7: 53.

Mimori, K., N. Sadanaga, Y. Yoshikawa, K. Ishikawa, M. Hashimoto, F. Tanaka, A. Sasaki, H. Inoue, K. Sugimachi, and M. Mori. 2006. Reduced tau expression in gastric cancer can identify candidates for successful paclitaxel treatment. Br. J. Cancer 94: 1894–1897.

Mohapatra, B., G. Ahmad, S. Nadeau, N. Zutshi, W. An, S. Scheffe, L. Dong, D. Feng, B. Goetz, P. Arya, T.A. Bailey, N. Palermo, G.E.O. Borgstahl, A. Natarajan, S.M. Raja, M. Naramura, V. Band, and H. Band. 2013. Protein tyrosine kinase regulation by ubiquitination: Critical roles of Cbl-family ubiquitin ligases. Biochimica et Biophysica Acta (BBA) - Molecular Cell Research 1833: 122–139.

Moritz, M., and D.A. Agard. 2001. Gamma-tubulin complexes and microtubule nucleation. Current Opinion in Structural Biology 11: 174–181.

Muz, B., P. de la Puente, F. Azab, and A.K. Azab. 2015. The role of hypoxia in cancer progression, angiogenesis, metastasis, and resistance to therapy. Hypoxia (Auckl) 3: 83–92.

Oberlies, N.H., and D.J. Kroll. 2004. Camptothecin and taxol: Historic achievements in natural products research. Journal of Natural Products 67: 129–135.

Pettersson, F., S.V. del Rincon, and W.H. Miller. 2014. Eukaryotic translation initiation factor 4E as a novel therapeutic target in hematological malignancies and beyond. Expert Opinion on Therapeutic Targets 18: 1035–1048.

Pommier, Y., Y. Sun, S.N. Huang, and J.L. Nitiss. 2016. Roles of eukaryotic topoisomerases in transcription, replication and genomic stability. Nature reviews Molecular Cell Biology 17: 703–721.

Potier, P. 1980. Synthesis of the antitumor dimeric indole alkaloids from *Catharanthus* species (vinblastine group). Journal of Natural Products 43: 72–86.

Regad, T. 2015. Targeting RTK signaling pathways in cancer. Cancers (Basel) 7: 1758–1784.

Sangai, T., A. Akcakanat, H. Chen, E. Tarco, Y. Wu, K.-A. Do, T.W. Miller, C.L. Arteaga, G.B. Mills, A.M. Gonzalez-Angulo, and F. Meric-Bernstam. 2012. Biomarkers of response to Akt inhibitor MK-2206 in breast cancer. Clin. Cancer Res. 18: 5816–5828.

Sangwan, V., and M. Park. 2006. Receptor tyrosine kinases: Role in cancer progression. Curr. Oncol. 13: 191–193.

Schroeder, C., J. Grell, C. Hube-Magg, M. Kluth, D. Lang, R. Simon, D. Höflmayer, S. Minner, E. Burandt, T.S. Clauditz, F. Büscheck, F. Jacobsen, H. Huland, M. Graefen, T. Schlomm, G. Sauter, and S. Steurer. 2019. Aberrant expression of the microtubule-associated protein tau is an independent prognostic feature in prostate cancer. BMC Cancer 19: 193.

Seca, A.M.L., and DG.C.A. Pinto. 2018 Plant secondary metabolites as anticancer agents: Successes in clinical trials and therapeutic application. Int. J. Mol. Sci. 19: 263.

Sehrawat, A., S.H. Kim, E.-R. Hahm, J.A. Arlotti, J. Eiseman, S.S. Shiva, L.H. Rigatti, and S.V. Singh. 2017. Cancer-selective death of human breast cancer cells by leelamine is mediated by bax and bak activation. Mol. Carcinog. 56: 337–348.

Sonenberg, N., and A.-C. Gingras. 1998. The mRNA 5′ cap-binding protein eIF4E and control of cell growth. Current Opinion in Cell Biology 10: 268–275.

Sonenberg, N. 2008. eIF4E, the mRNA cap-binding protein: From basic discovery to translational research. Biochemistry and Cell Biology 86: 178–183.

Sonenberg, N., and A.G. Hinnebusch. 2009. Regulation of translation initiation in eukaryotes: Mechanisms and Biological Targets. Cell 136: 731–745.

Stuckey, R., N. García-Rodríguez, A. Aguilera, and R.E. Wellinger. 2015. Role for RNA:DNA hybrids in origin-independent replication priming in a eukaryotic system. Proceedings of the National Academy of Sciences of the United States of America 112: 5779–5784.

Sundquist, W.I., and K.S. Ullman. 2015. An ESCRT to seal the envelope. Science 348: 1314–1315.

Svejstrup, J.Q., K. Christiansen, H. Gromova, A.H. Andersen, and O. Westergaard. 1991. New technique for uncoupling the cleavage and religation reactions of eukaryotic topoisomerase I. The mode of action of camptothecin at a specific recognition site. Journal of Molecular Biology 222: 669–678.

Tanida, S., T. Mizoshita, K. Ozeki, H. Tsukamoto, T. Kamiya, H. Kataoka, D. Sakamuro, and T. Joh. 2012. Mechanisms of cisplatin-induced apoptosis and of cisplatin sensitivity: Potential of BIN1 to act as a potent predictor of cisplatin sensitivity in gastric cancer treatment. Int. J. Surg. Oncol. 2012: 862879–862879.

Witherup, K.M., S.A. Look, M.W. Stasko, T.J. Ghiorzi, G.M. Muschik, and G.M. Cragg. 1990. *Taxus* spp. needles contain amounts of taxol comparable to the bark of Taxus brevifolia: Analysis and Iisolation. Journal of Natural Products 53: 1249–1255.

Xie, Y., B. Li, W. Bu, L. Gao, Y. Zhang, X. Lan, J. Hou, Z. Xu, S. Chang, D. Yu, B. Xie, Y. Wang, H. Wang, Y. Zhang, X. Wu, W. Zhu, and J. Shi. 2018. Dihydrocelastrol exerts potent antitumor activity in mantle cell lymphoma cells via dual inhibition of mTORC1 and mTORC2. International Journal of Oncology 53: 823–834.

Yang, H., A. Ganguly, and F. Cabral. 2010. Inhibition of cell migration and cell division correlates with distinct effects of microtubule inhibiting drugs. Journal of Biological Chemistry 285: 32242–32250.

CHAPTER 4

In silico and Computational Analysis of Plant Secondary Metabolites from African Medicinal Plants

Smith B Babiaka,[1] *Pascal Amoa Onguéné*,[2] *Boris Davy Bekono*[3] and *Fidele Ntie-Kang*[1,4,*]

Introduction

Natural Products in Drug Discovery

Natural products (NPs) are known to play an important role in drug discovery, as they often provide scaffolds as starting points for hit/lead discovery (Harvey et al. 2015, Rodrigues et al. 2016, Newman et al. 2020). They constitute huge sources of biologically active metabolites for the development of drugs (Newman et al. 2016, Beller et al. 2015, Pye et al. 2017). Several known drugs, e.g., the anticancer compounds (1 to 5, Fig. 1), are known to be derived from natural sources (Wani et al. 1971). Of major importance is that NPs continue to play a role as drugs (Newman et al. 2016), as biological probes, and

[1] Department of Chemistry, Faculty of Science, University of Buea, P. O. Box 63, Buea, Cameroon.
Email: babiaka.smith@ubuea.cm
[2] Department of Chemistry, University Institute of Wood Technology Mbalmayo, University of Yaoundé I, BP 50, Mbalmayo, Cameroon.
Email: amoapascal2@gmail.com
[3] Department of Physics, Ecole Normale Supérieure, University of Yaoundé I, P.O. Box 47, Yaoundé, Cameroon.
Email: borisbekono@gmail.com
[4] Department of Pharmaceutical Chemistry, Martin-Luther University of Halle-Wittenberg, Kurt-Mothes Str. 3, 06120, Halle (Saale), Germany.
* Corresponding author: ntiekfidele@gmail.com (FNK)

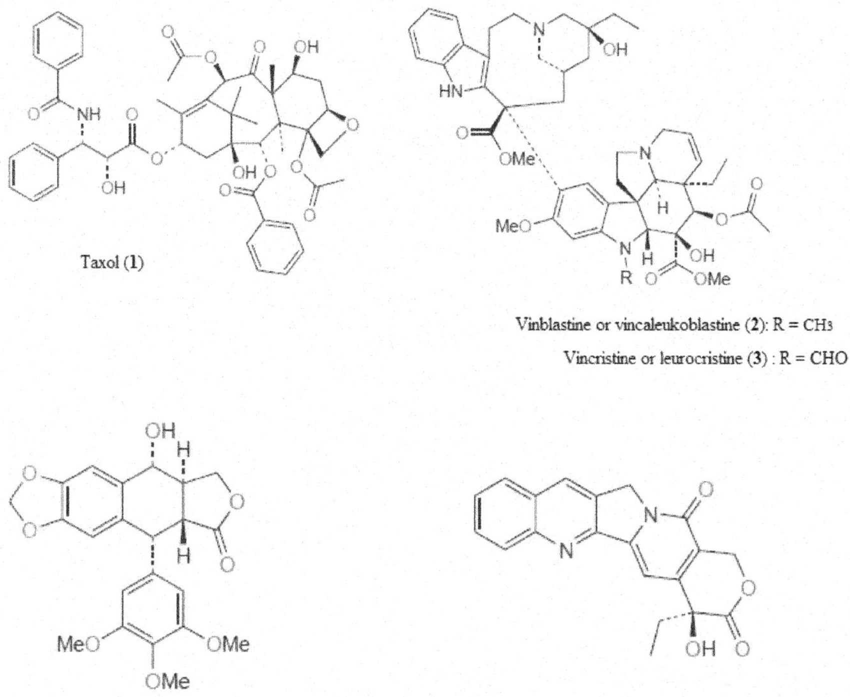

Taxol (1)

Vinblastine or vincaleukoblastine (2): R = CH_3

Vincristine or leurocristine (3) : R = CHO

Podophyllotoxin (4)

Camptothecin (5)

Figure 1: 2D structures of selected naturally occurring NP anticancer drug leads.

as study targets for synthetic and analytical chemists (Walsh et al. 2010). In fact, it has been shown that about half of all approved drugs between 1981 and 2010 were NP-based (Stratton et al. 2015). What makes NPs unique, when compared with synthetic drugs (SDs) is that they often contain more complex scaffolds and chiral centres, with more oxygen atoms and aromatic groups (Grabowski et al. 2007). In addition, a study involving a comparison of SDs versus NPs showed that drugs derived from NP-based structures display greater chemical diversity and occupy wider regions of chemical space (Newman et al. 2016). This is because drugs which are synthesized based on NP pharmacophores often exhibit lower hydrophobicity and greater stereochemical content when compared with drugs which are completely of synthetic origin. The aforementioned study showed that, of all approved drugs, NPs constituted 6% (unaltered), 26% (NP derivatives), 32% (NP mimics) or from NP pharmacophores, 73% of small molecule antibacterials and 50% of anticancer drugs (including taxol, vinblastine, vincristine, topotecan, etc.). This implies that if the structural features provided by nature are successfully incorporated into SDs, this would increase the chemical diversity available for small-molecule drug discovery (Newman et al. 2016). However, the reasons for the decline of interest by the pharmaceutical industry during the last two decades range from the time factor involved in the search for NP lead compounds to the labour intensiveness of the whole process (Li et al. 2009). This has now been rendered much easier within industrial settings by streamlined screening procedures and enhanced organism sourcing mechanisms (Pan et al. 2013).

Despite their evolving role in drug discovery (Koehn et al. 2005, Harvey et al. 2008), a recent chemoinformatic study involving a dataset of all published microbial and marine-derived compounds since the 1940s (comprising 40,229 NPs) showed that most NPs being published today bear close similarity[1] to previously published structures, with a plateau being observed since the mid-1990s (Pye et al. 2017). This study had, thus, suggested that the range of scaffolds readily accessible from nature is limited, i.e., scientists are now close to having described all of the chemical space covered by NPs, even though appreciable numbers of NPs with no structural precedents continue to be discovered. A reproduction of the previous study on another dataset of 32,380 NPs, also showed the same trend (Skinnider et al. 2017). By carrying out a similar analysis on a dataset of randomly selected compounds from the ZINC database, these compounds having overall lower structural similarity, the authors of the latter study further proved that such trends may be a feature of any growing database of chemical structures, rather than reflecting trends specific to NP discovery. Besides, it was also shown that since 1990, the rate of structurally novel compound discovery has dramatically outpaced random expectation[2] (Skinnider et al. 2017). This means that NPs discovered within the last three decades have been characterized by unprecedented chemical diversity,[3] suggesting that the dream of continuously discovering new chemical structures from nature remains positive.

Natural Product Databases

Databases have played a key role in drug discovery as they allow systematic annotation and storage of data for both basic and advanced applications (van Santen et al. 2020). Renewed interest in drug discovery from natural sources has ignited the development of several database resources and compound libraries, some of which are more comprehensive, including compounds from terrestrial, marine and microbial organisms, while other databases focus on particular disease types or on compounds from specific geographical regions or organism types. Moreover, several companies specialized in the commercialisation of NP compound samples are currently in the market (Chen et al. 2017), while other general sample suppliers include both NPs and SDs in their catalogues. A comprehensive list of useful databases for NP lead discovery and dereplication has been made available by the OMICs group (https://omictools.com/natural-products-category). A very recent study carried out by Chen et al. on 25 virtual libraries and 41 physical sample collections, revealed that of the ~ 250,000 NPs included in the investigated virtual library collections, about 10% have readily purchasable samples (Chen et al. 2017). In our study, we will focus on NP resources from African medicinal plants (Table 1).

Drug-likeness versus Natural Product-likeness Assessment

According to Lipinski's "Rule of Five" (ro5), an orally bioavailable drug should respect a set of rules (MW < 500 Da; log P < 5; HBD < 5; HBA < 10), and must not violate more than 2 of the "rules" (Lipinski et al. 1997). While Lipinski initially hypothesized that NPs do

[1] Two compounds were considered to be dissimilar by taking a Tanimoto cutoff of $T_c < 0.4$.
[2] By use of the Kolmogorov–Smirnov test, $P = 6.2 \times 10^{-14}$.
[3] The median maximum T_c has declined relative to random expectation, with $P = 7.6 \times 10^{-11}$.

Table 1: Summary of currently available natural product database resources in Africa.

Library name	Library size	Source organism	Web Accessibility	References
CamMedNP	1,859	224	http://african-compounds.org/about/cammednp/	Ntie-Kang et al. 2013a
ConMedNP	3,177	376	http://african-compounds.org/about/conmednp/	Ntie-Kang et al. 2014a
AfroDb	986	NM	http://african-compounds.org/about/afrodb/	Ntie-Kang et al. 2013b
AfroCancer	390	NM	http://african-compounds.org/about/afrocancer/	Ntie-Kang et al. 2014b
AfroMalariaDb	511	131	http://african-compounds.org/about/afromalariadb/	Onguéné et al. 2014
Afrotryp	321	NM	http://african-compounds.org/about/afrotryp/	Ibezim et al. 2017
p-ANAPL	534	NM	http://african- compounds.org/about/p-anapl/	Ntie-Kang et al. 2014c
NANPDB	4,928	751	http://african-compounds.org/nanpdb/	Ntie-Kang et al. 2017
EANPDB	1870	300	http://african-compounds.org.	Simoben et al. 2020
ETM-DB	4285		http://biosoft.kaist.ac.kr/etm	Bultum et al. 2019
SANCD	600	143	https://sancdb.rubi.ru.ac.za/	Hatherley et al. 2015

NM-not mentioned

not generally comply with the popular ro5 (Lipinski et al. 1997) it has recently been shown that a large majority of known approved drugs are either NPs or NP mimics (Newman et al. 2014). Thus, NPs have the huge potential to be developed into drugs, often ascribed to as 'lead-like' molecules (Newman et al. 2008). Several metrics have been used to assess the 'drug-likeness' (Daina et al. 2017), the 'lead-likeness' (respecting the "Rule of 3.5": $150 < MW < 350$; $\log P < 4$; $HBD < 3$; $HBA < 6$), (Teague et al. 1999) and the 'natural product-likeness' (Ertl et al. 2008) of a molecule. The concept of 'natural product-likeness' (NP-likeness), a Bayesian measure which allows for the determination of how molecules are similar to the structural space covered by natural products, was originally developed by Ertl et al. (2008) and has now been implemented in several open-source, open-data tools (Jayaseelan et al. 2012). The NP-likeness score is an efficient approach to separate NPs from synthetic molecules (SMs), with possible applications in virtual screening, prioritization of compound libraries toward NP-likeness, and the design of building blocks for the synthesis of 'NP-like' libraries (Ertl et al. 2008). The NP-likeness score (ranging from –5 to 5) is computed for a whole molecule, as a sum of contributions of fragments, f_i, (considered to be independent of each other, Eqn. 1) in the molecule, normalized relative to the molecule size:

$$f_i = \log\left(\frac{A_i}{B_i} \cdot \frac{B_{tot}}{A_{tot}}\right) \qquad (1)$$

where A_i is the number of NPs which contain fragment i, B_i is the number of SMs which contain fragment i, A_{tot} is the total number of NPs, and B_{tot} is the total number of SMs in the training set.

Chemical Space of Natural Products

Several investigations of the three-dimensional (3D) chemical space, occupied by compounds of synthetic and natural origins, using principal component analysis (PCA) have been published (Feher et al. 2003, Koch et al. 2005, Larsson et al. 2007, Singh et al. 2009, Rosén et al. 2009, López-Vallejo et al. 2012, Lachance et al. 2012, Gu et al. 2013, Newman et al. 2016). It was generally observed that, when compared with FDA-approved drugs and SDs, the distribution of NPs in chemical space cover regions that lack representation in synthetic medicinal chemistry compounds (Fig. 2), thus showing that NPs have a much wider coverage of chemical space. The unexplored areas appear to contain lead-like NPs that could subsequently be of interest in drug discovery. This indicates that these areas have not yet been investigated in drug discovery.

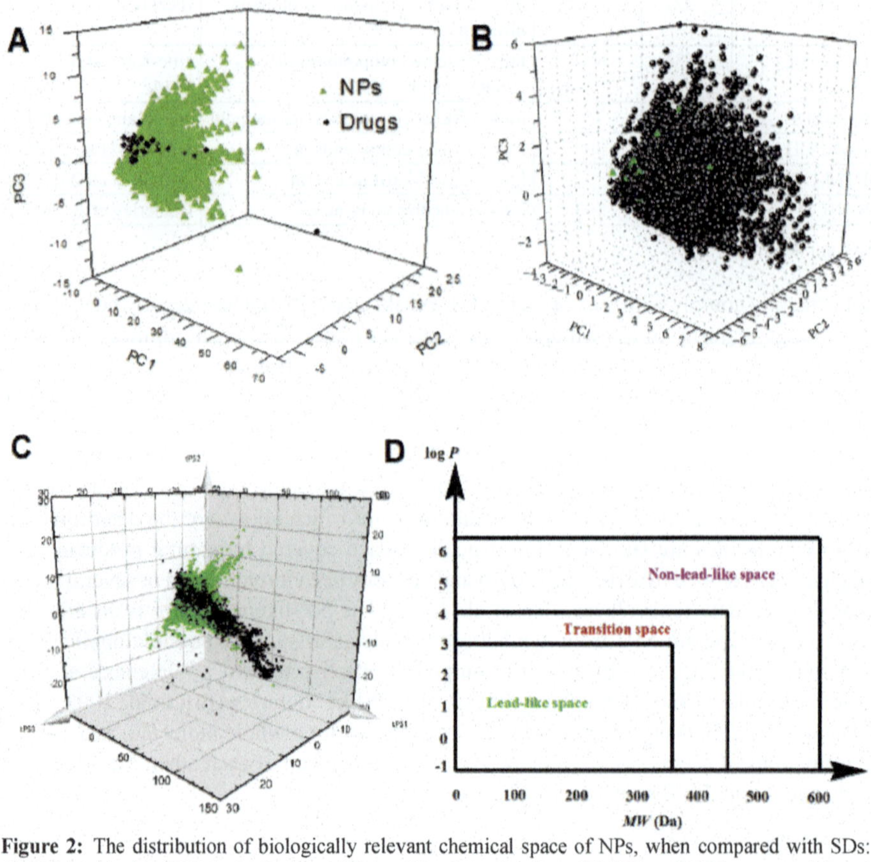

Figure 2: The distribution of biologically relevant chemical space of NPs, when compared with SDs: (A) PCA analysis of NPs in the Universal Natural Products Database (UNPD) and FDA-approved drugs. The green triangles and black dots represent natural products and FDA-approved drugs, respectively (Lachance et al. 2012); (B) PCA analysis of NPs contained in medicinal plants and 25 FDA-approved drugs for the treatment of type II diabetes mellitus (T2DM). The black dots and green triangles represent natural products and FDA-approved drugs, respectively (Rosén et al. 2009); (C) Predicted score (tPS) plots of NPs (in green) and bioactive medicinal chemistry compounds from the World of Molecular Bioactivity (WOMBAT) database (in black) (Feher et al. 2003); (D) Property space representation for lead-like molecules of some selected chemical libraries (Pascolutti et al. 2012).

Virtual Screening

In the quest to identify new and/or promising lead compounds from the plants, *in silico* (computer based) modeling is often employed since this approach has proven to accelerate the process and cuts down the cost of identifying lead compounds (López-Vallejo et al. 2011). Virtual screening methods are useful because, in principle, they narrow down the number of compounds to be actually tested in biological assays. This is practicable when the *in silico* scoring methods are sufficiently able to discriminate between active compounds and inactive ones. The approaches of structure-based methods, e.g., docking (Liu et al. 2017, Carregal et al. 2017, Wang et al. 2017), or ligand-based methods, e.g., quantitative structure-activity relationship (QSAR) and pharmacophore searching (Sun et al. 2016, Karhu et al. 2017, Gao et al. 2017) or a combination of several approaches within the same workflow (Svensson et al. 2012) have been recently employed with relative success within several lead generation programs from NP libraries.

Computer-based Prediction of Drug Metabolism and Pharmacokinetics

Many drugs often fail to enter the market as a result of poor pharmacokinetic profiles (Darvas et al. 2002). Thus, it has become imperative nowadays to design lead compounds which can be easily orally absorbed, easily transported to their desired site of action, not easily metabolised into toxic metabolic products before reaching the targeted site of action and easily eliminated from the body before accumulating in sufficient amounts that may produce adverse side effects. The sum of the above-mentioned properties is often referred to as ADME (absorption, distribution, metabolism and elimination) properties, or better still ADMET, ADME/T or ADMETox (when considerations are given to toxicity issues). The inclusion of pharmacokinetic considerations at earlier stages of drug discovery programs (Navia et al. 1996, Hodgson et al. 2001) using computer-based methods has become sufficiently popular (Lombardo et al. 2003, Gleeson et al. 2011, Irwin et al. 2012, Sterling et al. 2015). The rationale behind *in silico* approaches are the relatively lower cost and the time factor involved, when compared to standard experimental approaches for ADMET profiling (Darvas et al. 2002, DiMasi et al. 2003). As an example, it only takes a minute in an *in silico* model to screen 20,000 molecules, when compared with 20 weeks in the "wet" laboratory to do the same exercise (http://www.eadmet.com/en/ochem.php). Due to the accumulated ADMET data in the late 1990s, many pharmaceutical companies are now using computational models that, in some cases, are replacing the "wet" screens (http://www.eadmet.com/en/ochem.php). This paradigm shift has therefore spurred up the development of several theoretical methods for the prediction of ADMET parameters. A host of these theoretical models have been implemented in a number of software programs currently available for drug discovery protocols (http://www.eadmet.com/en/ochem.php, Cruciani et al. 2002, Meteor software, 2010, QikProp software 2011), even though some of the predictions could be disappointing (Tetko et al. 2006). The software tools currently used to predict the ADMET properties of potential drug candidates often make use of QSAR (Hansch et al. 2004, Tetko et al. 2006) or knowledge-based methods (Greene et al. 1999, Button et al. 2003, Cronin 2003). A promising lead compound may, therefore, be defined as one which combines potency with an attractive ADMET profile. As such, compounds with uninteresting predicted ADMET profiles may be completely dismissed from the list of potential drug candidates early enough (even if these prove to be highly potent). Otherwise, the DMPK properties are "fine-tuned" to improve their chances of making it to clinical trials (Hou et al. 2008).

This study intends to valorise African flora as both a rich source of secondary metabolites and a valuable starting point for nature-inspired drug discovery. To the best of our knowledge, there is no write-up on the cheminformatics analysis of plant secondary metabolites from African medicinal plants database, despite the long history of the use of their source organisms in traditional medicine, which dates back to prehistoric and pharaonic times (Shahat et al. 2001, Abdel-Azim et al. 2011).

NP Databases in Africa

Natural products libraries/databases of compounds from African medicinal plants that have been published recently include: (CamMedNP, ConMedNP, NANPDB, p-ANAPL, AfroDb, AfroCancer, AfroMalariaDb, Afrotryp, Ethiopia (ETM-DB), and South Africa (SANCDB). A huge effort has been made to gather traditional information-such as taxonomy, common names, location, medicinal uses, and part use,about plant species from the continent that was scattered in literature by these authors. In this section, we summarize the content and relevance of each of these databases that have emerged from each region of the continent.

CamMedNP and ConMedNP

Content of CamMedNP

The first version of the Cameroonian medicinal chemistry and natural products database (CamMedNP) released in 2013 contained more 2,500 compounds of natural origin, along with some hemisynthetic derivatives (Ntie-Kang et al. 2013). The compounds in CamMedNP were retrieved from peer-reviewed publications, Ph.D., M.Sc. theses and conference presentations by several research teams in Cameroon. The database reported a total of 224 distinct medicinal plant species belonging to 55 plant families. After a year, the same group published an update expanding the database to cover compounds from 10 countries in Central Africa, with the acronym; Congo Basin Medicinal Plant and Natural Products (ConMedNP) database library (Ntie-Kang et al. 2014). Compounds in CamMedNP had 3D optimized structures, annotated with chemical, biological, pharmacological data as well as literature references. Chemical information includes chemical structure, drug-like physicochemical properties. The biological information comprises species, geographical location and biological activities. A diversity analysis was also carried out in comparison with the ChemBridge diverse database.

Impact of the Central African natural product databases

Virtual screening has been performed on the CamMedNP library which has led to the identification of kinase inhibitors (Fig. 3, Table 2), examples, the protein kinase C (PKC), which plays a key role in neurotransmission in the central nervous system (Zhao et al. 2015). Other *in vitro* hits identified include promiscuous kinase inhibitor staurosporine (**6**, IC_{50} = 64 nM) (Islam et al. 2011), together with fisetin (**7**) (Sundarraj et al. 2017)

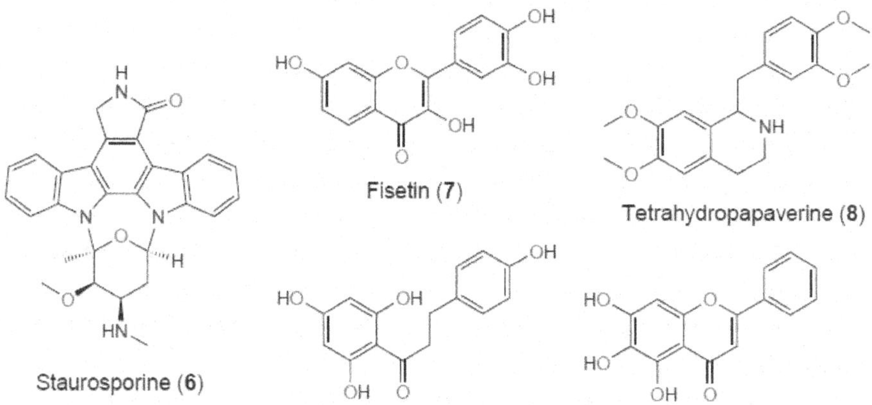

Staurosporine (**6**) Fisetin (**7**) Tetrahydropapaverine (**8**) Phloretin (**9**) Baicalein (**10**)

Figure 3: Kinase inhibitors identified by *in silico* and *in vitro* screening campaigns in which CamMedNP is included in the screening set.

Table 2: Summary of kinase inhibitors identified from virtual campaigns in which CamMedNP was included in the screening set.

Compound name (#)	PubChem ID	Source organism	Target	IC_{50} (mM)
Staurosporine (**6**)	44259	*Streptomyces* sp. B5136	PKC	0.06
Fisetin (**7**)	5281614	diverse fruits and vegetables	PKC	0.37
Tetrahydropapaverine (**8**)	5418	*Glaucium flavum* GFLOMT2	PKC	0.19
Phloretin (**9**)	4788	*Afzelia bipendensis*	Rho	0.22
Baicalein (**10**)	5281605	*Scutellaria lateriflora*	Rho	0.95

and tetrahydropapaverine (**8**) (Ehrenworth et al. 2017), which both showed inhibitory potencies within the nanomolar range (IC_{50} = 370 and 190, respectively). Another virtual screening campaign has also led to the identification of Rho kinase inhibitors, with a potential for the development of drugs against pulmonary hypertension (Su et al. 2015). These are phloretin (**9**) (Mpondo 1994) and baicalein (**10**) (Wang et al. 2014), with IC_{50} values of 0:22 and 0:95 mM, respectively.

Content of ConMedNP

ConMedNP was an extention of the CamMedNP with ~ 3200 compounds of natural origin, along with some of NP derivatives, from 9 other countries of the Central African sub-region. For each compound, the optimized 3D structure was used to calculate physico-chemical properties to determine the oral availability on the basis of the ro5. A comparative analysis was carried out with the "drug-like", "lead-like", and "fragment-like" subsets, containing 1726, 738 and 155 compounds, respectively. ConMedNP was also compared with their previously published CamMedNP library and the Dictionary of NPs. A diversity analysis was carried out in comparison with the DIVERSet™ Database (containing 48,651 compounds) from Chem-Bridge. The results in the database have proven that drug discovery, beginning with natural products from the Central African flora, could be a promising endeavor (Table 3).

Table 3: Summary of property distributions and comparison of the ConMedNP library, with it's various subsets.

Library name	Library size	Totaumers	Mw (Da)	log P	HBA	HBD	NRB
ConMedNP	3,177	7,838	426.70	4.18	5.85	2.39	5.31
Drug-like	1,726	3,900	326.16	2.87	4.97	1.79	2.96
Lead-like	738	1,610	269.58	2.48	4.17	1.49	2.01
Fragment-like	155	355	192.12	1.74	3.31	1.08	1.14
CamMedNP	1,859	5,286	421.63	4.07	6.00	2.40	5.51

$\overline{Mw}$: average molar weight; $\overline{lop\ P}$: average logarithm of the octan-1-ol/water partition coefficient; $\overline{HBA}$: average number of hydrogen bond acceptors; $\overline{HBD}$: average number of hydrogen bond donors; $\overline{NRB}$: average number of rotatable bonds.

AfroDb

AfroDb is a select highly potent and diverse natural product library containing ~ 1,000 potent compounds from African medicinal plants with a wide range of biological activities (Ntie-Kang et al. 2013). For each isolated compound in the dataset, the 3D structure was used to calculate physico-chemical properties used in the prediction of oral bioavailability on the basis of Lipinski's ro5. A comparative analysis was carried out with the "drug-like", "lead-like", and "fragment-like" subsets, as well as with the Dictionary of Natural Products (DNP, Chapman and Hall/CRC Press, Boca Raton, FL, USA). A diversity analysis was done in comparison with the ChemBridge diverse database. Furthermore, descriptors related to absorption, distribution, metabolism, excretion and toxicity (ADMET) were used by the authors to predict the pharmacokinetic profile of the compounds within the dataset. Their results prove that drug discovery, beginning with natural products from the African flora, could be highly promising. AfroDb has now been included as a subset in the ZINC database (http://zinc. docking.org/catalogs/afronp).

AfroCancer

The library includes "drug-likeness" of ~ 400 compounds from African medicinal plants that have shown *in vitro* and/or *in vivo* anticancer, cytotoxic, and antiproliferative activities (Ntie-Kang et al. 2014). Docking and binding affinity calculations were carried out in the dataset, in comparison with known anticancer agents comprising ~ 1,500 published naturally occurring plant-based compounds from around the world. Authors reported that, for five scoring methods, a significant number of the identified compounds within their dataset had docking scores comparable with those of the bound inhibitors within the X-ray structures of the drug targets, implying the presence of potential binders within the Afrocancer dataset. In addition, the docking pose of the antiproliferative luteolin-7-β-glucopyranoside, isolated from the Egyptian medicinal plant, *Livistona australis* (Kassem et al. 2012), showed similar binding interactions as the native ligand (9α-fluorocortisol), co-crystallized in the androgen receptor (PDB code: 1GS4, Fig. 4).

After two years, the same group performed a related study, pharmacophore models were generated and validated in virtual screening protocols for eight known anticancer

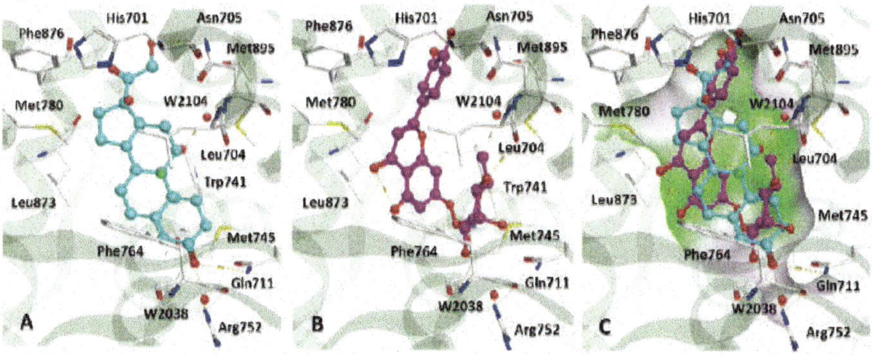

Figure 4: (A) crystal structure of the androgen receptor (1GS4) in complex with co-crystallized 9α-fluorocortisol; (B) in complex with docked luteolin-7-*O* β-glucopyranoside; (C) docking pose of the luteolin-7-β-glucopyranoside (C-atoms colored pink) compared with the co-crystallized 9α-fluorocortisol (C-atoms colored cyan). The molecular surface of the binding pocket is displayed and colored according to the hydrophobicity. Polar Regions are shown in magenta, hydrophobic regions in green.

drug targets, including tyrosine kinase, protein kinase B β, cyclin-dependent kinase, protein farnesyltransferase, human protein kinase, glycogen synthase kinase, and indoleamine 2,3-dioxygenase 1 (Ntie-Kang et al. 2016). The validated pharmacophore models was used as 3D search queries for virtual screening of the AfroCancer database (~ 400 compounds from African medicinal plants), along with the Naturally Occurring Plant-based Anticancer Compound-Activity-Target dataset (comprising ~ 1,500 published naturally occurring plant-based compounds from around the world). Additionally, an in silico assessment of toxicity of the two datasets was carried out by the use of 88 toxicity end points predicted by the Lhasa's expert knowledge-based system (Derek), showing that only an insignificant proportion of the promising anticancer agents would be likely showing high toxicity profiles. A diversity study of the two datasets was carried out using the analysis of principal components from the most important physico-chemical properties often used to access drug-likeness of compound datasets, showed that the two datasets do not occupy the same chemical space.

AfroMalariaDb

In this study, 3D molecular models were generated for antimalarial compounds of African origin (from 'weakly' active to 'highly' active), which were identified from literature sources (Onguéné et al. 2014). Selected computed molecular descriptors related to the ADMET properties of the phytochemicals were analysed and compared with those of known drugs to assess the 'drug-likeness' of the compounds in their dataset. More than 500 antimalarial compounds identified from 131 distinct medicinal plant species belonging to 44 plant families from the African flora were considered. On the basis of Lipinski's ro5, about 70% of the compounds were predicted to be orally bioavailable, while on the basis of Jorgensen's 'Rule of Three' (ro3), a corresponding > 80% were compliant. An overall drug-likeness parameter indicated that approximately 55% of the compounds could be potential leads for the development of drugs. From the authors analyses, it could be estimated that > 50% of the compounds exhibiting antiplasmodial/

antimalarial activities, derived from the African flora, could be starting points for drug discovery against malaria. The 3D models of the compounds could be employed in virtual screening.

Afrotryp

This was a relatively small dataset (~ 350 compounds) of NPs from African flora with known activities against the human African trypanosomiasis (HAT) parasite (Ibezim et al. 2017). Apart from the computer-based predictions for pharmacokinetic properties of the library a docking study was conducted, using 3 docking/scoring methods, to assess the affinity of the library dataset towards the binding site of 6 selected validated anti-*Trypanosoma* drug targets. Authors observed that about 42% of the compounds contained in the Afrotryp dataset were predicted to show a good overall performance in terms of predicted parameters for absorption, distribution, metabolism, elimination and toxicity properties. Furthermore, they performed docking calculations which identified 15 compounds having lowest theoretical binding energies toward the studied proteins, 9 of which are suited for the treatment of stage 2 HAT, due to their low polar surface area.

p-ANAPL

The pan-African natural products library (p-ANAPL) is a consortium of natural product (NP) collections isolated from African biota and owned by scientists and/or groups of scientists working in African institutions. The library is the largest physical collection of NPs derived from African medicinal plants directly available for screening purposes. A virtual library of 3D structures of compounds was generated and Lipinski's ro5 was used to evaluate likely oral availability of the samples. A majority of the compound samples are made of flavonoids and about two thirds (2/3) are compliant to the ro5. The pharmacological profiles of thirty six (36) selected compounds in the collection were described in the article (Ntie-Kang et al. 2014).

Northern African Natural Products Database (NANPDB)

The Northern African Natural Products Database (NANPDB) is the most comprehensive database containing ~ 5000 NPs, covering literature data for the period from 1962 to 2017 (Ntie-Kang et al. 2017). Computed physico-chemical properties, often used to predict drug metabolism and pharmacokinetics, as well as predicted toxicity information, were carried out for each compound in the data set. Also, for each compound, the details of its chemical structure, biological and physico-chemical properties, source species, and literature information are provided. The importance of the database was demonstrated by the low number of compounds already annotated in the PubChem database and the relatively few references common with PubMed. NANPDB has a web accessibility (http://african-compounds.org/nanpdb/) and a chemistry aware interface for exploitation by other research groups.

Eastern Africa Natural Products Database (EANPDB)

The Eastern Africa Natural Products Database (EANPDB) recently published in 2020 contains structural and information of 1870 unique molecules isolated from about 300 source species from the Eastern African region (Simoben et al. 2020). The compound SMILES, computed physicochemical properties and toxicity profiles of each compound was computed by the authors. Furthermore, the EANPDB was combined with the Northern African Natural Products Database (NANPDB), to form an African Natural Products Database (ANPDB), containing ~ 6500 compounds (http://african-compounds.org). Latrunculins A and B isolated from the sponge *Negombata magnifica* (Podospongiidae) with antitumour activities were identified through substructure searching from the dataset as potential putative binders of histone deacetylases (HDACs).

The integrated Ethiopian traditional herbal medicine and phytochemicals database (ETM-DB) is a comprehensive online (http://biosoft.kaist.ac.kr/etm) relational database which contains 4285 compounds from Ethiopian medicinal plants (Bultum et al. 2019). The ADMET and drug-likeness properties of all the compounds in the dataset was predicted. The authors reported that, out of the 4285 compounds in the database, 4080 of them passed the FAF-Drugs4 input data curation stage, with 876 having acceptable druglikeness properties.

South African Natural Compound Database (SANCD)

The SANCDB (Hatherley et al. 2015) contains compound information of 600 NPs from plants and marine life native to South Africa which have been published in journal articles, book chapters and theses. The databases also include information on molecular structure, name, structural class, source organism, and physicochemical properties of the compounds. The web-base NP database provides a useful resource for the *in silico* screening of South African NPs for virtual screening. SANCDB comprises a MySQL database, integrated into a Django application freely available at https://sancdb.rubi.ru.ac.za/.

Conclusions and Future Directions

The African continent comprises a wide range of medicinal plants that have been screened for their secondary metabolites and potential biological and therapeutic effects. Although it has been reported that under-exploited plants from African biodiversity offer untold medical and economic promise that should be pursued (Gurib-Fakim 2017). This review describes for the first time the *in silico* and computational analysis of plant secondary metabolites from African medicinal plants. The main highlight describes the contents of CamMedNP, ConMedNP, NANPDB, p-ANAPL, AfroDb, AfroCancer, AfroMalariaDb, Afrotryp, Ethiopia (ETM-DB), and South Africa (SANCDB). The report also includes the identification of NP lead molecules with anti-HIV and sirtuin inhibitory activities by virtual screening followed by *in vitro* assays, along with synthetic NP mimics, whose drug metabolism and pharmacokinetic profiles were predicted using computer-based methods. While a lot of effort has been put in to establish these libraries, further research and collaboration is required to establish a unified chemoinformatic natural product database from African medicinal plants.

Acknowledgments

We thank other authors for inspiration and for their published databases used in this review.

Abbreviations

ADMET	:	absorption, distribution, metabolism, excretion and toxicity
Afrotryp	:	African trypanosomiasis
CamMedNP	:	Cameroonian Medicinal chemistry and Natural Products database
ConMedNP	:	Congo Basin Medicinal plants and Natural Products database
DMPK	:	drug metabolism and pharmacokinetics
DNP	:	Dictionary of Natural Products
FDA	:	United States Food and Drug Administration
HBA	:	hydrogen bond acceptors
HBD	:	hydrogen bond donors
HIV	:	human immunodeficiency virus
log P	:	logarithm of the n-octanol/water partition coefficient
MMFF	:	Merck Molecular Forcefield
MOE	:	Molecular Operating Environment
MW	:	molecular weight
NP	:	natural product
NRB	:	number of rotatable bonds
P-ANAPL	:	pan-African Natural Products Library
PBMs	:	plant-based metabolites
PCA	:	principal component analysis
PDB	:	protein databank
QSAR	:	quantitative structure-activity relationship
RMSD	:	root mean square deviation
ro5	:	"Rule of Five"
ro3	:	"Rule of Three"
SDs	:	synthetic drugs
SDs	:	synthetic drugs
3D	:	Three dimensional
2D	:	Two dimensional

References

Abdel-Azim, N.S., K.A. Shams, A.A.A. Shahat, M.M. El Missiry, S.I. Ismail, and F.M. Hammouda. 2011. Egyptian herbal drug industry: Challenges and future prospects. Res. J. Med. Plant 5: 136–141.

Beller, H.R., T.S. Lee, and L. Katz. 2015. Natural products as biofuels and bio-based chemicals: Fatty acids and isoprenoids. Nat. Prod. Rep. 32: 1508–1526.

Bultum, L.E., A.M. Woyessa, and D. Lee. 2019. BMC Complement. Altern. Med. 19: 212.

Button, W.G., P.N. Judson, A. Long, and J.D. Vessey. 2003. Using absolute and relative reasoning in the prediction of the potential metabolism of xenobiotics. J. Chem. Inf. Comput. Sci. 43: 1371–1377.

Camp, D., R.A. Davis, M. Campitelli, J. Ebdon, and R. J. Quinn. 2012. Drug-like properties: Guiding principles for the design of natural product libraries. J. Nat. Prod. 75: 72–81.

Carregal, A.P., F.V. Maciel, J.B. Carregal, B. Dos Reis Santos, A.M. da Silva, and A.G. Taranto. 2017. Docking-based virtual screening of Brazilian natural compounds using the OOMT as the pharmacological target database. J. Mol. Model. 23: 111.

Chapman and Hall/CRC Press. 2005. Dictionary of Natural Products on CD-Rom, London.

Chen, Y., C. de Bruyn Kops, and J. Kirchmair. 2017. Data resources for the computer-guided discovery of bioactive natural products. J. Chem. Inf. Model. 57: 2099–2111.

Cronin, M.T.D. 2003. Computer-assisted prediction of drug toxicity and metabolism. pp. 259–278. *In*: Hillisch, A., and R. Hilgenfeld (Eds.). Modern Methods of Drug Discovery. Basel: Birkhäuser.

Cruciani, C., P. Crivori, P.A. Carrupt, and B. Testa. 2000. Molecular fields in quantitative structure-permeation relationships: The VolSurf approach. J. Mol. Struc.-Theochem 503: 17–30.

Daina, A., O. Michielin, and V. Zoete. 2017. SwissADME: A free web tool to evaluate pharmacokinetics, drug-likeness and medicinal chemistry friendliness of small molecules. Sci. Rep. 7: 42717.

Darvas, F., G. Keseru, A. Papp, G. Dormán, L. Urge, and P. Krajcsi. 2002. *In Silico* and *Ex silico* ADME approaches for drug discovery. Top. Med. Chem. 2: 1287–1304.

DiMasi, J.A., R.W. Hansen, and H.G. Grabowski. 2003. The price of innovation: New estimates of drug development costs. J. Health. Econ. 22: 151–185.

Ehrenworth, A.M., and Peralta-Yahya, P. 2017. Accelerating the semisynthesis of alkaloidbased drugs through metabolic engineering. Nat. Chem. Biol. 13: 249–258.

Ertl, P., S. Roggo, and A. Schuffenhauer. 2008. Natural product-likeness score and its application for prioritization of compound libraries. J. Chem. Inf. Model. 48: 68–74.

Feher, M., and J.M. Schmidt. 2003. Property distributions: Differences between drugs, natural products, and molecules from combinatorial chemistry. J. Chem. Inf. Comput. Sci., 43: 218–227.

Gao, Q., Y. Wang, J. Hou, Q. Yao, and J. Zhang. 2017. Multiple receptor-ligand based pharmacophore modeling and molecular docking to screen the selective inhibitors of matrix metalloproteinase-9 from natural products. J. Comput. Aided Mol. Des. 31: 625–641.

Gleeson, M.P., A. Hersey, and S. Hannongbua. 2011. *In-silico* ADME models: A general assessment of their utility in drug discovery applications. Curr. Top. Med. Chem. 11: 358–381.

Grabowski, K., and G. Schneider. 2007. Properties and architecture of drugs and natural products revisited. Curr. Chem. Biol. 1: 115–127.

Greene, N., P.N. Judson, and J.J. Langowski. 1999. Knowledge-based expert systems for toxicity and metabolism prediction: DEREK, StAR and METEOR. SAR QSAR Environ. Res. 10: 299–314.

Gu, J., L. Chen, G. Yuan, and X. Xu. 2013. A drug-target network-based approach to evaluate the efficacy of medicinal plants for type II diabetes mellitus. Evid. Based Complement. Alternat. Med. 203614.

Gu, J., Y. Gui, L. Chen, G. Yuan, H.Z. Lu, and X. Xu. 2013. Use of natural products as chemical library for drug discovery and network pharmacology. PLoS ONE 8: e62839.

Gurib-Fakim, A. 2017. Capitalize on African biodiversity. Nature 548: 7.

Hansch, C., A. Leo, S.B. Mekapatia, and A. Kurup. 2004. QSAR and ADME. Bioorg. Med. Chem. 12: 3391–3400.

Harvey, A.L. 2008. Natural products in drug discovery. Drug Discov. Today 13: 894–901.

Harvey, A.L., R. Edrada-Ebel, and R.J. Quinn. 2015. The re-emergence of natural products for drug discovery in the genomics era. Nat. Rev. Drug Discovery 14: 111–129.

Hatherley, R., D.K. Brown, T.M. Musyoka, D.L. Penkler, N. Faya, K.A. Lobb, and Ö.T. Bishop. 2015. SANCDB: A South African natural compound database. J. Cheminform. 7: 29.

Hodgson, J. 2001. ADMET – turning chemicals into drugs. Nat. Biotechnol. 19: 722–726.

Hou, T., and J. Wang. 2008. Structure-ADME relationship: Still a long way to go? Expert Opin. Drug Metab. Toxicol. 4: 759–770.

Ibezim, A., B. Debnath, F. Ntie-Kang, C.J. Mbah, and N.J. Nwodo. 2017. Binding of anti-Trypanosoma natural products from African flora against selected drug targets: A docking study. Med. Chem. Res. 26: 562–579.

Irwin, J.J., T. Sterling, M.M. Mysinger, E.S. Bolstad, and R.G. Coleman. 2012. ZINC: A free tool to discover chemistry for biology. J. Chem. Inf. Model. 52: 1757–1768.

Islam, M.T., A. Von Tiedemann, and H. Laatsch. 2011. Protein kinase C is likely to be involved in zoosporogenesis and maintenance of flagellar motility in the *peronosporomycete zoospores*. Mol. Plant Microbe Interact 24: 938–947.

Jayaseelan, K.V.C. Steinbeck, P. Moreno, and A. Truszkowski. 2012. Ertl, P. Natural product-likeness score revisited: An open-source, open-data implementation, BMC Bioinformatics 13: 106.
Karhu, E., J. Isojärvi, P. Vuorela, L. Hanski, and A. Fallarero. 2017. Identification of privileged antichlamydial natural products by a ligand-based strategy. J. Nat. Prod. 80: 2602–2608.
Kassem, M.E.S., S. Shoela, M.M. Marzouk, and A.A. Sleem. 2012. A sulphated flavone glycoside from *Livistona australis* and its antioxidant and cytotoxic activity. Nat. Prod. Res., 26: 1381–1387.
Koch, M.A., A. Schuffenhauer, M. Scheck, S. Wetzel, M. Casaulta, A. Odermatt, P. Ertl, and H. Waldmann. 2005. Charting biologically relevant chemical space: A structural classification of natural products (SCONP). Proc. Natl. Acad. Sci. USA. 102: 17272–17277.
Koehn, F.E., and G.T. Carter. 2005. The evolving role of natural products in drug discovery. Nat. Rev. Drug Discov. 4: 206–220.
Lachance, H., S. Wetzel, K. Kumar, and H. Waldmann. 2012. Charting, navigating, and populating natural product chemical space for drug discovery. J. Med. Chem. 55: 5989–6001.
Larsson, J., J. Gottfries, S. Muresan, and A. Backlund. 2007. ChemGPS-NP: Tuned for navigation in biologically relevant chemical space. J. Nat. Prod. 70: 789–794.
Li, J.W.H., and J.C. Vederas. 2009. Drug discovery and natural products: End of an era or an endless frontier? Science 325: 161–165.
Liu, Y., Y. Ren, Y. Cao, H. Huang, Q. Wu, W. Li, S. Wu, and J. Zhang. 2017. Discovery of a low toxicity O-GlcNAc transferase (OGT) inhibitor by structure-based virtual screening of natural products. Sci. Rep. 7: 12334.
Lipinski, C.A., F. Lombardo, B.W. Dominy, and P.J. Feeney. 1997. Experimental and computational approaches to estimate solubility and permeability in drug discovery and development settings. Adv. Drug Delivery Rev. 23: 3–25.
Lombardo, F., E. Gifford, and M.Y. Shalaeva. 2003. *In silico* ADME prediction: Data, models, facts and myths. Mini Rev. Med. Chem. 3: 861–875.
López-Vallejo, F., T. Caulfield, K. Martínez-Mayorga, M.A. Giulianotti, A. Nefzi, R.A. Houghten, and J.L. Medina-Franco. 2011. Integrating virtual screening and combinatorial chemistry for accelerated drug discovery. Comb. Chem. High Throughput Screen 14: 475–487.
López-Vallejo, F., M.A. Giulianotti, R.A. Houghten, and J.L. Medina-Franco. 2012. Expanding the medicinally relevant chemical space with compound libraries. Drug Discov. Today 17: 718–726.
Meteor software, version 13.0.0, Lhasa Ltd, Leeds, UK, 2010.
Mpondo, T.N. 1994. Phytochemical studies on *Afzelia bipendensis* from Cameroon. Annales Fac. Sci. UYI 3: 107–113.
Navia, M.A., and P.R. Chaturvedi. 1996. Design principles for orally bioavailable drugs. Drug Discov. Today 1: 179–189.
Newman, D.J. 2008. Natural products as leads to potential drugs: An old process or the new hope for drug discovery? J. Med. Chem. 51: 2589–2599.
Newman, D.J., and G.M. Cragg. 2016. Natural products as sources of new drugs from 1981 to 2014. J. Nat. Prod., 79: 629–661.
Newman, D.J., and G.M. Cragg. 2020. Natural products as sources of new drugs over the nearly four decades from 01/1981 to 09/2019. J. Nat. Prod. 83: 770–803.
Ntie-Kang, F., D. Zofou, S.B. Babiaka, R. Meudom, M. Scharfe, L.L. Lifongo, J.A. Mbah, L.M. Mbaze, W. Sippl, and S.M.N. Efange. 2013a. AfroDb: A select highly potent and diverse natural product library from African medicinal plants. PLoS ONE 8: e78085.
Ntie-Kang, F., J.A. Mbah, L.M. Mbaze, L.L. Lifongo, M. Scharfe J. Ngo Hanna, F. Cho-Ngwa, P. Amoa Onguéné, L.C. Owono Owono, E. Megnassan, W. Sippl, and S.M.N. Efange. 2013b. CamMedNP: Building the Cameroonian 3D structural natural products database for virtual screening. BMC Complement. Altern. Med. 13: 88.
Ntie-Kang, F., P.A. Onguéné, G.W. Fotso, K. Andrae-Marobela, M. Bezabih, J.C. Ndom, B.T. Ngadjui, A.O. Ogundaini, B.M. Abegaz, and L.M. Mbaze. 2014a. Virtualizing the p-ANAPL compound library: A step towards drug discovery from African medicinal plants. PLoS ONE 9: e90655.
Ntie-Kang, F., J.N. Nwodo, A. Ibezim, C.V. Simoben, B. Karaman, V.F. Ngwa, W. Sippl, M.U. Adikwu, and L.M. Mbaze. 2014b. Molecular modeling of potential anticancer agents from African medicinal plants. J. Chem. Inf. Model. 54: 2433–2450.

Ntie-Kang, F., P.A. Onguéné, M. Scharfe, L.M. Mbaze, L.C.O. Owono, E. Megnassan, W. Sippl, and S.M.N. Efange. 2014c. ConMedNP: A natural product library from Central African medicinal plants for drug discovery. RSC Adv. 4: 409–419.

Ntie-Kang, F., C.V. Simoben, B. Karaman, V.F. Ngwa, P.N. Judson, W. Sippl, and L.M. Mbaze. 2016. Pharmacophore modeling and *in silico* toxicity assessment of potential anticancer agents from African medicinal plants. Drug Des. Devel. Ther. 10: 2137–2154.

Ntie-Kang, F., K.K. Telukunta, K. Döring, C.V. Simoben, A.F.A. Moumbock, Y.I. Malange, L.E. Njume, J.N. Yong, W. Sippl, and S. Günther. 2017. NANPDB: A resource for natural products from Northern African sources. J. Nat. Prod. 80: 2067–2076.

OCHEM - A platform for the creation of *in silico* ADME/Tox prediction models (http://www.eadmet.com/en/ochem.php).

Onguéné, P.A., F. Ntie-Kang, J.A. Mbah, L.L. Lifongo, J.C. Ndom, W. Sippl, and L.M. Mbaze. 2014. The potential of anti-malarial compounds derived from African medicinal plants, part III: An *in silico* evaluation of drug metabolism and pharmacokinetics profiling. Org. Med. Chem. Lett. 4: 6.

Pascolutti, M., and R.J. Quinn. 2014. Natural products as lead structures: Chemical transformations to create lead-like libraries. Drug Discov. Today 19: 215–221.

Pan, L., H.B. Chai, and A.D. Kinghorn. 2013. Discovery of new anticancer agents from higher plants. Front. Biosci. (Schol. Ed.) 4: 142–156.

Pye, C.R., M.J. Bertin, R.S. Lokey, W.H. Gerwick, and R.G. Linington. 2017. Retrospective analysis of natural products provides insights for future discovery trends. Natl. Acad. Sci. USA 114: 5601–5606.

QikProp software, version 3.4, Schrödinger, LLC, New York, NY, 2011.

Rodrigues, T., D. Reker, P. Schneider, and G. Schneider. 2016. Counting on natural products for drug design. Nat. Chem. 8: 531–541.

Rosén, J., J. Gottfries, S. Muresan, A. Backlund, and T.I. Oprea. 2009. Novel chemical space exploration via natural products. J. Med. Chem. 52: 1953–1962.

Shahat, A.A., L. Pieters, S. Apers, N.M. Nazeif, N.S. Abdel-Azim, D.V. Berghe, and A.J. Vlietinck. 2001. Chemical and biological investigations on *Zizyphus spina-christi* L. Phytother. Res. 15: 593–597.

Simoben, C.V., Qaseem, A., Moumbock, A.F.A., Telukunta, K.K., Günther, S., Sippl, W., and Ntie-Kang, F. 2020. Pharmacoinformatic investigation of medicinal plants from East Africa. Mol. Inf. 39: 2000163.

Singh, N., R. Guha, M.A. Giulianotti, C. Pinilla, R.A. Houghten, and J.L. Medina-Franco. 2009. Chemoinformatic analysis of combinatorial libraries, drugs, natural products, and molecular libraries small molecule repository. J. Chem. Inf. Model. 49: 1010–1024.

Skinnider, M.A., and N.A. Magarvey. 2017. Statistical reanalysis of natural products reveals increasing chemical diversity. Proc. Natl. Acad. Sci. USA. 114: E6271–E6272.

Sterling, T., and J.J. Irwin. 2015. ZINC 15—Ligand Discovery for Everyone. J. Chem. Inf. Model. 55: 2324–2337.

Stratton, C.F., D.J. Newman, and D.S. Tan. 2015. Cheminformatic comparison of approved drugs from natural product versus synthetic origins. Bioorg. Med. Chem. Lett. 25: 4802–4807.

Su, H., J. Yan, J. Xu, Z.-X. Fan, X.-L. Sun, and K.-Y. Chen. 2015. Stepwise high-throughput virtual screening of Rho kinase inhibitors from natural product library and potential therapeutics for pulmonary hypertension. Pharmaceut. Biol. 53: 1201–1206.

Sun, Y., H. Zhou, H. Zhu, and S.W. Leung. 2016. Ligand-based virtual screening and inductive learning for identification of SIRT1 inhibitors in natural products. Sci. Rep. 6: 19312.

Sundarraj, K., A. Raghunath, and E. Perumal. 2017. A review on the chemotherapeutic potential of fisetin: *In vitro* evidences. Biomed. Pharmacother. 97: 928–940.

Svensson, F., A. Karlén, and C. Sköld. 2012. Virtual screening data fusion using both structure and ligand-based methods. J. Chem. Inf. Model. 52: 225–232.

Teague, S.J., A.M. Davis, and P.D. Leeson. 1999. Oprea, T. The design of leadlike combinatorial libraries. Angew. Chem. Int. Ed. Engl., 38: 3743–3748.

Tetko, I.V., P. Bruneau, H.W. Mewes, D.C. Rohrer, and G.I. Poda. 2006. Can we estimate the accuracy of ADMET predictions? Drug Discov. Today 11: 700–707.

van Santen, J.A,, S.A. Kautsar, M.H. Medema, and R.G. Linington. 2020. Microbial natural product databases: Moving forward in the multi-omics era. Nat. Prod. Rep. https://doi.org/10.1039/D0NP00053A.

Walsh, C.T., and M.A. Fischbach. 2010. Natural products version 2.0: Connecting genes to molecules. J. Am. Chem. Soc. 132: 2469–2493.

Wang, Y.-C., H.-Y. Yang, L.-T. Kong, and F.-X. Yu. 2014. *In vitro* evidence of baicalein's inhibition of the metabolism of zidovudine (AZT). Afr. Health Sci. 14: 173–177.

Wang, W., M. Wan, D. Liao, G. Peng, X. Xu, W. Yin, G. Guo, F. Jiang, W. Zhong, and J. He. 2017. Identification of potent chloride intracellular channel protein 1 inhibitors from traditional Chinese medicine through structure-based virtual screening and molecular dynamics analysis. Biomed. Res. Int. 4751780.

Wani, M.C., H.L. Taylor, M.E. Wall, P. Coggon, and A.T. McPhail. 1971. Plant antitumor agents. VI. Isolation and structure of taxol, a novel antileukemic and antitumor agent from *Taxus brevifolia*. J. Am. Chem. Soc., 93: 2325–2327.

Zhao, J., and C. Zhou. 2015. Virtual screening of protein kinase C inhibitors from natural product library to modulate general anaesthetic effects. Nat. Prod. Res. 29: 589–591.

ZINC12 subsets: AfroDb Natural Products (http://zinc. docking.org/catalogs/afronp). Published by permission from the authors of ref 100. Accessed June 27, 2016.

CHAPTER 5

Botany, Phytochemistry, Pharmacology, and Toxicity of the Southern African *Strychnos* Species

Willie Marenga, Abdullah Makhzoum and *Gaolathe Rantong**

Introduction

Strychnos are medicinally important plants from the *Loganiaceae* family with about 200 plant species that include shrubs, climbers, and large trees that produce edible and non-edible fruits (Mwamba 2006, Rajesh et al. 2012). The word *Strychnos* was coined because of the family's poisonous properties which resemble the poisonous nightshade known as *strukhnos* in Greek. *Strychnos* species are mainly found in the central tropical and sub-tropical parts of Central America, Australia, Asia, and Africa (Mwamba 2006). There are 75 *Strychnos* species identified in Africa. Twenty of these species produce edible fruits. In Africa, *Strychnos* species are mainly found in semi-arid and dry regions (Philippe et al. 2004, Mwamba 2006, Van Wyk 2013, Adebowale et al. 2014, Ngadze et al. 2017a, Ngadze 2018). According to Leeuwenberg (1963), there are 33 *Strychnos* species in the Southern Africa region. The *Strychnos* species's plant parts that are mostly used in traditional medicine are the leaves, barks, and roots. The investigation on ethnobotanical uses of medicinal plants has led to the discovery of about 74% of plant derived drugs. The discovery of these new biologically active compounds starts in the field where detailed, accurate, and specific ethnobotanical information is obtained from traditional medical practitioners and herbalists (Asuzu and Nwosu 2017).

Department of Biological Sciences & Biotechnology, Botswana International University of Science & Technology, Botswana.
* Corresponding author: rantongg@biust.ac.bw

The different plant parts of the *Strychnos* species produce bioactive compounds commonly known as *Strychnos* alkaloids (SAs), a class of dihydroindole-type alkaloids, and these are perceived as the main bioactive compounds of the *Strychnos* species. Strychnine and brucine are the most studied *Strychnos* bioactive compounds and have been reported to possess anti-cancer, anti-inflammatory, antimicrobial, and analgesic activities (Tian et al. 2013, Bribi 2018). The SAs interact with molecular targets inside the cells and tissues. The chemical structures of SAs resemble endogenous substrates, hormones, and neurotransmitters, and this triggers a response at the corresponding molecular targets. Plants have a diversity of biosynthetic pathways that have produced a wide range of chemical structures which are used in drug development, and these account for more than 50% of the current drugs available. Plants remain a treasure trove of new alkaloids with therapeutic potential. However, the SAs have some side effects that are caused by their potential toxicities (Tian et al. 2013, Bribi 2018).

The *Strychnos* species can adapt and thrive in harsh conditions such as soil with very low fertility and dry climatic conditions (Mwamba 2006, Ngemakwe et al. 2017, Ngadze et al. 2017b, Ngadze 2018). Due to this, they are extremely vital to the survival of humans and animals during the years of drought. Even though some of the *Strychnos* species are found in human settlements, the majority of them still grow in the wild and are at the risk of extinction due to their exposure to rapid climate change, unregulated harvesting and unregulated use by humans (Ngemakwe et al. 2017, Ngadze et al. 2017b, Ngadze 2018). Due to their risk of extinction, it is important that the different medicinal uses of *Strychnos* species in Southern African countries are documented. It is also important to identify present species and evaluate the population dynamics of the different species that exist. This information is important in the conservation process and commercialization of the *Strychnos* species that are found in Southern Africa.

Distribution of *Strychnos* spp. in Southern Africa

Strychnos species are distributed among the Miombo and Savannah woodlands of tropical Southern Africa and Malagasy (Mwamba 2006, Orwa et al. 2009). They are adaptable to riverine forests, warm to hot tropical savannas, rocky hills, deep valleys, coastal bushlands, and other woodland regions that receive approximately 300–1500 mm rainfall per annum and grow best at an altitude of 0–2000 m above sea level (Leeuwenberg 1963, Mwamba 2006). Table 1 depicts the habitat, distribution, and altitude of different *Strychnos* species found in Southern Africa.

Ethno Pharmacological Use of the Southern African *Strychnos* Species

Different *Strychnos* species have been used in folk medicine for centuries to treat many ailments and conditions. The different anatomical parts of the *Strychnos* species have specific uses in the treatment of various ailments such as diabetes mellitus, snakebites, gonorrhea, wounds, fever, malaria, and other conditions in different parts of the world (Mwamba 2006). Table 2 below shows the ethnobotany of different anatomical parts of Southern African *Strychnos* species.

Table 1: Distribution and habitat of *Strychnos* species in the Southern African region.

Species	**Section**	**Distribution**	**Habitat**	**Altitude**
1. S. cocculoides	Spinosae	Angola, Namibia, Botswana, Zambia, Malawi, Zimbabwe, Eswatini (Swaziland) South Africa, and Mozambique	Savanna (Dry and Open woodlands), bushveld, gallery, and sand forests	400–2000 m
2. S. spinosa	Spinosae	Angola, Namibia, Botswana, Zambia, Malawi, Mozambique, Zimbabwe, South Africa, Comoro Islands, Madagascar, Seychelles, and Mauritius	Savanna (Dry and Open woodlands), bushveld, gallery, and sand forests	0–2200 m
3. S. pungens	Densiflorae	Angola, Namibia, Botswana, Zambia, Malawi, Zimbabwe, South Africa, and Mozambique	Savanna (Dry and Open woodlands), rocky slopes, and rocky koppies	0–2000 m
4. S. innocua	Densiflorae	Angola, Botswana, Zambia, Malawi Zimbabwe, and Mozambique	Dry and Open evergreen woodlands and rocky koppies	0–1600 m
5. S. madagascariensis	Densiflorae	Angola, Botswana, Zambia, Malawi Zimbabwe, Madagascar, and Mozambique	Open evergreen woodlands, rocky koppies, and coastal bush	0–1800 m
6. S. gerrardii	Densiflorae	Madagascar, Mozambique, Eswatini, and South Africa	Dry coastal or dune forests sometimes on Wet coastal forests	0–500 m
7. S. lucens	Densiflorae	Angola and Zambia	Gallery forests and rocky hills in woodlands	0–1700 m
8. S. henningsii	Breviflorae	Angola, Zambia, Malawi, Zimbabwe, Mozambique, Eswatini, Madagascar, and South Africa	Dense Forest, woodlands, and dry areas	0–2000 m
9. S. angolensis	Breviflorae	Angola and Mozambique	Gallery forests	0–1500 m

Table 1 contd. ...

...Table 1 contd.

Species	Section	Distribution	Habitat	Altitude
10. S. campicola	Breviflorae	Angola	Rain forests, secondary forests, gallery forests, or thickets on rocky slopes	0–900 m
11. S. mitis	Breviflorae	Angola, Zimbabwe, Mozambique, Eswatini, South Africa and Comoros Islands	Upland and lowland rain forests, gallery Forests and coastal bush	0–2300 m
12. S. icaja	Breviflorae	Angola	Rain Forests and Secondary forests	0–800 m
13. S. decussata	Rouhamon	Zambia, Zimbabwe, Mozambique, Malawi, Madagascar, Eswatini, and South Africa	Wet coastal forests	0–1600 m
14. S. potatorum	Rouhamon	Zambia, Malawi, Zimbabwe, Mozambique, Namibia, Botswana, South Africa and Madagascar	Gallery Forests, Dry Forest, and Woodlands	0–1500 m
15. S. usambarensis	Rouhamon	Botswana, Zambia, Zimbabwe, Mozambique, Eswatini, and South Africa	Rain forests, secondary Forests on riverbanks, deep valleys, gallery forest, semi-evergreen and coastal evergreen bushland	0–2000 m
16. S. floribunda	Rouhamon	Angola	Rain forests, secondary Forests on the riverbanks, gallery forest, bushland	0–800 m
17. S. trichoneura	Rouhamon	Madagascar	Riverine forest	400 m
18. S. dale	Rouhamon	Angola	Rain forests or secondary forests	0–750 m
19. S. aculeata	Aculeatae	Angola	Rain forests or secondary forests	0–700 m

Table 1 contd. ...

...Table 1 contd.

Species	Section	Distribution	Habitat	Altitude
20. S. matopensis	Penicillatae	Zambia, Zimbabwe, and Mozambique	Semi-evergreen bushland or Brachystegia-woodland, on rocky hills and termite mounds, and gallery forest	900–1600 m
21. S. diplotricha	Penicillatae	Madagascar	Montane Rainforests	800–1600 m
22. S. mostueoides	Penicillatae	Madagascar	Low forest near the riverbanks	0–500 m
23. S. bifurcata	Penicillatae	Madagascar	Woodlands	Not provided
24. S. myrtoides	Penicillatae	Mozambique and Madagascar	Brachystegia-woodlands or light forests	0–600 m
25. S. pentantha	Penicillatae	Madagascar	Calcareous hills	50–400 m
26. S. moandaensis	Lanigerae	Angola	Rain forests, on riverbanks	Low elevation
27. S. scheffleri	Lanigerae	Angola	Rain forests, secondary Forests mostly on riverbanks or gallery forest	0–1100 m
28. S. panganensis	Lanigerae	Madagascar	Lowland rain forests, gallery forests, or thickets in coastal evergreen bushland	0–500 m
29. S. kasengaensis	Lanigerae	Angola and Zambia	Riverine forests or in crevices of granitic rocks in woodlands	0–1500 m
30. S. mellodora	Brevitubae	Zimbabwe and Mozambique	Rain forests	800–1200 m
31. S. johnsonii	Brevitubae	Angola	Rain forests, secondary Forests mostly on riverbanks or gallery forest	0–750 m
32. S. xantha	Dolichanthae	Zambia and Mozambique	Gallery forests	0–1800 m
33. S. gossweileri	Dolichanthae	Angola and Zambia	Gallery forests, groves on granitic rocks, or woodlands	0–700 m

Sources: (Van Wyk and Gericke 2000, Ruffo et al. 2002, Setshogo and Venter 2003, Mwamba 2006, Orwa et al. 2009, Van Wyk 2013, Adebowale 2014, Leeweunberg 1963)

Table 2: Ethnobotany of *Strychnos* species found in Southern Africa.

Plant name	Parts used and uses	References
S. cocculoides	**Roots:** extracts used as an aphrodisiac, and remedy for eczema, hydrocele, male infertility, sore throat, abdominal pains, amenorrhea, heavy menstrual flow, and stomach pains. **Root, Stem bark and leaves:** extracts drunk as medicine for pains, gonorrhea, and other venereal diseases. **Stem bark:** powder added to soup and extracts drunk as an aphrodisiac and remedy for male organ disorders, infertility, and stomach pain. **Leaves:** extracts drunk as a remedy for snakebites, neck pain, and the paste are applied on sores and wounds. **Ripe pulp:** mixed with honey and used as a cough syrup. It is also used as washing soap due to the presence of saponins. **Unripe pulp:** mixed with milk as a laxative and to induce vomiting; grounded unripe pulp extracts are used as a remedy for eardrops, ear pains, swelling, snakebites dye, and as a dye that controls insects.	(Ruffo et al. 2002, Mwamba 2006, Orwa et al. 2009, Maroyi 2011, Elago and Tjaveondja 2015) (Orwa et al. 2009) (Ruffo et al. 2002, Mwamba 2006, Orwa et al. 2009, Maroyi 2011, Elago and Tjaveondja 2015) (Ruffo et al. 2002, Mwamba 2006, Motlhanka et al. 2008, Orwa et al. 2009) (Mwamba 2006, Ngadze et al. 2017a, Ngadze 2018)
S. innocua	**Roots:** extracts used as an aphrodisiac, fly repellants, and a remedy for snakebites, gonorrhoea, and other venereal diseases. **Barks and twigs:** extracts drunk by expectant mothers to facilitate childbirth. Small twigs are used as toothpicks and toothbrushes across Africa. **Leaves:** extracts used as a remedy for malaria and as an effective anthelmintic against both *Pheretima posthuma* and *Ascaris ouum.* **Pulp:** extracts from unripe pulp is a remedy for dysentery and as an eardrop. **Pulp and seeds:** produce oil that is used to make soap. **Seeds:** extracts induce vomiting.	(Ruffo et al. 2002, Mwamba 2006) (Ruffo et al. 2002) (Asase et al. 2005) (Mwamba 2006) (Mwamba 2006)
S. madagascariensis	**Roots:** extracts used as medicine for inflammation and wounds. **Stem Bark:** infusion is used in obstetrics in East Africa and to make herbal tea which is a remedy for venereal diseases. **Stem bark and leaves:** infusion is used as a remedy for malaria. **Leaves:** Infusion used to control ticks. **Pulp:** extracts from unripe pulp is used as an eardrop, a remedy for dysentery and to control jigger flies. **Seeds:** extracts used as a remedy for hypertension and induce vomiting.	(Ruffo et al. 2002, Mwamba 2006) (Ruffo et al. 2002, Mwamba 2006, Rahman 2016) (Rahman 2016) (Ruffo et al. 2002, Mwamba 2006, Mawela 2009, Orwa et al. 2009) (Ruffo et al. 2002, Mwamba 2006, Orwa et al. 2009) (Mwamba 2006, De Wet et al. 2016)

S. pungens	**Roots:** Extracts used as a remedy for stomach aches, fever, inflamed eyes, bronchitis, and bone diseases that affect both hands and feet. **Stem Bark:** infusion used in obstetrics in East Africa. **Leaves:** extracts are used as a remedy for cough and the powder is used as an adorate. **Pulp:** extracts from unripe pulp is used as a remedy for snakebites. **Seeds:** extracts induce vomiting. **Rind:** used as one of the ingredients to make burned powder.	(Loeb et al. 1956, Ruffo et al. 2002, Mwamba 2006, Orwa et al. 2009, Adebowale 2014) (Loeb et al. 1956, Mwamba 2006) (Mwamba 2006) (Mwamba 2006) (Loeb et al. 1956)
S. spinosa	**Roots:** extracts mixed with milk to treat oedema. It is also mixed with coconut oil and applied on the feet to kill jigger fleas, flies. Extracts are used as a remedy for diabetes mellitus, leprosy, fever, earache, colds, inflamed eyes, headache, gastric and intestinal problems as well as diarrhoea. It is also used to make analgesics. **Root, stem bark, and leaves:** extracts used as a remedy for male organ dysfunction. **Roots and leaves:** extracts drunk as a medicine for pain, gonorrhoea, cuts, wounds, and other venereal diseases. Extracts were reported to exhibit some anti-plasmodial, anti-trypanosomal, and anthelmintic properties. **Roots and unripe fruits:** extracts used as a purgative, emetic, remedy for earache and snake bite antidote by the Zulus of South Africa. It is also used as alternatives to synthetic pesticides and ascaricides. **Stem bark:** extracts drunk as a remedy for fever, male organ dysfunction, infertility, and stomach pains. **Stem barks and leaves:** extracts used as a tea that is remedial for blenorragia and the infusion is applied on the body as medicine for scabies. **Leaves:** extracts drunk as a remedy for malaria, diabetes mellitus, snakebites, conjunctivitis, liver damage, syphilis, and the paste are applied on sores and wounds. It is also used to make analgesics. It is mixed with lion excrement and used for mental illnesses. It possesses some anthelmintic activity, and it was effective against jigger fleas, *Pheretima posthuma* and *Ascaris suum*. **Leaves and ripe fruits:** used as food for the lactating mothers for increased milk production. It is also used as feed for domestic livestock and wild animals. **Ripe pulp:** used as a remedy for hypertension.	(Ruffo et al. 2002, Mwamba 2006, Orwa et al. 2009, Isa et al. 2014, Isa et al. 2016) (Mwamba 2006, Orwa et al. 2009, Karou et al. 2011, Isa et al. 2014) (Hoet et al. 2007, Isa et al. 2014, Isa et al. 2016) (Sitrit et al. 2003, Madzimure et al. 2013, Isa et al. 2014) (Kabine et al. 2015, Issa et al. 2018) (Rahman 2016) (Mwamba 2006, Orwa et al. 2009, Isa et al. 2014) (Orwa et al. 2009) (Kabine et al. 2015, Issa et al. 2018)

Table 2 contd. ...

...Table 2 contd.

Plant name	Parts used and uses	References
S. potatorum	**Roots:** extracts are used as an aphrodisiac and their vapour is used as a remedy for colds and venereal diseases. **Roots and Leaves:** extracts used as medicine for epilepsy and cough. **Root, Stem bark and leaves:** ground into powder to make a fish poison. **Leaves:** ground into powder and used as a remedy for painful and watering eyes. **Pulp:** used as medicine for coughs and as a detergent. **Seeds:** extracts used to treat many ailments such as liver diseases, anorexia, kidney diseases, eye diseases, stomach disorders, gonorrhoea, malaria, leucorrhoea, bronchitis, epilepsy, chronic diarrhoea, urolithiasis, strangury, dysuria, polyuria, kidney and bladder stones, and diabetes mellitus. It is also used in the purification and clarification processes of muddy water.	(Van Wyk and Gericke 2000, Mallikharjuna et al. 2007, Schmelzer et al. 2008, Biswas et al. 2012, Rajesh et al. 2012, Packialakshmi et al. 2014, Ngarivhume et al. 2015, Gangwar and Choubey 2017) (Gangwar and Choubey 2017) (Gangwar and Choubey 2017) (Gangwar and Choubey 2017) (Mwamba 2006) (Sanmuga Priya and Venkataraman 2007, Mallikharjuna et al. 2007, Schmelzer et al. 2008, Rajesh et al. 2012, Biswas et al. 2012, Packialakshmi et al. 2014, Ngarivhume et al. 2015, Gangwar and Choubey 2017)
S. usambarensis	**Roots:** extracts are rubbed on the nostrils of hunting dogs to improve their scent. **Roots and leaves:** extracts used to poison arrows by hunters of the Banyambo tribe in Rwanda and Tanzania and it possessed anti-tumour, antimitotic, antiplasmodial, and antiamoebic activities.	(Schmelzer et al. 2008) (Wright et al. 1991, Frédérich et al. 2003, Schmelzer et al. 2008, Omosa et al. 2016)
S. aculeata	**Roots:** extracts used in the treatment of gonorrhea and pneumonia. **Stem bark:** usually mixed with the fruits of *Piper guineese* Schumach to make some infusion for treating edema and rubbed on the genital area as a remedy for gonorrhea and swelling. Extracts used as a remedy for oedema, pulmonary complaints, food poisoning, trypanosomiasis, scrotal elephantiasis, insanity, cough expectorant, and emetic. **Stem bark and leaves:** sap used as a remedy for worm infections and fever. **Leaves:** extracts used as a remedy for pulmonary tuberculosis by the Badala people of Congo. **Fruit and seeds:** used as a fish poison and a detergent for washing clothes. **Pulp:** infusion is applied on the head of the patient as a remedy for insanity and macerated fruit pulp is used as an abortifcient. **Seeds:** crushed into powder that is used as clysters for treating abdominal edema, also made into a paste to poison arrows.	(Schmelzer et al. 2008) (Schmelzer et al. 2008) (Schmelzer et al. 2008) (Schmelzer et al. 2008) (Schmelzer et al. 2008) (Schmelzer et al. 2008)

S. angolensis	**Roots:** crushed into powder is used as a remedy for large ulcers. **Stem bark:** extracts used to poison arrows by the Mbuti and Efe people.	(Schmelzer et al. 2008) (Schmelzer et al. 2008)
S. floribunda	**Roots:** extracts used as a remedy for edema. **Roots and Stem bark:** extracts used as a remedy for kidney diseases, edema, and diarrhea. **Stem bark:** decoction used as a remedy for palpitations, tachycardia, and other coronary heart diseases complications.	(Schmelzer et al. 2008) (Schmelzer et al. 2008) (Schmelzer et al. 2008)
S. henningsii	**Roots:** decoction used as a remedy for chest pain, internal injuries, and hookworm infections. Fresh roots are chewed as a remedy for snake bites. **Roots and Stem bark:** extracts used as rodents poison by the Malagasy people. **Roots, stem, and bark:** boiled in soup and used as a remedy for fitness, painful joints, and general body pains. The soup is also used as an aphrodisiac, a remedy for colic, nausea, and syphilis. **Stem bark:** decoction used as a remedy for malaria, colic, internal parasites in children, stomach-ache, dizziness, and it is purgative and anthelmintic. It is used for treating diarrhoea and heart water in cattle **Branches:** decoctions are added to the milk and meat-based soups, and it is a remedy for rheumatism, gynaecological complaints, gastrointestinal disorders, and a general tonic. **Fruit:** used as a beer flavorant by Mbeere people.	(Maundu and Tengnäs 2005, Schmelzer et al. 2008, Rajesh et al. 2012, Kipkemoi et al. 2013) (Schmelzer et al. 2008) (Kipkemoi et al. 2013) (Schmelzer et al. 2008) (Schmelzer et al. 2008, Kipkemoi et al. 2013) (Kipkemoi et al. 2013)
S. icaja	It is only administered under the supervision of traditional practitioners as it's regarded as toxic. **Roots:** cold infused with palm wine or palm oil to alleviate gastrointestinal complaints, hernia, and skin conditions. The macerated root bark is consumed as an abortifacient, anthelmintic, and applied on the body to serve as a snake repellent. It is used as a remedy for sterility through the cleansing of the rectum and forms part of the main ingredients of the arrow poison. **Roots and twig:** burnt ashes are used as a remedy for insanity and malaria. **Root bark and Stem bark:** infusion and extracts have been used as an ordeal poison. **Stem bark:** alcoholic extracts are used as a remedy for haemorrhoids. **Whole plant, root bark, and fruits:** extracts used as a fish poison.	(Schmelzer et al. 2008) (Schmelzer et al. 2008) (Schmelzer et al. 2008, Tchinda et al. 2012) (Schmelzer et al. 2008) (Schmelzer et al. 2008) (Schmelzer et al. 2008)

Table 2 contd. ...

...Table 2 contd.

Plant name	**Parts used and uses**	**References**
S. myrtoides	**Roots:** infusion or decoction strongly enhances the effects of chloroquine in the treatment of malaria. **Leaves:** used to prepare tea that serves as a remedy for colic.	(Rasoanaivo et al. 2001, Schmelzer et al. 2008, Rahman 2016) (Rahman 2016)
S. panganensis	**Roots:** decoction used as a remedy for chest complaints.	(Nuzillard et al. 1996)
S. mostueoides	**Aerial parts:** decoction used as a remedy for malaria.	(Rasoanaivo et al. 2001, Rahman 2016)
S. scheffleri	**Stem bark:** decoction used by nursing mothers as a remedy for cleansing the womb.	(Tchinda et al. 2012)
S. diplotricha	**Root:** decoction used as a remedy for malaria. **Stem bark:** decoction used as a remedy for leprosy and wounds.	(Rasoanaivo et al. 2001, Rajesh et al. 2012, Rahman 2016) (Rahman 2016)
S. moandaensis	**Root:** decoction used as a remedy for malaria.	(Verpoorte et al. 2010)
S. decussata	**Roots:** decoction or infusion is taken as a remedy for stomach disorders, wounds, abscesses, snakebites, and the root powders are usually applied externally on the site of snakebites.	(Maundu and Tengnäs 2005)
S. mitis	**Roots and wood:** extracts used as an ordeal poison by the Mbuti people. **Root bark:** extracts used as a remedy for malaria and fever.	(Maundu and Tengnäs 2005) (Rajesh et al. 2012)
S. mellodora	**Roots:** extracts used as a remedy for malaria. **Stem bark:** decoction used as a remedy for malaria.	(Rajesh et al. 2012) (Maundu and Tengnäs 2005)
S. pentantha	**Roots:** extracts used as a remedy for malaria.	(Rajesh et al. 2012)
S. lucens	**Roots:** extracts used as a remedy for malaria and powdered roots are consumed as a remedy for hook worm infestations.	(Schmelzer et al. 2008, Rajesh et al. 2012)
S. gossweileri	**Root:** extracts used as a remedy for malaria.	(Rajesh et al. 2012)
S. johnsonii	**Root:** extracts used as a remedy for malaria.	(Rajesh et al. 2012)

Pharmacology and Phytochemistry of the Southern African Region *Strychnos* Species

There are more than 300 different alkaloids isolated from various *Strychnos* species that possess a wide range of biological activities (Frédérich et al. 2003, Adebowale 2014). The most renowned alkaloids from *Strychnos* species are strychnine and brucine. Some plants contain strychnine-like alkaloids that are highly concentrated within the seeds, stem bark, and roots. Recent studies have shown that some of the leaves of *Strychnos* species are highly concentrated with alkaloids that can cause fatalities if consumed by humans. There is a lot of diversity in the pharmacological activity of the different alkaloids which is hard to understand (Frédérich et al. 2002, Mwamba 2006, Adebowale 2014). Various studies on indole alkaloids from the *Strychnos* species have shown that compounds like sungucine are slightly active whilst strychnogucine B and 18-hydroxyisosungucine are highly active against quinine and chloroquine resistant strains of *Plasmodium falciparum* (Wright et al. 1991, Frédérich et al. 1999, Frédérich et al. 2001, Frédérich et al. 2002, Frédérich et al. 2003, Aniszewski 2007, Omosa et al. 2016).

Different compounds were extracted from the different plant parts of the various *Strychnos* species found in Southern Africa (Fig. 1). Isobutyl acetate, 2-methylbutyl acetate, ethyl-2-methylbutyrate, 2, 6-ditetrabuty l-4-methyl-phenol, butyl-2-methyl butyrate, 5-(1-hydroxyethyl)-4- H-494 methyl-3-pyridinecarboxylic acid, 5-methyl-2,7-naphthyridine-M-489 4-carboxylic acid, 3-(1-hydroxyethyl)-5-methoxycarbonylpyridine, and geranyl acetate were extracted from ripe fruit pulp of *S. cocculoides* using solid phase microextraction (SPME) and their identity was established by the use of GC–FID and GC–MS systems (Shoko et al. 2013).

Bis-iridoid glucoside, 6-O-nicotinoyltetrahydrocantleyine, phenolic apioglucoside, iridoid monoterpene, iridoid glucoside, 2-hydroxy-3-O-β-glucopyranosyl-benzoic acid, iridoid glucosides, loganic acid, morroniside (5a), morroniside (5b), sarracenin, cocculoside, and 3,4,5-trimethoxyphenol β-D-apiofuranosyl-(1, 6)-β-D-glucopyronoside, were identified from the root and stem bark methanol extracts of *S. cocculoides* that were chromatographed through Vacuum Liquid Chromatography and their chemical structures were identified with NMR (Sunghwa and Koketsu 2009). Neozeylanicine (4-methoxycarbonyl-1-methyl-2,7-naphthyridine), condensamine, 3-(1-hydroxyethyl)-4-methyl-5-methoxycarbonylpyridine, 5-methyl-2,7-naphthyridine-4-carboxylic acid, and 6-O-nicotinoyltetrahydrocantleyine were extracted from the unspecified parts of the *S. cocculoides* (Buckingham et al. 2010). Strychnovoline was extracted from the leaves of *S. cocculoides* (Buckingham et al. 2010).

Neozeylanicine (4-methoxycarbonyl-1-methyl-2, 7-naphthyridine), 3-(1-hydroxyethyl)-4-methyl-5-methoxycarbonylpyridine, akagerine, 5-methyl-2,7-naphthyridine-4-carboxylic acid, and 6-O-nicotinoyltetrahydrocantleyine were extracted from unspecified parts of the *S. spinosa* (Buckingham et al. 2010). *Trans-â*-ocimene, chavicol, indole, *p-trans*-anol, dihydroeugenol, vanillin, 2,6-3-(1-hydroxyethyl)-5-methoxycarbonylpyridine, *trans*-isoeugenol, 5-methyl-2,7-naphthyridine-4-carboxylic acid, and 5-(1-hydroxyethyl)-4-methyl-3-pyridinecarboxylic acid were extracted from the ripe yellow fruit of *S. spinosa* and determined by atomic absorption spectroscopy with an ICP-OES Optima 3000 (Sitrit et al. 2003, Buckingham et al. 2010, Rodrigues et al. 2018). Scaevodimerine A was extracted from the aerial parts of *S. spinosa* whilst akagerine, kribine, 10-hydroxyakagerine, *β*-amyrin, lupeol, uvaol, *β*-sitosterol,

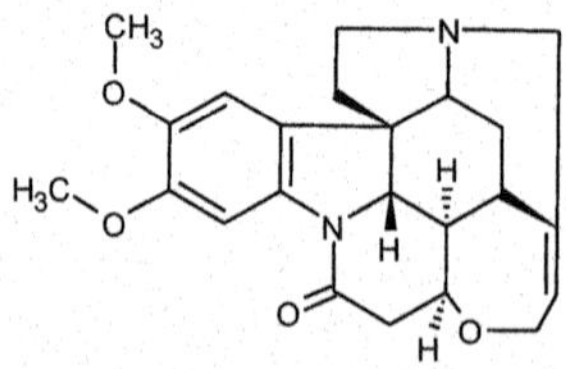

Brucine

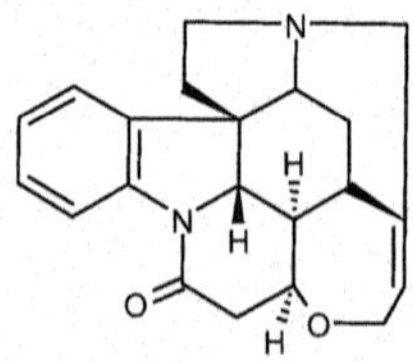

Strychnine

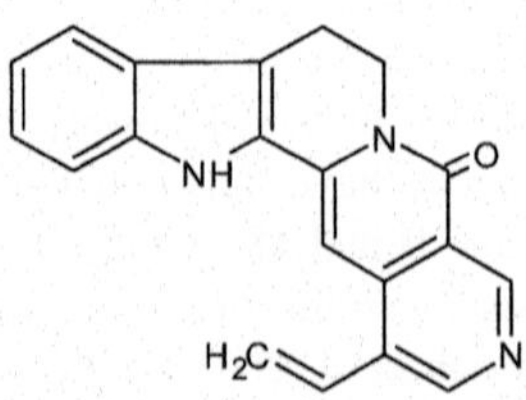

Angustine

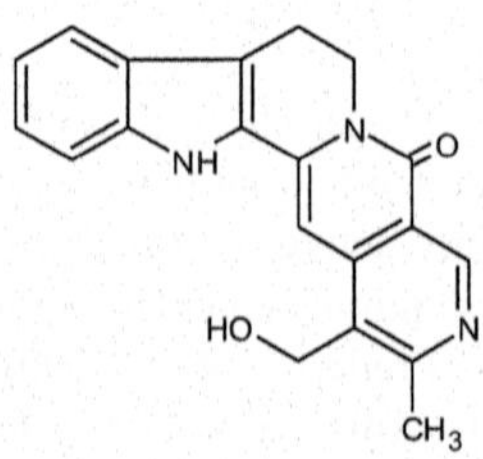

Angustoline

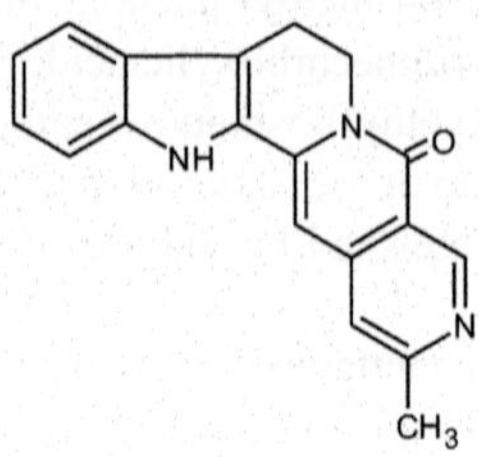

Angustidine

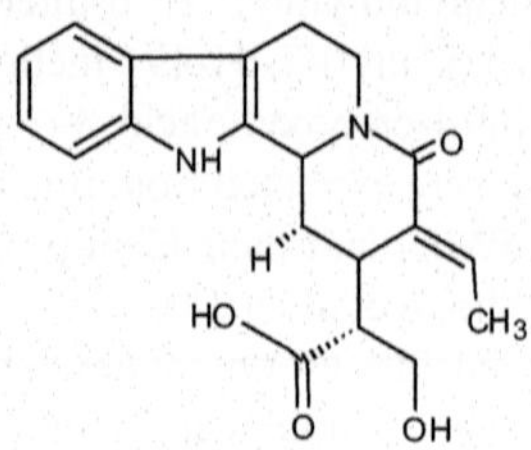

Isositsirikine

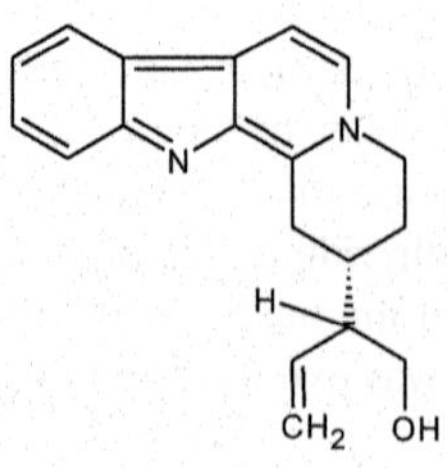

Matadine

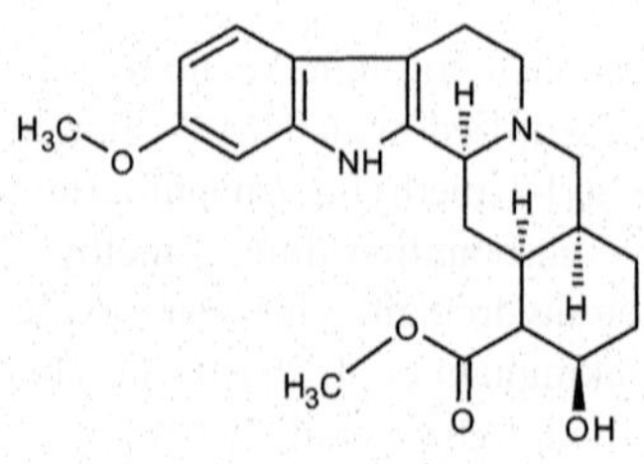

Dihydroxyapogeissoschizine

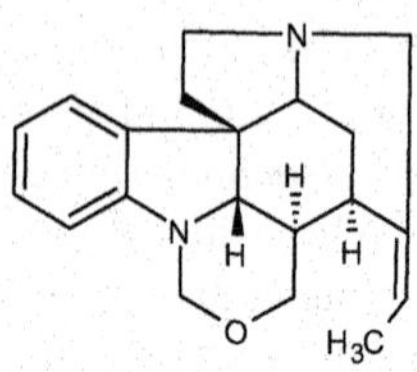

Rosibiline

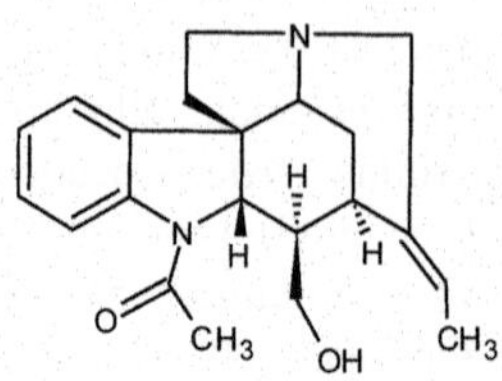

Isoretuline

Figure 1 contd. ...

...Figure 1 contd.

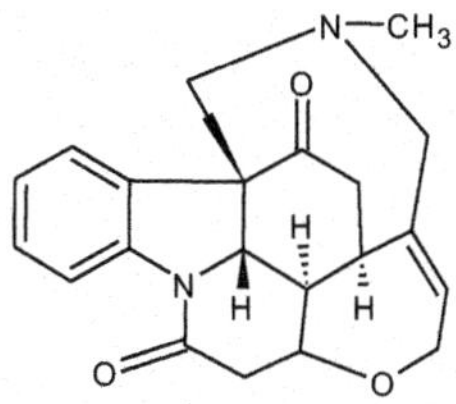

Icajine

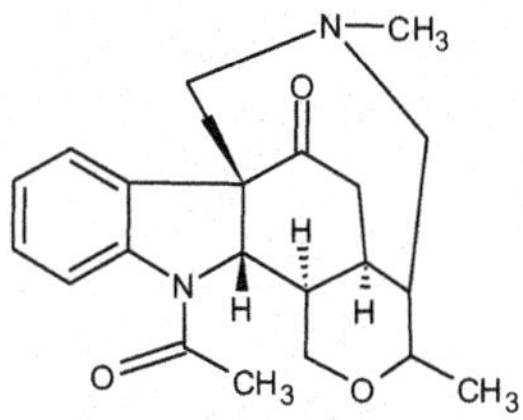

Strychnobrasiline

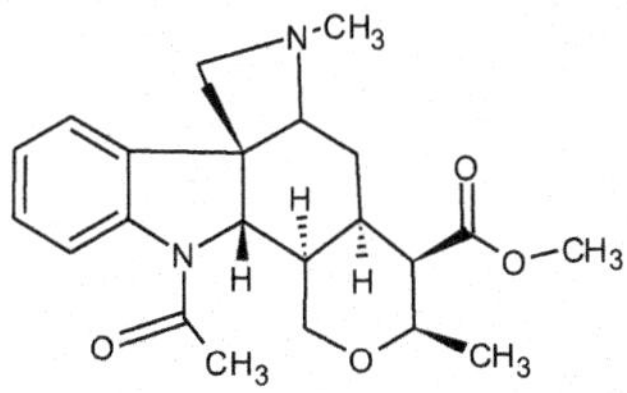

Malagashanine

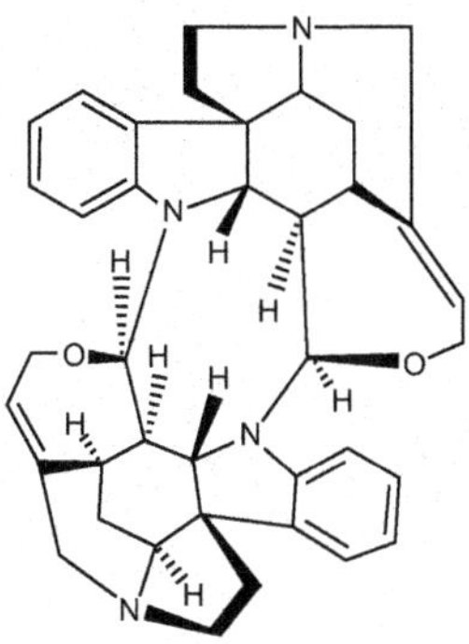

Caracurine

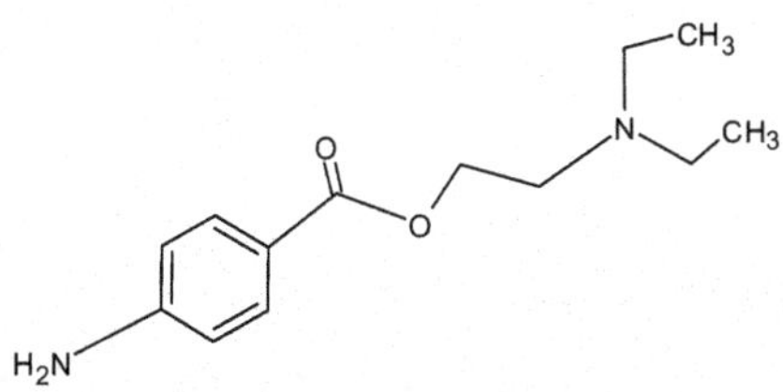

Novacine

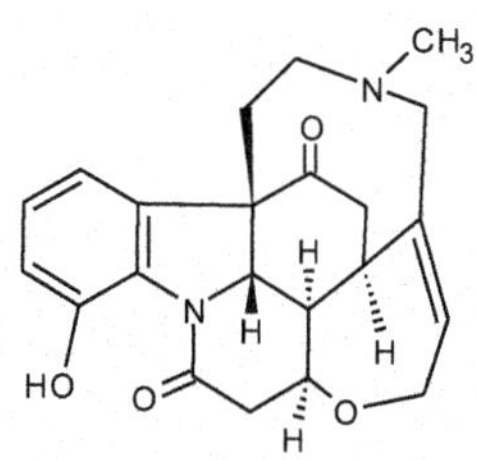

Vomicine

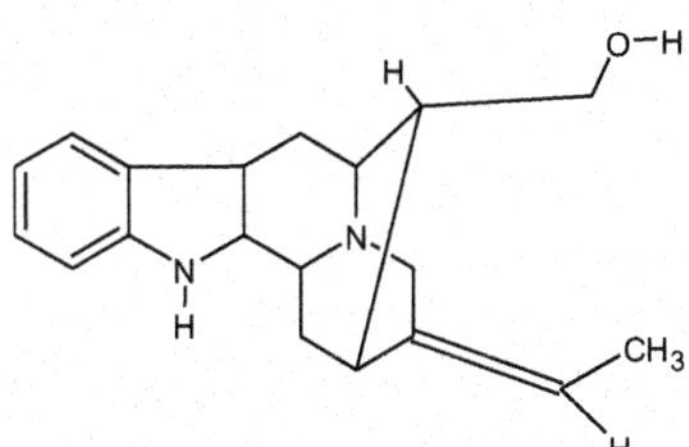

Normacusine B

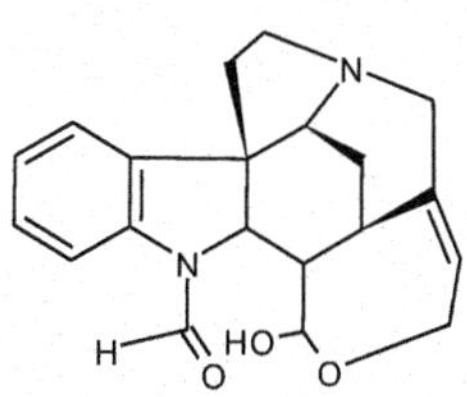

Diaboline

Figure 1 contd. ...

...Figure 1 contd.

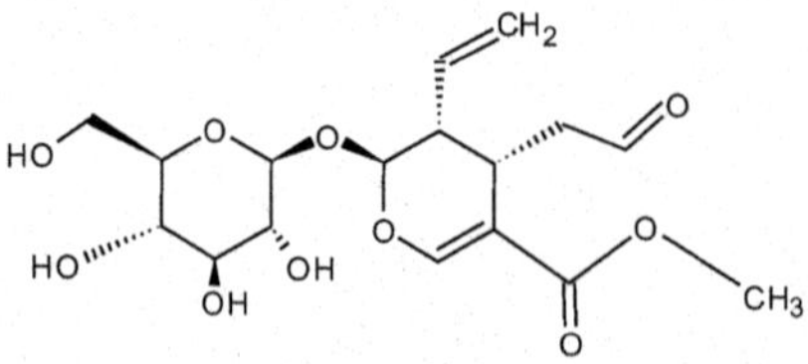

Secologanin

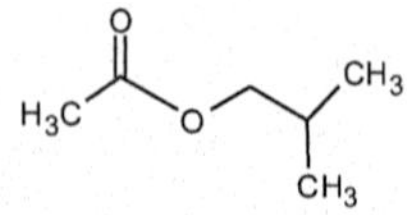

Isobutyl acetate

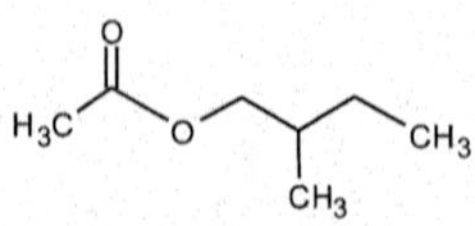

Methylbutyl acetate

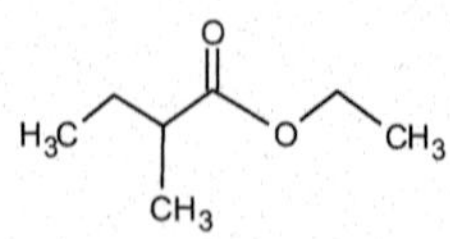

Ethyl-2-methylbutyrate

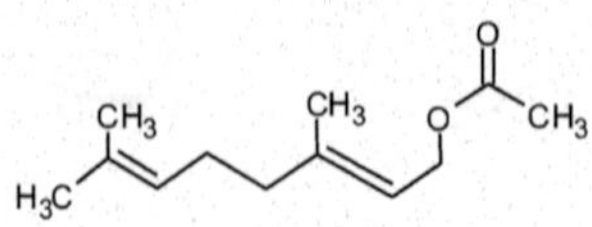

Geranyl acetate

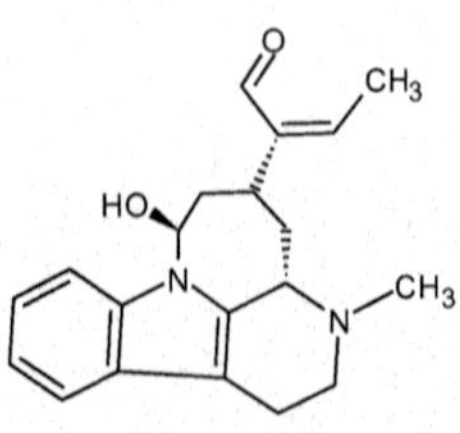

Akagerine

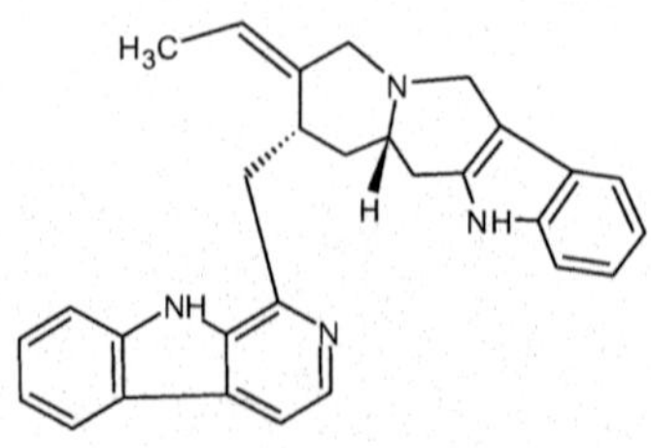

Usambarensine

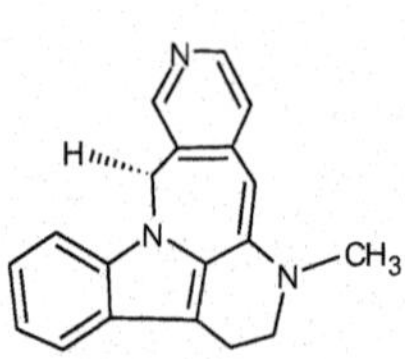

Decussine

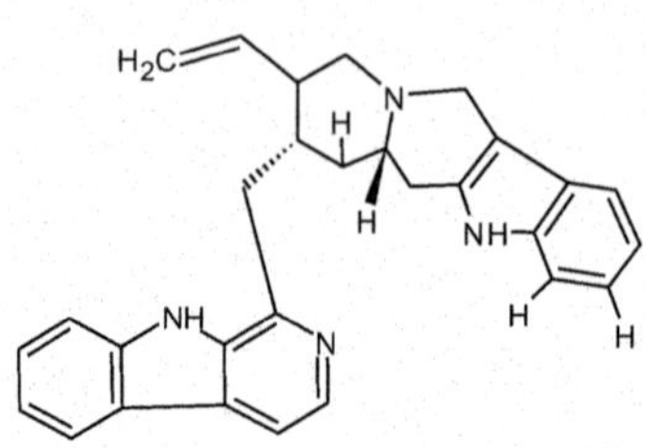

Usambarine

Figure 1 contd. ...

...Figure 1 contd.

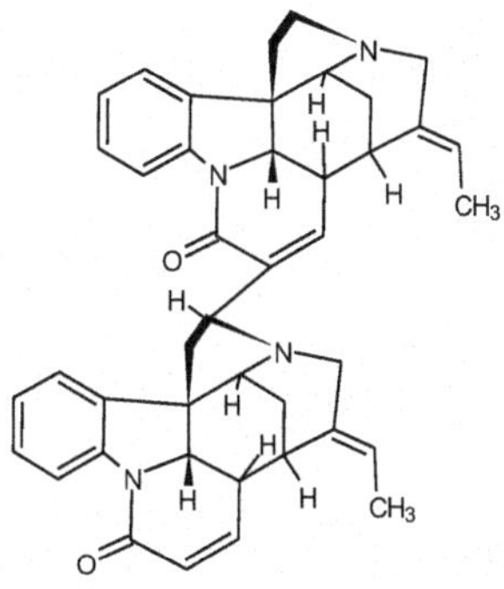

Sungucine

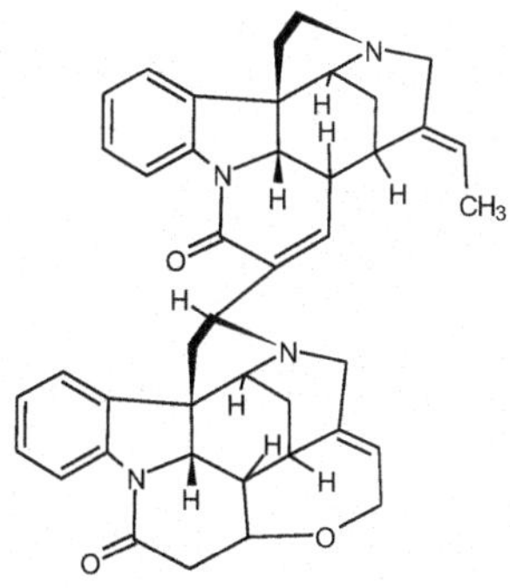

Strychnogucine B

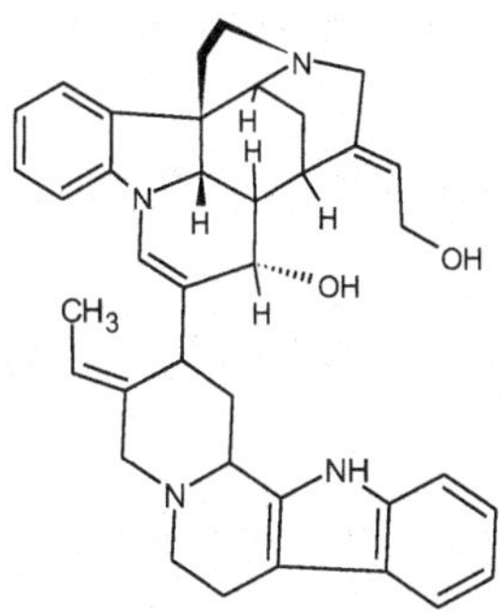

Longicaudatine F

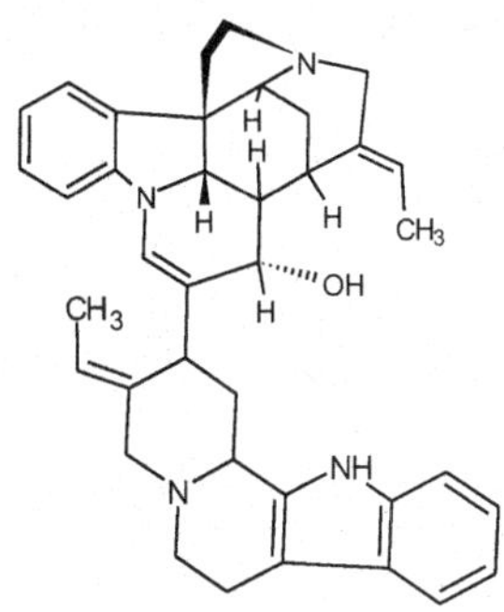

Longicaudatine Y

Figure 1: Structures of some chemical compounds isolated from Southern African *Strychnos* Species.

clerosterol, vinyl cholesterol, dimethoxyphenol, saringosterol, 24-hydroperoxy-24-11-methoxy-diaboline, and 12-hydroxy-11-methoxydiaboline were extracted from the leaves and the methods used to determine their chemical structures were not stated (Ohiri et al. 1983, Hoet et al. 2004, Buckingham et al. 2010). Henningsoline, 11-methoxy-diaboline,12-hydroxy-11-methoxydiaboline, akagerine, kribine, scaevodimerine A, 6-O-nicotinoyltetrahydrocantleyine, loganin, secologanin, sweroside, secoxyloganin, dimethyl acetal, secologanoside, secologanoside 7-methyl ester, secologanoside dimethyl ester, secologanic acid, vogeloside, epi-vogeloside, (5S)-5-carboxystrictosidine, cantleyoside, triploside A, (E)-aldosecologanin, 2,4,6-trimethoxyphenol 1-O-β-D-glucopyranoside, benzyl alcohol O-R-L-arabinopyranosyl-(1, 6)-β-D glucopyranoside, trifolin, geraldol, hyperin, astragalin, hirsutrin nicotiflorin, 10-hydroxyakagerine and 11-methoxy-diaboline were extracted from the stem bark of *S. spinosa* (Ohiri et al. 1983, Itoh et al. 2005, Buckingham et al. 2010, Tor-Anyiin et al. 2015, Rahman 2016). Cantleyine, tetrahydrocantleyine, 4-carbomethoxy-l-methyl-2,7-naphtyridine,5-carbomethoxy-4-methyl-2,7-naphtyridine, scaevodimerine, O-nicotinoyl-7-tetrahydrocantleyine, 3-carbomethoxy-5-(l-hydroxyethyl)-pyridine, 3-carbomethoxy-4-methyl-5-(1-

hydroxyethyl)-pyridine, 11-methoxy-diaboline, 11-methoxy-12-hydroxy-diaboline, 11–methoxy-henningsamine, 12-hydroxy-11-methoxyhenningsamine, and 6-O-nicotinoyltetrahydrocantleyine were extracted from the root barks of the *S. spinosa* (Buckingham et al. 2010, Rahman 2016). Kingiside aglucone and secoiridoid were extracted from the unripe fruits of *S. spinosa* (Msonthi et al. 1985).

Sitsirikine16(S)-isositsirikine, 16(R)-isositsirikine, diaboline, 12-hydroxy-11-methoxyhenningsamine,12-hydroxy,11-methoxydiaboline, venoterpine, henningsamine, gentianine lactam, 11-methoxyneooxydiaboline, O-acetylretuline, and scaevodimerine D were extracted from the leaves, stem bark, and root bark of *S. pungens* (Thépenier et al. 1990, Frédérich et al. 1999, Buckingham et al. 2010). O-acetylretuline was isolated from the stem bark, gentianine lactam was isolated from the unspecified parts whilst O-acetylvenoterpine was extracted from the seeds of *S. pungens* (Buckingham et al. 2010). Galactomannan (2, 3, 4, 6-tetra-omethylgalactopyranose and 2, 3, 6-tri-O-methylmannopyranose) and galactan (2, 3-di-omethylmannopyranose) were extracted from the pulp of *S. innocua* (Rajesh et al. 2012). Bakankoside was isolated from the seeds (Buckingham et al. 2010). 16(S)-E-isositsirikine, 9-methoxy-16(S)-E-isositsirikine, 9-methoxy-16(R)-E-isositsirikine, strychnorubigine normacusine B and 16(R)-E-isositsirikine were extracted from leaves, root bark, and stem bark of *S. madagascariensis* (Rahman 2002).

Angustine, angustidine, diaboline, acetyl-diaboline, isomotiol, sitosterol, stigmasterol, campesterol, and chemical composition of polysaccharide fractions such as 2, 3, 4, 6-tetra-O-methylgalactopyranose, 2, 3, 6-tri-O-methylmannopyranose and 2, 3-di-O-methylmannopyrannos were extracted from seeds, stem barks and leaves of *S. potatorum* (Ohiri et al. 1983, Sanmuga Priya and Venkataraman 2007, Schmelzer et al. 2008, Sanmuga and Venkataraman 2010, Rajesh et al. 2012, Gangwar and Choubey 2017). Normacusine B, diaboline, brucine, loganin, akuammidine, norharmane, 20-epiantirhine, (20S)-dihydroantirhine, henningsamine, diaboline N-oxide, β-carboline, 1-carbamoyl-β-carboline, 11-methoxy-12-hydroxydiaboline, 11-methoxy-diaboline, harmane carboxamide, cantleyine, dihydrolongicaudatine, dihydrolongicaudatine Y, normavacurine 3S, ochrolifuanine A, 18,19-dihydrousambarensine (nigritanine), polyneuridine, nor-C-fluorocurarine, deacetylretuline, bisnordihydrotoxiferine, ochrolifuanine E, normacusine B, 11-methoxy-henningsamine, antirhine, (20R)- and (20S)-dihydro-antirhine were isolated from the root bark of *S. potatorum* (Massiot et al. 1992, Frederich et al. 1999, Schmelzer et al. 2008, Sanmuga and Venkataraman 2010, Buckingham et al. 2010, Rahman 2016).

Afrocurarine, akagerine, angustine, antirhine β-N-methosalt (quartenary alkaloid), normelinonine F, dihydrousambarensine (nigritanine), usambarensine, N_b-methylusambarensine, 109-hydroxyusambarensine, nalindine, usambarine, usambaridine Br, dihydrousambarine, strychnopentamine, isostrychnopent-amine, strychnofoline, isostrychnofoline, strychnophylline, isostrychnophylline 10-hydroxyusambarine, usambaridine Br, 11-hydroxyusambarine, 10-hydroxydihydrousambarine, 11-hydroxydihydrousambarine, 3',4',5',6'-tetradehydrolongicaudatine Y, 10-hydroxy-Nb-methylcorynantheol, chrysopentamine, melinonine F,N_4-methyl-10-hydroxyusambarine, 18,19-dihydrousambaridine Br, 18,19-dihydrousambaridine Vi, strychnobaridine, 6,7-dihydroflavopereirine, strychnambarine, strychnobaridine, strychnofoline, isostrychnofoline, strychnophylline, isostrychnophylline, N_b-methylusambarensine were isolated from the leaves of *S. usambarensis* (Ohiri et al. 1983, Frédérich et al. 1999, Aniszewski 2007, Schmelzer et al. 2008, Buckingham et al. 2010). Kribine,

N_4-methyl-10-hydroxyusambarine, guianensine, 3',4',5',6'-tetradehydrolongicaudatine Y, longicaudatine Y, isoretuline and tchibangensine were extracted from the stem bark of *S. usambarensis* (Frédérich et al. 2002, Buckingham et al. 2010) whilst decarbomethoxydihydrogambirtannine was isolated from the fruits (Ohiri et al. 1983). Akagerine, macusine B, harmane, 5', 6'-dihydro-flavopereirine, malindine, isomalindine and O-methylmacusine B, dihydro-O-methylmacusine B, usambarensine, N_b-methyl usambarensine, N_b-methyl-norharmane, 5',6'-dihydrousambarine, C-curarine, C-calebassine, dihydrotoxiferine, afrocurarine, flurocurarine, melinonine F and tchibangensine were isolated from the root bark of *S. usambarensis* (Ohiri et al. 1983, Kolm 1984, Frédérich et al. 2002, Aniszewski 2007, Schmelzer et al. 2008, Buckingham et al. 2010).

Strychnofendlerine N^a-acetylstrychnosplendine, spermostrychnine, strychnofendlerine, isosplendine, N^a-acetylstrychnosplendine, and N^a-acetyl-O-methylstrychnosplendine were isolated from the root bark of *S. aculeata* (Ohiri et al. 1983, Schmelzer et al. 2008, Buckingham et al. 2010). Spermostrychnine, N_a-acetyl-isostrychnosplendine and N^a-acetyl-O-methylstrychnosplendine were extracted from the stem bark of *S. aculeata* (Ohiri et al. 1983, Schmelzer et al. 2008, Buckingham et al. 2010).

Gentianine, aspidospermine, akuammicine, tubifoline, condylocarpine, akuammidine, 16(R)-epi-isositsirikine, 16(S)-epi-isositsirikine, tubifolidine, normavacurine, antirhine (anthirine), strychnofluorine (18-hydroxy-nor-C-fluorocurarine), 11-Methoxy-Wieland and Gumlich aldehyde, 11-methoxy-17-methoxy-Wieland-Gumlich aldehyde, 11-methoxy-epi-17-methoxy-Wieland-Gumlich aldehyde, 11-methyl diaboline, curacurine V, flavopereirine, caracurine VII-N-oxide (WGA N-oxide) and tubotaiwine were extracted from the root bark of *S. angolensis* (Ohiri et al. 1983, Schmelzer et al. 2008). Gentianine, aspidospermine, akuammicine, tubifoline, condylocarpine, akuammidine, 16(R)-epi-isositsirikine, 16(S)-epi-isositsirikine, tubifolidine, normavacurine, antirhine (anthirine), strychnofluorine (18-hydroxy-nor-C-fluorocurarine), flavopereirine, caracurine VII-N-oxide (WGA N-oxide), 11-methoxy-Wieland-Gumlich aldehyde, 11-methoxy-17-methoxy-Wieland-Gumlich aldehyde, 11-methoxy-epi-17-methoxy-Wieland-Gumlich aldehyde, 11-methyl diaboline, curacurine V tubotaiwine, and tertiary dimeric alkaloid caracurine V were isolated from the stem bark of *S. angolensis* (Ohiri et al. 1983, Schmelzer et al. 2008) whilst angustine and angustidine were extracted from the leaves (Ohiri et al. 1983).

Bisnordihydrotoxiferine, akagerine, decussine, rouhamine, strychnocarpine, N_a-desacetylisoretuline, desacetylisoretuline, isorosibiline, the common sterols β-sitosterol, stigmasterol, and campesterol were extracted from the stem bark of *S. floribunda* (Ohiri et al. 1983, Schmelzer et al. 2008, Buckingham et al. 2010, Rahman 2016) whilst angustine was extracted from the leaves (Ohiri et al. 1983, Rahman 2016).

Holstiine, holstiline, rindline, retuline, condensamine, diaboline, henningsamine, henningsoline, 0-acetyl-henningsoline, 11-methoxydiaboline, 2,16-dehydro-diaboline, 2,16-dehydro-11-methoxydiaboline and spermostrychnine class were isolated from the stem bark of *S. henningsii* (Ohiri et al. 1983, Schmelzer et al. 2008, Rajesh et al. 2012). Holstiine, splendoline, 23-hydroxy-spermostrychnine, 19-epi-23-hydroxyspennostrychnine, retuline, henningsiine, deshydroxyacetyl-henningsiine, 0-acetyl-henningsiine, 3-hydroxy-henningsiine, henningsiine-N (4)-oxide, diaboline, 23-hydroxy-spermostrychnine-N(4)-oxide,

17,23-dihydroxy-spennostrychnine, Cyclostrychnine, spermostrychnine, henningsamide, 0-acetyl-henningsamide, deshydroxyacetyl-henningsamide were isolated from the root bark of *S. henningsii* (Ohiri et al. 1983, Schmelzer et al. 2008, Rajesh et al. 2012, Rahman 2016).

Alkaloids of the tsilanine class, retuline class, and spermostrychnine class, tsilanine, 10-methoxy-tsilanine, 0-demethyltsilanine, O-acetyl-retuline, 0-demethyl, and 10-methoxy-tsilanine were isolated from leaves and seeds of *S. henningsii* (Ohiri et al. 1983, Schmelzer et al. 2008, Rajesh et al. 2012, Rahman 2016) whilst alkaloids of the retuline class, tsilanimbine, N_a-deacetyl-isoretuline, N(a)-acetyl-11-methoxystrychnosplendine, 18-hydroxy-isoretuline, N_a-deacetyl, 18-hydroxy-isoretuline and Na-deacetyl-18-hydroxy-17-0-methoxy-isoretuline were extracted from the twigs (Ohiri et al. 1983, Rajesh et al. 2012, Rahman 2016).

Tertiary indole alkaloids which include strychnine and pseudostrychnine (12-hydroxystrychnine), 4-hydroxystrychnine, colubrine and brucine, the pseudo series pseudostrychnine, pseudocolubrines, pseudobrucine, the N-methyl-sec-pseudo series: icajine, 19,20α-epoxy-vomacine, 19,20α-epoxy-11-methoxyvomacine, 19,20α-epoxy-15-hydoxyicajine, vomicine, 19,20α-epoxy-15-hydroxynovacine, 19,20α-epoxy-15-hydroxyvomacine, 15-hydroxyicajine, 19,20α-epoxy-15-hydroxy-11-methoxyvomacine, 19,20α-epoxy-11,12-dimethoxyicajine and novacine were isolated from the stem and leaves of *S. icaja* (Ohiri et al. 1983, Schmelzer et al. 2008, Rajesh et al. 2012, Tchinda et al. 2012). 19,20α-epoxy-12-methoxyicajine, 19,20α-epoxy-15-hydroxynovacine, vomacine, icajine, 19,20α-epoxy-11-methoxyvomacine, 19,20α-epoxy-15-hydroxyvomacine, 19,20α-epoxy-15hydroxy-12-methoxyicajine and 19,20α-epoxy-15-hydroxy-12-methoxyicajine were extracted from the fruits of *S. icaja* (Ohiri et al. 1983, Tchinda et al. 2012).

Bisnordihydrotoxiferine, protostrychnine, 12-hydroxystrychnine, 19,20α-epoxy-vomacine, strychnine, strychnogucine B, guiaflavine, icajine, sungucine, N-strychninium and strychnohexamine were isolated from the roots of *S. icaja* (Schmelzer et al. 2008, Rajesh et al. 2012). Strychnine is the main alkaloid component of the fruits and seeds of *S. icaja* (Schmelzer et al. 2008).

Two major indole alkaloids, strychnobrasiline and malagashanine, 4 minor alkaloids: malagashanol, 12-hydroxy-19-epimalagashanine, myrtoidine, 11-demethoxymyrtoidine, N_b-C(21)-secocuran alkaloids, viz., 3-epi-myrtoidine, 11-demethoxy-3-epi-myrtoidine, and 11-demethoxy-12-hydroxy-3-epi-myrtoidine and malagashanine were isolated from stem bark of *S. myrtoides* (Schmelzer et al. 2008, Rajesh et al. 2012) whilst strychnobrasiline, malagashanine, 3-epi-myrtoidine, 11-demethoxy-12-hydroxy-3-epimyrtoidine and 11-demethoxy-3-epi-myrtoidine were extracted from the leaves (Schmelzer et al. 2008, Rajesh et al. 2012, Rahman 2016).

Normacusine B, caracusine VII, strychnofendlerine, W.G. aldehyde, spermostrychnine, strychnobrasiline, deacetylstrychnobrasiline, malagashine, and malagashanine were isolated from the root bark and stem bark of *S. mostueoides* (Schmelzer et al. 2008, Rajesh et al. 2012, Rahman 2016). Moandaensine was isolated from the root bark of *S. moadensis* and was observed on the TLC and its chemical structure was determined using a 3D NMR (Verpoorte et al. 2010). Angustine, angustidine, angustoline, N_a-acetyl-strychnosplendine, N^a-acetyl-O-methylstrychnosplendine, N_a-acetyl-O-methylstrychnosplendine, strychnobrasiline, and strychnofendlerine were extracted from the leaves of *S. moadensis* (Ohiri et al. 1983, Buckingham et al. 2010).

N^a-deacetylmalagashine, N^a-desacetylisoretuline, desactyl-isoretuline, bisnordihydrotoxiferine, mavacurine, fluorocurine, and malagashine were isolated from the stem bark of *S. scheffleri* and their chemical structures were elucidated with 1D and 2D NMR and MS (Ohiri et al. 1983, Buckingham et al. 2010, Tchinda et al. 2012).

Gluco-alkaloid was extracted from the leaves of *S. decussata* (Ohiri et al. 1983, Rahman 2016) whilst rouhamine (5,6-dehydro-decussine), 3,14-dihydro-decussine, decussine, malindine, macusine B, 10-hydroxy-3,14-dihydro-decussine, 0-methyl-macusine B, macusine A, macusine C, akagerine, akagerine lactone, 17-0-methyl-akagerine, 10-hydroxy-akagerine, 10-hydroxy-17-0-methyl-akagerine, 10-hydroxy-17-0-methyl-kribine, 10-hydroxy-epi-17-(9-methyl-kribine, bisnordihydrotoxiferine, and mostueine were isolated from the stem bark (Ohiri et al. 1983, Buckingham et al. 2010, Rahman 2016).

Three new N_b-C(21)-secocuran alkaloids, namely 3-epi-myrtoidine, 11-demethoxy-3-epi-myrtoidine, and 11-demethoxy-12-hydroxy-3-epi-myrtoidine, myrtoidine, strychnobrasiline, strychnofendlerine, malagashanine, and 11-demethoxymyrtoidine were isolated from the stem bark of *S. diplotricha* (Rasoanaivo et al. 2001, Rajesh et al. 2012, Rahman 2016).

3,4,5,6-tetradehydropalicoside, 3,4,5,6-tetradehydrodolichantoside, Desoxycordifoline (bcarboline-3-carboxylate glucoalkaloid), melinonine F(Nb-methylatedharmanium cation), strictosidine b-glucosidase, dolichantoside, and palicoside were extracted from the stem bark of *S. mellodora* (Aniszewski 2007, Rajesh et al. 2012). Matopensine, 16-(S)-E-isositsirikine, 12-hydroxy-11-methoxydiaboline, N-desacetylretuline, N-desacetylisoretuline, N-desacetylspermostrychnine, 12-hydroxy-11-methoxy-nor-C-fluorocurarine, 12-hydroxy-11-methoxy-N-acetylmon-Cfluorocurarimine and four dimeric alkaloids, panganensines 19R, panganensines 19 S, panganensines X, and panganensines Y were isolated from the root bark of *S. panganensis* (Rajesh et al. 2012). Normalindine, ajmalicinial, tetrahydroalstonial, 17-O-ethylakagerine, akagerine lactone, O-ethylakagerine lactone, β-carboline, dihydrocycloakagerine, janussine A, janussine B, normalindine, norepimalindine, oxojanussine, and tetrahydroakagerine were isolated from the root bark of *S. johnsonii* (Buckingham et al. 2010, Rajesh et al. 2012) whilst isoanthirine, antirhine lactone, dihydrocycloakagerine, janussine A, janussine B, tetrahydroakagerine were extracted from the stem bark (Buckingham et al. 2010). Cantleyine, tetrahydrocantleyine, strychnovohne, tubotaiwine, tubotaiwine N-oxide nicotinoyl-7-tetrahydrocantleyine, 16(S)-E-isositsirikine, 16(R)-E-isositsirikine, strychnovoline, and 6-O-nicotinoyltetrahydrocantleyine were isolated from the stem bark, root bark, and leaves of *S. mitis* (Buckingham et al. 2010, Rahman 2016). Bisnor-C-alkaloid D and longicaudatine F, desoxy-Wieland-Gumlich aldehyde, Na-formyloxydeoxy-Wieland-Gumlich aldehyde, bisnor-C-curarine, longicaudatine N-oxide, longicaudatine Z, longicaudatine Y, longicaudatine F, matopensine, matopensine mono-N-oxide, 18-hydroxymatopensine, 18,18'-dihydroxymatopensine, 16-methoxyisomatopensine, 16-ethoxyisomatopensine, desacetylisoretulinal, rosibiline, strychnofuranine, and caracurine VI were isolated from the root bark and stem bark of *S. matopensis* (Buckingham et al. 2010).

O-acetylretuline was isolated from the barks and leaves of *S. kasengaensis* (Buckingham et al. 2010). Desoxy-Wieland-Gumlich aldehyde, diol, caracurine VII, matopensine, matopensine mono-N-oxide, retuline, desacetylretuline, O-acetylretuline, 11-methoxyretuline, isoretuline, Na-desacetylisoretuline, O-acetylisoretuline, isoretulinal, N(1)-desacetyl-18-hydroxyisoretuline, 18-acetoxy-

Na-desacetylisoretulin, 11-methoxyisoretuline and 16,17-dehydroisostrychnobiline were extracted from the root bark of *S. kasengaensis*, and separated through the use of medium pressure liquid chromatography and their chemical structure was determined by high resolution mass spectrometry and H NMR (Thepenier et al. 1984, Buckingham et al. 2010). 10,10'-dimethoxytetrahydrousambarensine and 10,10'-dimethoxy-N4-methyltetrahydrousambarensine were isolated from the leaves of *S. dale* (Buckingham et al. 2010) whilst mostueine, 17-O-methylakagerine, decussine, kribine, epi-17-O-methylkribine, 3,4-dihydro-1-methyl-b-carboline, usambarensine, usambarensine-N_b-oxide, tchibangensine, 5',6'-dihydrousambarensine-N-oxide, 4',17-dihydro-17α-tchibangensine, 10,10'-dihydroxytetrahydrousambarensine, 10,10'-dihydroxy-N4-methyltetrahydrousambarensine, 10-hydroxy-10'-methoxytetrahydrousambarensine and 10-hydroxy-10'-methoxy-N4-methyltetrahydrousambarensine were extracted from the stem and root bark (Buckingham et al. 2010).

Diploceline, 16-epi-diploceline, matadine, 18-hydroxynorfluorocurarine, dolichantoside, isodolichantoside, strychnochromine, strychnoxanthine, strychnofluorine, alstonine, and 2,7-dihydroxyapogeissoschizine were isolated from the root bark of *S. gossweileri* (Ohiri et al. 1983, Buckingham et al. 2010). (16S)-E-isositsirikine and Strychnorubigine were extracted from the root bark of *S. lucens* (Buckingham et al. 2010). Dehydrogentianine lactam was isolated from an unspecified part of *S. xantha* (Buckingham et al. 2010). Angustine, angustidine, and angustoline were isolated from the leaves of *S. xantha* and *S. trichoneura* (Ohiri et al. 1983, Rahman 2016).

Antimicrobial Activity of Compounds from the Southern African *Strychnos* Species

Different compounds, mainly indolomonoterpenoid alkaloids and other alkaloids contain four types of bisindole skeletons that are strongly effective against *Plasmodium*. In other studies, *S. spinosa* root bark extracts exhibited anti-parasitic alkaloid activity against *Leishmania* spp. (Frederich et al. 2004, Hoet et al. 2004, Zirihi et al. 2005, Frederich et al. 2008). *S. angolensis, S. diplotricha, S. gossweileri, S. henningsii, S. innocua, S. johnsonii, S. lucens, S. mellodora, S. mitis, S. potatorum, S. cocculoides, S. pungens,* and *S. spinosa* produced some crude extracts that have an antiplasmodial activity against a chloroquine-susceptible strain of *Plasmodium falciparum* in an *in-vitro* assay (Philippe and Angenot 2005). *S. cocculoides* contains iridoid glycosides that have potential biological activities (Ngadze et al. 2018). The leaf and stem extracts of *S. spinosa* have shown good inhibitory activities against *E. coli* and *S. aureus* (Ugoh and Bejide 2013).

The stem bark extracts of *S. floribunda* also contain decussine and rouhamine that have exhibited antimicrobial activity against *Bacillus subtilis* and *Staphylococcus aureus*. The roots of *S. usambarensis* contain usambarine, usambarensine, and 18, 19-dihydrousambarine which are less effective against *P. falciparum in vitro*, but strongly active against *Entamoeba histolytica in vitro*. Akagerine is slightly effective against protozoa (Schmelzer et al. 2008). Usambarensine and 5',6'-dihydrousambarensine are highly effective against the gram-positive bacteria *S. aureus, B. subtilis,* and *Mycobacterium smegmatis* (Schmelzer et al. 2008). The stem bark alkaloids of *S. henningsii* contain holstiine and holstiline which have low to average antimalarial activity against chloroquine-resistant *P. falciparum* (Schmelzer et al. 2008). However, they do not possess any anti-amoebic activity. The leaves and root extracts of *S. angolensis* have shown moderate anti plasmodial activity against *P. falciparum in vitro* (Frédérich et al. 1999, Frederich et al. 1999, Frédérich et al. 2003, Frederich et al. 2004, Schmelzer et al. 2008). The root bark isolates of *S. mostueoides* contain malagashine that has a significant effect activity against chloroquine resistant *P. falciparum* strains. The root bark extracts of *S. potatorum* contain ochrolifuanine A, which was tested *in vitro* and showed significant effect against chloroquine sensitive and chloroquine resistant *P. falciparum* strains (Schmelzer et al. 2008). The leaves of *S. icaja* possess monomers such as protostrychnine, genostrychnine, and pseudostrychnine, the bisindolic alkaloid strychnogucine, and the trimeric indolomonoterpenic alkaloid strychnohexamine that have shown antiplasmodial activity against *P. falciparum* (Schmelzer et al. 2008).

Pharmacology

Strychnine and brucine are the most important alkaloids of the *Strychnos* species. Both alkaloids are used medicinally for stimulating the central nervous system (CNS), to treat chronic heart diseases, stomach pains, and as a remedy for asthma and epilepsy. Furthermore, strychnine and brucine speed up the secretion of gastric juices and increase sensory awareness. The *Strychnos* species are associated with two different types of toxic effects that are attributed to the presence of strychnine and its monomeric derivatives. They cause tetanization of the body's muscles and paralysis of the body by its' various quaternary alkaloids that form part of the curare 3 poisons. Unripe fruits and seeds of *Strychnos* species are traditionally used as a remedy for snakebites across Southern Africa as they induce vomiting. This is attributable to the presence of strychnine or strychnine-like alkaloids that stimulate the central nervous system and prevent respiratory depression which could ultimately cause sudden death after bites from cobra or mamba snakes. Strychnine obstructs the postsynaptic receptors of the inhibitory neurotransmitter glycine which are in the spinal cord and motor neurons (Van Wyk and Gericke 2000, Coates Palgrave et al. 2002, Philippe et al. 2004, Philippe and Angenot 2005, Ngemakwe et al. 2017, Guo et al. 2018).

According to Mwamba (2006), the seeds of *S. cocculoides* bear some minute amount of strychnine ($C_{21}H_{22}N_2O_2$) and are therefore regarded as nontoxic. Strychnine was reported in *S. spinosa* plants grown in Florida, but no alkaloids were reported (Mwamba 2006). Brucine is not present in the seeds of *S. cocculoides, S. spinosa, S. madagascariensis,* and *S. pungens*. In Tanzania, *S. spinosa* leaves were reported to be toxic whereas in Mauritius they are nontoxic, but cause narcotic effects (Mwamba 2006).

The root barks of *S. aculeata* contain N-acetyl-O-methylstrychnosplendine which is responsible for intense muscle paralysis both *in vitro* and *in vivo*. It also contains spermostrychnine that is responsible for clonic convulsions but not tonic seizures in mice (Schmelzer et al. 2008).

The stem bark of *S. floribunda* contains bisnordihydrotoxiferine that possesses antimicrobial and antidiarrheal activity *in vivo* in mice. The stem bark extracts of *S. floribunda* also contain decussine and rouhamine that have a muscle-relaxant effect (Schmelzer et al. 2008). The stem bark alkaloids of *S. henningsii* have shown convulsive, hypotensive, and cardiac tranquilizer effects due to the stimulation of the central nervous system (CNS) and possess anticancer properties. It also contains holstiine and holstiline, which have a low to average antimalarial activity against chloroquine-resistant *P. falciparum*. However, they do not possess any anti-amoebic activity. The mixture of stem and bark extracts did not show muscle relaxant or convulsive activity. Other alkaloids from *S. henningsii* have shown additional pharmacological activities such as the anti-inflammatory effects of retuline on edema, whilst isoretuline, acetylisoretuline, and desacetylisoretuline demonstrated analgesic effects in mice. Isoretuline and O-acetylisoretuline, in guinea pigs, have shown antispasmodic response. The leaves and root extracts of *S. angolensis* have shown moderate antiplasmodial activity against *P. falciparum in vitro* (Schmelzer et al. 2008).

S. icaja is a good source of strychnine, brucine, 12-hydroxystrychnine and other tertiary alkaloids. Strychnine enhances adrenaline levels, elevates blood pressure, and results in violent convulsions. Strychnine is widely used in medicine as an anesthetic and stimulator of the central nervous system. *S. icaja* also contains sungucine and isosungucine that possess antiplasmodial activities and exhibit cytotoxic activities against human cancer cells. Pharmacological experiments have proven that the quaternary alkaloid can function as a muscle relaxant and is also cardiotoxic, resulting in irreversible cardiac arrest (Frederich et al. 2000, Frédérich et al. 2001, Schmelzer et al. 2008).

The root bark and stem bark extracts of *S. myrtoides* have shown no inherent antimalarial activity against the chloroquine resistant strain of *P. falciparum in vitro* and *in vivo*, but significantly enhanced the effect of chloroquine in mice with no reports of cytotoxicity (Schmelzer et al. 2008). The root bark isolates of *S. mostueoides* contain about eight indole alkaloids namely: normacusine B, caracusine VII, strychnofendlerine, spermostrychnine, strychnobrasiline, deacetylstrychnobrasiline, malagashine, and malagashanine (Schmelzer et al. 2008). The root bark extracts also contain the monomeric alkaloid of the corynanthe class known as normacusine B, which is also found in *Rauvolfia, Tabernaemontana,* and *Vinca* species. It is sympatholytic and its hypotensive activity is stronger than that of reserpine (derived from *Rauvolfia serpentine* and *vomitonia*), which is a medication for high blood pressure (Schmelzer et al. 2008). Malagashine has shown significant activity against chloroquine resistant *P. falciparum* strains (Schmelzer et al. 2008).

The root bark isolates of *S. potatorum,* stem bark, and leaves of *S. mitis*, the root bark of *S. spinosa* contain the monoterpene alkaloid cantleyine, which has shown a relaxing effect on the isolated smooth muscles of the trachea of guinea-pigs (Massiot et al. 1992, Rahman 2016). The root bark extracts also contain normacusine B, which is sympatholytic, and has a hypotensive activity which is stronger than that of reserpin. The root bark extracts contain ochrolifuanine A, a dimeric alkaloid of the ß-carboline class, which was *in vitro* tested and showed significant activity against chloroquine sensitive and chloroquine resistant *P. falciparum* strains (Schmelzer et al. 2008). The total alkaloidal extracts of the seeds, bark, and leaves of *S. potatorum* have shown strychnine-like activity *in vivo*, elevated hypotensive effects, depressant activity and contractility of the heart (Schmelzer et al. 2008). *In vivo* tests of seed powder and aqueous extracts of the seeds are effective against ulcer formation by decreasing acid secretory activity, elevated mucin activity, and marked hepatoprotective activity in rats (Schmelzer et al. 2008). Methanol extracts of the seeds of *S. potatorum* have shown diuretic effects and antidiarrheal activity on castor oil induced diarrhea in rats (Schmelzer et al. 2008). The seeds of *S. potatorum* possess polyelectrolytes that can be used as coagulants to purify turbid water (Schmelzer et al. 2008). Laboratory tests have shown substantial improvement in the esthetic and microbiological quality of the surface water that has undergone direct filtration with the *S. potatorum* seeds as a coagulant (Schmelzer et al. 2008).

Quaternary alkaloids of the root bark of *S. usambarensis* have a muscle paralyzing, a curare-like effect that prevents the excitation of the skeletal muscles which are antagonized by acetylcholine. The higher dosage of the alkaloids causes various side effects such as a drop in blood pressure, blocking of the vagus nerve, change in heart rate, and change in the respiration rate (Schmelzer et al. 2008). The root bark *of S. usambarensis* also contains malindine which has been shown to have intense muscle relaxant activity, and which is not antagonized by acetylcholine and is not of the curare type. The root bark of *S. usambarensis* contains akagerine and its derivatives, which are potent convulsant agents and 100 times less active than strychnine (Schmelzer et al. 2008). The root bark of *S. usambarensis* contains usambarensines, which do not paralyze the skeletal muscles but does have an atropine-like and spasmolytic activity on smooth muscles (Schmelzer et al. 2008). It also contains harmane which induces the enrichment of biogenic amines such as serotonine in the brain. In small dosages, harmane results in hallucinations, and high doses cause convulsions and respiratory paralysis (Schmelzer et al. 2008). The root bark extracts of *S. usambarensis* contain strychnopentamine and 5',6'-dihydrousambarensine which are effective against *P. falciparum in vitro* and inactive against *Plasmodium berghei in vivo* (Schmelzer et al. 2008). Isostrychnopentamine has shown antiplasmodial activity both *in vitro* against various chloroquine resistant and chloroquine-sensitive strains of *P. falciparum* and *in vivo* against chloroquine-sensitive strains of rodent infecting *P. berghei* and *Plasmodium vinckei* (Schmelzer et al. 2008). However, harmane does not possess any antimicrobial activity (Schmelzer et al. 2008). Numerous *S. usambarensis* alkaloids are toxic to several tumour cells (Schmelzer et al. 2008). The chloroform extract of the leaves of *S. usambarensis* are effective against lymphatic leukaemia *in vivo* in mice (Schmelzer et al. 2008). Strychnopentamine, chrysopentamine, and isostrychnopentamine are perceived as potential anticancer agents (Schmelzer et al. 2008). The curare alkaloids play a vital role in decreasing the risk of anesthesia due to low doses applied during anesthetic procedures. The anesthetic makes it easy to attain adequate muscle relaxation during clinical or surgical procedures (Schmelzer et al. 2008).

Toxicology

The ripe pulp of *S. cocculoides, S. spinosa, S. pungens, S. innocua, S. madagascariensis, S. gerrardii, S. gossweileri, S. decussata,* and *S. lucens* are reported as nontoxic agents whereas the pulp of ripe fruits from *S. usambarensis, S. icaja, S. campicola, S. aculeata, S. angolensis* and *S. potatorum* are considered to be toxic (Leeuwenberg 1963, Ohiri et al. 1983, Mwamba 2006, Adebowale 2014). The ripe pulp of *S. potatorum* is detoxified through the cooking process due to the deactivation of the toxins by heat. The seeds of *S. potatorum* have found use as a fish poison at the Kruger National Park of South Africa and this is attributed to the presence of saponin (Mwamba 2006). The seeds of *S. cocculoides* bear some minute amount of strychnine ($C_{21}H_{22}N_2O_2$), which renders them nontoxic (Mwamba 2006). Brucine is not present in the seeds of *S. cocculoides, S. spinosa, S. madagascariensis* and *S. pungens* (Mwamba 2006). The leaves of *S. spinosa* in Tanzania have been reported to be toxic whereas in Mauritius they were reported to cause narcotic effects. The presence of strychnine was reported on *S. spinosa* plants grown in Florida, but no alkaloids were reported (Mwamba 2006).

The lethality of *Strychnos* species has enabled them to be used as a poison on arrow tips or curare poison by the ancient hunters of South America and the rest of Africa. The hunters of the Banyambo tribe from Rwanda and Tanzania use the roots and leaf extracts of *S. usambarensis* to poison their arrows. The isolation of compounds from *S. usambarensis* showed the presence of the usambarine type of alkaloids, such as isostrychnopentamine, which forms part of the main ingredient of the African curare for poisoning arrows (Kolm 1984, Wright et al. 1991, Frédérich et al. 2003, Philippe and Angenot 2005, Schmelzer et al. 2008).

Fruits and seeds of *S. aculeata* are very toxic (Schmelzer et al. 2008). The fruit is a good source of saponins (Schmelzer et al. 2008). The leaves of *S. angolensis* produce a considerable amount of mucilage. The pounded bark of *S. angolensis* has been used by the Efe and Mbuti tribes of Congo to make poisoned arrows (Schmelzer et al. 2008). The root bark of *S. icaja* has been used as an arrow tip and ordeal poison in central African countries such as Gabon, Central African Republic, and the Democratic Republic of Congo. The principal compounds responsible for the toxicity are strychnine and 12-hydrostrychinine (Schmelzer et al. 2008, Buckingham et al. 2010).

Conclusions and Future Perspective

Strychnos species have been used for medicinal and nourishment purposes for centuries. However, their medicinal potential has not yet been explored and commercialised in Southern Africa. Thirty three *Strychnos* species have been reported in Southern Africa and their medicinal exploration is necessary, especially as some of them are closely related to the already explored Asian *Strychnos* species. *S. icaja* is closest to *S. nux-vomica* in terms of alkaloids composition. Both are a good source of strychnine, brucine, colubrine, icajine, novacine, vomicine, normacusine B and many other compounds that exhibit some pharmacological activities such as hepatoprotection, immunomodulation, antimutagenic, cardiovascular protective, hypoglycaemic, anti-spasmodic, anticancer, anti-inflammatory, antimicrobial and anti-tumour properties (Schmelzer et al. 2008, Hiruntad et al. 2015).

References

Adebowale, A. 2014. Biosystematic studies in southern African species of *Strychnos* L. (Loganiaceae).

Adebowale, A., Y. Naidoo, J. Lamb, and A. Nicholas. 2014. Comparative foliar epidermal micromorphology of Southern African *Strychnos* L. (Loganiaceae): Taxonomic, ecological and cytological considerations. Plant Systematics and Evolution 300: 127–138.

Aniszewski, T. 2007. Alkaloids-Secrets of Life:: Aklaloid Chemistry, Biological Significance, Applications and Ecological Role. Elsevier 2: 19–21.

Asase, A., A.A. Oteng-Yeboah, G.T. Odamtten, and M.S. Simmonds. 2005. Ethnobotanical study of some Ghanaian anti-malarial plants. Journal of Ethnopharmacology 99: 273–279.

Asuzu, C.U., and M.O. Nwosu. 2017. Studies on morphology and anatomy of *Strychnos spinosa* Lam. (Loganiaceae) 4: 71–7.

Biswas, A., S. Chatterjee, R. Chowdhury, S. Sen, D. Sarkar, M. Chatterjee, and J. Das. 2012. Antidiabetic effect of seeds of *Strychnos potatorum* Linn. in a streptozotocin-induced model of diabetes. Acta Pol. Pharm. 69: 939–943.

Bribi, N.J.A.J.B. 2018. Pharmacological activity of alkaloids: A review. 1: 1–6.

Buckingham, J., K.H. Baggaley, A.D. Roberts, and L.F. Szabo. 2010. Dictionary of Alkaloids with CD-ROM. CRC 2: 9–2377.

Coates Palgrave, K., R. Drummond, and E.J.C.T. Moll. 2002. Trees of Southern Africa. Struik Publishers. 1212.

De Wet, H., M. Ramulondi, and Z.J.S.A.J.o.B. Ngcobo. 2016. The use of indigenous medicine for the treatment of hypertension by a rural community in northern Maputaland, South Africa 103: 78–88.

Elago, S.N., and L.T. Tjaveondja. 2015. A comparative evaluation of the economic contributions and uses of Strychnos cocculoides and Schinziophyton rautanenii fruit trees to poverty alleviation in mile 20 village of Namibia. Agriculture and Food Sciences Research 2: 25–31.

Frederich, M., M.-P. Hayette, M. Tits, P. De Mol, and L. Angenot. 1999. *In vitro* activities of strychnos alkaloids and extracts against *Plasmodium falciparum*. Antimicrobial Agents and Chemotherapy 43: 2328–2331.

Frédérich, M., M. Tits, M.-P. Hayette, V. Brandt, J. Penelle, P. DeMol, G. Llabres, and L. Angenot. 1999. 10 '-Hydroxyusambarensine, a new antimalarial bisindole alkaloid from the roots of Strychnos u sambarensis. Journal of Natural Products 62: 619–621.

Frederich, M., M.-C. De Pauw, G. Llabres, M. Tits, M.-P. Hayette, V. Brandt, J. Penelle, P. De Mol, and L. Angenot. 2000. New antimalarial and cytotoxic sungucine derivatives from Strychnos icaja roots. Planta Medica 66: 262–269.

Frédérich, M., M.-C. De Pauw, C. Prosperi, M. Tits, V. Brandt, J. Penelle, M.-P. Hayette, P. DeMol, and L. Angenot. 2001. Strychnogucines A and B, two new antiplasmodial bisindole alkaloids from Strychnos icaja. Journal of Natural Products 64: 12–16.

Frédérich, M., M.-J. Jacquier, P. Thépenier, P. De Mol, M. Tits, G. Philippe, C. Delaude, L. Angenot, and M. Zèches-Hanrot. 2002. Antiplasmodial activity of alkaloids from various *Strychnos* species. Journal of Natural Products 65: 1381–1386.

Frédérich, M., M. Bentires-Alj, M. Tits, L. Angenot, R. Greimers, J. Gielen, V. Bours, and M.-P. Merville. 2003. Isostrychnopentamine, an Indolomonoterpenic Alkaloid from Strychnos usambarensis, induces cell cycle arrest and apoptosis in human colon cancer cells. Journal of Pharmacology and Experimental Therapeutics 304: 1103–1110.

Frederich, M., M. Tits, E. Goffin, G. Philippe, P. Grellier, P. De Mol, M.-P. Hayette, and L. Angenot. 2004. *In vitro* and *in vivo* antimalarial properties of isostrychnopentamine, an indolomonoterpenic alkaloid from Strychnos usambarensis. Planta Medica 70: 520–525.

Frederich, M., M. Tits, and L. Angenot. 2008. Potential antimalarial activity of indole alkaloids. Transactions of the Royal Society of Tropical Medicine and Hygiene 102: 11–19.

Gangwar, M., and A. Choubey. 2017. Phytochemical screening and total flavonoid content assays of various solvent extracts of fruits of strychnos potatorum linn. 6: 69–75.

Guo, R., T. Wang, G. Zhou, M. Xu, X. Yu, X. Zhang, F. Sui, C. Li, L. Tang, and Z. Wang. 2018. Botany, phytochemistry, pharmacology and toxicity of *Strychnos nux-vomica* L.: A review. The American Journal of Chinese Medicine 46: 1–23.

Hiruntad, Y., C. Palanuvej, and N. Ruangrungsi. 2015. Pharmacognostic specification and quantitative analysis of strychnine and brucine in Strychnos nux-vomica seeds. Chulalongkorn University 1: 21–29.

Hoet, S., F. Opperdoes, R. Brun, V. Adjakidjé, and J. Quetin-Leclercq. 2004. *In vitro* antitrypanosomal activity of ethnopharmacologically selected Beninese plants. Journal of Ethnopharmacology 91: 37–42.

Hoet, S., L. Pieters, G.G. Muccioli, J.-L. Habib-Jiwan, F.R. Opperdoes, and J. Quetin-Leclercq. 2007. Antitrypanosomal activity of triterpenoids and sterols from the leaves of Strychnos spinosa and related compounds. Journal of Natural Products 70: 1360–1363.

Isa, A.I., M.D. Awouafack, J.P. Dzoyem, M. Aliyu, R.A. Magaji, J.O. Ayo, and J.N.J.B.c. Eloff. 2014. Some *Strychnos spinosa* (Loganiaceae) Leaf Extracts and Fractions have Good Antimicrobial Activities And Low Cytotoxicities. 14: 456.

Isa, A.I., J.P. Dzoyem, S. Adebayo, M. Suleiman, and J.N. Eloff. 2016. Nitric oxide inhibitory activity of Strychnos spinosa (Loganiaceae) leaf extracts and fractions. African Journal of Traditional, Complementary and Alternative Medicines 13: 22–26.

Issa, T.O., Y.S. Mohamed, S. Yagi, R.H. Ahmed, T.M. Najeeb, A.M. Makhawi, and T.O. Khider. 2018. Ethnobotanical investigation on medicinal plants in Algoz area (South Kordofan), Sudan. Journal of Ethnobiology and Ethnomedicine 14: 31.

Itoh, A., N. Oya, E. Kawaguchi, S. Nishio, Y. Tanaka, E. Kawachi, T. Akita, T. Nishi, and T.J.J.o.n.p. Tanahashi. 2005. Secoiridoid Glucosides from Strychnos s pinosa 68: 1434–1436.

Kabine, O., B. Mamadou, B. Fatoumata, K. Namagan, H.N. Luopou, and B.A. Mamadou. 2015. Anti-oxidative activity of fruit extracts of some medicinal plants used against chronic diseases (diabetes, hypertension) in Kankan, Guinea. J. Plant Sci. 3: 1–5.

Karou, S.D., T. Tchacondo, M.A. Djikpo Tchibozo, S. Abdoul-Rahaman, K. Anani, K. Koudouvo, K. Batawila, A. Agbonon, J. Simpore, and C.J.P.b. de Souza. 2011. Ethnobotanical study of medicinal plants used in the management of diabetes mellitus and hypertension in the Central Region of Togo. 49: 1286–1297.

Kipkemoi, M.N.R., N.P. Kariuki, N.V. Wambui, O. Justus, and K.J.I.J.o.M.P.R. Jane. 2013. Macropropagation of an endangered medicinal plant, Strychnos henningsii (gilg), (Loganiaceae) for sustainable conservation. 2: 247–253.

Kolm, H.J.T. 1984. Effects of three alkaloids isolated from Strychnos usambarensis on cancer cells in culture (in French): Bassleer, R., M.-C. Depauw-Gillet, B. Massart, J.-M. Marnette, P. Wiliquet, M. Caprasse, and L. Angenot (eds.). Histology and Cytology Service, Faculty of Medicine and Laboratory of Pharmacognosy, Institute of Pharmacy, University of Liège, Belgium. Planta Med. 45: 123.

Leeuwenberg, A.J.A.b.n. 1963. The Loganiaceae of Africa V. Usteria Willd. 12: 112–118.

Loeb, E.M., C. Koch, and E.-M.K. Loeb. 1956. Kuanyama Ambo magic. 6. Medicinal, cosmetical, and charm flora and fauna. The Journal of American Folklore 69: 147–174.

Madzimure, J., E.T. Nyahangare, H. Hamudikuwanda, T. Hove, S.R. Belmain, P.C. Stevenson, and B.M. Mvumi. 2013. Efficacy of *Strychnos spinosa* (Lam.) and *Solanum incanum* L. aqueous fruit extracts against cattle ticks. Tropical Animal Health and Production 45: 1341–1347.

Mallikharjuna, P., L. Rajanna, Y. Seetharam, and G. Sharanabasappa. 2007. Phytochemical studies of Strychnos potatorum Lf-A medicinal plant. Journal of Chemistry 4: 510–518.

Maroyi, A. 2011. An ethnobotanical survey of medicinal plants used by the people in Nhema communal area, Zimbabwe. Journal of Ethnopharmacology 136: 347–354.

Massiot, G., P. Thepenier, M.-J. Jacquier, L. Le Men-Olivier, and C.J.P. Delaude. 1992. Alkaloids from roots of Strychnos potatorum. 31: 2873–2876.

Maundu, P., and T. Tengnäs. 2005. Useful trees and shrubs for Kenya. Techninal handbook no.35.

Mawela, K.G. 2009. The toxicity and repellent properties of plant extracts used in ethnoveterinary medicine to control ticks. University of Pretoria 1: 18–54.

Motlhanka, D., P. Motlhanka, and T. Selebatso. 2008. Edible indigenous wild fruit plants of Eastern Botswana. International Journal of Poultry Science 7: 57–460.

Msonthi, J., C. Galeffi, M. Nicoletti, I. Messana, and G. Marini-Bettolo. 1985. Kingiside aglucone, a natural secoiridoid from unripe fruits of Strychnos spinosa. Phytochemistry 24: 771–772.

Mwamba, C.K. 2006. Monkey orange: Strychnos cocculoides. Crops for the Future 1: 1–111.

Ngadze, R.T., A.R. Linnemann, L.K. Nyanga, V. Fogliano, and R. Verkerk. 2017a. Local processing and nutritional composition of indigenous fruits: The case of monkey orange (*Strychnos* spp.) from Southern Africa. Food Reviews International 33: 123–142.

Ngadze, R.T., R. Verkerk, L.K. Nyanga, V. Fogliano, and A.R. Linnemann. 2017b. Improvement of traditional processing of local monkey orange (*Strychnos* spp.) fruits to enhance nutrition security in Zimbabwe. Food Security 9: 621–633.

Ngadze, R.T. 2018. Value addition of Southern African monkey orange (*Strychnos* spp.): Composition, utilization and quality. Wageningen University 1: 1–179.

Ngadze, R.T., R. Verkerk, L.K. Nyanga, V. Fogliano, R. Ferracane, A.D. Troise, and A.R. Linnemann. 2018. Effect of heat and pectinase maceration on phenolic compounds and physicochemical quality of Strychnos cocculoides juice. PloS one 13: e0202415.

Ngarivhume, T., C.I. van't Klooster, J.T. de Jong, and J.H. Van der Westhuizen. 2015. Medicinal plants used by traditional healers for the treatment of malaria in the Chipinge district in Zimbabwe. Journal of Ethnopharmacology 159: 224–237.

Ngemakwe, P.N., F. Remize, M. Thaoge, and D. Sivakumar. 2017. Phytochemical and nutritional properties of underutilised fruits in the southern African region. South African Journal of Botany 113: 137–149.
Nuzillard, J.-M., P. Thépenier, M.-J. Jacquier, G. Massiot, L. Le Men-Olivier, and C.J.P. Delaude. 1996. Alkaloids from root bark of Strychnos panganensis 43: 897–902.
Ohiri, F., R. Verpoorte, and A.B. Svendsen. 1983. The African *Strychnos* species and their alkaloids: A review. Journal of Ethnopharmacology 9: 167–223.
Omosa, L.K., J.O. Midiwo, V.M. Masila, B.M. Gisacho, R. Munayi, K.P. Chemutai, G. Elhaboob, M.E. Saeed, S. Hamdoun, and V. Kuete. 2016. Cytotoxicity of 91 Kenyan indigenous medicinal plants towards human CCRF-CEM leukemia cells. Journal of Ethnopharmacology 179: 177–196.
Orwa, C., Mutua, A., Kindt, R., Jamnadass, R. and Anthony, S. 2009. Agroforestree Database: A Tree Reference and Selection Guide, Version 4.0. 15.
Packialakshmi, N., C. Suganya, and V.J.I.J.R.P.N.S. Guru. 2014. Studies on Strychnos potatorum seed and screening the water Quality assessment of drinking water 3: 380–396.
Philippe, G., L. Angenot, M. Tits, and M. Frederich. 2004. About the toxicity of some *Strychnos* species and their alkaloids. Toxicon 44: 405–416.
Philippe, G., and L. Angenot. 2005. Recent developments in the field of arrow and dart poisons. Journal of Ethnopharmacology 100: 85–91.
Rahman, A.-u. 2016. Studies in natural products chemistry: Bioactive Natural Products (Part XII). Elsevier 50: 13–434.
Rajesh, P., V.R. Kannan, S. Latha, and P. Selvamani. 2012. Phytochemical and pharmacological profile of plants belonging to Strychnos genus: A review. Carbon 250, 4.
Rasoanaivo, P., G. Palazzino, M. Nicoletti, and C.J.P. Galeffi. 2001. The co-occurrence of C (3) epimer Nb, C (21)-secocuran alkaloids in Strychnos diplotricha and Strychnos myrtoides. 56: 863–867.
Rodrigues, S., E.S. de Brito, and E. de Oliveira Silva. 2018. Maboque/Monkey Orange—Strychnos spinosa, Exotic Fruits. Elsevier, pp. 293–296.
Ruffo, C., A. Birnie, and B. Tengnas. 2002. Edible wild plants of tanzania (Vol. 27). Regional Land Management Unit/Sida 27: 19–799.
Sanmuga Priya, E., and S.J.P.b. Venkataraman. 2007. Anti-inflammatory effect of Strychnos potatorum. Seeds on Acute and Subacute Inflammation in Experimental Rat Models 45: 435–439.
Sanmuga, P.E., and S.J.P.J. Venkataraman. 2010. Pharmacognostical and phytochemical studies of *Strychnos potatorum* Linn seeds 2: 190–197.
Schmelzer, G., A. Gurib-Fakim, and R. Arroo. 2008. Plant resources of tropical Africa: Medicinal plants 1. Wageningen/Leiden, Netherlands: PROTA Foundation/Backhuys Publisher 11: 1–790.
Setshogo, M., and F. Venter. 2003. Trees of Botswana: Names and distribution, Southern African Botanical Diversity Network Report No 18. Cape Town, South Africa 18: 19–125.
Shoko, T., Z. Apostolides, M. Monjerezi, and J. Saka. 2013. Volatile constituents of fruit pulp of Strychnos cocculoides (Baker) growing in Malawi using solid phase microextraction. South African Journal of Botany 84: 11–12.
Sitrit, Y., S. Loison, R. Ninio, E. Dishon, E. Bar, E. Lewinsohn, and Y. Mizrahi. 2003. Characterization of monkey orange (*Strychnos spinosa* Lam.), a potential new crop for arid regions. Journal of Agricultural and Food Chemistry 51: 6256–6260.
Sunghwa, F., and M. Koketsu. 2009. Phenolic and bis-iridoid glycosides from Strychnos cocculoides. Natural Product Research 23: 1408–1415.
Tchinda, A.T., V. Tamze, A.R. Ngono, G.A. Ayimele, M. Cao, L. Angenot, and M.J.P.L. Frédérich. 2012. Alkaloids from the stem bark of Strychnos icaja 5: 108–113.
Thepenier, P., M.-J. Jacquier, G. Massiot, L. Le Men-Olivier, and C.J.P. Delaude. 1984. Dehydroisostrychnobiline, matopensine and other alkaloids from Strychnos kasengaensis 23: 2659–2663.
Thépenier, P., M.-J. Jacquier, J. Hénin, G. Massiot, L. Le Men-Olivier, and C.J.P. Delaude. 1990. Alkaloids from Strychnos pungens 29: 2384–2386.
Tian, J.X., C. Peng, L. Xu, Y. Tian, and Z.J.J.B.C. Zhang. 2013. *In vitro* metabolism study of Strychnos alkaloids using high-performance liquid chromatography combined with hybrid ion trap/time-of-flight mass spectrometry 27: 775–783.

Tor-Anyiin, T., J. Igoli, J. Anyam, and J. Anyam. 2015. Isolation and antimicrobial activity of sarracenin from root bark of Strychnos spinosa. Journal of Chemical Society of Nigeria 40.

Ugoh, S.C., and O.S.J.N.S. Bejide. 2013. Phytochemical screening and antimicrobial properties of the leaf and stem bark extracts of Strychnos spinosa 11: 123–128.

Van Wyk, B.-E., and N. Gericke. 2000. People's plants: A guide to useful plants of Southern Africa. Briza Publications 1: 30–352.

Van Wyk, B. 2013. Field guide to trees of southern Africa. Penguin Random House South Africa 3: 298–299.

Verpoorte, R., M. Frédérich, C. Delaude, L. Angenot, G. Dive, P. Thépenier, M.-J. Jacquier, M. Zèches-Hanrot, C. Lavaud, and J.-M.J.P.L. Nuzillard. 2010. Moandaensine, a dimeric indole alkaloid from Strychnos moandaensis (Loganiaceae) 3: 100–103.

Wright, C.W., D.H. Bray, M.J. O'Neill, D.C. Warhurst, J.D. Phillipson, J. Quetin-Leclercq and L. Angenot. 1991. Antiamoebic and antiplasmodial activities of alkaloids isolated from Strychnos usambarensis. Planta Medica 57: 337–340.

Zirihi, G.N., L. Mambu, F. Guédé-Guina, B. Bodo, and P. Grellier. 2005. *In vitro* antiplasmodial activity and cytotoxicity of 33 West African plants used for treatment of malaria. Journal of Ethnopharmacology 98: 281–285.

Chapter 6

Preference of *Agrobacterium rhizogenes* Mediated Transformation of Angiosperms

Kamogelo M Mmereke,[1,*] *Aaqib Javid,*[2] *Sukanya Majumdar,*[3] *Ipshita Ghosh,*[3] *Sonia Malik,*[4] *Abdullah Makhzoum*[1] and *Sumita Jha*[3,*]

Introduction

Angiosperms are the largest and most diverse group in the kingdom plantae. These are vascular plants with stems, roots and produce seeds in fruits (Basile et al. 2017). Plants are exposed to various environmental stresses that trigger accumulation of reactive oxygen species. The accumulated reactive oxygen species have an effect on proteins, nucleic acids, enzymes and other molecules within the plant (Bartwal et al. 2013). To counteract molecular degradation, plants have developed antioxidative defense systems made up of both enzymatic and non-enzymatic molecules known as plant secondary metabolites or more recently as plant specialized metabolites (Akula and Ravishankar 2011, Bartwal et al. 2013). Secondary metabolites are compounds not primarily involved in the plant growth and development but rather essential for survival and overall fitness of plants, produced in response to stress conditions, and are responsible for the odor,

[1] Department of Biological Sciences & Biotechnology, Botswana International University of Science & Technology, Botswana.
[2] Department of Plant Biotechnology, Institute of Genetics and Biotechnology, Hungarian University of Agriculture and Life Sciences, 1118 Budapest, Hungary.
[3] Department of Botany, Calcutta University, 35 Ballygunge Circular Road, Kolkata 700 019.
[4] Laboratory of Woody Plants and Crops Biology (LBLGC), University of Orleans, 1 Rue de Chartres-BP 6759, 45067 Orleans, France.
* Correspondence authors: kamogelo.mmereke@studentmail.biust.ac.bw; sumitajha.cu@gmail.com

taste and color of the plant (Akula and Ravishankar 2011). Secondary metabolites are also referred to as phytochemicals and have a vast area of applications in pharmaceutical, nutraceutical, cosmetics and agriculture fields (Akula and Ravishankar 2011, Tiwari and Rana 2015, Malik et al. 2021). The potential of some of the medicinal plant species have been explored due to their properties in treating various ailments and their application in other fields for the benefit of human welfare (Sharma et al. 2020, Joshi et al. 2011, Dapar et al. 2020).

Secondary metabolite production and accumulation depends on rate of plant propagation and cultivation as well as climatic conditions during harvesting (Lucchesini and Mensuali-Sodi 2010, Nartop 2018). Plants that provide secondary metabolites are not adapted to agricultural cultivation but grow naturally in the environment. Secondary metabolites are generally produced in plants under stress conditions and in some cases, plants may require 5–7 years of growth to obtain a substantial amount of a specific product (Akula and Ravishankar 2011, Yang et al. 2018, Murthy et al. 2014).

In this regard plant tissue culture is proposed to curb these disadvantages. Plant tissue culture is an advanced approach that provides an ideal research tool to study various plant science and other related aspects. Studies on primary and secondary metabolism, cytodifferentiation, morphogenesis, plant tumour physiology and formation of plant hybrids via protoplast fusion are examples of research that plant tissue culture can give valuable insight into (Smith 2006, Roberts et al. 1998). This technique is improved by novel methods such as gene editing and stress induction by biotic or abiotic factors. Current day pharmaceuticals are typically based on plant derived metabolites. With the exponential demand for pharmaceuticals, uniform supply is compromised. Plant tissue culture is an alternative to avert this as it assures independence from geographical conditions by eliminating the need to rely on wild plants. Plant cell and tissue culture techniques are indispensable for the production of disease-free plants, and rapid multiplication of rare plant genotypes (Malik et al. 2014, Malik et al. 2016).

Plant genetic transformation allows the use of plants for the production of engineered products such as vaccines and multiple pharmaceuticals as well the production of plant derived metabolites of important commercial value (Makhzoum et al. 2013). Micropropagation, hairy root culture, cell suspension cultures, callus culture are some of the commonly explored *in vitro* culture techniques. Identification and isolation of useful bioactive compounds requires a method that permits its continuous production. Usually, the compounds are extracted from their natural sources. This however implies the over exploitation of the natural plants due to overharvesting. Further limitations include the slow growth rates of many plants, low concentration of the active compounds of interest and the frequent need for biotic and abiotic stresses to induce their biosynthesis. All these factors make yield of secondary metabolites from highly inefficient sources and emphasize the need for novel approaches for secondary metabolites production. *In vitro* culture under controlled conditions offers a profound technology system for the natural products extraction and study of plants.

Elicitors or precursors are added to plant growth media to modify it and enhance production of important plant defense compounds. Several types of elicitors have been used as chitin, pectin cellulose, yeast extract and nanoparticles, and precursors that are intermediary compounds of the natural compounds are also used to enhance the biosynthesis of secondary metabolites (Yue et al. 2014). Exogenous sterols, and stigmasterol are examples of precursors used in plant biotechnology. When searching for appropriate precursors, it is important to look at the entire biosynthetic pathway and include several molecules involved in different steps of the process. In *Rhodiola rosea*, the accumulation of different secondary metabolites increased by many folds after feeding the *in vitro* grown plants with different precursors such as trans-cinnamic acid, cinnamaldehyde and cinnamyl alcohol (Javid et al. 2020).

In vitro cultivation is an option to propagate plants resulting in increased biomass accumulation within a short period of time and increased secondary metabolite content (Lucchesini and Mensuali-Sodi 2010). However, in certain plant species, the yield of secondary metabolites are very low or completely lacking when grown under *in vitro* conditions (György 2004, Isah et al. 2018). Therefore, different strategies need to be evaluated for these plant species to increase the accumulation of important metabolites. These include callus, cell suspension, hairy root cultures and somatic embryogenesis.

Induction of hairy root cultures in plants is one of the popular plant tissue culture techniques adopted to increase the secondary metabolite production and accumulation. Hairy root culture is an *Agrobacterium rhizogenes* mediated transformation strategy where transfer DNA (T-DNA) from the root inducing plasmid (Ri) is stably integrated into the plant's genome (Chandra 2012, Chilton et al. 1982, Stougaard et al. 1987). Hairy root phenotype is expressed upon the successful transformation of the plant by *A. rhizogenes* (Furze et al. 1987, Makhzoum et al. 2013). Successful infection is characterized by the appearance of a white stubble at the wounded site (Bhadra et al. 1993). Neoplastic transformed or hairy roots are characterized by genetic stability, lateral branching, high proliferation and growth in hormone-free media (Chandra 2012, Yoshikawa and Furuya 1987). Symptoms of hairy or transformed roots commonly start to appear after 5–7 days (Kifle et al. 1999), 2–4 weeks (Kamada et al. 1986), 6–8 weeks (Pythoud et al. 1987) 3–6 weeks after infection, depending upon explant or plant species.

Hairy roots are induced using a specific strain of *Agrobacterium rhizogenes*. It is a gram negative, non-spore forming, short rod shaped, soil born bacteria associated with plants, which infect mostly dicotyledonous plants and some mono-cotyledonous species. They can use glucose as a carbon source, growing aerobically. *Agrobacterium* are usually found in association with roots, tubers, and or underground stems. *Agrobacterium rhizogenes* contains root inducing plasmids, also called Ri-plasmid. Upon infection, the plasmid is integrated in the genome of the host plant, forcing continuous root growth, which produces a food source for the bacterium, i.e., opines, and causes it to grow abnormally resulting in hairy roots. The abnormal roots are particularly easy and economical to culture in artificial media because hormones are not needed in contrast to adventitious roots, and they are neoplastic with indefinite growth (Makhzoum et al. 2015, Makhzoum et al. 2013). *Agrobacterium rhizogenes* has the root inducing plasmid (Ri plasmid) that harbors the T-DNA, an important factor in transformation that directs opine synthesis and the differentiation and growth of plant cells (Cardarelli et al. 1987, Jung and Tepfer 1987, Nepovím et al. 2004). The T-DNA copy number and positional effects determine the rate of transformation and the morphology of hairy roots established hence different clones can be established from the same site of infection (Taya et al. 1992). T-DNA is conserved; remains essentially unchanged throughout the evolution of plants. Therefore, even after a long-time the plant may regenerate and display characteristics of *A. rhizogenes* infection: hairy root regenerants with wrinkled leaves, short inter nodal lengths and adventitious roots though opines will not be detectable (David et al. 1984, Kamada et al. 1986).

Agrobacterium is a useful tool in molecular biology in such that it can be modified into a transformation vector and be applied in metabolic engineering. *Agrobacterium* binary vectors are a standard tool in plant genetic engineering studies for the transformation of higher plants. Binary vectors are accessorized with T-DNA borders, multiple cloning sites, and replication functions of both *E. coli* and Agrobacterium, selectable marker genes, promoter, terminator, and reporter genes to improve the efficiency of and give further capabilities to the system. The binary vector system contains two plasmids, the one carrying virulence genes helps in DNA transfer (helper plasmid) and the second one carrying T-DNA border sequences, is harboring the gene of interest and selectable markers are used as vectors to transfer DNA into plants via Agrobacterium. Common binary vectors used in plant transformation are pCAMBIA, pBIN19, pGREEN and gateway.

The hairy root syndrome in plant biotechnology is purposed to study metabolic processes, increased production of valuable secondary metabolites as well as a genetic engineering tool to express recombinant proteins in plants. Hairy root induction stimulates fast biomass accumulation with increased metabolite production (Deno et al. 1987, Jung and Tepfer 1987, Le Flem-Bonhomme et al. 2004). Hairy roots possess high genetic and metabolite production stability as they carry the same chromosome number and karyotype as that of their parent plant (Häkkinen et al. 2016). The fact that these cultures grow in hormone free media and display no somaclonal variation enables their ability to produce secondary metabolites that are identical to that of the intact plant (Bulgakov et al. 1998, Flores and Filner 1985, Yoshikawa and Furuya 1987). Undifferentiated cultures that do not carry the same chromosome number and karyotype and undergo soma clonal variation however will produce chemically diverse metabolites from that of the intact plants. Those metabolites varying from the intact plant usually come complexed with other metabolites which makes their extraction difficult. Fast growth and proliferation of hairy roots is induced by the Ri plasmid, and therefore does not have chemical impact on secondary metabolites but only increases metabolite accumulation.

It has been revealed that hairy roots biosynthesis capacity reflects that of normal roots (Cardarelli et al. 1987, Constabel and Towers 1988, Flores and Filner 1985, Hamill et al. 1989), this is advantageous over conventional cell cultures in a sense that secondary metabolite accumulation is higher, while the chemical spectrum of the synthesized secondary metabolites is not any different. High proliferation is due to the continuous cell division of the hairy root cells throughout the culture period (Le Flem-Bonhomme et al. 2004). Hairy root cultures can only produce secondary metabolites that are correlated with the organization of cells as roots (Flores and Filner 1985). *Calystegia sepium* leaves did not produce alkaloids but the roots produced the alkaloid while *Atropa belladonna* leaves produced alkaloids though in small amounts (Deno et al. 1987).

Moreover, hairy roots not only produce secondary metabolites but can also remove some undesired compounds and heavy metals from the environment by a process known as phytoremediation. Phytoremediation is a green and an eco-friendly technology, in relation to traditional de-contamination methods, and thus has gained importance. Earlier on, it was applied for the removal of inorganic pollutants from soils, but gradually, it has evolved in more efficient way, to be able to treat organic pollutants. This is also as roots are the main contact organ and are also the site where the first reactions against pollutants take place. Hairy roots are able to metabolize hazardous compounds by common metabolic pathways (Le Flem-Bonhomme et al. 2004), which constitute an additional advantage to this system.

Role of *rol* Genes

Root inducing plasmid T-DNA of *Agrobacterium rhizogenes* has segments encompassing open reading frames which are assigned to specific transcripts. These open reading frames encode oncogenic genes, root locus (*rol*), that are virulent factors to induce hairy root expression and specifically affect root morphology (Constabel and Towers 1988, Mano et al. 1989). Each *rol* gene is specific to a particular phenotype and together these genes synergistically induce hairy roots to produce stronger effects as compared to when they are expressed individually (Bonhomme et al. 2000).

rolA

This is a 300bp ORF encoding 100 amino acid binding proteins and is a nucleic acid with a not fully understood biochemical function. Plants expressing a single *rol A* gene show wrinkled leaves, stunted growth and extremely shortened internodes and rounded leaves (Roychowdhury et al. 2013).

rolB

The *rol B* gene contains ORF of 777bp encoding 316 amino acid protein. It has been proposed that *rolB* gene is neither directly nor indirectly involved in hormone metabolism as there is no difference between wild plants and transformed metabolic profiles of radio-labelled indole acetic acid or its conjugates. The other reason supporting the notion is that plants which are transformed with 35S *rolB* had no free Indole acetic acid (IAA), amides of which were found in decreased levels and its ester conjugates were unaltered (Mano et al. 1989). This therefore concludes that *rolB* does not recognize this substrate *in vivo*. The *RolB* gene increases sensitivity of the cells to auxin where it is expressed. This gene has been reported to possess β-glucosidase activity which is hydrolysis bound auxins thereby bringing an increase in auxin in the cultures (Bonhomme et al. 2000, Mano et al. 1989). Tobacco protoplasts expressing *rolB* from the endogenous 35S promoter are more sensitive to auxin as compared to the untransformed ones (Estruch et al. 1991).

In addition, *rolB* stimulates flower and root formations. In plants transformed with *rolB* gene root and flower formations occur sooner and are more numerous as compared to the wild plant (Delbarre et al. 1994). Transformed tobacco plants expressing *rolB* from its endogenous promoter display alterations in leaf and flower morphology and also show increased formation of adventitious roots in the stem. However, expressing *rolB* under the control of 35S promoter makes transformation less efficient, this implies that localized *rolB* expression is important to stimulate root initiation (Estruch et al. 1991). In addition, when *Vitis amurensis* was transformed with the *rolB* gene, Resveratrol content was increased by more than hundred-fold to the control level. Resveratrol, a stilbene, can prevent fungal infections (Kiselev et al. 2007). Similarly, when the calli of *Maackia amurensis* was transformed with the *rolB* gene, high accumulation of isoflavonoids were observed upon high expression of this *rolB* gene (Grishchenko et al. 2016).

rolC

This gene contains 540 bp ORF encoding 180 amino acid protein. Transformed plants expressing *rolC* have an increased leaf length to leaf width ratio. Comparing them to the wild type *rolC* the transformed plants are shorter, with smaller flowers, which flower early and have reduced pollen production (Kodahl et al. 2016). Strong expression of *rolC* results in dwarfed bushy plants with a reduced apical dominance (Kodahl et al. 2016). *RolC* also has β-glucosidase activity that releases cytokinin from their glucosidase conjugates (Estruch et al. 1991). The decrease in apical dominance suggests that *rolC* expression is coupled to increase in cytokinin activity. *In vivo*, this gene has β-glucosidase activity capable of releasing free active cytokinins from their inactive conjugates (Kodahl et al. 2016). In comparison to other *rol* genes, *rolC* increased ginsenosides by 1.8–3 fold as compared to control plants, while other *rol* genes accumulated less ginsenosides (Constabel and Towers 1988).

rolD

The rolD gene contains an ORF of 1032 bp encoding a 344-amino acid protein and is a late-auxin induced gene like *rolB*, but at higher auxin levels the promoter induction is low as compared to *rolB* (Grichko and Glick 2001). The *rolD* gene was not seen in all *A. rhizogenes* strains like the other *rol* genes, but is only present in the T_L DNA of agropine strain Ri plasmids (Pavlova et al. 2013). The *rolD* gene has a Dof (a transcription factor)-binding element in its promoter region, which probably has a role in auxin induction, the characteristic feature of *rolB* gene, but *rolD* is the only *rol-gene* incapable of inducing root formation. It is the least studied gene, but is the only *rol* gene whose biochemical function has been determined, i.e., as a functional ornithine cyclodeaminase which produces proline by reducing ornithine (Trovato et al. 2001). Maintenance of hairy root growth and increase in flowering are among the major effects of *rolD* on the morphology of transformed plants (Trovato et al. 2001). Overexpressing of *rolD* results in an increased production of the pathogenesis-related protein (PR-1) in transformed plants, which is produced as a defense response (Bettini et al. 2003).

ORF8

The ORF8 gene encodes a protein of 780 amino acids and has the longest sequence among the genes present on the T_L-DNA. The N-terminal and C-terminus domains of ORF8 showed some similarity to the RolB protein and to the iaaM proteins found in *A. tumefaciens* respectively, was reported by (Levesque et al. 1988). Hairy root regenerated tobacco plants showed a difference amongst various research groups in Phenotypical analysis of the ORF8 transgenic plants under the CaMV-35S promoter, Lemcke, reporting no changes in morphology while Ouartsi and his colleagues in 2004 observed some divergence in cotyledon morphology and attributed it to auxin-induced cell division and expansion (Lemcke and Schmülling 1998, Ouartsi et al. 2004). Meanwhile, Umber, Clément and Otten, reported the significant morphological differences like hampered growth and rough, mottled leaves having thick and fleshy midribs, among the transformed and untransformed plants (Umber et al. 2002, Clément et al. 2006, Otten and Helfer 2001).

ORF13

The ORF13 promoter is known to be wound-inducible, composed of approximately 600 base pairs, and is conserved in different *A. rhizogenes* strains. The expression begins after 5 hours from the wound infection, and it is maintained for 17 hours where it reaches a maximum point. The ORF13 promoter consists of an 11-base pair motif repeats, which may have a role in wound induction. While conducting studies on tobacco plants Hansen and collegues reported ORF13 genes are wound-inducible and show organ specific expression (Hansen et al. 1997). Lemcke and Schmülling observed morphological changes such as dwarfing, wrinkled leaves, shortened internodes and roots, in transgenic tobacco plants harboring ORF13 (Lemcke and Schmülling 1998). For example, in *A. thaliana* approximately 1% biomass production was reduced due to extreme dwarfing caused by the overexpression of ORF13 (Kodahl et al. 2016).

ORF14

The *ORF14* gene on the Ri-plasmid has been conserved over time among different strains of *Agrobacterium rhizogenes*, and belongs to the *rolB, rolC, ORF8* and *ORF13* gene family (Levesque et al. 1988). Lemcke reported that no change in plant morphology was recorded, where ORF14 was overexpressed (Lemcke and Schmülling 1998). Capone and his co-workers concluded that *rol* genes and *ORF13, ORF14* co-act in synergy, improving root induction in *N. tabacum* and *Daucus carota* (Capone et al. 1989). So far, no morphological changes have been reported in the studies that have been conducted on the expression of *ORF14*. It is very uncertain to assign the range of effects on transformed plants, as a narrow range of species have been studied for transformation with *ORF14*.

Factors that Influence Hairy Root Transformation

There are various factors that affect the transformation frequency of hairy roots viz., the source, type and age of explant, the media used and solidifying agent, type of agrobacterium strain, co-cultivation period, and temperature are few among them. The age at which the seedling, stem, tuber, leaf, etc., is used also has a significant effect on the transformation rate. Young explants transform more efficiently and rapidly than the older ones and/or in some species, with age the plant becomes recalcitrant to transformation. In some cases root tips show a higher transformation efficiency than even hypocotyls, but co-cultivation of Triphysaria root tips with *A. rhizogenes* produced very few transformed roots (Handa et al. 1994), so it should wisely should be chosen as the type of explant. Overgrowth of the bacterium is one of the recurring problems affecting the hairy root transformation process, such that in most cases it depends on the type of media used for bacterial overgrowth, it particularly becomes problematic in Hoagland's medium as compared to MS medium (Sandal et al. 2007), and can also occur if the co-cultivation period is increased. It is widely known that the bacterium is most virulent at the temperature below or at 20°C. In addition, the solidifying agent, Phytagel had higher transformation rates than using phyto agar in Triphysaria. In some species, co-cultivation in the dark promotes the transformation rate. In conclusion, for specific species each parameter needs to be optimized. It is also possible to select hairy root clones that grow faster and produce higher metabolites content as compared to their normal plant roots. Hairy root clones vary in their secondary metabolites content, growth rate and productivity (Furze et al. 1987).

Clone Selection and Medium

Assessment of growth patterns in cultures is important for evaluation of growth kinetics of plant cultures. Grown index (GI) is one of the growth determining criteria used to select clones for further studies and is defined as harvest fresh or dry weight, normally at 21 days of culture (Giulietti et al. 1993). GI is calculated by the ratio of the total mass transferred and final volume accumulated during the propagation of culture. Clones from *Brugmansia candida* hairy roots obtained by excision of single root tips showed different patterns of growth and classification as follows; roots with a high degree of lateral branching and a profusion of root hairs on growing laterals and tips, roots with branches only on one side and linear unbranched roots (Giulietti et al. 1993). Media preparation requires precaution to avoid contamination, supplementation and nutritional composition of media determines the growth and development of cultures. Clones are selected after passages on solid media and liquid media, the attractive ones that satisfy different parameters are selected for further analysis and exploitation. In some clone's fresh weight increases about 20 times in 21 days of culture, while with other clones the fresh weight increases only 4 times. Some of the clones regenerate shoots with wrinkled leaves (Giulietti et al. 1993). Among clones there are variations in growth and metabolite spectrum and accumulation. These variations can be attributed to phenomena linked to transformation, a secondary consequence of variation in copy number, size and chromosomal location of Ri T-DNA fragments integrated in the plant genome (Ambros et al. 1986, Jouanin et al. 1987, Vilaine and Casse-Delbart 1987).

Differences in the morphologies of hairy roots grown on media can exist between the same cultivar which may be linked to divergence in growth rates and adaptation in liquid media of different clones (Bhadra et al. 1993). Clones cultured in solid medium display different morphologies and characteristics; thin/thick, straight/coil-like, regular/cluster type branches and thin/bulbous root tips long and short. To illustrate the possible links that exist between adaptability to liquid media and morphology, two clones of a thin and coiled nature from solid media were transferred to liquid media and behaved differently, with the persistent culture the coiled clone fragments, while the thin clone was cultures successful and appeared white or pale forming tight and dense bunches that fill the cross sections of the culture flask during growth in liquid culture. These roots did not thicken with age in the core of the root bunch and grew long roots at the periphery of roots that branched and aligned regularly with the flow direction (Bhadra et al. 1993).

Strain Selection

There are different strains of *A. rhizogenes*, depending on the type of opines, a carbon source, that they catabolise. *A. rhizogenes* classified as; Agropine-type strains [A4, 15834] bring about roots containing agropine, mannopine, mannopinic acid, and agropinic acid and mannopine-type strains [8196, TR7] elicit roots containing only mannopine, mannopinic acid and agropinic acid (Petit et al. 1983). *A. rhizogenes* strain

MAFF 03-01724 is a mikimopine strain and has successfully transformed *Ajuga reptans var. atropurpurea* to enhance steroid accumulation (Matsumoto et al. 1991).

Different strains have different capabilities to induce hairy roots. Specific virulence genes also affects the transformation efficiency of strains (Parr and Hamill 1987). Low transformation efficiency is characterized by the difficulty to establish root cultures and slow growth of the established hairy roots. Irregular T-DNA structure leads to failed transformation, therefore intact T-DNA copies in the plant genome, *rol* genes and auxin biosynthetic genes are key features of a strain required to establish successful hairy root cultures (Parr and Hamill 1987). Growth rates and growth cycle of hairy root cultures can be influenced by manipulating inherent factors, e.g., *rol* genes, which influence cell division in transformed hairy roots (Mehrotra et al. 2015, Vincent et al. 2015). Sterile leaf segments of *Withania somniferum* inoculated with *A. rhizogenes* strains A4, LBA 9402 and LBA 9630 by wounding with sterile needles exhibit different transformation frequencies; 0.9 ± 0.07 and 0.70 ± 0.05 respectively, while LBA 9360 failed to induce transformation and root initiation occurred within 14–17 days with LBA 9402 with A4 giving the best hairy root growth (Banerjee et al. 1994).

Host specificity is determined by the T-DNA structure within the plasmid. Strain A4, with *BamH1* restriction fragments corresponding to those found in carrot roots, transformed *Atropa belladonna* while strain AR8196, with *BamH1* restriction fragments similar to the ones found in *Convolvulus arvensis*, potato and carrot roots better transformed *Calystegia sepium* (Jung and Tepfer 1987). In *Althaea officinalis* the content of secondary metabolites varied among the strains as well, among four different strains A4, A13, ATCC15834 and ATCC15834 GUS. AR15834 GUS strain possesses a higher potential for hairy root development, especially in liquid MS medium in comparison to other strains involved (Nandini et al. 2017). Moreover, major tea crops of Asia and Africa were recalcitrant to *Agrobacterium* infection, due to the bactericidal effects of polyphenols that are exuded by its leaves *in vitro*. These hindrances were overcome by media manipulations and involvement of antioxidants (Tavassoli and Afshar 2018). This indicates that strain and variety selection is important in maximizing biomass accumulation and secondary metabolite production.

Variations in T-DNA copy number and location in the Chromosome also gives rise to distinct clones, as has been reported in protoplast-derived hairy root clones of *Nicotiana rustica* (Furze et al. 1987). Auxin genes coded for by the T_L of T-DNA (*tms* homologous loci) in Ri plasmid is important in root induction, in its absence no transformation occurs as is evident in the failed hairy root induction of *Duboisia myoporoides* (Nepovím et al. 2004).

Plant Variety

Plant variety is important in hairy root culture establishment as the infection rate and transformation efficiency; ability to grow in hormone free medium and ability to adapt to liquid medium are quite dependent on this factor (Bhadra et al. 1993). Two *Catharanthus roseus* plant varieties; flowerless variety and a little delicate normal plant, were investigated for the hairy root transformation efficiency. Transformation was easy in *C. roseus cv* little delicate, a total 30% of tested roots started to grow while that of the flowerless variety only 11% started growing (Toivonen et al. 1989). Leaf, petiole and roots of red beet (*Beta vulgaries*) were infected with *Agrobacterium rhizogenes* strain A4, and many adventitious roots appeared mainly in the cut ends of leaf sections.

Woody plants are difficult to induce hairy roots (Parr and Hamill 1987). Plants produce aromatic compounds that activate the *A. rhizogenes* plasmid to induce transformation. If those compounds are produced in low content or not synthesized at all, transformation does not occur (Toivonen et al. 1989). Some plants can be easily transformed, while others cannot be transformed by a particular strain. Supplying the exogenous Aceto-syringone at desired concentration helps in the efficient transformation process, by inducing vir-genes of the bacterium.

Plant parts used also affects the extent of hairy root transformation, *Agrobacterium rhizogenes LBA 9402* induced root cultures from *Brugmansia candida* sterile seedlings where stems exhibit a high root initiation frequency of 90% as compared to woody Solanaceous plants such as *Duboisia* (Giulietti et al. 1993). *Atropa belladonna* hairy root analysis at a stationary phase reveals increased contents of hydroxyhyoscyamine and scopolamine at various degrees, leaves at two months of *A. belladona* whole plant regenerant contains, on dry weight basis, 0.02% of hyoscyamine, 0.02% of hydroxyhyoscyamine and 0.45% of scopolamine. This further conversion to scopolamine in leaves is observable in other transgenic *A. belladona* plants which show moderate conversion in branch roots, indicating highly efficient conversion of hyoscyamine to scopolamine occurs during translocation of alkaloids from root to aerial plant parts (Hashimoto et al. 1993). *Perezia cuernavacana* explants; leaves, roots and shoots infected with *AR12* did not express hairy roots despite the addition of nopaline, acetosyringone and glucose however hairy roots formed in internodal explants (Arellano et al. 1996). Plant origin is also a factor that influences the extent of hairy root transformation. From hairy roots established with two natural *Chaenactis douglasii* that are morphologically identical but from different populations, in one amongst them doubled thiarubrine content as was compared to the other (Parr et al. 1988). Different alkaloid contents are recorded in various species of *Nicotiana* intact plants (Flores and Filner 1985). This can be used to assume that the hormonal environment of a normal plant is not the same as that of hairy roots. Hairy root cultures can produce their own auxin and there is a level of auxin control by the bacterial genes (Parr et al. 1988). Hairy root cultures have also been revealed to express a biochemical variability between strains initiated from different plants and between strains issued from the same plant (Hamill et al. 1989).

During the transformation process, it is very common that many plants species show the browning effect, which could hinder the normal transformation process, the browning could be attributed to many issues involving recalcitrant plant, oxidation phenolic exudates, pH changes bacterial overgrowth and many others, the important point here to mention is that in addition some amino acids have proven well in reducing browning in these explants. In addition, treating the explant first with ascorbic acid and citrate at desired ratio have minimized the effect, nevertheless the incorporation of some L-form of amino acids have enhanced the root formation ability.

Media and Nutrients

The nutrient composition of growth medium is a key factor in successful induction and establishment of hairy roots. Nutrients in the media like vitamins, carbons source and hormones modify the medium for maximum biomass production (Hamill et al. 1989,

Parr and Hamill 1987). The addition of acetosyringone to media at the optimal concentration increases the efficiency of transformation by attracting the bacteria to the wounded plant, and induces the virulence genes, thus facilitating the transfer of the T-DNA region. Acetosyringone binds to *VirA* protein in the *Agrobacterium* membrane. This activates *VirG*, which in turn switches on the other *vir* genes, including *virD* and *virE2*.

Carbon sources are provided at either full, half, or quarter strength to know the optimal concentration for maximum biomass accumulation (Le Flem-Bonhomme et al. 2004, Parr and Hamill 1987). Commonly used carbon sources in hairy root media are sucrose and glucose added to Gamborg's, Murashinge and Skoog, White's or Nitsch's basal growth medium.

Aljamicine and Serpentine accumulated in *C. roseus* hairy root cultures were found to enhance with increase in sucrose content (Le Flem-Bonhomme et al. 2004). Similarly, scopolamine accumulation in *Datura stramonium* was affected by sucrose content. Addition of sucrose at 30 g/l (w/v) increased scopolamine content from 0.015 to 0.15% in this plant species. Murashige and Skoog medium showed scopolamine accumulation of 0.11% while 0.32% was recorded in Street medium (Hamill et al. 1989). Different concentrations of media and sucrose were investigated for their effects in stimulating hairy root growth and valepotriates production and accumulation in *Centranthus ruber*. Quarter and half strength B5 media stimulated biosynthesis of 7-desisovaleroyl-7-acetylvaltrate (DIA-VAL) in the hairy roots of *C. ruber* whereas it was only detected in small amounts in the non-transformed roots. Isovaltrate (IVAL) content was unaffected by various culture media tested and remained very low. The fastest hairy root growth of 4.2 g was observed with half strength B5-3 and B5-3 medium but the latter medium led to a poor valepotriate content (0.6% dry wt.). In the quarter B5 medium, 2–5% sucrose had no significant effect on growth and the valepotriate content, but 7% sucrose decreased substantially the valepotriate content. Except for B5 medium, dilution of other investigated media to half or quarter strength had no effect on the growth and valepotriate content. The highest concentration of valeopotriates was observed in quarter or half strength B5-3 media and reached 3.0% dry wt. This was the same range as the valeopotriates content of non-transformed roots (3.4% dry wt). Media pH is an important factor in determining the successful establishment of hairy root cultures.

Exogenous Growth Regulators

Exogenous growth regulators in the medium play a significant role in hairy root growth and secondary metabolite accumulation. Auxin as Indole acetic acid was shown to have little or no effect on *Duboisia myoporoides* hairy root cultures, while combination of cytokinin benzyladenine and auxin 2, 4-dichlorophenoxyacetic acid increased cell yield. In *Linum album* hairy root line (LYR2i), addition of indole 3 acetic acid (IAA) induced thicker root tips, however a compact green callus was induced when 2,4-dichlorophenoxy acetic acid (2,4-D) was added. The addition of IAA and 2,4-D to the basal medium, resulted in an increased amount of podophyllotoxin and 6-methoxypodophyllotoxin by 1.86-fold and 1.45-fold as that of control, respectively (Farkya and Bisaria 2008). NAA was investigated for its effect on thiarubrine production in *Crataegus douglassii* hairy root cultures. It was revealed that this auxin reduced thiarubrine accumulation as its level started to degrade. Hairy root cultures produce their own auxin so the addition of

NAA increased the concentration that subsequently induced callus formation (Parr et al. 1988). *Amsonia elliptica* hairy roots grown in B5 containing 0.5 mg/L NAA showed rapid growth doubling fresh weight every four days either in the dark or light (16 hr period/day). Hairy roots in hormone free medium showed faster growth as compared to those in the dark (Sauerwein et al. 1991).

Plant growth and biochemical productivity can be enhanced by using plant bioregulators. These are biochemical compounds added in small quantities in plants to enhance their growth and maximize their productivity. Various concentrations of N,N-dimethylmorpholmium iodide (DMI) adeed to *C. ruber* hairy root culture of 40 days posed an unfavorable effect on the growth of *C. ruber* and strongly reduced valeopotriate production as compared to control,valeopotriate content decreased with increase of DMI (Gränicher et al. 1995).

Incubation Parameters

Light

Normally hairy root cultures are maintained in the dark at 25°C, varying this parameter has effects on growth and productivity of the cultures. When maintained in light, hairy roots developed morphologically distinct plant regenerants. Normal types resembled non-transformed plants and leaves were not wrinkled; the wrinkled type had wrinkled leaves observed in a number of regenerants from hairy roots of a number of plant species and the rooty type had abundantly differentiated adventitious roots from the shoots (Saitou et al. 1991). *A. reptans* var. atropurpea hairy root culture increased rapidly 250 times more in thirty days and reached a stationary phase in comparison to cultures maintained in light. In this condition growth was inhibited and steroid content decreased to one third/fourth, this indicates that production of 20-hydroecdysone is related to growth of roots (Matsumoto et al. 1991). Hairy root cultures of *Amsonia elliptica* grown in darkness grew slowly doubling fresh weight every 14 days, while those cultured in 16 hr light period/day showed rapid growth doubling every 7 days. Under both conditions alkaloid accumulation was slow. 17α-o-methylhimbine detected at 0.3–1.9 ug/flask with pleiocarpamine produced at any time point yield recorded between 1.7 and 13.4 ug/flask. In hairy root cultures maintained in light, vallesiachotamine constituted 138.0 ug/flask (Sauerwein et al. 1991).

Shaking

Beet *A. rhizogenes* (A4) induced hairy roots released pigments into the culture broth upon cessation of culture shaking as supported by diminishing transparency of culture broth and reached a stationary phase after 40–48 hours after shaking ceased. At 48 hours after cessation of culture hairy roots in the early cultures (culture times: 11 and 15 days), about 10% of both pigments in the cells were released into the medium. For hairy roots in the early phase (culture times: 25 days), the intracellular pigments were released into the medium at 21–25% (Taya et al. 1992).

Oxygen Limitation

In static condition, where culture shaking is ceased, pigments were released into the medium. This phenomenon can be explained by dissolved oxygen (DO) levels that rapidly dropped to zero within two hours and gradually increased to 1.3 ppm at 48 hours. The increase in DO is due to lowered oxygen uptake by hairy root cells and diffusion from the gas phase. In jar fermenter cultures to detect effects of DO level on pigment leakage of betanin and vulgaxanthin-I were detected at minimal when DO was controlled at 1 and 6 ppm at 50 hours. On the other hand, at 26 hours maximal pigment concentration was recorded at 2.8 and 6.4 mg/dm3 betanin and vulgaxanthin-I respectively, which corresponded to 63% of total intracellular pigments. Therefore pigment release into the medium was considered to occur when the cells were subjected to oxygen limitation under a very low DO level, which may lead to partial disruption or relaxation of cell membranes (Taya et al. 1992).

Upon investigation of hairy root growth after cessation of shaking for 48 hours it was found there is no difference in root tip elongation. However, cell mass growth was attained in liquid cultures in early cultures; 11 and 15 days although low as compared to the control. Late cultures; 25 days showed no cell growth implying that the cells had become fragile due to oxygen starvation and were sensitive to hydraulic stress (Taya et al. 1992).

Duration of Culture

Time course allocated to culture also affects the performance of culture. *C. ruber* hairy root culture time course of growth and valeopotriate production in half strength B5 medium supplemented with 3% sucrose at 5-day interval in flasks was investigated. Fresh weight of roots increased from original inoculum to 50 mg to reach 4.2 g with a growth index of 84, with maximum biomass reached by 35 days of culture. On a mg/g dry weight basis, the overall valeopotriate content reached 31.4 mg/g, representing a mean accumulation rate of 0.7 mg/g/day. Between day 10 and day 25 the growth rate was maximum and reached 80% of the final fresh weight. VAL (valtrate), DIA-VAL (7-desisovaleroyl-7-acetylvaltrate) and HVAL (7-homovaltrate) were the major constituents of the valeopotriate mixture synthesized by the hairy roots. The VAL content increased rapidly and regularly between the 10th and 45th day of culture to reach 1.5% dry wt. DIA-VAL and HVAL contents increased continually between the 20th and 40th day of culture to reach a constant level of about 0.5% and 0.4% dry wt. respectively. DI, IVHD and IVAL were also detected but their levels remained low (Gränicher et al. 1995). Lack of accumulation of valeopotriates could be due other biosynthetic pathways competing for common precursors or to a very low activity of one or more of the relevant biosynthetic enzymes, during the first 45 days of culture in medium. This suggests that all valeopotriates were retained within the tissues. After 45 days of culture the hairy roots turned brown and died and valeopotriates were released into the medium (Gränicher et al. 1995). *Atropa belladonna* hairy root analysis at the stationary phase reveals increased contents of hydroxy hyoscyamine and scopolamine at various degrees.

All the above-mentioned factors influence hairy root morphology and growth pattern as well as secondary metabolite accumulation. Distinct hairy roots result in varied secondary metabolite production. Thin and slow growing hairy roots record little biomass and consequently low secondary metabolite content accumulates. Secondary metabolites can also be low in the roots due to diminishing intracellular pools. This is a phenomenon whereby secondary metabolites released into the medium continue until the internal ratio of secondary metabolite equates that of external concentration, accumulation ratio is reached (Furze et al. 1987). Thick hairy roots produced higher ajmalicine content as compared to less vigorous roots (Toivonen et al. 1989). After accumulation in roots, secondary metabolites are transported to aerial parts of the plant. In hairy roots, the metabolites are released into the culture medium, therefore metabolites found in the aerial parts of the intact plant can also be noticed in the growth medium (Flores and Filner 1985). However, secondary metabolites released under *in vitro* conditions have not shown any correlation with *in vivo* transportation activity. For instance, some intact plants with high metabolite content in shoots secreted only a small amount in the growth medium. In addition, the pattern of metabolites found in hairy roots mirrors that found in culture medium, but that in roots of intact plant does not correlate the pattern found in aerial parts. Lack of correlation can be because in intact plants the transport system includes long distance and specific uptake mediated transportation along the vascular system coupled with biochemical modifications of the compounds during or after transportation (Flores and Filner 1985).

Chemical Classification of Bioactive Compounds Produced/Over-Produced

Different classification methods are used to analyze secondary metabolites. These techniques are categorized into both qualitative and quantitative analysis. With each technique, parameters such as sample volume, temperature, gas pressure are optimized.

Thin Layer Chromatography

Thin layer chromatography (TLC) is a separation technique based on a multistage distribution of compounds. TLC is a qualitative analysis method, which identifies and determines purity compounds in a sample mixture in a process that involves the stationary and mobile phases for compounds separation. The stationary phase is made up of a thin layer of an adsorbent coating usually silica gel, aluminum oxide or cellulose. Silica is the most used adsorbent material for TLC. Structurally silica gel is a matrix of Si-OH groups which can interact with molecules via hydrogen bonding and adsorption (Namir et al. 2019). Various compounds are carried up the plate at different speeds due to varying interactions with the adsorbent material (Rother et al. 2011). For example, polar molecules with groups such as hydroxy (OH) and amine (HN_2) will tend to form hydrogen bonds with the silica matrix groups of the silica gel and thus move slowly up the plate. Non-polar compounds will have fewer interactions with the matrix and be more soluble in the solvent phase and therefore rise up quickly with the solvent front (Namir et al. 2019, Rother et al. 2011).

The mobile is made up of a mixture of solvents usually ethanol, methanol, distilled water, hexane and ethylacetate at given ratios. The solvent is added into a glass jar in which the TLC plate is immersed, and the solvent ascends or migrate up the plate while separating the compounds. Small volumes, in microliters, are applied to the TLC plate in a process known as spotting. The spotted plate is placed in the solvent jar and the solvent mixture is drawn up via capillary action. Given the structure of the adsorbent and solvent, analytes of the sample mixture migrate at different rates therefore separation is achieved (Balammal and Kumar 2014). Once the solvent has moved or migrated a particular distance sufficient to separate the components of the spot, the plate is either visualized directly using ultraviolet light, usually 215 nm–450 nm. The plate is visualized by projecting ultraviolet light onto the sheet upon which the spot will appear on the sheet where the compounds absorb the light impinging on a certain area. The plate can also be developed by spraying with stains to check for specific types of molecules. Common staining agents used are anisaldehyde, phosphor and sulfuric acid as well as ninhydrin. After staining, the plate is baked in an oven at moderate heat for a short period of time, usually 5 minutes (Namir et al. 2019).

To quantify results, the distance travelled by the sample is divided by the total distance travelled by the mobile phase, a ratio known as retention factor. A substance whose structure resembles that of the stationery phase will have a low retention factor value while one that resembles the mobile phase will have high retention factor value (Namir et al. 2019). Retention factors are characteristic but will change depending on the exact condition of the mobile and stationary phase. For this reason a standard, known compound, is also spotted along with the analytes (Rother et al. 2011). TLC was used to confirm that betanin and vulgaxanthun-1 in the medium are the same as those in the hairy roots of red beet (Taya et al. 1992) and separation against authentic compounds on a precoated silica gel plate allowed for isolation of 17 alkaloids.

Two spots correlating to Scopolamine and Atropine were observed in alkaloid extracts of *Atropa belladonna* hairy root cultures and were comparable to *Atropa belladonna* grown in the field (Kamada et al. 1986).

Thin layer chromatography confirmed that betanin and vulgaxanthin-1 in the medium are the same as those in the hairy roots of red beet (Taya et al. 1992) and separation of plant extract on silica gel against authentic samples allowed for the isolation of 17 alkaloids. Using a pure standard of chlorogenic acid, TLC run on a pre-coated silica gel plate developed by ethyl-acetate-formic acid-glacial acetic acid-water (100:11:11:27), visualized under UV set at 254 nm and treated with ferric chloride reagent, identified hydroxycinnamic acid esters from methanolic extract of freshly dried roots of *Leontopodium alpinum* Cass (Hook 1994). *Swertia japonica* hairy root methanol extract subjected to TLC analysis on a plate developed with C_6H_6-EtOAC revealed the presence of bellidifolin and methylbellidifolin (Ishimaru et al. 1990). *Datura stramonium* and *Hyoscyamus niger* hairy root extract residue dissolved in methanol solution was separated along with hyoscyamine and scopolamine as standards. The TLC plate was developed with 1,1,1-$C_2H_3Cl_3$-$(C_2H_5)_2$-$C_2H_5)_2NH$ (9:1) and dried at 105°C for 2 hours, sprayed with 4-dimethylaminobenzaldehyde dissolved in EtOH-8NH_2SO_4 (1:1) and heated at 105°C for 1 hour and was visualized at 493 nm. *D. stramonium* had only scopolamine while *H. niger* had hyoscyamine (Jaziri et al. 1988). In *A. belladonna* and *C. sepium* hairy root extracts the presence of cuscohygrine, atropine, hyoscyamine and scopolamine was determined by the TLC assay, with draggendorf alkaloid as a standard. Similarly cuscohygrine and atropine, hyoscyamine were extracted from *C. sepium* and

A. belladonna hairy roots respectively (Jung and Tepfer 1987). *Atropa belladonna* alkaloid residue fraction extracted with chloroform dissolved in MeOH and run-on silica gel plate developed with two different solvent systems, EToAC-$_2$PrOH-10% NH_4OH (9:7:3) and $CHCl_3$-acetone-MeOH-28% NH_4OH (75:10:15:2). The spots were detected and identified with draggendorffs reagent to reveal atropine and scopolamine (Kamada et al. 1986). TLC analysis on silica gel of *Lactuca virosa* hairy roots methanol extract yielded 5 guanolide glycosides; Crepidiaside B, picriside A, macrodiniside A, ixerin F, Scorzoside and 3 other unkown compounds also found in the roots on intact plants. The mobile phase of Benzene-EtoAc (50%) followed by $CHCl_3$-MeOH up to 20% developed the plate (Mano et al. 1989). AcOEt layer of the *Cassia* extract was run on a TLC plate precoated with silica gel and developed on benzene-AcOEt eluent. Pinselin, germichrysone, chrysopanol, emodin, physcion, 1,8-di-o-methylchrysophanol and 8-O-methylchrysophanol were identified by direct comparison with authentic samples (KO et al. 1995). *Ajuga reptans var. atopurpurea* hairy root extract separated with TLC on solvent CHCl3:MeOH:H_2O (13:7:4) revealed four phyto-ecdysteroids; 20-hydroxy-ecdysteroids, norcyasterone, cyasterone and isocyasterone in hairy root clones (Matsumoto et al. 1991).

High Performance Liquid Chromatography

This technique is a modern application of liquid chromatography that provides versatility in separating and analyzing organic mixtures of compounds that are volatile and thermally unstable and have relatively high molecular weight (Galant et al. 2015, Klimczak and Gliszczyńska-Świgło 2015). The mobile phase carries the injected sample through a separation column packed with a stationary phase and to the detector, it is in this column that individual compounds are separated based on physicochemical interactions and the elution order is based on this interaction. The separated compounds are detected by the detector based on absorption of light or changes in the refractive index, electrochemical/ conductivity changes or simply the size distribution of eluting (Sahu et al. 2018). The detector, commonly a UV/Vis diode array detector, provides the required sensitivity in the shape of peaks having an area in direct proportion to the amount of the compound present (Brighenti et al. 2017, Lozano-Sánchez et al. 2018). The output is monitored and evaluated by the operating software which not only required calculations on the response but also supports operating parameters like injection volume and sequence, detection wavelength and wash cycles. Due to its high sensitivity, HPLC requires thorough sample clean up and sample handling before injection (Sahu et al. 2018). The mechanism responsible for distribution between phases include surface absorption, ion exchange, relative solubilities and steric effects (Sahu et al. 2018). The column is packed with microscale beads functionalized with chemical groups that as the mixture flows through the column, the component interacts with the stationery phase differently (Galant et al. 2015). Column chromatography happens at a higher flow rate, and therefore higher pressure is required to allow for an increased interaction of the stationary phase and components in the mobile phase. Compounds interact with the stationery phase differently and therefore travel down the length of the column to the detector at a different rate (Galant et al. 2015, Lozano-Sánchez et al. 2018). The time required for a component to exit the column or elute is called retention time. The result is a plot for retention time against intensity or a chromatogram. The retention time is used to identify the component, the peak size is used to quantify the amount of the compound in the initial solution (Lozano-Sánchez et al. 2018).

Caranthine and ajmalicine alkaloids from *Catharanthus roseus* were quantified by HPLC (Toivonen et al. 1989). FC like components from ethanolic extracts of horse radish hairy roots quantified by HPLC fractionation revealed several compounds positive in a competitive radio immune assay (Babakov et al. 1995). HPLC analysis of metabolites from *C. roseus* hairy roots was performed with alkaloidal fractionated by (72:25) mixture of MeOH:5 mM (NH4) 2HPO4 at a flow rate of 0.8 ml/min. The detection wavelength set at 254 nm, and peak identification was based on comparison of retention times of authentic standards of ajmalicine, serpentine, caranthine, vindoline and vinblastine. Alkaloids were identified in the hairy root clones but not in the medium. Vindoline was determined by HPLC by comparison of peak retention times to a standard. Compounds eluting at the same time as the standard medium (Bhadra et al. 1993).

Hairy roots of *Withaneria somniferum* showed that *Withaferin A* was detected at all growth stages though in trace amounts with HPLC analyses using MeOH:H2O (55:45) as the mobile phase with rate (ml/min) with reference compounds *Withaferin A, withcristine* and *iochromolide*. Ten-week-old culture medium, ten-week-old hairy roots and control roots showed different chromatograms depicting profiles in withanolides. Ten-week-old hairy roots and the corresponding medium showed distinguishable peak shapes of *withaferin A*, while the rest of the roots and their medium contained maximum *withaferin A*. This implies concomitant release of withanolide into the medium. The control root extracts exhibited the presence of *withaferin A* in traces while the other two compounds were not identified (Banerjee et al. 1994). HPLC detected low amounts of scopolamine in the range of 1 *ug/g* fresh weight of *B. candida* hairy roots (Giulietti et al. 1993).

The quantification of valeopotiates from *C. ruber* was based on the simultaneous estimation of the monoene and diene derivatives in a single HPLC run. An isocratic methanol-water (69:31) mixture was used as mobile phase at a flow rate of 0.7 ml/min for 10 min then at 1.4 ml/min for 30 min. The detection was performed at 208 nm. VAL(valtrate), DIA-VAL(7-desisovaleroyl-7-acetylvaltrate) and HVAL(7-homovaltrate) were the major constituents of the valeopotriate mixture synthesized by the hairy roots (Gränicher et al. 1995). A wide range of alkaloids were generated in *Brugmansia candida* hairy root clones as per gas chromatography (GC) analysis in which content of alkaloids and the spectrum changed during the culture. Scopolamine was found in levels sufficient to allow detection under the better resolution of GC from in the order of 0.2–0.5 *ug/g* (Giulietti et al. 1993).

Gas Chromatography/Mass Spectroscopy

This is a hybrid tandem technique that couples' the separation technique of gas chromatography with detection properties of mass spectroscopy (MS, also used for structure elucidation) to provide high efficiency of sample analysis. GC/MS covers identification and quantification of a large variety of volatile and non-volatile metabolites (Lelevic et al. 2020). Gas chromatography is a method of choice for chemical analysis and because of its high resolution it provides the highest overall efficiency and performance of all the separation methods and is readily operated in tandem with MS (Lorenzo and Pico 2017). GC/MS has a high resolution open tubular columns with bonded phases that are capable of separating hundreds of volatile constituents in a single run (Lorenzo and Pico 2017, Santos and Galceran 2003). Injection technique and analytical columns' stationary phase profoundly influence the results of GC. Polarity of the stationery phase is the most important parameter and should be matched as closely as possible with the polarity of the analytes (Lelevic et al. 2020). For identification purposes its best to determine retention indices for each analyte on two columns of differing polarity (Santos and Galceran 2003). A mass spectroscopy detector is preferred in analysis of complex volatile mixtures since it offers both qualitative and quantitative information (Mbughuni et al. 2016). GC/MS allows for a mass spectral library matching for the identification of unknown chromatographic peaks where MS helps to fragment the component and identify components on bases of their mass (Lorenzo and Pico 2017, Mbughuni et al. 2016). This technique also provides enhanced range identification, sensitivity, increased analyzable samples and faster results (Mbughuni et al. 2016, Santos and Galceran 2003).

To identify scopolamine and Atropine in the Alkaloid extract in *Datura stramonium* and *Datura metel* sample extracts were subjected to GC-MS, using desired buffer solution for extraction which elucidated the presence of these two compounds in the samples (Bandaranayake and Yoder 2018). For the first time in Algerian *Datura* species the production of 13 potential alkaloids by selected hairy root line was detected (Harfi et al. 2018), homatropine was used as a standard for alkaloids. GC/MS analysis of the alkaloidal methanol extract from *C. roseus* hairy roots was done with temperature maintained at 150°C for 2 minutes, then raised to 300°C for another 10 minutes. This technique was used as an independent method for detection of vindoline and ajmalicine. Sample molecular ion (m/z = 456) of vindoline could not be detected but fragment ions (m/z = 188) and (m/z = 174) were present. The absence of (m/z = 456) was due to a limitation in sensitivity since the sample could only be concentrated to a vindoline level of approximately 0.1–02 mg/ml and the molecular ion for vindoline is fairly weak in the 1mg/ml standard (Bhadra et al. 1993). Analysis of hairy root ethanolic extract of horseradish revealed the presence of fusicoccins A (Babakov et al. 1995).

Infrared Spectrum

This technique is used for structure elucidation and for quality control purposes, infrared light that interacts with molecules and is analyzed in three ways: absorption, emission and reflection (Dutta 2017). When exposed to infrared light in the ranges of 1280 cm^{-1}–10 cm^{-1}, sample molecules selectively absorb radiation of specific wavelengths which causes the change of dipole moment of sample molecules and consequently the vibrational energy levels of sample molecules transfer from ground state to excited state (Lopes et al. 2018). The frequency of absorption peaks is determined by the vibrational gap. Functional groups in molecules are determined by measuring the vibration of atoms (Lopes et al. 2018). With this technique it is possible to identify the functional groups of a molecule by comparing its vibrational frequency on an IR spectrum or an IR stored data bank (Dutta 2017).

This technique can also be applied in the field of quantitative analysis based on Beer-Lambert law. IR spectroscopy monitors the interaction of functional groups in chemical molecules with infrared light resulting in predictable vibrations that provides fingerprint characteristics of chemical or biochemical components in a sample (Bunaciu et al. 2015, Dutta 2017). It provides a sensitive probe for specific functional groups in chemical polymers. Infrared spectroscopy analyzed hexane extract of *Perezia Cuernavaca* hairy root concentrated at reduced pressure (20 mm Hg) to 25% of its original volume. Perezone was catalyzed by quick chilling in an ice bath and dried at a low pressure (5 mmHg) for one hour. The nature of the crystals obtained from the hexane extract from transformed roots were confirmed by IR spectroscopy, the IR spectrum of the sample extracted from transformed roots was identical to those of purified perezone. The spectrum showed characteristic relative positions of peaks at 3285, 1655, 1645, 1620 and 1620 cm^{-1} (Arellano et al. 1996).

Novel Compounds Produced Perturbing Secondary Metabolites Using *A. rhizogenes*

Plant secondary metabolites have a myriad of applications as they are the primary ingredients in the food industry to make food flavoring, pharmaceuticals industry as herbal medicines and synthesis of drug components, cosmetics industry to make fragrances as well as in textile industry to make dyes. *Calystegia sepium* produced cuscphygrine alkaloid with an 11-fold increase of transformed hairy roots as compared to untransformed roots. Alkaloid content production in *Convolvulus sepium* transformed under optimal conditions improved by a factor of 53 as compared to untransformed roots producing 2.3 mg/l/day in the 2 liter fermenter (Jung and Tepfer 1987). Normal *in vitro* cultures of *Atropa belladonna* produced low tropane alkaloid, Scopolamine and atropine contents as compared to the field plant. Scopolamine and Atropine content in hairy root cultures was comparable to that in field plants with recorded quantities of 0.024% and 0.371% dry weight (Yoshikawa and Furuya 1987). In *Duboisia myoporoides* transformed roots, low content of scopolamine and high hyoscyamine content was produced as compared to the untransformed roots. It was suggested that as hairy roots culture is prolonged enzyme activity decreases hence why hyoscyamine could not be effectively transformed to scopolamine. This can be averted by controlling cell division and growth by limiting nutrients uptake and addition of growth regulators and thus increase the scopolamine accumulation (Nepovím et al. 2004). DL-34 clone of *D. myoporoides* cultured in HF medium produced low scopolamine content at the earliest stages but increased during the stationary phase while hyoscyamine remained even low during the two stages. At weeks 4–5 scopolamine increased to 3.3–3.6 mg/l. *Atropa belladonna* transformed roots produced a higher content of atropine and hyoscyamine by 16 fold as compared to the normal roots and production increased by 5.8 fold as compared to the normal roots. Transformed roots produced alkaloids at a rate of 53 mg/l/day in an agitated flask (Jung and Tepfer 1987). *Datura Strymonium* strains showed a marked difference in scopolamine content ranging from 0 to 0.56% while *Hyoscyamus niger* produced hyoscyamine (0.07%) to a level comparable with that of the intact plant (Hamill et al. 1989). *Catharanthus roseus* produced between 1 mg/g–2 mg/g dry weight aljamicine and caranthine which is comparable to content derived from the normal soil grown plant roots. The TLC analysis revealed that the normal roots and *in vitro* hairy roots have qualitatively similar more than 20 alkaloids but different quantities per HN-MR analysis. The hairy root culture HN-MR spectra revealed lochnericine, pericalline, yohimbine and O-acetyl vallesamine (Toivonen et al. 1989). Quinoline alkaloids are important N-based aromatic compounds that have been isolated as far back as 200 years ago. These alkaloids are the bases of anticancer drugs that are being developed. Quinoline alkaloids also have antimalarial, anti-tumour, antibacterial, insecticidal activities hence the significant interest from researchers around the world. Common quinoline alkaloid are quinine isolated from Cinchona tree in 1820 and camptothecin isolated from a Chinese tree *Camptotheca acuminate. Cinchona ledgeriana* hairy root cultures produced quinine, cinchonidine, quinamine and quinidine as major components. Total quinoline alkaloid content in *C. ledgeriana* hairy root cultures was 2 to 3 fold higher than in other dispersed cultures, with the major quinolone alkaloids being quinine and quinodine amounting to 50% and 70% respectively (Parr and Hamill 1987). Phenylpropanoids synthesis was investigated with *Coreopsis tinctoria* and *Coreopsis lanceolata.* Derivatives of 1' hydroxy *C. tinctoria* root culture yielded a ten-fold increase in biomass fresh weight while *C. lanceolata* biomass weight increased by 5 fold. Root *in vitro* cultures accumulated Phenylpropanoids compounds in comparable amounts as the intact plant. The cultures revealed that the occurrence of Phenylpropanoids is closely linked to root tissue differentiation. Saponins were produced in *Panax ginseng* hairy roots such as Gingsenosides Rb and Rg. Hairy roots in hormone free medium produced same content and constituents of saponins similar to those of the callus culture but produced more Saponins when the medium was supplemented with Indole Butyric Acid (Yoshikawa and Furuya 1987). Root alkaloid content in Nicotiana species, compared between soil grown normal roots and *in vitro* hairy roots, varied in secondary metabolite *N. rustica* and *N. tabacum* to principally contain nicotine and anatabine while *N. hesperis* contained principally nicotine and anabasine while those of *N. africana* nicotine and nornicotine. In all the species hairy root cultures had more of the metabolites as compared to the normal plant roots (Constabel and Towers 1988). *Duboisia leichhadartii* hairy roots produced a high content of scopolamine, 47 times more as compared to the parent plant. Clones that did not produce scopolamine produced hyoscyamine (Jaziri et al. 1988).

Peroxidase accumulation was enhanced in hairy root cultures of horseradish. Three hairy root regenerants were established and classified into three types; normal, wrinkled and root type according to their morphology. Peroxidase activity of the wrinkled and normal regenerates was comparable to the wild type plants. While that of the root regenerates was higher two times higher than of the untransformed shoots. Shoots and roots of the root type were three to ten times higher than those of the other types. Isoelectric focusing gel electrophoresis confirmed that there is no difference in peroxidase isoenzyme pattern in all tested plants. The only significant difference was noted between shoots and roots of regenerants and several isoenzymes could be detected specifically in roots/shoots (Saitou et al. 1991). Hairy roots from root explant grew faster than the other types and peroxidase activity increased per flask during HR growth and reached a maximum value of about 30 units/flask after 4 weeks of culture in the rooty type while the normal type took 6 weeks to reach a maximum (Saitou et al. 1991). Hairy root lines were established with phytoecdysone producing plants; *Aja reptans var. atropurpurea, Archyranthes faurei, Pfaffia iresinoides* and *Vitex stricken.* Different clones were isolated and four phyecdysone compounds were detected in the hairy root clones and the proportions of steroids were relative to that of original roots. 20-hydroxyecdysone was the most abundant, most of it retained in the tissues and less than 1 ug/ml was released into the medium. Best clone produced the highest content up to 0.14% which is more than four times that of original roots. Clones that grow rapidly show/contained large amounts of steroids compared to slow growing clones suggesting that steroid production is related to the growth of roots (Matsumoto et al. 1991). Indole alkaloids were isolated from hairy roots of *Amsonia elliptica.* 17α-o-methylyohimbine was only detected in trace amounts, with pleiocarpamine was the main alkaloid produced. Vllesiachotamine was the main alkaloid in light cultured hairy roots (Sauerwein et al. 1991).

Adventitious roots were established with *A. rhizogenes* (A4) to establish axenic pigment producing hairy root culture enhance production of betalains, red violet betacyanin's and yellow betaxanthins. Betanin and vulgaxanthin-I, the two major components of betanins, were released into the culture broth. Betanin and vulgaxanthin-I in hairy roots were produced in contents comparable to the original plant; hairy roots accumulated 2.8–6.1 mg/g dry cells while the original plant accumulated 4.4.–6.3 mg/g dry cells betanins. Vulgaxanthin-I accumulated 5.3–9.3 mg/g dry cells while the original plant accumulated 2.9–3.8 mg/g dry cells (Taya et al. 1992). The highest perezone production in *P. cuernavacana* was observed in the stationery growth phase. The yield of perezone obtained from the stationery growth-phase root culture was 1.3% of the dry weight and yield of perezone from field plants ranges from 2–8% dry weight (Arellano et al. 1996).

Fungus *fusiococcum amygdali* Del. Endogenous fusicoccin (FC) or related substances were found in horseradish (*Armoracia rusticana*) hairy roots. Presence of FC like substances in ethanolic extracts from roots were established in a radioreceptor binding assay with plasmalemmal FC receptors and with radio immune analysis with an anti-serum specific for FC A. FC like ligands were found in the tissue and medium of aseptically grown culture. Quantification of FC related substances in roots of natural plants yielded about 30 nmol kg, which is comparable to the values for the transformed root cultures (Babakov et al. 1995).

Phytochemical analyses with A4 mediated hairy roots live to determine production of withanolides in the hairy roots of *Withania somniferum* as well as in the culture medium at different growth phases. *Withaferin A* was detected at all growth stages though in trace amounts (Banerjee et al. 1994).

Alkaloids were extracted from *C. roseus* hairy roots with methanol and concentrated with (NH4)2HPO4 solution. Vinblastine was not detected. Specific yields of the alkaloids in leaves were comparable to those reported in previous research. Vinblastine was identified in the leaf sample at levels less than 0.1 mg/g dry weight. Four clones had identical ajmalicine yields approximately 1.5 mg/g dry weight while only one clone produced significantly higher levels of ajmalicine of approximately 4 mg/g dry weight. Specific yield of serpentine 98%, carnthine levels of 1–2 mg/g dry weight, vindoline levels of 0.2–0.4 mg/g dry weight were observed (Bhadra et al. 1993). Hairy root clones of *B. candida* generated 20 tropane alkaloids of which the spectrum was dominated by hyoscyamine (56%), scopolamine (11%), tropine (10%). The contribution of 3α-acetoxytropane is low at 2% and the contribution of apo-derivatives (apo-scopolamine and apo-hyoscyamine) is at 3% and 4% respectively. Clones of the hairy root cultures showed a substantial utilization of tropane nucleus in the formation of 3α-acetoxytropane (58%) rather than hyoscyamine and scopolamine (24%). While in some clones tropine is esterified to form normal tropane alkaloids (hyoscyamine and scopolamine:68%) (Giulietti et al. 1993). *Centranthus ruber DC* sterile plantlets were infected with *Agrobacterium rhizogenes* strain R1601. The established roots produced a spectrum of valepotriates which quantitively mirrors that on non-transformed. VAL (valtrate), DIA-VAL (7-desisovaleroyl-7-acetylvaltrate) and HVAL(7-homovaltrate) were the major constituents of the valeopotriate mixture synthesized by the hairy roots. Valtrate and 7-homovaltrate were the major valepotriates of the hairy roots and of the non-transformed roots (Gränicher et al. 1995). The amount of tropane alkaloids varied in clones of transformed Atropa belladonna. In control hairy roots of the wild type hyoscyamine content ranged from 0.05% to 0.3% dry weight. Compounds hydroxy-hyoscyamine and scopolamine were present but in lower levels, each less than 0.1% dry weight. In hairy one clone induced from regenerated plants, hydroxy-hyoscyamine and scopolamine increased significantly and established as the main alkaloids (Hashimoto et al. 1993).

Determination of *Agrobacterium rhizogenes* Transformation

Southern Blot Hybridization of T-DNA

This is a technique employed to determine the presence of T-DNA fragments in the genomic DNA of transformed hairy roots that are homologous to the plasmid DNA. Southern blot analysis involves detecting specific DNA fragments by gel electrophoresis (Green and Sambrook 2019, Tarik et al. 2018). The sample DNA is digested by restriction enzymes such as *BamHI* and *EcORV* and separated on gel by electrophoresis, the fragments are then blotted on a nylon membrane. The membrane is then transferred to an alkali solution with probes, usually purified DNA fragments of the Ri plasmid sequence that are specific to the target DNA sequence on the nylon membrane. Phosphorus labelled [P^{32}] T_R and T_L DNA fragments of the Ri plasmid are used as the probe that will hybridize with the sample DNA to confirm transformation of the plant genome (Mitiouchkina and Dolgov 1998). The presence of T-DNA is interpreted by the presence of bands that are of the same size as the control probes and a DNA marker or ladder is used as a ruler that gives the band sizes in bases (Green and Sambrook 2019, Tarik, et al. 2018).

Opine Detection by High Voltage Paper Electrophoresis

Opines are amino acid derivatives synthesized in *Agrobacterium rhizogenes* transformants by enzymes encoded on the T-DNA of the Ri plasmid (Taya et al. 1992). Opines are detected to confirm root transformation by the Ri plasmid (Yoshikawa and Furuya 1987). Agropine, cumcopine and mannopine are the most commonly detected opines in hairy roots. High voltage paper electrophoresis followed by silver staining is a technique that detects these compounds (Kärkönen et al. 2017). In fresh hairy root cultures opines are very much detectable but are absent in prolonged HRCs and also vary among members of the same clone (Kamada et al. 1986). Absence of opines in HRCs has also been evident in tobacco and carrot.

Visual Observation of Phenotypic Characteristics

Visual observations can also be used to confirm hairy roots: lateral branching, active proliferation and lack of geotropism (Toivonen et al. 1989).

PCR Analysis of insertion of T-DNA

Polymerase chain reaction is a molecular technique that exponentially amplifies small segments of DNA into millions of copies. PCR works with small quantities of reagents therefore it can achieve more sensitive detection and higher levels of amplification of specific sequences in less time than other methods (Navarro et al. 2015). Theoretically PCR amplifies DNA exponentially doubling the template with every cycle, so that the relative difference between samples can be measured as intensity of bands on the gel (Chhalliyil et al. 2020).

PCR is a thermal cycling process, and each cycle includes steps of template denaturation, primer annealing and primer extension. During denaturation the reaction is heated strongly to separate the strands and provide a single strand for the next step. Annealing cools the reaction mixture so that the primer can bind to the complementary sequences on the single stranded template DNA. Extension raises the reaction temperature so that Taq polymerase extends the primer synthesizing new strands of DNA (Ho Huu 2015, Kadri 2019, Levin et al. 2018). This steps or conditions on average can be optimized as follows; initial denaturation 94°C 7 minutes, 30 cycles of denaturation at 94°C for 1 minute, annealing at 55°C for 2 minutes, extension at 72°C for 3 minutes and final extension at 72°C for 10 minutes (Jaziri et al. 1994). The cycle is repeated 20–40 times in a typical PCR reaction which generally takes 2–4 hours depending on the length of the DNA region being copied (Kadri 2019).

DNA template, primers, DNA oligonucleotide bases, DNA polymerase and buffers are the core ingredients of the reaction (Chhalliyil et al. 2020, Ho Huu 2015). Primers are short segments of DNA that are custom designed to complement the DNA template. A pair of primers, forward and reverse, are designed to match the segment of the target DNA through complementary base pairing. DNA polymerase fragments of the same length form a band on the gel. The DNA band contains many copies of the target DNA region (Macao et al. 2015).

Gel electrophoresis is a common technique employed to visualize PCR results. This is a technique in which fragments of DNA are pulled through a gel matrix by an electric current, and it separates DNA fragments according to size. A standard DNA ladder is typically included so that the size of the fragments in the PCR sample can be determined. When DNA polymerase bumps on a primer with a longer piece of DNA it attaches near the end of the primer and starts adding nucleotides. DNA nucleotide bases are the building blocks of DNA that are added by the primers to the template strand to amplify it. The PCR mixture consist of all the types of nucleotides in the mixture (Ho Huu 2015, Levin et al. 2018).

Oligonucleotides ROLB-TL1, ROLB-TR1, ROLB-TL3R and ROLB-TR3R are used to identify the insertion of TL-DNA or TR-DNA of the plasmid into the plants' genome by PCR amplification. The sequence of the oligonucleotides has to respond to the terminal coding regions T-DNA. ROLB-TL1 and ROLB-TR1 have to correspond to the N-terminal coding region of TL-DNA *rol B* gene rolBTR. ROLBTL3R or ROLBTR3R have to correspond to complementary sequences of the C-terminal coding region of TL-DNA *rol B* gene or rolBTR. With the ROLB-TL1 and ROLBTL3R primer combination a band occurring at 670 bp is expected from hairy root clones, while 673 bp is obtained with ROLB-TR1 and ROLB-TR3R Oligonucleotide primers TL-DNA (*rol A-1* and *rol B-2*) and TR-DNA (*Ags-1* and *Ags-2*) for T-DNA of MAFF 03-01724 and ATCC. PCR mixture electrophoresed on agarose gel and visualized by ethidium bromide (Arellano et al. 1996, Hamill et al. 1991).

P. cuervanacava hairy roots were induced in internodal segments upon infection with *Agrobacterium rhizogenes AR12* carrying the GUS gene fused to the *35S CaMV* constitutive promoter in the Ri plasmid T-DNA region. The integration and expression of the GUS marker gene was confirmed by PCR. The size of the amplified DNA fragment correspond to the gus gene sequence according to the primers used (Arellano et al. 1996).

Co-integration of the *rol* genes in the pRi15834 and the engineered genes in the T-DNA of *pHY* into the genome of *A. belladona* produced kanamycin resistant hairy roots having the 35S-H6H chimeric transgene that was analyzed by PCR. Genomic DNA isolated and amplified with two primers that would anneal the first and third exons of the *H. niger* H6H gene (Hashimoto et al. 1993).

A. rhizogenes Transformation has changed the Role of Conventional Plant Tissue Culture in Secondary Metabolites Accumulation and Synthesis

Callus cultures have attempted to produce alkaloids but were unsuccessful as upscaling was not possible. The cell lines only produced high content alkaloids during selection but the content decreased with several subcultures under non-selective conditions (Kamada et al. 1986). Hairy roots however continue to produce atropine and scopolamine under non-selective conditions and prolonged time in the amount higher than or same as of the roots of plants grown in the field (Kamada et al. 1986).

Ever since secondary products production was reported in 1986 from different laboratories (Hamill et al. 1990, Mano et al. 1989, Ryoichi et al. 1989), it was established that hairy roots are able to synthesize and accumulate secondary metabolites (SMs) characteristic of the parent plant. It was also established in *Nicotiana rustica*, that HRs can be cultured for scale up of growth and nicotine production in a fermenter as a batch/continuous-flow system (Rhodes et al. 1986). Since that time this controllable, sustainable and large scalable HRC platform has been immensely utilized for its multi-faceted approaches in the field of research and in pharmaceutical industries such as genetic improvement of medicinal plants, production of a wide range of valuable bioactive SMs at laboratory scale as well as at the commercial level by scale-up using bioreactors, and they provide an easier, efficient and quicker way for extraction, isolation and quantification of phytochemicals which are important for their ethnobotanical and pharmacological properties.

The recent trends on the progress of this technology have been presented in Table 1 (2010–2020) via up-to date research applications in context to the role HRCs which is developed by transformation with wild type strains of *Agrobacterium rhizogenes*, in the production of valuable SMs. Several reports clearly indicate that the hairy root cultures harboring *rol* genes of *A. rhizogenes* offer a potential alternative means of production of different potential pharmaceutical compounds, such as scopolamine, anisodamine and hyoscyamine in *Brugmansia candida* (Cardillo et al. 2010a), scopolamine and hyoscyamine in *Przewalskia tangutica* (Lan and Quan 2010) and in *Hyoscyamus muticus*

Table 1: Secondary metabolites accumulated in cultures transformed with wild type strains of *Agrobacterium rhizogenes* (2010–2020).

Sl. No.	Plant material	Family	Target metabolites	Effect on secondary metabolites in transformed hairy root/plants	References
1	*Brugmansia candida*	Solanaceae	Scopolamine, Anisodamine, Hyoscyamine	HRC's in bioreactor produced the predominant alkaloid Anisodamine to a maximum concentration of 10.05 ± 0.76 mg/g DW.	Cardillo et al. 2010
2	*Arachis hypogaea*	Fabaceae	Resveratrol, Arachidin-1, Arachidin-3	HRCs after elicitation at exponential growth phase showed accumulation of Resveratrol, Arachidin-1 and Arachidin-3 accumulated in the medium.	Condori et al. 2010
3	*Harpagophytum procumbens*	Pedaliaceae	Harpagoside, Verbascoside, Isoverbascoside	Selected hairy root clones produced high amounts of Verbascoside and Isoverbascoside comparable to those found in root tubers.	Grabkowska et al. 2010
4	*Przewalskia tangutica*	Solanaceae	Hyoscyamine, Scopolamine	First report of HRC in *P. tangutica* producing tropane alkaloids.	Lan and Quan 2010
5	*Plumbago indica*	Plumbaginaceae	Plumbagin	Dry root biomass and Plumbagin accumulation were found to be maximum in hairy roots induced in leaf explants infected with *A. rhizogenes* ATCC 15834.	Gangopadhyay et al. 2010
6	*Rubia akane.*	Rubiaceae	Alizarin, Purpurin	Selection of hairy root lines following infection with different strains resulted in high anthraquinone yielding root lines.	Lee et al. 2010
7	*Panax quinquefolium*	Araliaceae	Ginsenosides	The crude Ginsenosides content transformed roots was about 0.2 g/g dry wt.	Mathur et al. 2010
8	*Fagopyrum esculentum* M.	Polygonaceae	Rutin	The content of Rutin (flavonol glycoside) was around 2.4 times more than that of wild type root.	Kim et al. 2010
9	*Glycyrrhiza uralensis*	Fabaceae	Licochalcone A	Licochalcone A and total flavonoid production in hairy roots enhanced by treatment with Tween 20.	Zhang et al. 2011

Table 1 contd. ...

...Table 1 contd.

Sl. No.	Plant material	Family	Target metabolites	Effect on secondary metabolites in transformed hairy root/plants	References
10	*Bacopa monnieri*	Scrophulariaceae	Bacopasaponin D, Bacopasaponin F, Bacopaside II, Bacopaside V, Bacoside A3 and Bacopasaponin C	Ri transformed plants derived from transformed roots showed enhanced accumulation of Bacosides.	Majumdar et al. 2011
11	*Gentiana cruciata*	Gentianaceae	Gentiopicroside	Gentiopicroside, Loganic acid, Swertiamarin and Sweroside were identified in HRCs. Maximum Gentiopicroside (1.08%) in a root clone.	Hayta et al. 2011
12	*Glehnia littoralis*	Apiaceae	Furanocoumarin	Hairy roots produced enhanced Xanthotoxin and Bergapten production than wild type roots.	Terato et al. 2011
13	*Hyoscyamus muticus*	Solanaceae	Tropane alkaloids	Tetraploid hairy root clone could produce more scopolamine than the diploid root clone.	Dehghan et al. 2012
14	*Lithospermum canescens*	Boraginaceae	Shikonin and its derivatives Acetyl shikonin and Isobutyryl shikonin	Shikonin not detected in HRCs. Enhanced accumulation of Acetylshikonin and Isobutyryl Shikonin was achieved.	Syklowska-Baranek, 2012
15	*Clitoria ternatea*	Fabaceae	Pentacyclic triterpenoid compound Taraxerol	Taraxerol yield in HRCs was nearly 4-fold as compared to natural roots.	Swain et al. 2012
16	*Decalepis arayalpathra*	Apocynaceae	2-hydroxy-4-methoxy benzaldehyde (MBALD)	0.22 % dry weight in HRCs	Sudha et al. 2013
17	*Nicotiana tabacum* L.	Solanaceae	Alkaloid Nicotine	Stimulation of branching in HRC resulted in enhanced nicotine yields in the media.	Zhao et al. 2013
18	*Artemisia vulgaris*	Asteraceae	Camphor, Camphene, α-thujone, Germacrene D, 1,8-cineole and β-caryophyllene	HRCs accumulated substantial amount of essential oils as compared to non-transformed roots.	Sujatha et al. 2013

19	*Cichorium intybus* L.	Asteraceae	A new neolignan glucoside; Caffeic acid and its three derivatives; sesquiterpene lactone Crepidiaside A	A new neolignan glucoside was isolated from hairy roots and its structure determined.	Malarz et al. 2013
20	*Dracocephalum moldavica* L.	Lamiaceae	Rosmarinic acid	A transformed root line with significantly higher content of RA than that of roots of field-grown plants of *D. moldavica* was established.	Weremczuk-Jezyna et al. 2013
21	*Tribulus terrestris* L.	Zygophyllaceae	*β-carboline alkaloids; Harmine*	HRCs synthesized secondary products at levels comparable to the wild-type roots	Sharifi et al. 2014
22	*Gentiana scabra*	Gentianaceae	Iridoids, Secoiridoids	Enhanced accumulation of Gentiopicroside, Swertiamarin and Loganic acid content in HRCs	Huang et al. 2014
23	*Polygonum multiflorum* Thunb.	Polygonaceae	Phenolics, flavonoids, anthraquinones (Emodin and Physcion)	Emodin and physcion content was 3.7-fold and 3.5-fold higher in the hairy roots after 20 day cultivation.	Thiruvengadam et al. 2014
24	*Rauvolfia serpentina*	Apocynaceae	Reserpine	High accumulation of reserpine was obtained in two HRC clones.	Ray et al. 2014
25	*Bacopa monnieri* (L.) Wettst.	Plantaginaceae	Bacoside A	Maximum bacoside A' content (10.02 mg g^{-1} DW) was obtained in a high yielding HRC.	Bansal et al. 2014
26	*Plumbago zeylanica*	Plumbaginaceae	Plumbagin	Plumbagin contents in HRCs (4.81 ± 0.16–6.69 ± 0.34 mg/g DW) were higher than that reported earlier.	Basu et al. 2015
27	*Cucumis anguria*	Cucurbitaceae	Phenolic compounds	Higher content of phenolic compounds in HRCs as compared to *in vitro* grown roots.	Yoon et al. 2015
28	*Isatis tinctoria*	Brassicaceae	Flavonoids (Rutin, Neohesperidin, Buddleoside, Liquiritigenin, Quercetin, Isorhamnetin, Kaempferol and Isoliquiritigenin	Under the optimal conditions the total FL accumulation in HRCs achieved was 438.10 μg/g DW.	Gai et al. 2015
29	*Plumbago zeylanica*	Plumbaginaceae	Plumbagin	A4-induced rhizoclone HRA2B5 was identified as the most superior clone with a higher Plumbagin yield.	Nayak et al. 2015

Table 1 contd. ...

...Table 1 contd.

Sl. No.	Plant material	Family	Target metabolites	Effect on secondary metabolites in transformed hairy root/plants	References
30	*Fagopyrum tataricum* Gaertn.	Polygonaceae	Rutin, Anthocyanin	Higher transcript levels for most metabolic pathway genes for the synthesis of Rutin (22.31, 15.48, and 13.04 μg/mg DW, respectively), Cyanidin 3-*O*-glucoside (800, 750, and 650 μg/g DW, respectively), and Cyanidin 3-*O*-rutinoside (2410, 1530, and 1170 μg/g DW, respectively.	Thwe et al. 2016
31	*Arachis hypogea*	Fabaceae	*Trans*-resveratrol	High yielding HRC clones identified	Halder and Jha 2016
32	*Oldenlandia umbellata* L.	Rubiaceae	Purpurin	Purpurin content in hairy roots was about 3.6 times higher than wild growing roots.	Krishnan and Siril 2016
33	*Linumusita tissimum*	Linaceae	Secoisolariciresi-noldiglucoside (SDG), Secoisolariciresi-nl (SECO) and Matairesinol (MAT)	HRCs accumulated SDG, SECO and MAT with a total lignan concentration of 1.057 l mol g^{-1}–1.227 and 1.057 l mol g^{-1}.	Gabr et al. 2016
34	*Lopezia racemosa* Cav	Onagraceae	Triterpenes and campesterol derivative	Identification of the triterpenes Ursolic and Oleanolic acids, and (23*R*)-2α,3β,23,28-tetrahydroxy-14, 15-dehydrocampesterol showing anti-inflammatory and cytotoxic activities.	Moreno-Anzúrez et al. 2017
35	*Sphagneticola calendulacea* (L.) Pruski	Asteraceae	Rutin Caffeic acid Kaempferol Wedelolactone	Accumulation of Wedelolactone in HRCs was enhanced by precursor feeding with PAL	Kundu et al. 2018
36	*Swertia chirayita* (Roxb.) H. Karst	Gentianaceae	Secoiridoids	The Swertiamerin content varied from 0.042 to 0.207% in the transformed root lines.	Samaddar et al. 2019
37	*Plumbago europaea* L.	Plumbaginaceae	Plumbagin	3.2 mg g^{-1} DW Plumbagin accumulation in HRCs.	Beigmohamadi et al. 2020
38	*Echium plantagineum* L.	Boraginaceae	Shikonins, and derivatives	Acetylshikonin content was 36.25 mg/L on average.	Fu et al. 2020
39	*Trachyspermum ammi* L.	Apiaceae	Thymol	Nearly 4.9-fold and 5.3-fold Enhanced accumulation of Thymol accumulation (5.3-fold) in comparison with the untransformed control roots.	Vamenani et al. 2020

(Dehghan et al. 2012), resveratrol, arachidin-1 and arachidin-3 in peanut (Cardillo et al. 2010b), harpagoside, verbascoside and isoverbascoside in *Harpagophytum procumbens* (Grąbkowska et al. 2010), anthraquinone in *Rubia akane* (Lee et al. 2010), rutin in hairy root lines of *Fagopyrum esculentum* (Kim et al. 2010), different bacosides and bacopasaponins in *Bacopa monnieri* (Mazumdar et al. 2011), bacoside A in *Bacopa monnieri* (Bansal et al. 2014), xanthotoxin and bergapten *in Glehnia littoralis* (Terato et al. 2011), acetylshikonin and isobutyrylshikonin in *Lithospermum canescens* (KatarzynaSykłowska-Baranek 2012), taraxerol in *Clitoria ternatea* (Swain et al. 2012), camphor, camphene, α-thujone, germacrene D, 1,8-cineole and β-caryophyllene in *Artemisia vulgaris* (Sujatha et al. 2013), 2-hydroxy-4-methoxy benzaldehyde in *Decalepis arayalpathra* (Sudha et al. 2013), rosmarinic acid A in *Dracocephalum moldavica* (Weremczuk-Jezyna et al. 2013), reserpine in *Rauvolfia serpentina* (Ray et al. 2014), gentiopicroside, swertiamarin and loganic acid in *Gentiana scabra* (Huang et al. 2014), plumbagin in *Plumbago zeylanica* (Basu et al. 2015, Nayak et al. 2015) and *P. europaea* (Beigmohamadi et al. 2020), flavonoids in *Isatis tinctoria* (Gai et al. 2015), phenolic compounds and anthocyanins in *Fagopyrum tataricum* (Thwe et al. 2016), phenolic and flavonoids in *Althea officinalis* (Tavassoli and Safipour Afshar 2018), swertiamerin in *Swertia chirayita* (Samaddar et al. 2019), acetylshikonin in *Echium plantagineum* (Fu et al. 2020, Hu et al. 2020), thymol in *Trachyspermum ammi* (Vamenani et al. 2020), etc. Cardillo et al. (2010) reported successfull production of tropane alkaloids, i.e., scopolamine, anisodamine, hyoscyamine in HRC of *Brugmansia candida* plants in a modified 1.5 L stirred tank. Accumulation of scopolamine and hyoscyamine in HRC of *Przewalskia tangutica*, an important and rare medicinal plant in Tibet Plateau of China, was reported by Lan and Quan (2010). Similarly, higher accumulation of the alkaloid reserpine in selected *LBA 9402* transformed hairy root lines of the endangered species, *Rauvolfia serpentina* was also reported (Ray et al. 2014).

Improvement of resveratrol, arachidin-1 and arachidin-3 production was also reported in peanut HRC by adapting strategies like sodium acetate-mediated elicitation, medium optimization (Condori et al. 2010). Optimization of different factors of genetic transformation of *Harpagophytum procumbens* using *Agrobacterium rhizogenes* strains led to establishment of fast-growing root clones of *H. procumbens*. The growth and anthraquinone (alizarin and purpurin) production in HRCs of *Rubia akane*, one of the Indigenous medicinal and natural dye plants in Korea, was established following infection with five different *A. rhizogenes* (Lee et al. 2010). Enhanced accumulation of an important flavonol glycoside, rutin in hairy root lines of *Fagopyrum esculentum* developed by transformation with *strain15834* was documented (Kim et al. 2010).

Ri transformed plants of *Bacopa monnieri* derived from transformed roots showed enhanced accumulation of Bacosides like Bacopasaponin D, Bacopasaponin F, Bacopaside II, Bacopaside V, Bacoside A3 and Bacopasaponin C (Mazumdar et al. 2011). Bansal et al. (2014) demonstrated the establishment of hairy root lines of *Bacopa monnieri* from leaf explants using different strains of *A. rhizogenes*. Except strain A4 induced hairy roots, all lines showed higher growth rate (maximum 6.8 g l^{-1}) and enhanced production of bacoside A (maximum 10.02 mg g^{-1} DW) than the untransformed root (Bansal et al. 2014).

HRCs of *Glehnia littoralis* showed enhanced production of xanthotoxin and bergapten in comparison with wild type roots (Terato et al. 2011). In an interesting study by Dehghan et al. (2012), tetraploid hairy root lines of *Hyoscyamus muticus* were established to understand the effects of ploidy level and culture medium on the production of tropane alkaloids. Dehgan et al. (2012) concluded that both culture conditions and ploidy level effect growth, alkaloid accumulation and the scopolamine/hyoscyamine ratio in plants and transformed root cultures. Despite its lower biomass production, tetraploid hairy root clone produced more scopolamine than the diploid counterpart under similar growth conditions (Dehghan et al. 2012). Although shikonin was not detected in HRCs of *Lithospermum canescens* irrespective of the transgenic line and culture treatment used, an eightfold increase in acetylshikonin and isobutyrylshikonin accumulation was achieved (KatarzynaSykłowska-Baranek 2012). Swain et al. (2012) reported nearly 4 fold enhanced production of anti-cancerous phytochemical, taraxerol in genetically transformed root cultures of *Clitoria ternatea* as compared to natural roots. Sujatha et al. (2013) reported higher growth rate and accumulation of substantial amount of essential oils in hairy roots of *Artemisia vulgaris* than non-transformed roots. Moreover, 87 compounds were identified from transformed root essential oils instead of 77 compounds from non-transformed root essential oils, among which camphor, camphene, α-thujone, germacrene D, 1,8-cineole and β-caryophyllene were main (Sujatha et al. 2013). HRCs of *Decalepis arayalpathra* showed production of root specific compound, 2-hydroxy-4-methoxy benzaldehyde (Sudha et al. 2013). Production of significant amount of rosmarinic acid A in A4-transformed hairy roots of *Dracocephalum moldavica* was also reported (Weremczuk-Jezyna et al. 2013). Strong effect of the type of culture media on rosmarinic acid A production also noted during this research. A transformed root line with significantly higher content of RA than that of roots of field-grown plants of *D. moldavica* can be possible via medium optimization (Weremczuk-Jeżyna 2013).

An efficient transformation system has been reported for *Digitalis purpurea* L., using *A. rhizogenes* strain LBA 9402 to infect excised leaves (Basu and Jha 2014). The transformed roots grew in absence of phytohormone unlike non-transformed roots and biomass increased 6 fold in 21 days. The root lines were genetically stable and maintained their high growth rate over a period of three years. However, none of the cardinolides, characteristic of the species, were detected in HRCs (Basu and Jha 2014). The production of gentiopicroside, swertiamarin and loganic acid in HRCs of *Gentiana scabra* and the effect of age, physiology and different plant growth regulators were studied by Huang et al. (2014). Samaddar et al. (2019) reported swertiamerin production for the first time in hairy root cultures in a range of 0.042 to 0.207% in N/5 basal medium of *Swertia chirayita*. Moreover, Ri-transformed plants of *Swertia chirayita* showed enhanced accumulation of swertiamerin as compared to non-transformed plants of similar age (Samaddar et al. 2019). Production of plumbagin within the range of 4.81–6.69 mg/g DW in HRCs of *Plumbago zeylanica* was reported (Basu et al. 2015). Nayak et al. also reported plumbagin production in *Agrobacterium rhizogenes*-mediated hairy root cultures of *Plumbago zeylanica*. A4-induced rhizoclone HRA2B5 was identified as the most superior clone with a higher plumbagin yield in comparison with other hairy root clones, *in vitro*-grown non-transformed roots and *in vivo* roots of naturally occurring *P. zeylanica* (Nayak et al. 2015). Accumulation of 3.2 $mg\,g^{-1}$ DW plumbagin in HRCs of *Plumbago europaea* after optimization of medium was reported by Beigmohamadi et al 2020). Hairy root cultures of *Isatis tinctoria* showed production of 438.10 μg/g DW total flavonoids accumulation which constituents of rutin, neohesperidin, buddleoside, liquiritigenin, quercetin, isorhamnetin, kaempferol and isoliquiritigenin (Gai et al. 2015).

Establishment of transformed hairy root cultures of *Fagopyrum tataricum* using different strains of *A. rhizogenes*, quantification of *in vitro* synthesis of phenolic compounds and anthocyanins in these hairy roots and the expression levels of the polypropanoid biosynthetic pathway genes were the main investigation area (Thwe et al. 2016). *R1000* was the most effective bacterial strains for hairy root induction because it induced the highest growth rate, root number, root length, transformation efficiency, and total anthocyanin and rutin content (Thwe et al. 2016).

Among the different hairy root lines of *Althea officinalis* induced using a different strain of *A. rhizogenes* and different explants, the highest total phenolic and flavonoids content was found at 1.57 ± 0.1 mg g^{-1} DW in HRCs (Tavassoli and Safipour Afshar 2018). The hairy root lines of *Echium plantagineum* which were established using *Agrobacterium rhizogenes* strain ATCC15834 showed the production of nine shikonin related compounds and a two fold higher acetylshikonin accumulation in HRs when cultured in the $^{1}/_{2}$B5 medium in compared with in the M9 medium (Fu et al. 2020). The high potential of *Trachyspermum ammi* hairy roots for the biosynthesis of a bioactive compound thymol was recently reported by Vamenani et al. (2020). Secondary metabolite accumulation in HRCs can be increased by selection of variants *in vitro* hairy root morphology affects Secondary metabolite accumulation (Hamill et al. 1991, Kamada et al. 1986). Different clones show marked differences in growth and metabolite accumulation as was noted with different clones of *D. leichdaartii* (Jaziri et al. 1988). Biosynthetic capacity of a natural plant can be predicted through the hairy root culture technique where differences in secondary metabolite biosynthesis among plant species can be attributed to differences in the biosynthetic capacities of the parent plants (Toivonen et al. 1989). Root alkaloid content in Nicotiana species, compared between soil grown normal roots and *in vitro* hairy roots, varied in secondary metabolite *N. rustica* and *N. tabacum* to principally contain nicotine and anatabine while *N. hesperis* contained principally nicotine and anabasine while those of *N. africana* nicotine and nornicotine. In all the species hairy root cultures had more of the metabolites as compared to the normal plant roots (Flores and Filner 1985). High secondary metabolite accumulation in hairy roots as compared to normal roots can be explained by the following: root synthesised metabolites like alkaloids can be transported to the shoots in intact plants. Although hairy roots can release these alkaloids in the medium this may provide a weaker sink than the shoot tissues in plants. The transport process is interfered with, and hairy roots end up accumulating the metabolites. The growth pattern: lateral branching fasting growth of hairy roots can also possibly explain the extensive accumulation of metabolites. Age of culture can also be a factor that explains secondary metabolite content in hairy roots (Flores and Filner 1985).

Regeneration of Hairy Roots from Callus Cultures and Protoplast Colony Suspension Culture

Regeneration capacity in Plants depends on their totipotency, by which somatic cells inherit information and development of the whole plant in specific conditions. Differentiation of meristem occurs and gain structure and functions of specialized organs, for example roots, leaves, stem, etc. Plant tissue culture clubbed with genetic engineering more specifically transformation has not only improved but has opened a new field for the safe production of bioactive compounds in higher amounts, a much-needed alternative to exploitation of a plants natural habitat. *Berberis aristata* an important is the source of berberin, natural alkaloid which is a principle active compound found in this plant. The amount of berberin present in plant's root is higher than is found in its bark (Andola et al.

2010). *B. aristata* has a long history for being used in the treatment of various ailments including diabetes, jaundice, malarial fever; provide relief in eye and ear infection, wound healing, gastrointestinal disorders, and additionally to acting as an important source for treating rheumatism, various skin diseases, anticancer, diuretic, stomachic, and anti-convulsive (Rahman and Ansari 1983). Moreover, it has also been used in textile dye industry (Semwal et al. 2009). In *B. aristate* callus culture the transformation efficiency has been reported higher than from the leaf explant. Fifty days old mature callus culture has been used for transformation with *Agrobacterium rhizogenes* strain 532 for the regeneration of hairy roots to produce bioactive compounds. Callus culture was treated with bacterial suspension for one hour. It was reported that the hairy roots were more efficiently regenerated from callus than any other explant varying from low efficiency (leaves) to inefficient (stem) (Brijwal and Tamata 2015). *Capsicum annum* hairy roots developed the callus due to the active expression of rol genes responsible for the synthesis of endogeneous auxin and cytokinins (Md et al. 2014). In addition, hairy roots regenerated from callus based on secondary metabolites bears much importance (Bulgakov et al. 2008). Lioshina and Bulko reported plant regeneration from hairy roots and callus in periwinkle and foxgloves. The shoots formation occurred after the roots were illuminated with light and some percentage of callus was also reported which after subculturing in light conditions also lead to the shoot formation with low efficiency on B5 medium supplemented with phytohormones (Liosshina and Bulko 2014). However, the plants regenerated from transformed explants usually show phenotype differing from its parent plant typically short internodes, small leaves and sometimes curved (Ooms et al. 1985). Moreover, not only does the presence or absence and low or high amounts of secondary metabolites produced vary in root cultures from callus and but also the growth rate of hairy roots induced. *Ammi majus* known for the source of umbelliferon and furanocoumarins showed that in callus cultures and cell suspensions furanocoumarins was not detected while in hairy roots both compounds were highly synthesized. The growth rate was thirty times higher in hairy roots as compared to callus or cell suspension cultures (Królicka et al. 2001). The regeneration of hairy roots from callus produced after transformation at or near the infection site in *Alhagi pseudoalhagi.* The callus grew very slowly on callus induction medium and resulted in shooting on the regeneration medium. The hairy roots were produced on the rooting media without any such phenotype alteration to the parent plant, which can fix nitrogen more efficiently and to more extent resist drought (Wang et al. 2001). Hairy root cultures are genetically stable; the cell wall in hairy roots is mostly resistant to cell wall degrading enzymes. *Nicotiana rustica* hairy roots showed resistance to cell wall degrading enzymes and very few protoplasts were produced (Cardarelli et al. 1987). The alternative method for isolation of protoplast from callus rather than root cells and regeneration of hairy root from protoplast suspension was first described by Furze et al. 1987, however there are reports for protoplast isolation directly from hairy roots, but the yield has been low. The hairy root tips used for callus induction in *Hyoscyamus muticus* in MS medium supplemented with phytohormones and used for protoplast isolation. Hairy roots were produced from protoplasts which differ in growth rate and morphology. The potential of hairy root protoplast culture can be used for the isolation of different variants (Robins et al. 1987) and for the easy gene delivery into the single cell *in vitro* (Paszkowski et al. 1984). Regeneration of single cell derived may provide a beneficial insight towards the capability for the altered production of secondary metabolites.

Up-Scaling-HRCs in Bioreactor Technology

Hairy root culture systems are genetically stable and therefore are suitable for fermentation to up-scale biomass accumulation and secondary metabolite accumulation (Jung and Tepfer 1987). Establishment and selection of clones serves to isolate highly productive hairy root clones that would successfully transfer to bioreactors to scale it up to production. Callus culture have failed to up-scale to fermentation while selected cell lines can only produce alkaloids in high contents during selection time and the quantity decreased with several subcultures (Kamada et al. 1986). Hairy roots on the other hand produce high content of secondary metabolite which is kept constant during selection and also with several subcultures (Kamada et al. 1986). For large scale production of phytochemicals synthesized in hairy roots, different bioreactor combinations have been used for the scale-up process. Bioreactors for culturing of hairy roots have been widely classified into three main types: Gas phase, liquid phase and Hybrid reactors (combination of both) (Kim et al. 2002). Moreover, bioreactors can also be classified as agitated and bed reactors (Eibl and Eibl 2002). In liquid phase reactors the biomass is completely submerged into the medium, while as, in Gas phase root biomass is not submerged into the media but are exposed to air and media both. The supply of liquid media in these reactors is essential for the production of phytochemicals, e.g., Nutrient mist reactor, droplet reactors, etc. In the liquid phase reactors the gaseous exchange is rate limiting, liquid phase reactors include airlift bioreactor, bubble column, stirred tank bioreactors, etc.

Studies have shown that when excised from the parent hairy roots can be successfully cultured in large scale fermenters. Large scale biomass production needs to optimize conditions to maximize accumulation (Jung and Tepfer 1987). Given an increase in sugar transformed roots will grow fast and result in increased accumulation. Fast growing hairy root clones contain large amounts secondary metabolites as compared to the slow growing clones. This can be used to assume that steroid production is related to growth of roots. Hence given the case, time course of growth and secondary metabolite production can be examined by measuring dry weight at given time intervals. From the initial inoculum, fresh weight of the hairy roots should increase and enter a stationery phase (Matsumoto et al. 1991). However, Srivastava and Srivastava reported the comparison of hairy root growth from *Azadirachta indica* for the production of Azadirachtin in different liquid phase bioreactors. The high shear stress caused by impellers of stirred tank reactor resulted in no hairy roots and high phenolic content after 25 days (Srivastava and Srivastava 2013). Given that hairy roots can maintain their genetic stability throughout their growth period and during product formation it is therefore possible to use fermentation techniques to scale up hairy root cultures and enhance root biomass accumulation and productivity. Conditions like volume of media to be added, aeration, temperature, speed of agitation are taken into consideration and optimized to maximize the fermentation process (Jung and Tepfer 1987). Agitation of transformed hairy root cultures increased biomass accumulation by a factor of 4 in *C. sepium* and 24 in *A. belladonna.* Optimal growth was observed when roots are not abased against the wall. The production of artemisinin from hairy roots of *Artemisia annua* had been established using Stirred tank bioreactor, resulted in the production of 10.3 mg/L of artemisinin using MeJa as an elicitor (Patra et al. 2014). Various bioreactor designs have been reported for hairy roots, such as STR, BCR, ALR, TBR and NMR type have been successfully tested. The production location, whether intra or extra cellular, affects the design of the reactor. The tendency of formation of clumps among primary roots and lateral roots hinders the culturing of hairy roots in large volumes, irrespective of bioreactor type. The rheological properties of hairy roots vary from species to species, and among the clones, thus it's very uneasy to select the one best bioreactor type for hairy root cultures. The sensitivity of the root system, against the shear stress vary from plant to plant, needs to be minimized which highly depends on the sensitivity of root system. The hairy root cultures of *Catharanthus roseus* in an STR type bioreactor marked the essence of mesh isolated impeller system (Nuutila et al. 1994), whereas the hairy root culturing of *Catharanthus trichophyllus* hairy roots were successfully obtained in simple STR reactor, where 10 g FW was inoculated into the bioreactor and after 9.5 weeks of culturing, 2500 g FW was harvested (Mishra and Ranjan 2008). These studies suggest that STR could be used for the scaling up of hairy root cultivation.

Airlift fermenter was exploited in a preliminary scale up experiment. Liquid MS was the growth medium and supported by poly urethane foam. After 45 days of culture the fresh weight of final biomass was 210 g. The growth ratio was 230 fold comparable to that of flask culture. The content of 20-hydroxyecdysone in the final biomass was 0.12% by dry weight basis. More physiological conditions need to be explored to enhance steroid productivity (Matsumoto et al. 1991).

Root morphology is an important parameter in the design and operation of large-scale hairy root culturein bioreactors. While most bioreactors have been reported for roots in liquid nutrients, a mist bioreactor for betacyanin production from hairy roots may enable exploitation of root morphologies that are not suited for liquid culture. Different morphologies observed among clones of the same cultivar perhaps highlight the randomness of the processing and integration of the Ri plasmid T-DNA into the plant genomic DNA. Understanding the role and function of *A. rhizogenes* affecting root growth may allow one eventually to genetically engineer optimal growth characteristics of hairy roots (Bhadra et al. 1993).

Conclusions

Hairy root culture system continues to provide a possibility to produce and discover secondary metabolites from different plants. There are available techniques, some are recent and are improved to provide high sensitivity to detect and characterize these metabolites. Stability of hairy root cultures allows for up-scaling production to suit industrial application; this is achieved by designing bioreactor systems that will allow for the enhancement in yield of hairy root biomass thus increasing by-product accumulation. However, there is still an open road ahead to explore hairy root culture aspects to further optimize the production and accumulation of secondary metabolites.

References

Akula, R., and G.A. Ravishankar. 2011. Influence of abiotic stress signals on secondary metabolites in plants. Plant Signal. Behav. 6: 1720–1731.

Alsoufi, A.S.M., C. Pączkowskl, A. Szakiel, and M. Długosz. 2019. Effect of jasmonic acid and chitosan on triterpenoid production in *Calendula officinalis* hairy root cultures. Phytochem. Lett. 31: 5–11.

Ambros, P., M. Matzke, and A. Matzke. 1986. Detection of a 17 kb unique sequence (T-DNA) in plant chromosomes by *in situ* hybridization. Chromosoma 94: 11–18.

Andola, H.C., R.S. Rawal, M.S.M. Rawat, I.D. Bhatt, and V.K. Purohit. 2010. Analysis of berberine content using HPTLC fingerprinting of root and bark of three Himalayan Berberis species. Asian J. Biotechnol. 2: 239–245.

Arellano, J., F. Vázquez, T. Villegas, and G. Hernández. 1996. Establishment of transformed root cultures of *Perezia cuernavacana* producing the sesquiterpene quinone perezone. Plant Cell Rep. 15: 455–458.

Babakov, A.V., L.M. Bartova, I.L. Dridze, A.N. Maisuryan, G.U. Margulis, R.R. Oganian, V.D. Voblikova, and G.S. Muromtsev. 1995. Culture of transformed horseradish roots as a source of fusicoccin-like ligands. J. Plant Growth Regul. 14: 163–167.

Balammal, G., and S. Kumar. 2014. A review on basic chromatographic techniques. Indian J. Pharm. Sci. 4: 221–238.

Bandaranayake, P.C., and J.I. Yoder. 2018. Factors affecting the efficiency of *Rhizobium rhizogenes* root transformation of the root parasitic plant *Triphysaria versicolor* and its host *Arabidopsis thaliana*. Plant Methods. 14: 1–9.

Banerjee, S., A. Naqvi, S. Mandal, and P. Ahuja. 1994. Transformation of *Withania somnifera* (L.) Dunal by *Agrobacterium rhizogenes*: Infectivity and phytochemical studies. Phytother. Res. 8: 452–455.

Bansal, M., A. Kumar, and M. Sudhakara Reddy. 2014. Influence of *Agrobacterium rhizogenes* strains on hairy root induction and 'bacoside A' production from *Bacopa monnieri* (L.) Wettst. Acta Physiol. Plant. 36: 2793–2801.

Bartwal, A., R. Mall, P. Lohani, S. Guru, and S. Arora. 2013. Role of secondary metabolites and brassinosteroids in plant defense against environmental stresses. J. Plant Growth Regul. 32: 216–232.

Basile, A., M. Fambrini, and C. Pugliesi. 2017. The vascular plants: open system of growth. Dev. Genes Evol. 227: 129–157.

Basu, A., and S. Jha. 2014. Genetic transformation of *Digitalis purpurea* L. by *Agrobacterium rhizogenes*. J. Bot. Soc. Bengal 68(2): 89–93.

Basu, A., R.K. Joshi, and S. Jha. 2015. Genetic transformation of *Plumbago zeylanica* with *Agrobacterium rhizogenes* strain LBA 9402 and characterization of transformed root lines. Plant Tissue Cult. Biotech. 25(1): 21–35.

Beigmohamadi, M., A. Movafeghi, S. Jafari, and A. Sharafi. 2020. Potential of the genetically transformed root cultures of *Plumbago europaea* for biomass and plumbagin production. Biotechnol Prog. 36(2): e2905. doi: 10.1002/btpr.2905.

Bettini, P., S. Michelotti, D. Bindi, R. Giannini, M. Capuana, and M. Buiatti. 2003. Pleiotropic effect of the insertion of the *Agrobacterium rhizogenes rolD* gene in tomato (*Lycopersicon esculentum* Mill.). Theor. Appl. Genet. 107: 831–836.

Bhadra, R., S. Vani, and J.V. Shanks. 1993. Production of indole alkaloids by selected hairy root lines of *Catharanthus roseus*. Biotechnol Bioeng. 41: 581–592.

Biswas, T., and A. Mathur. 2017. Plant Anthocyanins: Biosynthesis, bioactivity and *in vitro* production from tissue cultures. Adv. in Biotech. Microbiol. 5(5): 555672.

Bonhomme, V., D. Laurain-Mattar, J. Lacoux, M.A. Fliniaux, and A. Jacquin-Dubreuil. 2000. Tropane alkaloid production by hairy roots of Atropa belladonna obtained after transformation with *Agrobacterium rhizogenes* 15834 and *Agrobacterium tumefaciens* containing *rolA, B, C* genes only. J. Biotechnol. 81: 151–158.

Brighenti, V., F. Pellati, M. Steinbach, D. Maran, and S. Benvenuti. 2017. Development of a new extraction technique and HPLC method for the analysis of non-psychoactive cannabinoids in fibre-type *Cannabis sativa* L. (hemp). J. Pharm Biomed. 143: 228–236.

Brijwal, L., and S. Tamta. 2015. *Agrobacterium rhizogenes* mediated hairy root induction in endangered *Berberis aristata* DC. Springer Plus 4: 443. https://doi.org/10.1186/s40064-015-1222-1.

Bulgakov, V.P., M.V. Khodakovskaya, N.V. Labetskaya, G.K. Chernoded, and Y.N. Zhuravlev. 1998. The impact of plant *rolC* oncogene on ginsenoside production by ginseng hairy root cultures. Phytochem. 49: 1929–1934.

Bulgakov, V.P., D.L. Aminin, Y.N. Shkryl, T.Y. Gorpenchenko, G.N. Veremeichik, P.S. Dmitrenok, and Y.N. Zhuravlev. 2008. Suppression of reactive oxygen species and enhanced stress tolerance in *Rubica cordifolia* cells expressing the *rol C* oncogene. Mol. Plant Microbe. Interact. 21: 1561–1570.

Bunaciu, A.A., H.Y. Aboul-Enein, and S. Fleschin. 2015. Vibrational spectroscopy in clinical analysis. Appl. Spectrosc. Rev. 50: 176–191.

Capone, I., L. Spano, M. Cardarelli, D. Bellincampi, A. Petit, and P. Costantino. 1989. Induction and growth properties of carrot roots with different complements of *Agrobacterium rhizogenes* T-DNA. Plant Mol. Biol. 13: 43–52.

Cardarelli, M., D. Mariotti, M. Pomponi, L. Spano, I. Capone, and P. Costantino. 1987. *Agrobacterium rhizogenes* T-DNA genes capable of inducing hairy root phenotype. Mol. Gen. Genet. 209: 475–480.

Cardillo, A.B., A.Á.M. Otálvaro, V.D. Busto, J.R. Talou, L.M.E. Velásquez, and A.M. Giulietti. 2010. Scopolamine, anisodamine and hyoscyamine production by *Brugmansia candida* hairy root cultures in bioreactors. Process Biochem. 45(9): 1577–1581.

Chandra, S. 2012. Natural plant genetic engineer *Agrobacterium rhizogenes*: Role of T-DNA in plant secondary metabolism. Biotechnol. Lett. 34: 407–415.

Chhalliyil, P., H. Ilves, S.A. Kazakov, S.J. Howard, B.H. Johnston, and J. Fagan. 2020. A real-time quantitative PCR method specific for detection and quantification of the first commercialized genome-edited plant. Foods 9: 1245.

Chilton, M.D., D.A. Tepfer, A. Petit, C. David, F. Casse-Delbart, and J. Tempé. 1982. *Agrobacterium rhizogenes* inserts T-DNA into the genomes of the host plant root cells. Nature. 295: 432–434.

Clément, B., S. Pollmann, E. Weiler, W.E. Urbanczyk, and L. Otten. 2006. *The Agrobacterium vitis* T-6b oncoprotein induces auxin-independent cell expansion in tobacco. Plant J. 45: 1017–27.

Condori, J., G. Sivakumar, J. Hubstenberger, M.C. Dolan, V.S. Sobolev, and F. Medina-Bolivar. 2010. Induced biosynthesis of resveratrol and the prenylated stilbenoids arachidin-1 and arachidin-3 in hairy root cultures of peanut: Effects of culture medium and growth stage. Plant Physiol. Biochem. 48(5): 310–8.

Constabel, C., and G. Towers. 1988. Thiarubrine accumulation in hairy root cultures of *Chaenactis douglasii*. J. Plant Physiol. 133: 67–72.

Dapar, M.L.G., G.J.D Alejandro, U. Meve, and S. Liede-Schumann. 2020. Quantitative ethnopharmacological documentation and molecular confirmation of medicinal plants used by the Manobo tribe of Agusan del Sur, Philippines. J. Ethnobiol. Ethnomedicine 16: 14. https://doi.org/10.1186/s13002-020-00363-7.

David, C., M.D. Chilton, and J. Tempé. 1984. Conservation of T-DNA in plants regenerated from hairy root cultures. Nat. Biotechnol. 2: 73–76.

Dehghan, E., S.T. Häkkinen, K.M. Oksman-Caldentey, and A.F. Shahriari. 2012. Production of tropane alkaloids in diploid and tetraploid plants and *in vitro* hairy root cultures of Egyptian henbane (*Hyoscyamus muticus* L.). Plant Cell Tiss Organ Cult. 110: 35–44.

Delbarre, A., P. Muller, V. Imhoff, H. Barbier-Brygoo, C. Maurel, N. Leblanc, C. Perrot-Rechenmann, and J. Guern. 1994. The *rolB* gene of *Agrobacterium rhizogenes* does not increase the auxin sensitivity of tobacco protoplasts by modifying the intracellular auxin concentration. Plant Physiol. 105: 563–569.

Deno, H., H. Yamagata, T. Emoto, T.Y. Yoshioka, Y. Yamada, and Y. Fujita. 1987. Scopolamine production by root cultures of *Duboisia myoporoides*: II. Establishment of a hairy root culture by infection with *Agrobacterium rhizogenes*. J. Plant Physiol. 131: 315–323.

Dutta, A. 2017. Fourier transform infrared spectroscopy. Spectroscopic Methods for Nanomaterials Characterization, 73–93.

Eibl, R., and D. Eibl. 2002. Bioreactors for plant cell and tissue cultures. Plant Biotechnol Transgen Plants. Marcel Dekker, Inc. 163–199.

Estruch, J.J., J. Schell, and A. Spena. 1991. The protein encoded by the *rolB* plant oncogene hydrolyses indole glucosides. EMBO J. 10: 3125–3128.

Farkya, S., and V.S. Bisaria. 2008. Exogenous hormones affecting morphology and biosynthetic potential of hairy root line (LYR2i) of *Linum album*. J. Biosci. Bioeng. 105: 140–146.

Flores, H., and P. Filner. 1985. Metabolic relationships of putrescine, GABA and alkaloids in cell and root cultures of Solanaceae. In Primary and secondary metabolism of plant cell cultures. Springer 174–185.

Fu, J.Y., H. Zhao, J.X. Bao, Z.L. Wen, R.J. Fang, A. Fazal, M.K. Yang, B. Liu, T.M. Yin, Y.J.. Pang, G.H. Lu, J.L. Qi, and Y.H. Yang. 2020. Establishment of the hairy root culture of *Echium plantagineum* L. and its shikonin production. 3 Biotech. 10(10): 429. doi: 10.1007/s13205-020-02419-7.

Furze, J.M., J.D. Hamill, A.J. Parr, R.J. Robins, and M.J.C. Rhodes. 1987. Variations in morphology and nicotine alkaloid accumulation in protoplast-derived hairy root cultures of *Nicotiana rustica*. J. Plant Physiol. 131(3-4): 237–246. doi:10.1016/s0176-1617(87)80163-3.

Gabr, A.M., H.B. Mabrok, K.Z. Ghanem, M. Blaut, and I. Smetanska. 2016. Lignan accumulation in callus and Agrobacterium rhizogenes- mediated hairy root cultures of flax (*Linum usitatissimum*). Plant Cell, Tissue and Organ Culture (PCTOC) 126(2): 255–267.

Gai, Q.Y., J. Jiao, M. Luo, Z.F. Wei, Y.G. Zu, E. Ma, and Y.J. Fu. 2015. Establishment of hairy root cultures by *Agrobacterium rhizogenes* mediated transformation of *Isatis tinctoria* L. for the efficient production of flavonoids and evaluation of antioxidant activities. PloS one 10(3): e0119022.

Galant, A., R. Kaufman, and J. Wilson. 2015. Glucose: Detection and analysis. Food Chem. 188: 149–160.

Gangopadhyay, M., S. Dewanjee, S. Bhattacharyya, and S. Bhattacharya. 2010. Effect of different strains of *Agrobacterium rhizogenes* and nature of explants on *Plumbago indica* hairy root culture with special emphasis on root biomass and plumbagin production. Nat. Prod. Commun. 5(12): 1913–1916.

Giulietti, A., A. Parr, and M. Rhodes. 1993. Tropane alkaloid production in transformed root cultures of *Brugmansia candida*. Planta Med. 59: 428–431.

Grąbkowska, R., A. Królicka, W. Mielicki, M. Wielanek, and H. Wysokińska. 2010. Genetic transformation of *Harpagophytum procumbens* by *Agrobacterium rhizogenes*: Iridoid and phenylethanoid glycoside accumulation in hairy root cultures. Acta Physiol Plant. 32: 665–673.

Gränicher, F., P. Christen, and I. Kapétanidis. 1995. Production of valepotriates by hairy root cultures of Centranthus ruber DC. Plant Cell Rep. 14: 294–298.

Green, M.R., and J. Sambrook. 2019. Analysis of DNA by agarose gel electrophoresis. Cold Spring Harbor Protocols (1), pp. pdb-top100388.

Grichko, V.P., and B.R. Glick. 2001. Flooding tolerance of transgenic tomato plants expressing the bacterial enzyme ACC deaminase controlledby the 35S, rolD or PRB-1b promoter. Plant Physiol. Biochem. 39(1): 19–25. ISSN 0981-9428. https://doi.org/10.1016/S0981-9428(00)01217-1.

Grishchenko, O.V., K.V. Kiselev, G.K., Tchernoded, S.A. Fedoreyev, M.V. Veselova, V.P. Bulgakov, and Y.N. Zhuravlev. 2016. *RolB* gene-induced production of isoflavonoids in transformed *Maackia amurensis* cells. Appl. Microbiol. Biotechnol. 100: 7479–7489. https://doi.org/10.1007/s00253-016-7483-y.

György, Z., A. Tolonen, M. Pakonen, P. Neubauer, and A. Hohtola. 2004. Enhancement of the production of cinnamyl glycosides in CCA cultures of *Rhodiola rosea* through biotransformation of cinnamyl alcohol. Plant Sci. 166: 229–236.

Häkkinen, S.T., E. Moyano, R.M. Cusidó, and K.-M. Oksman-Caldentey. 2016. Exploring the metabolic stability of engineered hairy roots after 16 years maintenance. Front. Plant Sci. 7: 1486.

Halder, M., and S. Jha. 2016. Enhanced trans-resveratrol production in genetically transformed root cultures of Peanut (*Arachis hypogaea* L.). Plant Cell Tiss Organ Cult. 124(3): 555–572.

Hamill, J.D., A.J. Parr, R.J. Robins, and M.J. Rhodes. 1986. Secondary product formation by cultures of *Beta vulgaris* and *Nicotiana rustica* transformed with *Agrobacterium rhizogenes*. Plant Cell Rep. 5(2): 111–114 https://doi.org/10.1007/BF00269247.

Hamill, J.D., R.J. Robins, and M.J. Rhodes. 1989. Alkaloid production by transformed root cultures of *Cinchona ledgeriana*. Planta Med. 55: 354–357.

Hamill, J.D., R.J. Robins, A.J. Parr, D.M. Evans, J.M. Furze, and M.J.C. Rhodes. 1990. Over-expressing a yeast ornithine decarboxylase gene in transgenic roots of *Nicotiana rustica* can lead to enhanced nicotine accumulation. Plant Mol. Biol. 15: 27–38. https://doi.org/10.1007/BF00017721.

Hamill, J.D., S. Rounsley, A. Spencer, G. Todd, and M.J. Rhodes. 1991. The use of the polymerase chain reaction in plant transformation studies. Plant Cell Rep. 10(5): 221–224.

Handa, T., T. Sugimura, E. Kato, H. Kamada, and K. Takayanagi. 1994. Genetic transformation of *Eustoma grandiflorum* with rol genes. Genetic Improvement of Horticultural Crops by Biotechnology 392: 209–218.

Hansen, G., D. Vaubert, D. Clerot, and J. Brevet. 1997. Wound-inducible and organ-specific expression of ORF13 from *Agrobacterium rhizogenes* 8196 T-DNA in transgenic tobacco plants. Mol. Gen. Genet. 254(3): 337–343.

Harfi, B., L. Khelifi, M. Khelifi-Slaoui, C. Assaf-Ducrocq, and E. Gontier. 2018. Tropane alkaloids GC/MS analysis and low dose elicitors effects on hyoscyamine biosynthetic pathway in hairy roots of Algerian Datura species. Sci. Rep. 8: 1–8.

Hashimoto, T., D.-J. Yun, and Y. Yamada. 1993. Production of tropane alkaloids in genetically engineered root cultures. Phytochem. 32: 713–718.

Hayta, S., A. Gurel, I. Akgun, F. Altan, M. Ganzera, B. Tanyolac, and E. Bedir. 2011. Induction of *Gentiana cruciata* hairy roots and their secondary metabolites. Biologia 66: 618–625. https://doi.org/10.2478/s11756-011-0076-4.

Ho, T.H. 2015. Development of amplification-based technologies for enrichment of nucleic acids with difficult sequences or low-abundance point mutations. http://urn.fi/URN:ISBN:978-951-51-0579-0.

Hook, I. 1994. Secondary metabolites in hairy root cultures of *Leontopodium alpinum* Cass. (Edelweiss). In Primary and Secondary Metabolism of Plants and Cell Cultures III. Springer. 321–326.

Huang, S.H., R.K. Vishwakarma, T.T. Lee, H.S. Chan, and H.S. Tsay. 2014. Establishment of hairy root lines and analysis of iridoids and secoiridoids in the medicinal plant *Gentiana scabra*. Bot. Stud. 55(1): 17. https://doi.org/10.1186/1999-3110-55-17.

Isah, T., S. Umar, A. Mujib, M.P. Sharma, P.E. Rajasekharan, N. Zafar, and A. Frukh. 2018. Secondary metabolism of pharmaceuticals in the plant *in vitro* cultures: Strategies, approaches, and limitations to achieving higher yield. Plant Cell Tiss Organ Cult. 132: 239–265. https://doi.org/10.1007/s11240-017-1332-2.

Ishimaru, K., H. Sudo, M. Satake, Y. Matsunaga, Y. Hasegawa, S. Takemoto, and K. Shimomura. 1990. Amarogentin, amaroswerin and four xanthones from hairy root cultures of *Swertia japonica*. Phytochem. 29: 1563–1565.

Javid, A., N. Gampe, F. Gelana, and Z. György. 2020. Enhancing the accumulation of rosavins in *Rhodiola rosea* L. plants grown *in vitro* by precursor feeding. Agronomy 11(12): 2531. https://doi.org/10.3390/agronomy11122531.

Jaziri, M., M. Legros, J. Homes, and M. Vanhaelen. 1988. Tropine alkaloids production by hairy root cultures of *Datura stramonium* and *Hyoscyamus niger*. Phytochem. 27: 419–420.

Jaziri, M., K. Yoshimatsu, J. Homès, and K. Shimomura. 1994. Traits of transgenic Atropa belladonna doubly transformed with different *Agrobacterium rhizogenes* strains. Plant Cell, Tissue and Organ Culture 38(2): 257–262.

Joshi, B., G.P. Sah, B.B. Basnet, M.R. Bhatt, D. Sharma, K. Subedi, J. Pandey, and R. Malla. 2011. Phytochemical extraction and antimicrobial properties of different medicinal plants: Ocimum sanctum (Tulsi), Eugenia caryophyllata (Clove), Achyranthes bidentata (Datiwan) and Azadirachta indica (Neem). J. Microbiol. Antimicrob. 3(1): 1–7. https://doi.org/10.5897/JMA.9000046.

Jouanin, L., P. Guerche, N. Pamboukdjian, C. Tourneur, F.C. Delbart, and J. Tourneur. 1987. Structure of T-DNA in plants regenerated from roots transformed by *Agrobacterium rhizogenes* strain A4. Mol. Gen. Genet. 206: 387–392.

Jung, G., and D. Tepfer. 1987. Use of genetic transformation by the Ri T-DNA of *Agrobacterium rhizogenes* to stimulate biomass and tropane alkaloid production in *Atropa belladonna* and *Calystegia sepium* roots grown *in vitro*. Plant Sci. 50: 145–151.

Kadri, K. 2019. Polymerase chain reaction (PCR): Principle and applications. pp. 146–163. *In*: Madan L. Nagpal, Oana-Maria Boldura, Cornell Balta, and Shymaa Enany (eds.). Synthetic Biology-New Interdisciplinary Science. IntechOpen. DOI: 10.5772/intechopen.86491. https://www.intechopen.com/chapters/67558.

Kamada, H., N. Okamura, M. Satake, H. Harada, and K. Shimomura. 1986. Alkaloid production by hairy root cultures in *Atropa belladonna*. Plant Cell Rep. 5: 239–242.

Kärkönen, A., R.A. Dewhirst, C.L. Mackay, and S.C. Fry. 2017. Metabolites of 2, 3-diketogulonate delay peroxidase action and induce non-enzymic H2O2 generation: Potential roles in the plant cell wall. Arch. Biochem. Biophys. 620: 12–22.

Kifle, S., M. Shao, C. Jung, and D. Cai. 1999. An improved transformation protocol for studying gene expression in hairy roots of sugar beet (*Beta vulgaris* L.). Plant Cell Rep. 18: 514–519.

Kim, Y.J., P.J. Weathers, and B.E. Wyslouzil. 2002. Growth of *Artemisia annua* hairy roots in liquid and gas-phase reactors. Biotechnol Bioeng. 80(4): 454–64.

Kim, Y.K., H. Xu, W.T. Park, N.I. Park, S.Y. Lee, and S.U. Park. 2010. Genetic transformation of buckwheat ('*Fagopyrum esculentum*' M.) with '*Agrobacterium rhizogenes*' and production of rutin in transformed root cultures. Aust. J. Crop Sci. 4(7): 485–490.

Kiselev, K.V., A.S. Dubrovina, M.V. Veselova, V.P. Bulgakov, S.A. Fedoreyev, and Y.N. Zhuravlev. 2007. The rolB gene-induced overproduction of resveratrol in Vitis amurensis transformed cells. J. Biotechnol. 128(3): 681–692. ISSN 0168–1656. https://doi.org/10.1016/j.jbiotec.2006.11.008.

Klimczak, I., and A. Gliszczyńska-Świgło. 2015. Comparison of UPLC and HPLC methods for determination of vitamin C. Food Chem. 175: 100–105.

Ko, K.S., Y. Ebizuka, H. Noguchi, and U. Sankawa. 1995. Production of polyketide pigments in hairy root cultures of Cassia plants. Chem. Pharm. Bull. 43(2): 274 278 https://doi.org/10.1248/cpb.43.274 https://www.jstage.jst.go.jp/article/cpb1958/43/2/43_2_274/_article/-char/en.

Kodahl, N., R. Müller, and H. Lütken. 2016. The *Agrobacterium rhizogenes* oncogenes *rolB* and *ORF13* increase formation of generative shoots and induce dwarfism in *Arabidopsis thaliana* (L.) Heynh. Plant Sci. 252: 22–29.

Krishnan, S.R., and E.A. Siril. 2016. Induction of hairy roots and over production of anthraquinones in *Oldenlandia umbellata* L.: A dye yielding medicinal plant by using wild type *Agrobacterium rhizogenes* strain. Indian J. Plant Physiol. 21: 271–278.

Królicka, A., I.I. Staniszewska, K. Bielawski, E. Malinski, J. Szafranek, and E. Lojkowska. 2001. Establishment of hairy root cultures of *Ammi majus*. Plant Sci. 160(2): 259–264. doi: 10.1016/s0168-9452(00)00381-2.

Lan, X., and H. Quan. 2010. Hairy root culture of *Przewalskia tangutica* for enhanced production of pharmaceutical tropane alkaloids. J. Med. Plants Res. 4(14): 1477–1481.

Le, F.-B.V., D. Laurain-Mattar, and M. Fliniaux. 2004. Hairy root induction of *Papaver somniferum* var. album, a difficult-to-transform plant, by *A. rhizogenes* LBA9402. Planta 218: 890–893.

Lee, S.Y., S.G. Kim, W.S. Song, Y.K. Kim, N. Park, and S.U. Park. 2010. Influence of different strains of *Agrobacterium rhizogenes* on hairy root induction and production of alizarin and purpurin in Rubia akane Nakai. Romanian Biotechnol. Lett. 15: 5405–5409.

Lelevic, A., V. Souchon, M. Moreaud, C. Lorentz, and C. Geantet. 2020. Gas chromatography vacuum ultraviolet spectroscopy: A review. J. Sep. Sci. 43: 150–173.

Lemcke, K., and T. Schmülling. 1998. Gain of function assays identify non-rol genes from *Agrobacterium rhizogenes* TL-DNA that alter plant morphogenesis or hormone sensitivity. Plant J. 15(3): 423–433.

Levesque, H., P. Delepelaire, P. Rouzé, J. Slightom, and D. Tepfer. 1988. Common evolutionary origin of the central portions of the Ri TL-DNA of *Agrobacterium rhizogenes* and the Ti T-DNAs of *Agrobacterium tumefaciens*. Plant Mol. Biol. 11: 731–744.

Levin, R.E., F.-G.C. Ekezie, and D-.W. Sun. 2018. DNA-based technique: Polymerase chain reaction (PCR). pp. 527–616. *In*: Da-Wen Sun (ed.). Modern Techniques for Food Authentication. Elsevier.

Lioshina, L.G., and O.V. Bulko. 2014. Plant regeneration from hairy roots and calluses of periwinkle *Vinca minor* L. and foxglove purple *Digitalis purpurea* L. Cytol. Genet. 48: 302–307. https://doi.org/10.3103/S009545271405003X.

Lopes, C.C.A., P.H.J.O. Limirio, V.R. Novais, and P. Dechichi. 2018. Fourier transform infrared spectroscopy (FTIR) application chemical characterization of enamel, dentin and bone. Appl. Spectrosc. Rev. 53: 747–769.

Lorenzo, M., and Y. Pico. 2017. Gas chromatography and mass spectroscopy techniques for the detection of chemical contaminants and residues in foods. pp. 15–50. *In*: Dieter Schrenk, and Alexander Cartus (eds.). Chemical Contaminants and Residues in Food. Elsevier.

Lozano-Sánchez, J., I. Borrás-Linares, A. Sass-Kiss, and A. Segura-Carretero. 2018. Chromatographic technique: High-Performance Liquid Chromatography (HPLC). pp. 459–526. *In*: Da-Wen Sun (ed.). Modern Techniques for Food Authentication. Elsevier.

Lucchesini, M., and A. Mensuali-Sodi. 2010. Plant tissue culture—an opportunity for the production of nutraceuticals. pp. 185–202. *In*: Giardi, M.T., G. Rea, and B. Berra (eds.). Bio-farms for Nutraceuticals. Springer.

Macao, B., J.P. Uhler, T. Siibak, X. Zhu, Y. Shi, W. Sheng, M. Olsson, J.B. Stewart, C.M. Gustafsson, and M. Falkenberg. 2015. The exonuclease activity of DNA polymerase γ is required for ligation during mitochondrial DNA replication. Nat. Commun. 6: 1–10.

Majumdar, S., S. Garai, and S. Jha. 2011. Genetic transformation of *Bacopa monnieri* by wild type strains of *Agrobacterium rhizogenes* stimulates production of bacopa saponins in transformed calli and plants. Plant Cell Rep. 30: 941–954.

Makhzoum, A., P. Sharma, M.A. Bernards, and J. Trémouillaux-Guiller. 2013. Hairy roots: An ideal platform for transgenic plant production and other promising applications. *In*: Gang, D. (ed.). Phytochemicals, Plant Growth, and the Environment. Springer. 42: 95–142.

Makhzoum, A., A. Bjelica, G. Petit-Paly, and M.A. Bernards. 2015. Novel plant regeneration and transient gene expression in *Catharanthus roseus*. The All Results Journals: Biol. 6: 1–9.

Malarz, J., A. Stojakowska, E. Szneler, and W. Kisiel. 2013. A new neolignan glucoside from hairy roots of *Cichorium intybus*. Phytochem. Lett. 6: 59–61.

Malik, S., S. Bhushan, M. Sharma, and P.S. Ahuja. 2014. Biotechnological approaches to the production of shikonins: A critical review with recent updates. Crit. Rev Biotechnol. 16: 1–14.

Malik, S., M. Sharma, and P.S. Ahuja. 2016. An efficient and economic method for *in vitro* propagation of *Arnebia euchroma* using liquid culture system. American J. Biotechnol. Med. Res. 1(1): 19–25.

Malik, S., S. Odeyemi, G.C. Pereira, L.M.F. Junior, H.A. Hamid, N. Atabaki, A. Makhzoum, E.B. Almeida, J. Dewar, and R. Abiri. 2021. New insights into the biotechnology and therapeutic potential of *Lippia alba* (Mill.) N.E.Br. ex P. Wilson. J. Essent Oil Res. DOI: 10.1080/10412905.2021.1936667.

Mano, Y., S. Nabeshima, C. Matsui, and H. Ohkawa. 1986. Production of tropane alkaloids by hairy root cultures of *Scopolia japonica*. Agri. Biol. Chem. 50(11): 2715–2722.

Mano, Y., H. Ohkawa, and Y. Yamada. 1989. Production of tropane alkaloids by hairy root cultures of *Duboisia leichhardtii* transformed by *Agrobacterium rhizogenes*. Plant Sci. 59: 191–201. https://doi.org/10.1016/0168-9452(89)90137-4.

Mathur, A., A. Gangwar, A.K. Mathur, P. Verma, G.C. Uniyal, and R.K. Lal. 2010. Growth kinetics and ginsenosides production in transformed hairy roots of American ginseng—*Panax quinquefolium* L. Biotechnol. Lett. 32(3): 457–461.

Mbughuni, M.M., P.J. Jannetto, and L.J. Langman. 2016. Mass spectrometry applications for toxicology. Ejifcc 27(4): 272–287.

Md, S.N., S.N. Jaafar, R.Z., Abdul, and C.R.C.M. Zain. 2014. Induction of hairy roots by various strains of *Agrobacterium rhizogenes* in different types of Capsicum species explants. BMC Res Notes 7: 414. https://doi.org/10.1186/1756-0500-7-414.

Mehrotra, S., V. Srivastava, L.U. Rahman, and A.K. Kukreja. 2015. Hairy root biotechnology-indicative timeline to understand missing links and future outlook. Protoplasma 252: 1189–1201. https://doi.org/10.1007/s00709-015-0761-1.

Mishra, B.N., and R. Ranjan. 2008. Growth of hairy-root cultures in various bioreactors for the production of secondary metabolites. Biotechnol. Appl. Biochem. 49: 1–10.

Mitiouchkina, T.Y., and S. Dolgov. 1998. Modification of chrysanthemum plant and flower architecture by *rolC* gene from *Agrobacterium rhizogenes* introduction. XIX International Symposium on Improvement of Ornamental Plants 508: 163–172.

Moreno-Anzúrez, N.E., S. Marquina, L. Alvarez, A. Zamilpa, P. Castillo-España, L.I. Perea-Arango, P.N. Torres, M. Herrera-Ruiz, E.R. Díaz García, J.T. García, and J. Arellano-García. 2017. A cytotoxic and anti-inflammatory campesterol derivative from genetically transformed hairy roots of *Lopezia racemosa* Cav. (Onagraceae). Molecules 22(1): 118. doi:10.3390/molecules22010118.

Murthy, H.N., E.J. Lee, and K.Y. Paek. 2014. Production of secondary metabolites from cell and organ cultures: Strategies and approaches for biomass improvement and metabolite accumulation. Plant Cell Tiss Organ Cult. 118: 1–16. https://doi.org/10.1007/s11240-014-0467-7.

Namir, H., R. Hadžić, and I. Malešević. 2019. Application of thin layer chromatography for qualitative analysis of gunpowder in purpose of life prediction of ammunition. Int. J. Biosen. Bioelectron. 5: 4–12.

Nandini, D., B.S. Ravikumar, and S.R. Amaraneni. 2017. Identification of atropine and scopolamine from *Datura* L. Plant and Market Samples by GC-MS. Asian J. Chem. 29(10): 2116–2118.

Nartop, P. 2018. Engineering of biomass accumulation and secondary metabolite production in plant cell and tissue cultures. pp. 169–194. *In*: Parvaiz Ahmad, Mohammad Abass Ahanger, Vijay Pratap Singh, Durgesh Kumar Tripathi, Pravej Alam, and Mohammed Nasser (eds.). Plant Metabolites and Regulation under Environmental Stress. Elsevier. https://doi.org/10.1016/C2016-0-03727-0.

Navarro, E., G. Serrano-Heras, M. Castaño, and J. Solera. 2015. Real-time PCR detection chemistry. Clinica Chimica Acta 439: 231–250.

Nayak, P., M. Sharma, S.N. Behera, M. Thirunavoukkarasu, and P.K. Chand. 2015. High-performance liquid chromatographic quantification of plumbagin from transformed rhizoclones of *Plumbago zeylanica* L.: Inter-clonal variation in biomass growth and plumbagin production. Appl. Biochem. Biotechnol. 175(3): 1745–70. doi: 10.1007/s12010-014-1392-2.

Nepovím, A., R. Podlipná, P. Soudek, P. Schröder, and T. Vaněk. 2004. Effects of heavy metals and nitroaromatic compounds on horseradish glutathione S-transferase and peroxidase. Chemosphere. 57(8): 1007–1015. doi: 10.1016/j.chemosphere.2004.08.030.

Nuutila, A.-M., L. Toivonen, and V. Kauppinen. 1994. Bioreactor studies on hairy root cultures of *Catharanthus roseus:* Comparison of three bioreactor types. Biotechnol. Tech. 8: 61–66.

Ooms, G., A. Karp, M.M. Burrell, D. Twell, and J. Roberts. 1985. Genetic modification of potato development using Ri T-DNA. Theoret. Appl. Genetics 70: 440–446. https://doi.org/10.1007/BF00273752.

Otten, L., and A. Helfer. 2001. Biological activity of the rolB-like 5′ end of the *A4*-orf8 gene from the *Agrobacterium rhizogenes* TL-DNA. Mol. Plant-Microbe Interact. 14: 405–411.

Ouartsi, A., D. Clérot, A.D. Meyer, Y. Dessaux, J. Brevet, and M. Bonfill. 2004. The T-DNA ORF8 of the cucumopine-type *Agrobacterium rhizogenes* Ri plasmid is involved in auxin response in transgenic tobacco. Plant Sci. 166(3): 557–567. ISSN 0168–9452. https://doi.org/10.1016/j.plantsci.2003.08.019.

Parr, A., A. Peerless, J. Hamill, N. Walton, R. Robins, and M. Rhodes. 1988. Alkaloid production by transformed root cultures of *Catharanthus roseus*. Plant Cell Rep. 7: 309–312.

Parr, A.J., and J.D. Hamill. 1987. Relationship between *Agrobacterium rhizogenes* transformed hairy roots and intact, uninfected Nicotiana plants. Phytochem. 26: 3241–3245.

Paszkowski, J., R.D. Shillito, M. Saul, V. Mandak, B. Hohn, and I. Potrykus. 1984. Direct gene transfer to plants. EMBO J. 3: 2717–2722.

Patra, N., and A.K. Srivastava. 2014. Enhanced production of artemisinin by hairy root cultivation of *Artemisia annua* in a modified stirred tank reactor. Appl. Biochem. Biotechnol. 174(6): 2209–22.

Pavlova, O.A., T.V. Matveyeva, and L.A. Lutova. 2013. Rol-genes of *Agrobacterium rhizogenes*. Ecol. Genet. 11: 59–68.

Petit, A., C. David, G.A. Dahl, J.G. Ellis, P. Guyon, F. Casse-Delbart, and J. Tempé. 1983. Further extension of the opine concept: Plasmids in *Agrobacterium rhizogenes* cooperate for opine degradation. Mol. Gen. Genet. 190: 204–214.

Pythoud, F., V.P. Sinkar, E.W. Nester, and M.P. Gordon. 1987. Increased virulence of *Agrobacterium rhizogenes* conferred by the *vir* region of pTiBo542: Application to genetic engineering of poplar. Biotechnol. 5: 1323–1327.

Rahman, A., and A.A. Ansari. 1983. Alkaloids of *Berberis aristata:* Isolation of aromoline and oxyberberine. J. Chem. Soc. Pak. 5: 283–284.

Ray, S., A. Majumder, M. Bandyopadhyay, and S. Jha. 2014. Genetic transformation of sarpagandha (*Rauvolfia serpentina*) with *Agrobacterium rhizogenes* for identification of high alkaloid yielding lines. Acta Physiol. Plant. 36(6): 1599–1605.

Rhodes, M.J.C., M. Hilton, A.J. Parr, J.D. Hamill, and R.J. Robins. 1986. Nicotine production by "hairy root" cultures of *Nicotiana rustica*: Fermentation and product recovery. Biotechnol. Lett. 8: 415–420. https://doi.org/10.1007/BF01026743.

Ricigliano, V., J. Chitaman, J. Tong, A. Adamatzky, and D.G. Howarth. 2015. Plant hairy root cultures as plasmodium modulators of the slime mold emergent computing substrate *Physarum polycephalum*. Frontiers in Microbiology 6: 720.

Roberts, L.W., P.B. Gahan, and R. Aloni. 1998. Vascular differentiation and plant growth regulators. Springer. 1: 1–89. Doi: 10.1007/978-3-642-73446-5.

Robins, R.J., J.D. Hamili, A.J. Parr, K. Smith, N.J. Walton, and M.J.C. Rhodes. 1987. Potential for use of nicotine acid as a selective agent for isolation of high nicotine producing lines of *N. rustica* hairy root cultures. Plant Cell Rep. 6(1): 22–176.

Rother, D., T. Sen, D. East, and I.J. Bruce. 2011. Silicon, silica and its surface patterning/activation with alkoxy-and amino-silanes for nanomedical applications. Nanomed. 6: 281–300.

Roychowdhury, D., B. Ghosh, B. Chaubey, and S. Jha. 2013. Genetic and morphological stability of six-year-old transgenic *Tylophora indica* plants. Nucleus (India) 56(2): 81–89. doi: 10.1007/s13237-013-0084-6.

Ryoichi, K., K. Hiroshi, and M. Asashima. 1989. Effects of high and very low magnetic fields on the growth of hairy roots of *Daucus carota* and *Atropa belladonna*. Plant Cell Physiol. 30(4): 605–608. https://doi.org/10.1093/oxfordjournals.pcp.a077782.

Sahu, P.K., N.R. Ramisetti, T. Cecchi, S. Swain, C.S. Patro, and J. Panda. 2018. An overview of experimental designs in HPLC method development and validation. J. Pharm. Biomed. Anal. 147: 590–611.

Saitou, T., H. Kamada, and H.JPS. Harada. 1991. Isoperoxidase in hairy roots and regenerated plants of horseradish (*Armoracia lapathifolia*). Plant Sci. 75(2): 195–201. doi: 10.1016/0168-9452 (91) 90234-y.

Samaddar, T., S. Sarkar, and S. Jha. 2019. *Agrobacterium rhizogenes* mediated transformation of the critically endangered species, *Swertia chirayita*. Plant Tiss. Cult. Biotechnol. 29(2): 231–244. https://doi.org/10.3329/ptcb.v29i2.44512.

Sandal, I., U. Saini, B. Lacroix, A. Bhattacharya, P.S. Ahuja, and V. Citovsky. 2007. Agrobacterium-mediated genetic transformation of tea leaf explants: Effects of counteracting bactericidity of leaf polyphenols without loss of bacterial virulence. Plant Cell Rep. 26(2): 169–176.

Santos, F.J., and M. Galceran. 2003. Modern developments in gas chromatography–mass spectrometry-based environmental analysis. J. Chromatogr. A 1000(1-2): 125–151.

Sauerwein, M., K. Ishimaru, and K. Shimomura. 1991. Indole alkaloids in hairy roots of *Amsonia elliptica*. Phytochemistry 30(4): 1153–1155.

Semwal, B.C., J. Gupta, S. Singh, Y. Kumar, and M. Giri. 2009. Antihyperglycemic activity of root of *Berberis aristate* DC. In alloxan induced diabetic rat. Int. J. Green Pharm. 3(3): 259.

Sharifi, S., T.N. Sattari, A. Zebarjadi, A. Majd, and H. Ghasempour. 2014. The influence of *Agrobacterium rhizogenes* on induction of hairy roots and ß-carboline alkaloids production in *Tribulus terrestris* L. Physiology and Molecular Biology of Plants 20(1): 69–80.

Sharma, P., R. Manchanda, R. Goswami, and S. Chawla. 2020. Biodiversity and therapeutic potential of medicinal plants. *In*: Shukla, V., and N. Kumar (eds.). Environmental Concerns and Sustainable Development. Springer, Singapore. https://doi.org/10.1007/978-981-13-6358-0_2.

Smith, R.H. 2006. Plant tissue culture techniques and experiments. Elsevier. 2: 1–175. ISBN: 978-0-12-415920-4.

Srivastava, M., S. Sharma, and P. Misra. 2016. Elicitation based enhancement of secondary metabolites in *Rauwolfia serpentina* and *Solanum khasianum* hairy root cultures. Pharmacogn. Mag. 12: 315–320. doi: 10.4103/0973-1296.185726.

Srivastava, S., and A.K. Srivastava. 2013. Production of the biopesticide azadirachtin by hairy root cultivation of *Azadirachta indica* in liquid-phase bioreactors. Appl. Biochem. Biotechnol. 171(6): 1351–61.

Stougaard, J., D. Abildsten, and K.A. Marcker. 1987. The *Agrobacterium rhizogenes* pRi TL-DNA segment as a gene vector system for transformation of plants. Mol. Gen. Genet. 207: 251–255.

Sudha, C.G., T.V. Sherina, T.P. Anu-Anand, J.V. Reji, P. Padmesh, and E.V. Soniya. 2013. *Agrobacterium rhizogenes* mediated transformation of the medicinal plant *Decalepis arayalpathra* and production of 2-hydroxy-4-methoxy benzaldehyde. Plant Cell Tiss. Organ Cult. 112: 217–226. https://doi.org/10.1007/s11240-012-0226-6.

Suprabuddha, K., S. Umme, N.A. Md., K.H. Alok, and M. Nirmal. 2018. Development of transgenic hairy roots and augmentation of secondary metabolites by precursor feeding in *Sphagneticola calendulacea* (L.) Pruski. Ind. Crops Prod. 121: 206–215.

Sujatha, G., S. Zdravković-Korać, D. Ćalić, G. Flamini, and B.R. Kumari. 2013. High-efficiency *Agrobacterium rhizogenes*-mediated genetic transformation in *Artemisia vulgaris*: Hairy root production and essential oil analysis. Ind. Crops Prod. 44: 643–652.

Swain, S.S., K.K. Rout, and P.K. Chand. 2012. Production of triterpenoid anti-cancer compound taraxerol in Agrobacterium-transformed root cultures of butterfly pea (*Clitoria ternatea* L.). Appl. Biochem. Biotechnol. 168(3): 487–503. doi: 10.1007/s12010-012-9791-8.

Syklowska-Baranek, K., A. Pietrosiuk, A. Gawron, A. Kawiak, E. Lojkowska, M. Jeziorek, and I. Chinou. 2012. Enhanced production of antitumour naphthoquinones in transgenic hairy roots lines of *Lithospermum canescens*. Plant Cell Tiss. Organ Cult. 108: 213–219. doi:10.1007/s11240-011-0032-6.

Takeshi Matsumoto, and Tanaka Nobukazu. 1991. Production of Phytoecdysteroids by Hairy Root Cultures of *Ajuga reptans* var. atropurpurea, Agricultural and Biological Chemistry, 55: 4, 1019–1025, DOI: 10.1080/00021369.1991.10870707.

Tarik, M., L. Ramakrishnan, H.S. Sachdev, N. Tandon, A. Roy, S.K. Bhargava, and R.M. Pandey. 2018. Validation of quantitative polymerase chain reaction with Southern blot method for telomere length analysis. Future sci. OA 4(4): FSO282. doi: 10.4155/fsoa-2017-0115.

Tavassoli, P., and A.S. Afshar. 2018. Influence of different *Agrobacterium rhizogenes* strains on hairy root induction and analysis of phenolic and flavonoid compounds in marshmallow (*Althaea officinalis* L.). Biotech. 3(8): 1–8. https://doi.org/10.1007/s13205-018-1375-z.

Taya, M., K. Mine, M. Kino-Oka, S. Tone, and T. Ichi. 1992. Production and release of pigments by culture of transformed hairy root of red beet. J. Ferment. Bioeng. 73(1): 31–36.

Terato, M., A. Ishikawa, K. Yamada, Y. Ozeki, and Y. Kitamura. 2011. Increased furanocoumarin production by *Glehnia littoralis* roots induced via *Agrobacterium rhizogenes* infection. Plant Biotechnol. 28: 317–321.

Thiruvengadam, M., N. Praveen, E.H. Kim, S.H. Kim, and I.M. Chung. 2014. Production of anthraquinones, phenolic compounds and biological activities from hairy root cultures of *Polygonum multiflorum* Thunb. Protoplasma. 251: 555–566.

Thwe, A., M.V. Arasu, X. Li, C.H. Park, S.J. Kim, N.A. Al-Dhabi, and S.U. Park. 2016. Effect of different *Agrobacterium rhizogenes* strains on hairy root induction and phenylpropanoid biosynthesis in tartary buckwheat (*Fagopyrum tataricum* Gaertn) Front. Microbiol. 7: 318. doi: 10.3389/fmicb.2016.00318. Tiwari, R., and C. Rana. 2015. Plant secondary metabolites: A review. Int. J. Eng. 3(5): 661–670.

Toivonen, L., J. Balsevich, and W.G.W. Kurz. 1989. Indole alkaloid production by hairy root cultures of *Catharanthus roseus*. Plant Cell Tiss. Organ Cult. 18: 79–93.

Trovato, M., B. Maras, F. Linhares, and P. Costantino. 2001. The plant oncogene *rolD* encodes a functional ornithine cyclodeaminase. Proc. Natl. Acad. Sci. 98(23): 13449–13453.

Umber, M., L. Voll, A. Weber, P. Michler, and L. Otten. 2002. The rolB like part of the *Agrobacterium rhizogenes* orf8 gene inhibits sucrose export in tobacco. Mol. Plant-Microbe Interact. 15: 956–62.

Vamenani, R., A. Pakdin-Parizi, M. Mortazavi, and Z. Gholami. 2020. Establishment of hairy root cultures by *Agrobacterium rhizogenes* mediated transformation of *Trachyspermum ammi* L. for the efficient production of thymol. Biotechnol. Appl. Biochem. 67(3): 389–395. doi: 10.1002/bab.1880.

Vilaine, F., and F. Casse-Delbart. 1987. Independent induction of transformed roots by the TL and TR regions of the Ri plasmid of agropine type *Agrobacterium rhizogenes*. Mol. Gen. Genet 206: 17–23.

Weremczuk-Jeżyna, I., I. Grzegorczyk-Karolak, B. Frydrych, A. Królicka, and H. Wysokińska. 2013. Hairy roots of *Dracocephalum moldavica*: rosmarinic acid content and antioxidant potential. Acta Physiol. Plant. 35: 2095–2103.

Yang, L., K.S. Wen, X. Ruan, Y.X. Zhao, F. Wei, and Q. Wang. 2018. Response of Plant Secondary Metabolites to Environmental Factors. Molecules 23(4):762. https://doi.org/10.3390/molecules23040762.

Yoon, J.Y., I.M. Chung, and M. Thiruvengadam. 2015. Evaluation of phenolic compounds, antioxidant and antimicrobial activities from transgenic hairy root cultures of gherkin (*Cucumis anguria* L.). S. Afr. J. Bot. 100: 80–86.

Yoshikawa, T., and T. Furuya. 1987. Saponin production by cultures of *Panax ginseng* transformed with *Agrobacterium rhizogenes*. Plant Cell Rep. 6: 449–453.

Yue, W., Q.L. Ming, B. Lin, K. Rahman, C.J. Zheng, T. Han, and L.P. Qin. 2014. Medicinal plant cell suspension cultures: Pharmaceutical applications and high-yielding strategies for the desired secondary metabolites, Crit. Rev. Biotechnol., 36(2): 215–232. doi: 10.3109/07388551.2014.923986.

Zhang, H.C., J.M. Liu, H.M. Chen, C.C. Gao, H.Y. Lu, H. Zhou, Y. Li, and S.L. Gao. 2011. Up-regulation of licochalcone A biosynthesis and secretion by Tween 80 in hairy root cultures of *Glycyrrhiza uralensis*. Fisch. Mol. Biotechnol. 47(1): 50–6. doi: 10.1007/s12033-010-9311-4.

Zhao, B., F.A.K.C.R. Agblevor, and J.G. Jelesko. 2013. Enhanced production of the alkaloid nicotine in hairy root cultures of *Nicotiana tabacum* L. Plant Cell Tiss. Organ Cult. 113: 121–129.

Zhong, J.J. 2001. Biochemical engineering of the production of plant-specific secondary metabolites by cell suspension cultures. Adv. Biochem. Eng. Biotechnol. 72: 1–26.

II-Plant Molecular Pharming

CHAPTER 7

Plant Molecular Pharming

Methods, Tools, Challenges Ahead for Production of Recombinant Proteins, and Potential Solutions

Shahram Shokrian Hajibehzad,[1,*] *Azadeh Mohseni,*[1] *Mohammad Dolati,*[1] *Saeid Malekzadeh Shafaroudi,*[1] *Kathleen Hefferon*[2] and *Abdullah Makhzoum*[3,*]

Introduction

The first report of a transgenic higher organism expressing a pharmaceutical was in 1987 when sheep β-lactoglobulin (BLG) was successfully produced in transgenic mice (Simons et al. 1987). Two years later, the primary idea of using plants as a potential platform for the production of recombinant proteins was sparked by Hiatt et al. They successfully expressed the catalytic IgG1 (6D4) monoclonal antibody in tobacco plants (Hiatt et al. 1989). After nearly ten years, the first report with the aim of expression, purification, and commercial sale of the glycoprotein avidin was performed in transgenic maize (Hood et al. 1997). Since that time, the word "plant molecular farming" has been frequently used to produce such components and several other proteins with the aim of medical or industrial applications in *planta* (Fischer et al. 2004, Hefferon 2013). Over the past two decades, this technology has been successfully applied for a wide range of production of some valuable proteins, such as antibodies (Hensel et al. 2015), antigens (Hajibehzad et al. 2016), growth factors, hormones (Giddings 2001), human blood products,

[1] Biotechnology and Plant Breeding Department, College of Agriculture, Ferdowsi University of Mashhad, Mashhad, Iran.
[2] Cornell University, Department of Cell Biology and Genetics, United States.
[3] Department of Biological Sciences & Biotechnology, International University of Science and Technology, Palapye, Botswana.
* Corresponding author: s.shokrianhajibehzad@uu.nl; makhzouma@biust.ac.bw

and valuable industrial components (i.e., enzymes, reagents, and biopolymers) in plants (Kempinski et al. 2015, Mielenz et al. 2015, Snell et al. 2015) (Fig. 1). Although these valuable components can be produced via traditional recombinant production systems, i.e., bacteria, yeast, mammalian cell cultures, the expression of recombinant proteins in plants as green bio-factories is generally thought to be more attractive than other expression platforms because, 1—the production of recombinant proteins in plants eliminates the potential for the contamination of final products with human pathogens such as viruses and prions, as well as endotoxins. As a result, vaccine delivery could be done for edible plants by feeding the plant tissues containing recombinant proteins as food (edible vaccines). Thus injection-related worries and the purification procedures would be eliminated (Davies 2010). 2—one of the significant advantages of plants over bacteria-based platforms is their capability to accomplish post-translational modification such as glycosylation and disulfide bond formation that, in contrast to bacteria, is similar to that of the mammalian cells. This enables plants to synthesize complex proteins with accurate folding and activity (Horn et al. 2004). 3—the production of recombinant proteins in plants could be restricted to the subcellular compartments such as Endoplasmic Reticulum (ER) or directed to the outside of cells, providing a safe environment that reduces degradation and thus increases yield of the final product (Llop-Tous et al. 2010). 4—plants only need to sunlight, water, and minerals for their growth and reproduction; therefore, these inexpensive and accessible inputs make them more economical than industrial facilities for the cost-effective production of recombinant proteins (Jamal et al. 2009, Twyman et al. 2003).

Here, we will first comparatively address several aspects of plant molecular farming, including the strategies that have been applied for the production of recombinant proteins in *planta*, such as transient expression and stable production in nuclei and chloroplasts. Moreover, the essential methods and tools applied by the mentioned strategies will be explained. In the second part, we will simultaneously discuss the bottlenecks ahead for production of recombinant proteins in plants and their potential solutions.

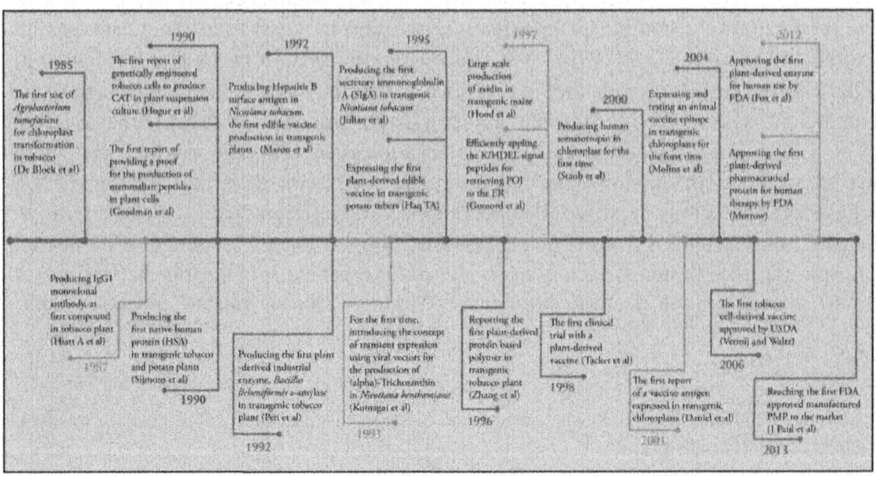

Figure 1: A brief history of plant molecular farming highlighting key reports.

Plant Nuclear genome Transformation

Plant genetic transformation is an effective tool in plant biotechnology and crop improvement. Specifically, plant nuclear genome transformation is one of the most common strategies employed for plant-based recombinant protein production. There are many potential plant transformation mechanisms; among them, *Agrobacterium* and gene gun-mediated transformation are the main strategies used to transfer a gene of interest (GOI) into the plant nuclear genome (Sanford 1990). Naturally, *Agrobacterium* species transform plants by transferring a large region of their large resident plasmid, referred to as transferred DNA (T-DNA), along with other virulence effector proteins, into the host plant genome. In *Agrobacterium*-mediated transformation, T-DNA is cut in single-strand form, escorted by some bacterial Vir proteins and plant-encoded proteins, exported through the plant cytoplasm using the type IV secretion system entered into the nucleus, and finally it is inserted into the plant genome. Integration of T-DNA into the plant chromatin, done either by a single or double strand T-DNA intermediate, generates a stable transgenic plant that can permanently express T-DNA encoded genes through the next generations (Tzfira et al. 2004). The precise mechanism by which T-DNA integrates into the plant genome has remained elusive. However, recently, van Kregten et al reported that *TEBICHI*, a plant homologous to polymerase θ (Pol θ), has a crucial role in T-DNA integration. Based on mutation analysis, they have proved that *TEBICHI* has neither role in plant infection nor T-DNA transfection into the plant nucleus; instead, it controls T-DNA integration and generates error-prone footprints at integration sites (Levy 2016, van Kregten et al. 2016). Nowadays, new strains of *Agrobacterium* harboring a disarmed plasmid (by the removal of tumor-inducing genes) are primarily used in plant biotechnology for producing transgenic plants. Indeed, owing to low copy numbers and the large size (ca. 200 kb) of *Agrobacterium* tumor-induced plasmid (Ti plasmid), the binary vector system has been employed over the native Ti-plasmid (Text Box 1).

The second methodology for plant transformation consists of a physical device known as the Gene gun. The process of particle bombardment used for stable transformation involves directly shooting DNA into both intact single and organized cells through the use of high-velocity microprojectiles. This method is extensively used for many plant species especially those which cannot be transformed by *Agrobacterium*-mediated transformation. Gene gun-mediated transformation needs a construct to proceed, while applied vectors in this procedure are different from common vectors which are mostly used for *Agrobacterium*-mediated transformation. As such, a gene construct that is commonly used in gene gun-mediated transformation is in the form of a linear expression cassette consisting of a promoter, GOI, and terminator. In contrast to common *Agrobacterium*-based constructs, the linear vector is coated by metal particles and employed with particle bombardment for transient or stable transformation of GOI. After bombardment, the expression cassette is used as the template for transcription either mostly with no integration into the plant genome (transient expression), or rarely by insertion into the plant chromosomes (Sanford 1990). For many years, until the middle of the 1990s, scientists mistakenly thought that due to host range restrictions, *Agrobacterium tumefaciens* would not be able to transform monocotyledons plants. Eventually, some of the most common monocotyledons crops including, wheat (Cheng et al. 1997), maize (Ishida et al. 1996), barley (Shrawat et al. 2007) and rice (Hiei et al. 1994) were successfully transformed

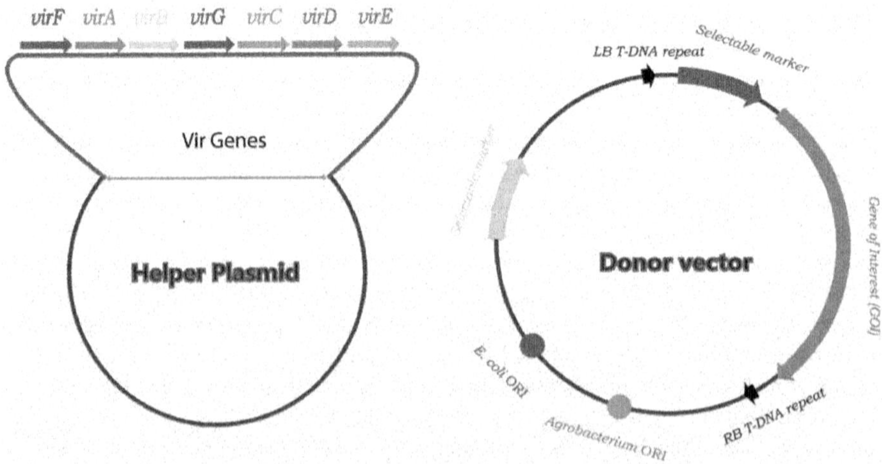

Text Box 1: Molecular mechanisms of the binary vector system

By the middle of the 1980s, the initial idea of binary vectors was established by Hoekema et al. They found that the *Vir* genes are not required to be in the same plasmid along with other regions. This led to the construction of a system for plant transformation in which the T-DNA and the Vir regions have been separated into distinct amplicons (Hoekema et al. 1983). The T-DNA binary system consists of two replicons, including (1) the helper plasmid—a region of *Agrobacterium* Ti-plasmid harboring the *Vir* genes, but lacking the T-DNA region. (2) The binary vector (donor plasmid)—this replicon generally contains the origin(s) of replication and antibiotic resistance gene(s), also known as non-T-DNA, as well as the transgene. It is notable that the origin(s) of replication and antibiotic resistance gene(s) are used to initiate replication in both *E. coli* and *Agrobacterium*, and select for the presence of the binary vector in bacteria, respectively. The T-DNA region contains the left and right borders (LB and RB), a transgene, a plant selectable marker, and promoters for the selectable marker and transgene. It is worth noting that separation of the *Vir* genes and T-DNA regions of Ti-plasmid would help to facilitate plasmid extraction, manipulation, and transformation using standard cloning procedures, whereas the helper plasmid is kept inside the *Agrobacterium* cells. Finally, disarmed (non-tumorigenic) *Agrobacterium* carrying the helper plasmid is transformed with the constructed binary vector. Although the term Binary vector refers to a complete binary vector system and consists of the two described plasmids, the replicon containing the T-DNA region is also referred to as a binary vector, as used in this review (Komari et al. 2006, Meyers et al. 2010). It should be noted that three of the aforementioned elements are essential for the transformation process, including A—T-DNA left and right borders (ca. 25 bp) which have surrounded the T-DNA in direct orientation and are necessary to cut the T-DNA region through the Vir proteins (Yadav et al. 1982), B—Vir genes which have resided in the helper plasmid (Lee and Gelvin 2008), and C—a group of genes which have been located within the Agrobacterium genome itself (O'Connell and Handelsman 1989).

by *Agrobacterium* species. The low level of expression and time-consuming procedures of transgenic plant production are the most important obstacles ahead for plant stable expression of recombinant proteins. Furthermore, the transformation of alien genes into the nuclear genome of plants has raised global concerns about the possibility of undesirable genetic pollution to other crops and closely related wild species.

Plant Chloroplast Transformation

Plastids are a group of organelles found in plant cells. Chloroplast transformation is an alternative platform for generating transgenic plants because of its potential for the high level production of recombinant proteins and the fact that the transgene is maternally inherited, which prevents the possibility of out-crossing of transgenes to related crops (Ahmad et al. 2016). The first successful chloroplast transformation was done in tobacco by De Block et al. who showed that *A. tumefaciens* harboring a modified Ti-plasmid vector containing a marker gene, is capable of being expressed in chloroplasts and can be used to introduce alien genes to the chloroplast genome (De Block et al. 1985). Since then, the chloroplast genome has been used for the production of vaccine antigens, antibodies, recombinant proteins, pharmaceuticals, hormones, and commercial enzymes. Although a number of higher plants including, *Arabidopsis* (Sikdar et al. 1998), *Oryza sativa* (Lee et al. 2006), wheat (Cui et al. 2011), lettuce (Lelivelt et al. 2005), cabbage (Liu et al. 2007), cotton (Kumar et al. 2004), and potato (Sidorov et al. 1999) have been successfully recruited for chloroplast transformation, this technology has not yet been applied for field based production.

The current procedure for chloroplast transformation starts with building a construct harboring the following elements that are required upstream and downstream of GOI: A—the promoter region which not only initiates transcription of GOI, but also determines the rate of transcription. B—the 5' untranslated region is transcribed from the corresponding DNA sequence to mRNA, but is not translated into protein sequence. This region is not only used as a ribosome binding site (also known as Shine-Dalgarno sequence), but also contains a sequence that is recognized by ribosomes. Similar to bacteria, the recognition of start codons in plastid is accomplished either by shine-Dalgarno or local minima of mRNA structure. Recently, Scharff et al. demonstrate that the plastid genes with weak structure in their start codon surrounding are independent of shine-Dalgarno. This evidence clearly confirms the presence of two district mechanisms in recognition of plastid start codons (Scharff et al. 2017). C—the 3' untranslated region begins immediately after the translation stop codon and plays crucial roles in mRNA stability, termination of translation, and post-transcriptional gene expression. D—the terminator region mediates transcriptional termination using signals that take part in a process that helps to release synthesized mRNA from the transcription complex (Bock 2014b). Despite the upmost importance of terminators in boosting gene expression, only a few attempts have been done on terminators in terms of their role in recombinant protein production. For a couple of decades, the most popular terminators which are mostly used for gene expression were nopaline synthase (NOS) and octopine synthase (OCS) terminators from *Agrobacterium tumefaciens* and 35S terminator from cauliflower mosaic virus. Recently, the terminator of *Arabidopsis* heat shock protein increased the transgene expression up to 2 times in

Table 1: Examples of proteins expressed in chloroplast of plants

Protein	Expression system	Applied vector to transform *A. tumefaciens*	The highest expression level	References
Recombinant protein				
gene encoding β-mannanase (man1) gene from *Trichoderma reesei*	*N. tabacum*	pLD (psbA promoter)	Maximum enzyme activity: 25 Units/g FW	Agrawal et al. 2011
xynA, encoding a bacterial xylanase from *Clostridium cellulovorans*	Tobacco plants (cv. 81V9 and I-64)	CEC (1-4) (mutated Prrn)	6% of TSP	Kolotilin et al. 2013
β-glucosidase family H3 (bgl1 gene) from *Aspergillus niger* cellulase type A (celA gene) and cellulose type B (celB gene) from *Thermotoga neapolitana* (celA-celB) celA celB	*N. tabacum*	pES4 (pES6, pHM4, pHM5 and pHM6) (rrn16 promoter)	Maximum enzyme activity: celA-celB: 58.21 ± 1.99 U/mg of TSP at PH 5 58.21 ± 1.99 U/mg of TSP at 65° C celA: 49.10 ± 1.84 U/mg of TSP at PH 5 49.10 ± 1.84 U/mg of TSP at 65° C celB: 48.72 ± 1.44 U/mg of TSP at PH 5 48.72 ± 1.44 U/mg of TSP at 65° C bgl1: 30.45 ± 2.01 U/mg of TSP at PH 5 30.45 ± 2.01 U/mg of TSP at 40° C	Espinoza-Sánchez et al. 2016
Antibody				
Factor IX (FIX) fused with CTB	*Lactuca sativa*	pLS (psbA promoter)	1.81 ± 0.10 mg/g dry weight [DW] 0.57% (±0.01%) of TLP 927.02 μg/gDW 0.57% TLP	Su et al. 2015 and Herzog et al. 2017

Vaccine				
cholera toxin-B subunit (CTB) fused to malarial vaccine antigens apical membrane antigen-1 (AMA1) (CTB-MSP1) CTB fused to merozoite surface protein-1(MSP1) (CTB- AMA1)	*Nicotiana tabacum* *Lactuca sativa*	pLD (psbA promoter) pLsDV (psbA promoter)	13.17% and 12.41% of the TSP in T0 and T1 respectively in tobacco 7.3% and 7.26% of the TSP in T0 and T1 respectively in lettuce 10.11% and 9.9% of the TSP in T0 and T1 respectively in tobacco 6.1% and 5.89% of the TSP in T0 and T1 respectively in lettuce	Davoodi-Semiromi et al. 2010
SAG1, the main surface antigen of the intracellular parasite *Toxoplasma gondii* SAG1 fused to HSP90 of *Leishmania infantum* (LiHsp83) (SAG1-LiHsp83)	*N. tabacum*	pAF (psbA promoter)	Up to ~ 100 µg per gram of FW in mature and old leaves in SAG1-LiHsp83	Albarracín et al. 2015
Pharmaceutical protein				
Native human coding sequence for the 13-kDa growth factor-β3 (TGFβ3) polypeptide Synthetic human coding sequence for the 13-kDa growth factor-β3 (TGFβ3) polypeptide	*N. tabacum*	pUM34 (*B. napus* rrn promoter)	0.11 ± 0.04% (n = 2) of TLP 8.2 ± 2% (n = 3) of TLP	Gisby et al. 2011
Retrocyclin-101 (RC101) fused with GFP Protegrin-1 (PG1) fused with GFP	Tobacco chloroplasts	pLD (psbA promoter)	32% ~ 38% of TSP 17% ~ 26% of TSP	Lee et al. 2011
exendin-4 (EX4) fused with cholera toxin B subunit (CTB)–protein (CTB-EX4)	Tobacco chloroplasts	EX4 5500 (psbA promoter)	14.3% of TLP	Kwon et al. 2013
Human Serum Albumin (HSA)	Tobacco chloroplasts	pLD (Prrn promoter and Shine-Dalgarno (SD) 5′ UTR) pLD (psbA promoter and 5′ UTR)	up to 11.1% TP (Using chloroplast psbA UTR)	Millán et al. 2003

comparison with the NOS terminator where transient expression in *Arabidopsis* leaves was employed (Nagaya et al. 2010).

More recently, Rosenthal et al. 2018 indicated that the intronless extension terminator enhances the transgene expression up to 13.6 folds in comparison to previously used standard terminators (Rosenthal et al. 2018). E—the expression cassette should be flanked by two fragments, typically 1–2 kb in size, which help to integrate the GOI in a pre-determined place within the genome by a process known as homologous recombination. These constructs are transferred to the chloroplast genome through either biolistic (biological ballistic) or polyethylene glycol (PEG) based methods. Of these, the gene-gun-mediated particle delivery system (biolistic), has been known as an efficient method of choice for stable plastid transformation as it is less laborious and time-consuming (Bock 2014a). It is worth noting that the choice of suitable selectable marker is extremely important for successful plastid transformation. Today, among the selection procedures which have been successfully engaged in plastid transformation, the *aadA* gene encoding an aminoglycoside 3"-adenylyl transferase that confers resistance to spectinomycin and streptomycin remains the most generally used selectable marker for chloroplast transformation (Jin and Daniell 2015, Svab and Maliga 1993).

The following advantages and disadvantages can be deduced for stable chloroplast transformation in comparison with nuclear genome transformation. A—in the chloroplast genome, transition-transversion nucleotide substitutions are nearly four times lower than in the nuclear genomes so that the protein of interest (POIs) could be expressed with minimal alteration in their structure and folding. B—integration of new genes of interest (GOIs) into the chloroplast genome occurs by homologous recombination. This accuracy and high efficiency of transferring genes to the chloroplast genome is integral to homologous recombination, such that the expression plasmid contains two homologous flanking sites (either ca. 0.5–1 kb). The GOI is replaced with a corresponding fragment within the chloroplast genome whose flanking sites have been designed upstream and downstream of the expression cassette through a double-crossover event. As a result, transgene expression in chloroplasts is more durable and uniform than nuclear transgene expression, which occurs via random insertion through non-homologous recombination and suffers from position effects that create variations in expression levels of different plant lines. C—in contrast to the nuclear genome, the expression of new genes is not affected by RNA silencing and/or other plastids epigenetic transgene silencing, suggesting that transgene expression in the chloroplast genome is stable over many generations. D—due to the polycistronic nature of the chloroplast genome, a cluster of transgenes could be arranged as synthetic operons and expressed under the control of single or multiple promoters. This phenomenon could not be implemented in a plant nuclear genome due to its monocistronic nature. In contrast to bacteria, plastid operons need additional promoters often located within operons as well as other post-transcriptional mechanisms such as intron splicing and other mRNA editing motifs (Hirose and Sugiura 1997). E—depending on the plant species, every cell has several plastids which each also has up to 100 copies or more of the plastid DNA. As a result, 10,000 to 50,000 identical copies of plastid DNA can be found in a single leaf cell. Due to the polyploidy nature of the chloroplast genome, extraordinary high levels of foreign protein expression (up to 50 percent of total soluble protein (TSP)) could be achieved, or nearly 10 to 100 times more than protein expression from a nuclear genome (Bock 2001). F—it has been shown that just one or a few chloroplastic genomes, out of several thousand copies are transformed during the transplastomic process. Therefore, for successful transformation, homoplasmic cells need to be generated from the initial heteroplasmic cells (which are consist of a mixed population of transformed and non-transformed genomic copies). G—while they have many advantages, chloroplasts are not a suitable place for the production of recombinant proteins requiring posttranslational modifications (PTM) such as glycosylation and specific phosphorylation patterns. However, recently, some chloroplast glycosylation have been recognized. It should be noted that disulfide bands are properly formed inside the chloroplasts (Lehtimäki et al. 2015).

Transient Expression

Since the production of stable transgenic plants is laborious and time-consuming, there is a need for further confirming whether the construct arrangement is accurate or not, and also whether the chosen host plant has a suitably high capacity to produce recombinant proteins. *Agrobacterium*-mediated transient expression is a powerful method to not only check the accuracy of expression constructs but also as an independent platform for producing recombinant proteins (Le Mauff et al. 2015, Li et al. 2016). There are two types of vector-mediated transient expression (i.e., virus-based replicons and classical T-DNA expression vectors) methods that have been implemented to produce recombinant proteins. Agroinfiltration and gene-gun methods could be employed for the introduction of these vectors to the plant cells (Hefferon 2016, Hefferon et al. 2004). Recombinant *A. tumefaciens* harboring GOI is co-infected with plant leaves for epichromosomal expression of pharmaceutical proteins via large-scale vacuum infiltration or bench-scale direct syringe injection. Transient expression via syringe could be applied using either leaf discs or intact leaves. For the last, the first three leaves from the top of the plants are usually used (Menassa et al. 2012). In this regard, along with creating a small nick using a needle on the abaxial side of the leaves, the agrobacteria culture are injected into the mesophylar air spaces of the cells using a needleless syringe (Sparkes et al. 2006). For the leaf disc syringe method, after producing leaf discs (with a radius of between 8 to 10 mm) commonly using cork borer, around 20 to 50 of them are poured into the plastic syringe (with the size of 20 mL) which is half-filled by *Agrobacterium* suspension. Immediately, by removing the extra air inside the syringe body, and also sealing the syringe tip, the plunger is pulled in to produce a small vacuum inside the syringe. This procedure could be applied up to three times and then the well-infiltrated leaf discs should be incubated on the MS medium and used for further molecular based analysis. Due to the use of whole plants in the intact leaf method, genetic pollution of agroinfiltrated plants could be considered a major disadvantage of the intact leaf method as compared with the leaf disc method (Matsuo et al. 2016, Piotrzkowski et al. 2012). The possibility of a vacuum-based expression system for the large scale production of pharmaceutical proteins is suitable for this purpose. This method consists of a desiccator which is connected to a vacuum pump in order to produce a large vacuum (around 100 mbar) pressure inside the desiccator. In this method, intact cut leaves and even whole plants could be applied. The whole plant is placed upside down in the desiccator which has been filled with *Agrobacterium* suspension media, so that the aerial regions of the plants are submerged in the mixture and the vacuum is applied for a period of 1 to 5 min and then released suddenly to produce well-infiltrated leaves (Chen et al. 2013). Although the vacuum infiltration method allows for greater levels of recombinant protein production, it usually needs 1 to 3 L of bacterial suspension to soak entire plants.

Table 2: Examples of proteins expressed in plants via agrobacterium-mediated stable transformation.

Protein	**Silencing suppressor**	**Expression system**	**Applied vector to transform *A. tumefaciens***	**The highest expression level**	**References**
Recombinant protein					
Soybean agglutinin (SBA)	P19	*Solanum tuberosum* tubers	pBI-101 ? promoter	0.31% TSP	Tremblay et al. 2011
human growth hormone (hGH) herbicide-resistant ahas gene	–	*Glycine max* (apical meristem of somatic embryonic axes)	pβcong3hGH for hGH (α′ subunit of the β-conglycinin promoter) pAC 321 for ahas gene (ahas promoter)	2.90 ± 0.069% of TSP	Cunha et al. 2011a
human adenosine deaminase (hADA)	–	*Pisum sativum* *N. benthamiana* *Lupinus mutabilis* (seed transformation for all)	pLPhADA (LegA2 promoter)	The highest activity of the enzyme: 3.23 U/g of tissue in pea 1.69 U/g of tissue in tobacco 4.26 U/g of tissue in tarwi	Doshi et al. 2016
Antibody					
mAb H10	p19	*N. benthamiana* leaves	pBI-Ω (CaMV 35S promoter)	1.1 mg/kg (FW)	Villani et al. 2009
Vaccine					
Rabies glycoprotein fused with B subunit of cholera toxin	–	*N. tabacum* leaves	pBI101 (CaMV 35S promoter)	0.4% of TSP	Roy et al. 2010
The Mycobacterial Ag85B/ESAT-6	–	*N. tabacum* leaves	pCB301-Kan (CaMV 35S promoter)	~ 4% of TSP	Floss et al. 2010
Synthetic COE gene of PEDV fused with the synthetic CTB (sCTB–sCOE)	–	*Lactuca sativa* cotyledons	pMYV514 (ubiquitin promoter)	0.0065% of TSP	Huy et al. 2011

cholera toxin B subunit and Domain III (amino acids 297–394) of dengue virus type 2 E glycoprotein (EIII) (CTB–EIII) vaccine	–	*N. tabacum* leaves	pMYV498 (CaMV 35S promoter)	0.0053% of TSP	Kim et al. 2010
F1-V fusion Vaccine	–	*Lactuca sativa* cotyledons	pBI121 (CaMV 35S promoter)	Up to 0.08% of TSP	Rosales-Mendoza et al. 2010
Taenia solium HP6/TSOL18	–	*Daucus carota*	pBin (CaMV 35S promoter)	14 µg/g dry-weight of carrot calli	Monreal-Escalante et al. 2016
Pharmaceutical protein					
Human coagulation factor IX (hFIX) with sub-cellular target acid-acetic-hydroxy-synthase gene (ahas)	–	*Glycine max* somatic embryonic axes form seeds	pβcong3 for hFIX (α′ subunit of the β-conglycinin promoter) pAC321 for ahas gene (ahas promoter)	0.158 mg/ ml (FW) (0.23% TSP)	Cunha et al. 2011b
B chain of human insulin-CTB fusion (CTB–InsB$_3$)	–	*N. tabacum* leaves	pBIN19 (duplicated CaMV 35S promoter)	0.11% of TSP	Li et al. 2006

Table 3: Examples of proteins expressed in plants via transient expression.

Protein	Silencing suppressor	Expression system	Applied vector to transform *A. tumefaciens*	Method	The highest expression level	References
Recombinant protein						
GFP GFP-HFBI fusion protein	p19	*N. benthamiana* leaves	pCaMterX (CaMV 35S promoter)	Vacuum agroinfiltration	51% GFP-HFBI of TSP	Joensuu et al. 2010
Soybean agglutinin (SBA)	p19	*N. benthamiana* leaves	pBI-101 (? promoter)	Syringe infiltration using intact leaves	4% of TSP	Tremblay et al. 2011
Antibody						
CMG2-Fc, the chimeric fusion protein (chimeric antibody) combining the Fc region of human IgG and the VWA domain of the CMG2 anthrax receptor	p1, p10, p19, p21, p24, p25, p38, 2b, and HCPro	*N. benthamiana* (intact plants and detached leaves)	pBIN for CMG2-Fc2/1, p1, p19, p25, and 2b pCB302 for p10, p21, p24, p38, and HcPro (CaMV 35S promoter for all)	Vacuum agroinfiltration	0.56 g/kg (FW) for p1 co-expression	Arzola et al. 2011
An aglycosylated version (Asn 297 was mutated for Gln 297) of TheraCIMR	–	*N. tabacum* leaves	pDEGF-R (CaMV 35S promoter)	Vacuum agroinfiltration	1.2 µg/gram (FW)	Rodríguez et al. 2005
hp-mAb^{PA}	–	*N. benthamiana* leaves	pBISfi (mutant pBI121) (CaMV 35S promoter)	Vacuum agroinfiltration	1 mg/kg plant tissue	Hull et al. 2005
mAb H10	p19	*N. benthamiana* leaves	pBI-Ω (CaMV 35S promoter)	Vacuum agroinfiltration	100 mg/kg (FW)	Villani et al. 2009

anti-CD20-human interleukin-2 (hIL-2) immunocytokine (2B8-Fc-hIL2)	p19	*N. benthamiana* leaves	pBI-Ω (CaMV 35S promoter)	Vacuum agroinfiltration	20 mg/kg (FW)	Marusic et al. 2016
C5-1 (a murine anti-human IgG) Co-expression of domain of human GalT fused to the N-terminal part (cytosolic tail and transmembrane domain) of *A. thaliana* N-acetylglucosaminyltransferase I (GNTI) (GalT/GNTI) with C5-1	HcPro	*N. benthamiana* Whole plant	pCAMBIA2300 for C5-1 (alfalfa plastocyanin promoter) pBLTI121 for GalT/GNTI with C5-1 (CaMV 35S promoter)	Vacuum agroinfiltration	1.5 g C5-1/kg (FW)	Vézina et al. 2009
Vaccine						
S1D epitope from porcine epidemic diarrhea virus S1D fused with cholera toxin B subunit	p19	*N. benthamiana* leaves	pMYV717 pMYV719 (duplicated CaMV 35S promoter for both vectors)	Vaccum agroinfiltration	0.07 % of TSP (for S1D fused with CTB)	Huy et al. 2016
Antigen						
Bacterial Antigen N-terminal domain of IpaD from *Shigella dysenteriae* conjugated to cholera toxin B subunit (CTB)	2B	*N. tabacum* *N. benthamiana* *Lactuca sativa* *Glycine max* (leafs in all plant species)	pBIEXCIpaD pBI-ZCIpaD pCAMBIA-ZCIpaD (CaMV 35S promoter for all)	Vacuum pump	0.224% of TSP *in N. benthamiana*	Hajibehzad et al. 2016

Agrobacterium is the natural enemy of plants. Because of their long interaction, they have been armed with different defense and counter defense mechanisms. As a host, plants have their own immune system which is also known as defense. Defense is aimed at protecting plant against a variety of pathogens like agrobacterium. On the other side, pathogens like agrobacterium have long designed a counter defense to suppress the plants defense system (Pumplin and Voinnet 2013). To put it briefly, the plant defense system is also active during transient expression which as a result alleviated the agrobacterium inoculation and thus final protein production. The suppression of the plant defense system by expressing a potent bacterial effector AvrPto to suppress PTI signaling enhance transgene expression. The same has also occurred through loss of function mutant for elongation factor Tu (EF-Tu) receptor, efr-1, which cannot sense EF-Tu MAMP (Wu et al. 2014). Recently, a new and robust transient expression system, called AGROBEST, that enables high levels of transgene expression has been employed. AGROBEST does not use a disarmed Agrobacterium or a mutant hose for protein production. It provides high levels of transgene expression by only using an acidic pH at 5.5. Low pH suppresses the plant defense pathway and thus enables the agrobacterium to grow and efficiently produce transgene expression (Wang et al. 2018).

Agroinfiltration is a rapid and flexible method for the production of recombinant proteins in plants. However, the versatility of this method is hampered by the activity of different proteases available in either the cytoplasm or apoplast of plant leaves. The protease activities degrade the protein of interest either inside the intact leaves or during the process of purification. In a broad attempt to inhibit the activity of proteases in transgene expression through agroinfiltration, more recently, Grosse-Holz et al. tested 29 different protease inhibitor candidates by cloning their corresponding genes of interest. The cloned constructs were simultaneously agroinfiltrated with three different proteins of interest, glycoenzyme α-Galactosidase; glycohormone erythropoietin (EPO); and IgG antibody VRC01r to enhance their accumulation. Of which, three unrelated protease inhibitors, the *N. benthamiana* NbPR4, NbPot1 and human HsTIMP which inhibit cysteine, serine and metalloproteases, respectively, were identified to improve the production of proteins of interest (Grosse-Holz et al. 2018).

Plant virus expression vectors have played an integral role in the successful development of modern plant-made pharmaceuticals. Distinct from transgenic crops, the expression of therapeutic proteins such as vaccines using recombinant plant viruses is rapid and avoids many of the regulatory restrictions associated with GMOs. In addition, the yield of desired protein can be vastly improved, facilitating the cost and improving the feasibility of oral immunization (Hefferon and Fan 2004). Initial, 'first generation' plant virus expression vectors were based on entire genomes of positive sense RNA viruses such as Tobacco mosaic virus (TMV), Cowpea mosaic virus (CPMV) and Potato virus X (PVX). However, 'second generation' vectors have advanced with respect to biological containment strategies, size restriction of vaccine proteins as well as yield improvements. Vaccines can be expressed as full-length proteins, as part of fusion proteins, or can be expressed on the surface of plant virus particles; as a collective repeat pattern is better recognized by the immune system. These 'deconstructed vectors' are composed solely of the minimum virus components required for replication. Removal of the movement and coat protein genes necessitates the delivery of the expression vector to the host plant by vacuum infiltration, and thus, the synchronous production of the vaccine protein in all plant tissues. A recent review on the use of plant virus vectors for molecular pharming can be found in (Hefferon 2014).

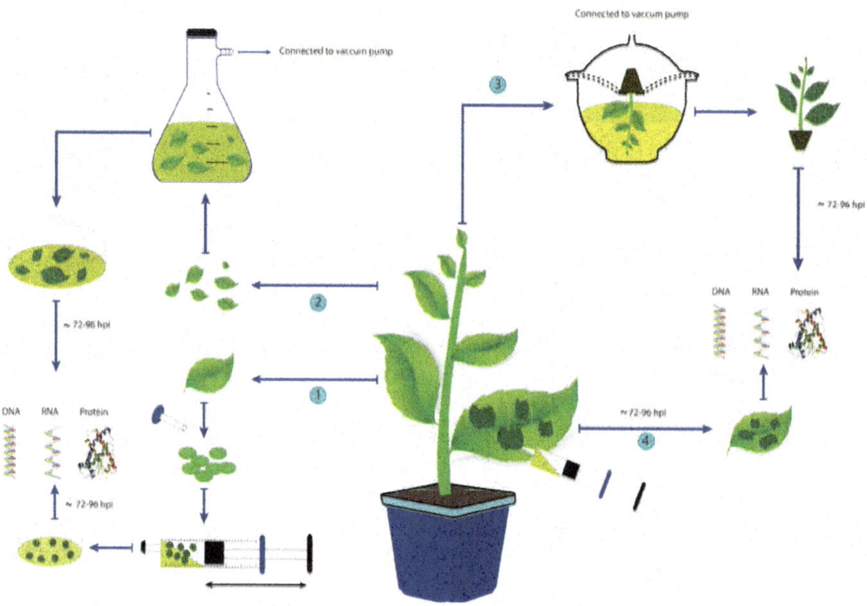

Figure 2: Schematic representation of the present transient expression procedures. 1—syringe-based infiltration method using leaf discs: in this method, leaves are cut into small pieces of approximately 8–10 mm using a cork borer and then added to the body of a 20 mL plastic syringe filled nearly to 10 mL with *Agrobacterium* suspension. Removing the extra air inside the syringe body and subsequently blocking the tip of the syringe using a plastic cap by pulling the plunger can create a small vacuum inside the syringe which is sufficient for infiltrating leaves. Agroinfiltration is achieved by releasing the plastic cap. 2—vacuum-based infiltration method using intact leaves: In this method, intact fresh leaves are cut and placed inside a covered Erlenmeyer flask which is half-filled with *Agrobacterium* suspension and connected to a vacuum pump to produce a large vacuum pressure. 3—whole plant body infiltration method: the entire plant leaf system is submerged into the infiltration buffer inside the desiccator which is connected to the vacuum pump. Notably, applying and releasing the vacuum through the pump will produce agroinfiltration. 4—syringe-based infiltration method using intact leaves: In this method, *A. tumefaciens* culture harboring GOI is resuspended in infiltration buffer and injected into the intact leaves using a needleless syringe. It is worth noting that, in all four procedures, the resulting infiltrated leaves will be used for molecular based analysis such as DNA, RNA, and protein extraction.

The Challenges ahead for Molecular Farming and their Potential Solutions

The significant advantage of plants over other traditional expression systems such as bacteria is cost-effectiveness. Based on reports, the cost of producing pharmaceutical proteins using plants is as little as 3–10 percent of the cost required for microbial fermentation (Chen et al. 2014). After all, the plants are often suffering from low expression levels of foreign genes along with the high cost of purification methods which are required to be resolved before commercialization (Sabalza et al. 2014). Although researchers recruit a series of cost-effective molecular-based purification methods, such as aqueous two-phase partitioning that is used to remove plant alkaloids and phenols, it seems there is a need to develop a new procedure which could be able to boost

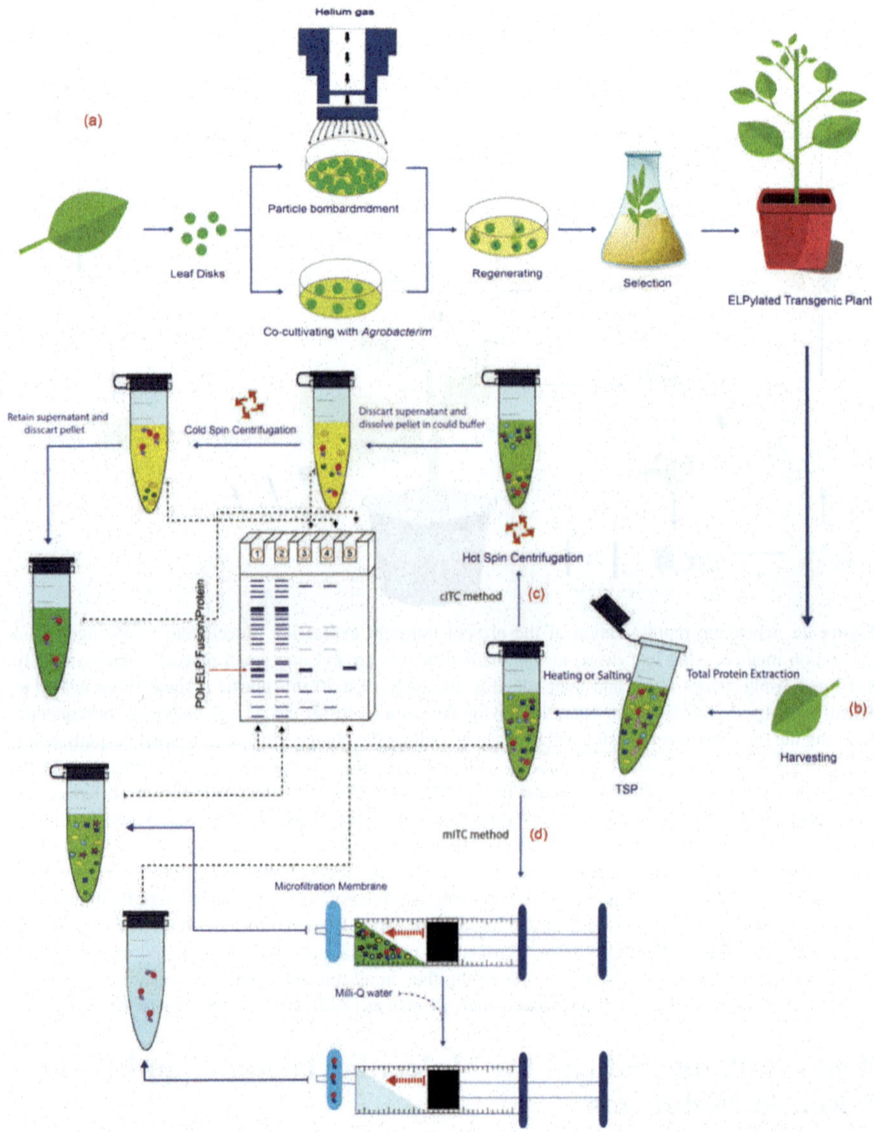

Figure 3: Schematic generation of transgenic plants using *Agrobacterium* and gene gun mediated transformation methods and purification of POI-ELP fusion proteins by inverse transition cycling (ITC). (a)—Leaves are cut into small pieces of approximately 8–10 mm using a cork borer and employed by *Agrobacterium* or Gene gun mediated transformation methods to produce transgenic plants. (b)—Leaves from the resulting transgenic plants are harvested and employed to extract Total Soluble Protein (TSP). Afterward, the inverse transition of ELPylated proteins is triggered by salting (i.e., 2.5 M NaCl) and/or heating a temperature above the T*t*, resulting in the formation of POI-ELP aggregates. After this step, the cITC and mITC methods could be employed for the purification of ELPylated proteins. Thus, in the cITC method (c)—the resulting phase inversed TSPs are centrifuged at a temperature above the T*t*, via hot spin centrifugation. After centrifugation, the supernatant containing the contaminating soluble proteins

Figure 3 contd. ...

foreign protein expression levels without any need for high cost purification techniques (Ma et al. 2005, Moustafa et al. 2016, Twyman et al. 2003). Recently, several approaches (i.e., transcription and/or translation-boosting, subcellular targeting of recombinant proteins to a suitable organelle, re-engineering protein structures to stabilize them, increasing a gene copy number of foreign genes, codon-optimization of the target gene, and choosing suitable hosts which are fitted to high level production of our POI) have been developed to increase foreign protein yields in plants. In the next section of this review, we will address some of the mentioned bottlenecks and their potential solutions.

Heterologous genes and importance of Codon Optimization

There are 61 different triplets or codons which correspond to produce 20 different amino acids. This means that most of the amino acids except for Tryptophan and Methionine are encoded by two to six codons. Therefore, because of the degenerate nature of the genetic code, the same amino acid sequence could be encoded and translated by different triplets. In addition, because the synonymous codons (those that encode the same amino acids) are not chosen randomly, the genomes of different organisms, and even different genomes of single organisms prefer to have distinct codon bias (Hershberg and Petrov 2008). It is worth noting that a number of studies have shown that gene expression levels not only could be affected by inside features (IFs), including codon usage bias, intron sequences, UTRs, putative transcription termination signals such as AAUAAA, cryptic splicing sites which may exist inside the heterologous genes and GC content which halt the transcription process and destabilize the mRNA, outside features (OFs) including host systems, culture conditions, expression cassettes and promoters could also alter GOI expression (Castillo-Davis and Hartl 2002, Coghlan and Wolfe 2000, Habibi et al. 2017, Ullrich et al. 2015, Wang and Roossinck 2006). Of these, as explained earlier, codon usage bias is the most important determinant of heterologous gene expression levels and is considered as a tool for optimization of foreign sequences. For instance, the codon-optimized *hACHE-S* gene resulted in an approximately 5–fold increase in hACHE-*S* accumulation using agroinfiltration based transient expression compared with the native hACHE-*S* protein, in which the level of the optimized version

---

...Figure 3 contd. from facing page

is discarded, while the pellet containing the insoluble POI-ELP coacervate along with other probable insoluble contaminants is retained and resuspended in low-salt buffer at a temperature below T*t*. The solution is again centrifuged as a cold spin, the supernatant containing solubilized POI-ELP is retained and the pellet containing the insoluble contaminants is removed. It should be noted that the centrifugation process could be repeated 3–5 times to obtain pure POI-ELP fusion protein. The pure ELPylated fusion protein could be treated with a specific protease, whose recognition site has been inserted between POI and ELP, and followed by one round of cITC, resulting in the separation of pure POI. (d)—since the aggregates of the ELPylated proteins attain up to 2 microns in diameter after inversion; inversed phase TSP is passed through a syringe, which is coupled to a 0.2 µm microfiltration membrane, to discard all contaminants and retain the ELPylated protein within the membrane. For resolubilization of retained ELPylated proteins, 3 mL of Milli-Q water is passed through the membrane at a temperature below the T*t*, resulting in ELPylated proteins that are also passed through the membrane and retained.

was 264 ± 40 mU/mg, while the non-optimized hACHE-S was 58 ± 3 mU/mg (Geyer et al. 2007). It seems, achieving high-level heterologous gene expression with optimized sequences compared with non-optimized sequences further show that codon usage bias has many effects on transgene expression, i.e., preventing premature polyadenylation and/or increasing translation rate of GOI and/or decreasing GOI susceptibility to RNA silencing (Heitzer et al. 2007, Tanaka et al. 2014). Today, optimization of codon usage of foreign genes to the host genome is the most important factor in the beginning of transgene expression studies. Because a reference set of highly expressed genes from a multitude of organisms and organelles are available, the codon optimization programs are able to analyze, compare and match the GOI sequence with the host's codon usage pattern (Liu and Xue 2005, Streatfield et al. 2001). The GOI sequence could be modified or *de novo* synthesized for incorporation to an expression vector (Egelkrout et al. 2012). Many available codon usage optimization programs such as Optimizer (http://genomes.urv.cat/OPTIMIZER/), DNAWorks version 3.2.3 (https://hpcwebapps.cit.nih.gov/dnaworks/), and Codon Optimization Tool (https://eu.idtdna.com/CodonOpt) could be used for this purpose. Codon Optimization Index value (CAI) is considered a quantitative tool for forecasting the correlation that has been observed between the codon bias of a gene and its expression level (Gustafsson et al. 2004). It is generally believed that increasing the CAI value of heterologous GOIs leads to its high-level expression compared to those with low CAI. However, increasing the CAI value inevitably brings an excess in A/T content which is generally considered as a marker for mRNA destabilization and reduction in protein production (De Rocher et al. 1998, Geyer et al. 2007).

Importance of Subcellular Localization of Foreign Proteins

By default, proteins of interest (POI) that lack targeting signal peptides are located into the cytoplasm (Spiegel et al. 1999). However, it seems that the cytoplasm is not a suitable location for the accumulation of recombinant proteins, as the resident ubiquitin/proteasome system may affect POI accumulation (Doran 2006, Robert et al. 2016). Furthermore, the cytoplasm lacks protein disulfide isomerase and specific chaperones which reinforce protein misfolding and thus contribute to the proteolysis of foreign proteins (Benchabane et al. 2008). For these reasons, in numerous cases, the accumulation level of foreign proteins in the cytoplasm is low and does not overstep 0.1 percent of TSP (Tavladoraki et al. 1993), however, in certain cases the cytoplasmic expression of antibodies is more desirable (Artsaenko et al. 1995). Therefore, co-expression of POI in conjugation with protease inhibitors is crucial for those proteins that need to be expressed and accumulated in the cytoplasm. To overcome these obstacles and the possibility of increasing the overall protein accumulation levels, recombinant proteins are purposefully targeted to subcellular compartments with a low level of protease activity and the capacity for post-translational modification, such as the endoplasmic reticulum (ER) and ER-derived protein bodies as well as to storage tissues such as seeds.

Directing Recombinant Proteins to the Endoplasmic Reticulum (ER)

Because the ER and its derivations such as proteins bodies, protein storage vacuoles, oil bodies, and lipid droplets have very low hydrolytic activity and high plasticity, they are considered safe locations for the accumulation of recombinant proteins in transgenic plants (Vitale and Pedrazzini 2005). In general, two types of proteins, which are typically accumulated inside the ER have been identified. 1—ER-resident proteins harboring signal peptides: in contrast to typical soluble proteins which ought to be secreted from the ER, some soluble proteins that carry tetrapeptide ER-retention motifs, i.e., KDEL or HDEL (Lys/His-Asp-Glu-Leu) will be directed to the ER lumen from the Golgi apparent by means of retrograde transport using the ERD2p receptor (Lewis et al. 1990, Semenza et al. 1990). Large amounts of those proteins act as helpers of protein folding such as chaperones, therefore, they will remain within the ER as long as they harbor functional ER-retention signal (Pelham 1990, Vitale and Denecke 1999). Hence, by fusion of KDEL or HDEL signal peptides to the POI's C-terminus, accumulation levels of foreign proteins would increase nearly by two or three orders of magnitude as compared to those which accumulate in the cytoplasm (Ram et al. 2002, Wandelt et al. 1992). However, it is well acknowledged that the K/HDEL based retrieval of foreign proteins into the ER could be a competitive mechanism that depends on the presence of other proteins harboring these signal peptides. As a result, the POI will either enter the secretion pathway or will be transported to lytic organelles, in the condition it fails to compete with different K/HDEL labeled proteins (Frigerio et al. 2001, Pimpl et al. 2006). To overcome this limitation, another strategy based on fusion proteins has been developed.

2—ER-resident aggregate form proteins: A group of seed storage proteins are translocated into the ER lumen and either accumulate principally within the ER as massive aggregates, also known as protein bodies (PBs) or are secreted into storage vacuoles (Herman and Larkins 1999, Vitale and Ceriotti 2004). Through an unknown mechanism the ER-derived PBs are kept close to the ER membrane. PBs direct the location for accumulation of recombinant proteins, as they supply a secure setting for folding and assembling of POIs, as well as increase the accumulation of foreign proteins (Galili 2004, Vitale and Ceriotti 2004). Prolamins, a species of plant storage proteins, accumulate within the lumen of ER in the form of secretions referred to as PBs. They are found in wheat (gliadin), barley (hordein), rye (secalin), corn (zein), and sorghum (kafirin) (Coleman and Larkins 1999, Esen 1987, Xu and Messing 2008). It is noteworthy that prolamins do not have the K/HDEL signal peptide at their sequences, and the molecular mechanism behind their accumulation inside the ER and finally into PB form appears to be a complex process that is not yet fully understood (Müntz 1998, Shewry and Halford 2002). However, recently, it has been demonstrated that the K/HDEL signal peptide is crucial for the accumulation and aggregation of β-glucosidase (PYK10) in the ER-derived protein bodies (Hara-Nishimura et al. 2004). It has been demonstrated that by genetically fusing the GOI with several fusion tags such as elastin-like polypeptides (ELPs), hydrophobins (HFBs), and Zera®, not only the ultimate expression levels of foreign proteins have been improved, the purification of recombinant proteins has been facilitated (Conley et al. 2009, Conley et al. 2011, Joseph et al. 2012, Khan et al. 2012). It is hypothesized that by inducing ER-derived PB, which results from the function of all three fusion tags, the POI are shielded from proteolysis, and consequently the final production yields are raised (Conley et al. 2009). Furthermore, in contrast to affinity chromatography-based systems that have been designed to simplify the purification of recombinant proteins through fusing them to carriers or peptides such as maltose binding protein, glutathione S-transferase, cellulose binding domain, and thioredoxin, ELPs, HFBs, and Zera tags are not dependent on chromatography (Maina et al. 1988, Makhzoum et al. 2014b, Nilsson et al. 1992, Ong et al. 1989, Tsao et al. 1996). Notably, affinity chromatography-based approaches for the purification of recombinant proteins suffer from two crucial limitations. First, they are not suitable for a high throughput scale or for purifying a large amount of POIs. The main cause is cost related, as multiple affinity columns have to be reused for every purification step. Second, in relation to metal affinity chromatography, there is a potential risk of metal toxicity which may exit the column and contaminate the pharmaceutical protein, thus preventing its use for recombinant protein purification.

N-terminal domain of γ Zein protein (Zera®)

The 27 KD γ Zein, a maize seed storage prolamin, along with β, δ, and ά zeins, accumulate in PBs (Lending and Larkins 1989). The γ and β Zeins, the major content of maize storage proteins, play distinguished roles in the formation of PBs, despite the absence of canonical retention signal H/KDEL. Nonetheless, δ and ά zeins do not accumulate in PBs unless, they co-express with γ and β Zeins (Ludevid et al. 1984, Prat et al. 1985). Zera® consists of three major subdomains namely an N-terminal signal peptide containing a Cys-Gly-Cys motif, a highly repetitive proline-rich sequence containing eight repeats of PPPVHL motif and a proline-X sequence containing four cysteines. Based on a mutational study, the latter two sub-domains have crucial roles in Zera multimerization and the formation of PBs (Llop-Tous et al. 2010). In detail, two intrinsic molecular properties of Zera participate in self-assembly, polymerization and PBs formation: 1—six cysteine residues present in the Zera sequence participate in the formation of disulfide bonds between Zera oligomers, a first pivotal step before the generation of PBs. Mutational studies of Zera® subdomains demonstrated that the two cysteine residues (Cys^7 and Cys^9) located within the N-terminal region are crucial for oligomerization. Nevertheless, the C-terminal cysteine residues (Cys^{64}, Cys^{82}, Cys^{84}, and Cys^{92}) do not participate in multimerization (Llop-Tous et al. 2010, Pompa and Vitale 2006). 2—the alcohol-soluble feature of Zera, which reflects its general hydrophobic nature, and also the amphipathic nature of its highly repetitive motif $(PPPVHL)_8$, facilitates Zera-Zera self-assembly and is probably responsible for its sustainability within the ER (Geli et al. 1994, Kogan et al. 2002). The peptide of 112 proline-rich amino acid at the N-terminal domain of γ Zein, known as Zera® is able to induce the formation of ER-derived PBs in seeds and even in vegetative tissues when is tagged to POIs (Coleman et al. 1996, Llop-Tous et al. 2010). Fusion of POIs with Zera can facilitate the purification and recovery of tagged recombinant proteins using iodixanol gradient centrifugation method, and consequently increase the final production level. As such, by fusing the cyan fluorescent protein (eCFP), human growth hormone (hGH) and human immunodeficiency virus negative factor (Nef) protein to the N-terminal domain of the γ Zein protein, the levels of fused recombinant proteins have accumulated by about 0.5 g/kg FW, 3.2 g/kg FW, and 1% of TSP, respectively (de Virgilio et al. 2008, Hajibehzad et al. 2016). Furthermore, Zera fused recombinant proteins could be dried at 37°C and stored up to 5 months at 25°C with no reduction in POIs stability. This property may be referred to PB-based protection of foreign proteins from any undesirable proteolytic activity (Torrent et al. 2009). Therefore, Zera-based generation of PBs possibly provides a promising avenue for storing a huge amount of recombinant proteins within the small limited area of the cell.

Elastin-Like Polypeptides (ELPs)

Elastin is a highly elastic extracellular protein that has been well-known as a plasticity and resilience supplier to vertebrate connective tissues. This is the most abundant protein in tissues or organs in which elasticity is the most important, including artery walls, lung, tendon, and skin (Daamen et al. 2007). In human beings, the *ELN* gene encodes tropoelastin protein, the soluble precursor of elastin (Curran et al. 1993). The encoded protein contains two major domains: 1—the hydrophilic domain that is choked with hydrophilic amino acids like Lys and Ala, and 2—the hydrophobic region that is rich in hydrophobic residues such as Val, Pro, Ala and Gly, which are arranged as tetra, penta, and hexa repeats. The elasticity and resilience of elastin protein is chiefly rooted in its hydrophobic domain (Urry et al. 1976). Numerous synthetic polypeptides based on elastin hydrophobic domain have been designed and employed as a replacement strategy to provide a straightforward, inexpensive, and effective method for increasing the POI's production, that facilitates and simplifies their purification (Floss et al. 2009, Kaldis et al. 2013, Phan et al. 2014). As such, an artificial elastin-like protein composed of a repetitive pentapeptide (Val-Pro-Gly-Xaa-Gly), where the Xaa residue could be any amino acid apart from proline (ELP) inspired by human elastin protein has been designed (Urry et al. 1991). Naturally, the physiologically reversible shifting of ELP from monomeric phase to self-aggregation micron-scale coacervates form is triggered by a minor alteration of temperature (2–3°C), which is also termed transition temperature (T*t*). Below T*t*, ELPs are in the monomeric form and soluble in water, and by increasing the temperature over T*t*, they form large hydrophobic aggregates and become insoluble (Li et al. 2001, Li and Daggett 2003). Self-aggregated and insoluble forms of ELPs could be returned to soluble phase by reducing the temperature to below T*t*; nevertheless, the reversion is influenced by several components, including buffer composition, ELP chain length, the guest Xaa residue, and salt concentration (Girotti et al. 2004, Meyer and Chilkoti 2004, Ribeiro et al. 2009). It should be noted that the transition could also be triggered by applying environmental conditions, in particular by the type and concentration of buffer salts. As an example, the Trx-ELP fusion protein demonstrated a T*t* of 60°C in PBS, 38°C in 1.0 M NaCl (added to PBS) and 18°C in 2.5 M NaCl (added to PBS). These results demonstrated that the inverse phase transition of any ELPylated fusion proteins could be applied at room temperature by adding for example 2.5 M NaCl to the PBS solution. This means that the inverse phase transition can be exerted not merely by increasing the temperature to above Tt, but by altering in PBS buffer composition (Ge et al. 2006).

Recently, ELPylation as a strategy by which GOIs are fused to ELP has been increasingly used for yield enhancement of several POIs including antigens, interleukins, single-chain Fv (scFv) antibodies, erythropoietin, spider silk proteins and several other proteins as well as for simplification of their purification procedures (Conley et al. 2009, Scheller et al. 2004). As such, by ELPylation, the temperature-dependent returnable characteristics of ELP make fusion partners able to be aggregated in the insoluble form and then allow them to be purified *via* a cost-effective non-chromatographic purification method referred to as Inverse Transition Cycling (ITC) (Meyer and Chilkoti 1999). Two different procedures are employed for the purification of POI-ELP fusion proteins. First, the centrifugation-based ITC method (cITC), involves the precipitation and purification of POI-ELP fusion proteins by employing a combination of heat and/or salt so as to obtain an inverse transition phase, followed by centrifugation and resolubilization at a temperature below T*t*. In detail, a hot spin (T > T*t*), centrifugation of isothermally inversed phase ELPylated proteins results in the formation of a pellet at the bottom of a micro tube consisting of POIs-ELPs and other insoluble contaminants. The resulting supernatant harboring other soluble contaminants is discarded while the pellet is collected and dissolved in a cold buffer. As a cold spin (T < T*t*), the resolubilized pellet containing POI-ELP is centrifuged again in a cold buffer to obtain a supernatant harboring the ELPylated protein and a pellet containing other contaminants. This combination of hot and cold spins represent one round of the ITC method (MacEwan et al. 2014, Meyer and Chilkoti 1999). Repeating the hot and cold spins will enhance the purity of POI yield (Fig. 3). The cITC method has been successfully recruited to purify glycoprotein 130 (Lin et al. 2006), erythropoietin (EPO) (Conley et al. 2009), human interleukin-10 (IL-10) (Kaldis et al. 2013) and several other proteins. To avoid the partial and/or full denaturation of ELPylated proteins which may happen by applying a temperature higher than the optimal range in cITS method, a new procedure called microfiltration-based ITC method (mITC) has been developed. Meyer and Chilkoti (2002) found as "upon inducing inverse phase transition, the aggregates of the ELPylated proteins attain up to 2 micron diameter", and is considered to be the basis for the main idea of the mITC method. Ge et al. 2006 found that the micron-scale coacervates could be easily sustained using a syringe coupled to a microfiltration membrane. In this method, a 0.2 μm hydrophilic microfiltration which is coupled to a syringe is used for hoarding of ELPylated proteins. The isothermally inversed phase ELPylated proteins is injected into the syringe and passed through the membrane in order to retain POI-ELP fusion aggregates and elute out other contaminant proteins. The resolubilization of ELPylated aggregates is done by adding Milli-Q water into the microinfiltration membrane at a temperature below T*t*. Finally, the resolubilized POI-ELP fusion proteins could be easily passed through the membrane and recovered. This method is able to retrieve up to 100% of POI present in the bacterial cell lysate after inducing phase transition by changing the physiological conditions and membrane recovery. The mITC method as an efficient, rapid and non-expensive procedure is more amenable not merely for lab-based purification of recombinant proteins, it could be employed for industrial scale applications (Ge et al. 2006, MacEwan et al. 2014, Meyer and Chilkoti 2004, Phan and Conrad 2011). As an example, up to now, ELPylated avian influenza virus haemagglutinin (AIVHA) and thioredoxin (Trx) proteins have been successfully purified and scaled up using the mITC method (Ge et al. 2006, Phan et al. 2013).

The cleavage of POI from ELP could be accomplished by treatment with a specific protease whose cleavage site has been inserted between them. However, protease-based purification of recombinant proteins also suffers from problems. For example, proteases can increase costs and add more steps to the purification procedure. Furthermore, the addition of a protease cleavage site may interfere with the structure of POI. Moreover, due to non-specific activity, proteases could interact with the ELP partner. Very recently, as an alternative technology, intein methodology has been designed and employed for recombinant protein purification via self-cleavage without the need for proteases (Banki et al. 2005).

In addition to low levels of production, transgenic plants also suffer from a high inter-transformant variation which may stem from various elements such as promoters, chromosomal position effects (Meyer 2000), and variability in expression levels derived from epigenetic procedures such as gene silencing phenomena (De Bolle et al. 2003). Of these elements, the promoter is not only an initiative sequence for starting gene transcription, it also affects transgene expression levels and more importantly, plays a role in expression variability among individual transformants (Butaye et al. 2004, Makhzoum et al. 2014a). The constitutive 35S promoter of the Cauliflower mosaic virus which directs gene expression uniformly in all tissues and cells is the most widely used promoter for the expression of recombinant proteins in plants. Theoretically, the constitutive promoters seem to confer high levels of transgene expression in all plant cells, while significant variability in expression levels has been practically observed when monocotyledonous or dicotyledonous have been used, arguing that the constitutive promoters are species-specific, making them essential for optimizing candidate promoters for specific groups (Dutt et al. 2014).

Viral Suppressors of RNA Silencing (VSRs) and their Potential Role in Molecular Faming

As previously mentioned, another cause for fluctuations in foreign gene expression in plants undoubtedly is the gene silencing mechanism which can act at both transcriptional (Transcriptional Gene Silencing, TGS) and post transcriptional (Post Transcriptional Gene Silencing, PTGS or RNA silencing) which levels to inhibit foreign gene production (Borges and Martienssen 2015). Inhibition of transgene expression at the TGS level is connected with promoter inactivation, methylation or chromatin remodeling, while suppression at the PTGS level mostly stems from sequence homology within transcribed regions (Aufsatz et al. 2002, Hernandez-Pinzon et al. 2007). Generally, multiple copies of foreign T-DNAs integrated into the plant genome and/or high level of gene expression derived from strong constitutive promoter act as a flag to trigger the host RNA silencing mechanism and to initiate transgene silencing. However, in some cases, it has been shown that transgene silencing could also be induced by a single copy of T-DNA without a higher level of expression (Meza et al. 2002, Muskens et al. 2000). Tissue-specific expression of recombinant proteins together with employing weak constitutive promoters with low expression potential can reduce problems related to RNA silencing (Depicker and Van Montagu 1997). Nonetheless, the latter seems not to be fruitful because the main objective of molecular farming, as mentioned earlier. Viruses, along host antiviral mechanisms, have evolved to inhibit RNA silencing-mediated inactivation (Díaz-Pendón and Ding 2008). The most common strategies recruited as a counter defense by viruses are virus-encoded suppressors of RNA silencing (VSRs), using great diversity and no obvious sequence homology (Burgyán and Havelda 2011, Gupta et al. 2014). VSRs are capable of suppressing RNA silencing in different silencing pathways such as viral recognition, dicing, RISC assembly and RNA targeting, as well as interference with viral genome epigenetic modification (Csorba et al. 2015). Viral suppressors of silencing (VSRs) is an effective strategy that has been widely used to prevent plant RNA silencing and confer high levels of expression. In this regard, co-expression of VSRs with GOI has been considered as a suitable strategy to confer high-levels of recombinant protein expression (Arzola et al. 2011). It is worth noting that the VSRs have been mostly employed to resolve a major drawback of the transient expression platform, while the transformation of PTGS mutants confers a high level of foreign gene expression in stably transgenic plants, even throughout consecutive generations (Voinnet et al. 2003). In this respect, the use of the PTGS *sgs2* mutant in *A. thaliana* plants as the recipient for transformation has demonstrated that the average GUS activity in primary transformants becomes enhanced nearly eightfold. This result is not unexpected because in sgs2 plants, RNA silencing has been deactivated (Butaye et al. 2004). In the following section, the strategies of several VSRs are described.

The most well-used VSR, P19 protein, has been identified as an expression enhancer and is able to increase expression yields of a range of proteins in *N. benthamiana* up to 50 fold (Voinnet et al. 2003). P19 protein binds to ds-siRNA, and inhibits HEN1 action, due to its higher affinity for si/miRNA than for HEN1 (Kontra et al. 2016). Sequestering duplex siRNAs and preventing RISC assembly are the consequences of these VSRs actions (Lakatos et al. 2004). It is noteworthy that P19 acts as a molecular caliper, selecting size-dependent siRNA duplexes and sequestering them in a sequence-independent manner (Pumplin and Voinnet 2013). The VSRs such as Pothoslatentaureus virus (PolV) P14, the p38 coat protein of TCV and CMV 2b protein have been shown to bind to dsRNA in a size-independent manner and interfere with vsiRNA maturation (Csorba et al. 2015). P19 interacts with AGO1 protein before RISC assembly to destabilize it (Kontra et al. 2016). Furthermore, other VSRs such as Poliovirus P0, PVX P25, and Tomato ringspot virus (ToRSV) suppressor CP have a function with AGO1. It is noteworthy that PVX P25, in addition to interacting with AGO1, interacts with AGO2, 3, and 4 (Csorba et al. 2015). AGO1 mRNA stability and translation are downregulated by miR168 accumulation (Rhoades et al. 2002). P19 VSR specifically promotes increasing the level of miR168 which reduces the antiviral action of AGO1 (Várallyay et al. 2013). Recently, it has been shown that a combination of p19, p1, and p0, as gene silencing suppressors, could boost gene expression through transient expression up to 400 $\mu g/g^{-1}$FW, which is close to three times higher than those only used a single gene silencing suppressor (Habibi et al. 2018).

CMV 2b protein is a kind of VSR that has multiple functions in the inhibition or suppression of RNA silencing. As mentioned above, it can bind to dsRNA and prevent dsRNA into siRNA. In addition it sequesters viRNA and its dsRNA precursor (González et al. 2010). The 2b protein of the Fny strain of CMV interacts with PAZ and PIWI domains of AGO1 and AGO4 proteins to inhibit the slicing activity of antiviral RISC complexes. AGO4-target hypomethylation, which is independent of CMV 2b catalytic potential, is another retainer of AGO4 (Hamera et al. 2012). CMV 2b has a role in transporting siRNA to the nucleus, and facilitates epigenetic modification (Kanazawa et al. 2011). Moreover, CMV 2b interferes in systemic silencing and blocks RNA silencing signals over long-distance but not in local infections (Roth et al. 2004).

The GW182 protein family contains a GW/WG motif as the component of the RISC which interacts with AGO proteins. Therefore, mimicking the GW-motif is another mechanism that VSRs such as TCV P38, and P1 in Sweet potato mild mottleipomovirus (SPMMV) utilize to function with unloaded and sRNA loaded AGO1 proteins, respectively, to interrupt with RNA silencing (Giner et al. 2010).

The host RDRs (RDR1, 2 and 6) contribute to RNA silencing amplification and systemic signal spread. VSRs are able to disrupt at this stage by interacting with RDR6 or a cofactor of RDR6, SGS3, and assist with virus replication and spread (Pumplin and Voinnet 2013). Notably, RDR1, which is considered to be the antagonist of RDR6, increases the inhibition of endogenous silencing. The Rice yellow stunt virus (RYSV) protein P6 has an acting mode with RDR6 to block secondary siRNA production, and suppress systemic signals. The V2 protein of Tomato yellow leaf curlovirus (TYLCV) either interacts with SGS3 or competes with it for binding to dsRNAs with 5′ overhangs (Fukunaga and Doudna 2009).

Potyviral helper-component protease (HC-Pro), in addition to protease activity, is involved in the binding and sequestration of sRNA duplexes, and impacts viral cell-to-cell and long-distance movement. HC-Pro also prevents endonuclease activity and protease activity of the 20S proteasome, increasing the spread of viral particles. HC-Pro interacts with host factors such as the endogenous suppressor of RNA silencing rgs–CaM (regulator of gene silencing-calmodulin-like protein), the ethylene-inducible transcription factor RAV2, translation initiation factors eIF4E/iso4E (Ivanov et al. 2016). Zucchini yellow mosaic virus (ZYMV) Hc-Pro interacts with HEN1 directly (Jamous et al. 2011). Ivanov and colleagues have shown that HC-Pro acts via two distinct mechanisms to suppress silencing. One mechanism involves the deactivation of two key enzymes, S-adenosyl methionine synthase (SAMS) and S-adenosyl- homocystein-hydrolase (SAHH) required for HEN1. The other mechanism involves direct interaction with AGO1 (Ivanov et al. 2016).

AL2 (also known as AC2, C2, or TrAP, transcriptional activator protein) suppressor of Tomato golden mosaic virus (TGMV) and L2 (also known as C2) suppressor of Beet curly top virus (BCTV) belong to the geminivirus family. Although AL2 and L2 are both involved in virus infection, AL2, but not L2 is required for viral late gene expression, and suppresses PTGS by encoding WEL1 (Werner exonuclease-like 1) and rgs-CaM domains as negative regulators of RNA silencing. Furthermore, in this mechanism, AL2 and L2 interact with each other, and target the activity of an adenosine kinase (ADK) enzyme, producing S-adenosyl methionine (SAM) as a cofactor of methyltransferase. Repression of ADK activity also interferes with TGS, thereby blocking defense against geminiviruses by restriction of viral genome methylation (Jackel et al. 2015). ßC1, VSR of Tomato yellow leaf curlChinavirus (TYLCCNV) geminivirus, inhibits the role of SAHH in SAM synthase by interacting with it, and thus affects TGS (Yang et al. 2011).

Using a completely different strategy, sweet potato chlorotic stunt virus (SPCSV) encodes a VSR which is an RNase III endonuclease, and cleaves siRNA duplexes of 21–24 nucleotides in length into approximately 14-bp products without any apparent function in mediating silencing (Cuellar et al. 2009).

The Role of Genetic Insulators in Boosting Transgene Expression

In 1928, Heintz defined heterochromatin and euchromatin based on their compaction at interphase. Basically, euchromatin is gene rich, less condensed and therefore more accessible and readily transcribed. By contrast, heterochromatin is generally gene poor, highly condensed and hence inaccessible for transcription machinery (Grewal and Jia

2007). Heterochromatin domains can be classified into two groups: facultative and constitutive. Facultative heterochromatin comprises developmentally regulated genes, and their expression and therefore level of compaction change in response to developmental and environmental signals. Constitutive heterochromatin, in contrast, includes large domains of the genome and contains repetitive DNA elements (e.g., satellite sequences and transposons) which remain condensed throughout the cell cycle (Feng and Michaels 2015, Wang et al. 2016).

Heterochromatin formation occurs in three main phases: establishment, spreading, and maintenance (Wang et al. 2014). Since spreading via positive feedback is an inherent feature of heterochromatin, restricting it to appropriate domains is critical for cells (Verrier et al. 2015). Similarly, cells need to limit the effect of enhancers (as distal regulatory elements) to specific regions of euchromatin. DNA sequences known as insulators or boundary elements, act as barriers (which prevent the spread of heterochromatin from one domain to the next) or enhancer-blockers (which prevent the communication between distant elements such as enhancers to influence gene expression) to constrain independent structural and functional chromatin domains (Wei et al. 2005).

Random integration of transgenes into the target genome can result in two types of problems. The first stems from the structure of the genomic sequence flanking the insertion site. If an integration event occurs within or close to heterochromatin, or the transgene becomes heterochromatinized due to the spreading of heterochromatin, a partial or complete loss of transgene expression (over time or followed by differentiation) could be expected. This phenomenon, referred to as chromosomal position effects can cause complete silencing of the transgene or position effect variation. The second problem, referred to as insertional mutagenesis, may arise within the regulatory elements of the transgene and potentially can alter the expression of nearby endogenous genes; e.g., activation of a proto-oncogene under the influence of a promoter or enhancer of a transgene (Emery 2011). Another type of problem can be addressed as enhancer-promoter interference. The mechanism by which enhancer elements regulate gene expression, is position- and orientation-independent. As in the case of 35S enhancer (which is widely used for constitutive expression of selectable markers), enhancer–promoter interference can constitutively activate tissue-specific promoters adjacent to the transgene (Singer et al. 2011).

To solve these obstacles, insulator sequences have been used frequently (Fig. 4). It has been demonstrated that barrier insulator elements flanking a transgene could potentially shield it from heterochromatinization. For example, the matrix attachment region (MAR) from *Phaseolus vulgaris* is a typical barrier element used in transgenic plants (Nandi and Khush 2015). MARs are A-T rich repeated DNA sequences which attach to the nuclear scaffold and thus contribute to the formation of rosette-like structures in metaphasic chromosomes. Since the chicken β-globin insulator element was the first vertebrate insulator to be identified and recovered as a MAR, we will take a quick look at its mechanism. cHS4 (chicken hypersensitivity site 4) insulator is located at the 5′ end of the chicken β-globin locus and functions as both barrier element and enhancer-blocker. This sequence contains strong binding sites for CTCF protein (Gaszner and Felsenfeld 2006). CTCF is a ubiquitous zinc finger protein and is described as the main insulator protein in vertebrates; nonetheless, any CTCF homologues have not been identified as of yet in yeast, *Caenorhabditis elegans* or plants. This protein contributes to the formation

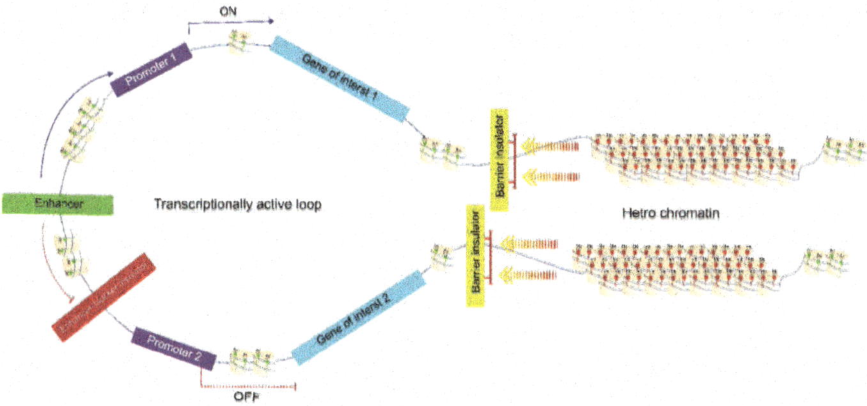

Figure 4: Ideal hypothetical representation of an insertional construct containing two types of insulators. Barrier insulators at the right and left borders of the construct could prevent it from heterochromatinization, hence forming and maintaining a transcriptionally active region. By using enhancer-blocker insulators between an enhancer and a promoter, enhancer activity could be constrained to a specific promoter. As is shown schematically, promoter 1 is expressed constitutively due to enhancer activity; however, promoter 2 is not affected because of the inhibitory effect of enhancer-blocker, so it could maintain, for instance, tissue-specific expression.

of chromatin loops which, in part, facilitate long-range chromosomal interactions and, on the other hand, separate antagonistic histone modification regions to form functional domains of gene expression (Ong and Corces 2014). A series of MARs have been used as barrier insulators in transgenic plants to investigate the improvement of transgene expression. These studies showed MARs effectiveness in the reduction of transgene silencing and an increase in expression levels. Based on these observations, MARs could act via increasing transformation efficiency, transgene expression levels (either absolutely or in proportion to copy number) or single gene insertion, resulting in the reduction of transgene rearrangement and transformant variability (Brouwer et al. 2002, Oh et al. 2005, Xue et al. 2005). Diamos et al. 2018 show that Rb7 and TM6, as MAR elements greatly increase the transgene expression in transient through agroinfiltration. They also indicate that the effect of Rb7 dependents on the type of promoter and terminator which is used along. For example, they introduced the combination of Rb7 with EU that was the strongest terminator as the strong MAR and terminator association. However, the combination of the second strongest terminator, NbACT3, with Rb7 showed the lowest expression ratio. They showed that the combination of terminators with MARs could boost the transgene expression up to 60% compared with NOS terminator alone (Diamos et al. 2018).

To prevent unwanted enhancer-promoter interactions and subsequent mis-expression of transgenes, enhancer-blocker insulators are employed (Fig. 4). These sequences may directly inhibit the enhancer by acting as a decoy, or function by forming chromatin loops by interacting with each other (Raab and Kamakaka 2010). In general, enhancer-promoter interference within transgenes in plants could be correlated through enhancer strength, promoter sensitivity and distance between enhancer and target promoter. It is observed that enhancer-blockers from humans, fungi and *Drosophila* can prevent CaMV35S enhancer interference in plants, to some extent. EXOB fragment (from *bacteriophage lambda*) and

TBS fragment (from petunia and as a MAR element) are identified as enhancer-blockers which can be used to remedy enhancer- or promoter-promoter crosstalk in transgenic plants (Hily et al. 2009, Yang et al. 2011).

Concluding Remarks

Plant molecular farming (PMF) is a branch of biotechnology where plants are employed as an expression platform to produce recombinant proteins is called plant molecular farming (PMF). During the last two decades, various pharmaceutical, industrial, and technical proteins have been made *in planta* as an alternative to other expression systems such as bacteria and mammalian cells. Several expression systems are currently employed for the production of valuable recombinant proteins. Among them, plants have many attractive advantages that make them suitable for the production of recombinant proteins. For example, plants offer an affordable expression platform to produce pharmaceutical materials. Hence, based on reports, the cost of producing pharmaceutical proteins using plants is as much as 3–10 percent of the cost required to produce them using microbial fermentation. Second, the final products of recombinant proteins produced in the plant are expected to be much safer than those produced in both microbes and animals in terms of contamination by pathogens, oncogenes, or viral DNA, which are dangerous to a human health.

Despite the advantages mentioned above, plant molecular farming has some challenges concerning the low expression level of foreign proteins and the high price of purification methods. These issues are required to be addressed before the commercialization of transgenic plants. Supporting reports, the production and accumulation of recombinant proteins in the cytoplasm is generally low and does not overstep 0.1% of TSP because of its robust ubiquitin/proteasome system, which may affect the accumulation of the protein of interest. Therefore, subcellular compartments such as the endoplasmic reticulum (ER) and ER-derived protein bodies have been considered since they have a low level of protease activity and are capable of post-translational modification. Other strategies have also been taken into account to overcome the low expression levels of recombinant proteins in plants. For example, the co-expression of VSRs and the GOI has been considered a suitable strategy to confer high-level recombinant protein expression. Furthermore, to prevent unwanted position effects and enhancer-promoter interference, which may affect heterologous gene expression, barrier insulator elements flanking a transgene and enhancer-blocker insulators have been used frequently (Fig. 4).

This review has described the rationale for designing genetic constructs for molecular pharming through a variety of sub-cellular organelles. Both transient and stable transgene expression has been discussed. How to increase expression of pharmaceutical proteins using viral suppressors of gene silencing and genetic insulators has been included, as have been solutions for potential challenges to the generation of plant-made pharmaceuticals. To date, several products have become commercialized, and more will enter the marketplace as the technology further matures. Many more products are currently undergoing clinical trials. The next decade will be a turning point for introducing molecular pharming into the mainstream of drug design, development, and delivery.

References

Agrawal, P., D. Verma, and H. Daniell. 2011. Expression of Trichoderma reesei β-mannanase in tobacco chloroplasts and its utilization in lignocellulosic woody biomass hydrolysis. PloS one 6(12): p.e29302.

Albarracín, R.M., M.L. Becher, I. Farran, V.A. Sander, M.G. Corigliano, M.L. Yácono, S. Pariani, E.S. López, J. Veramendi, and M. Clemente. 2015. The fusion of Toxoplasma gondii SAG1 vaccine candidate to Leishmania infantum heat shock protein 83-kDa improves expression levels in tobacco chloroplasts. Biotechnology Journal 10(5): 748–759.

Artsaenko, O., M. Peisker, U. Nieden, U. Fiedler, E.W. Weiler, K. Müntz, and U. Conrad. 1995. Expression of a single-chain Fv antibody against abscisic acid creates a wilty phenotype in transgenic tobacco. Plant J. 8(5): 745–750.

Arzola, L., J. Chen, K. Rattanaporn, J.M. Maclean, and K.A. McDonald. 2011. Transient co-expression of post-transcriptional gene silencing suppressors for increased in planta expression of a recombinant anthrax receptor fusion protein. Int. J. Mol. Sci. 12(8): 4975–4990.

Aufsatz, W., M.F. Mette, J. van der Winden, A.J. Matzke, and M. Matzke. 2002. RNA-directed DNA methylation in Arabidopsis. Proc. Natl. Acad. Sci. 99(suppl 4): 16499–16506.

Banki, M.R., F. Liang, and D.W. Wood. 2005. Simple bioseparations using self-cleaving elastin-like polypeptide tags. Nat. Methods 2(9): 659.

Benchabane, M., C. Goulet, D. Rivard, L. Faye, V. Gomord, and D. Michaud. 2008. Preventing unintended proteolysis in plant protein biofactories. Plant Biotechnol. J. 6(7): 633–648.

Bock, R. 2001. Transgenic plastids in basic research and plant biotechnology. J. Mol. Biol. 312(3): 425–438.

Bock, R. 2014a. Engineering chloroplasts for high-level foreign protein expression. Chloroplast biotechnology: Methods and Protocols, 93–106.

Bock, R. 2014b. Genetic engineering of the chloroplast: Novel tools and new applications. Curr. Opin. Biotechnol. 26: 7–13.

Borges, F., and R.A. Martienssen. 2015. The expanding world of small RNAs in plants. Nat. Rev. Mol. Cell Biol. 16(12): 727–741.

Brouwer, C., W. Bruce, S. Maddock, Z. Avramova, and B. Bowen. 2002. Suppression of transgene silencing by matrix attachment regions in maize a dual role for the maize 5′ ADH1 matrix attachment region. The Plant Cell 14(9): 2251–2264.

Burgyán, J., and Havelda, Z. 2011. Viral suppressors of RNA silencing. Trends Plant Sci. 16(5): 265–272.

Butaye, K.M., I.J. Goderis, P.F. Wouters, J.M.T. Pues, S.L. Delauré, W.F. Broekaert, A. Depicker, B. Cammue, and M.F. De Bolle. 2004. Stable high-level transgene expression in *Arabidopsis thaliana* using gene silencing mutants and matrix attachment regions. Plant J. 39(3): 440–449.

Castillo-Davis, C.I., and D.L. Hartl. 2002. Genome evolution and developmental constraint in Caenorhabditis elegans. Mol. Biol. Evol. 19(5): 728–735.

Chen, Q., H. Lai, J. Hurtado, J. Stahnke, K. Leuzinger, and M. Dent. 2013. Agroinfiltration as an effective and scalable strategy of gene delivery for production of pharmaceutical proteins. Advanced Techniques in Biology & Medicine 1(1).

Qiang Chen, Luca Santi, and Chenming Zhang. 2014. Plant-Made Biologics. BioMed Research International, vol. 2014, Article ID 418064, 3 pages.

Cheng, M., J.E. Fry, S. Pang, H. Zhou, C.M. Hironaka, D.R. Duncan, T.W. Conner, and Y. Wan. 1997. Genetic transformation of wheat mediated by *Agrobacterium tumefaciens*. Plant Physiol. 115(3): 971–980.

Coghlan, A., and K.H. Wolfe. 2000. Relationship of codon bias to mRNA concentration and protein length in Saccharomyces cerevisiae. Yeast 16(12): 1131–1145.

Coleman, C.E., E.M. Herman, K. Takasaki, and B.A. Larkins. 1996. The maize gamma-zein sequesters alpha-zein and stabilizes its accumulation in protein bodies of transgenic tobacco endosperm. The Plant Cell 8(12): 2335.

Coleman, C.E., and B.A. Larkins. 1999. The prolamins of maize, seed Proteins. Springer, 109–139.

Conley, A.J., J.J. Joensuu, A.M. Jevnikar, R. Menassa, and J.E. Brandle. 2009. Optimization of elastin-like polypeptide fusions for expression and purification of recombinant proteins in plants. Biotechnol. Bioeng. 103(3): 562–573.

Conley, A.J., J.J. Joensuu, A. Richman, and R. Menassa. 2011. Protein body-inducing fusions for high-level production and purification of recombinant proteins in plants. Plant Biotechnol. J. 9(4): 419–433.

Csorba, T., L. Kontra, and J. Burgyán. 2015. Viral silencing suppressors: Tools forged to fine-tune host-pathogen coexistence. Virology 479: 85–103.

Cuellar, W.J., J.F. Kreuze, M.-L. Rajamäki, K.R. Cruzado, M. Untiveros, and J.P. Valkonen. 2009. Elimination of antiviral defense by viral RNase III. Proc. Natl. Acad. Sci. 106(25): 10354–10358.

Cui, C., F. Song, Y. Tan, X. Zhou, W. Zhao, F. Ma, Y. Liu, J. Hussain, Y. Wang, and G. Yang. 2011. Stable chloroplast transformation of immature scutella and inflorescences in wheat (*Triticum aestivum* L.). Acta Biochim. Biophys. Sin. 43(4): 284–291.

Cunha, N.B., A.M. Murad, T.M. Cipriano, A.G.C. Araújo, F.J. Aragão, A. Leite, G.R. Vianna, T.R. McPhee, G.H. Souza, M.J. Waters, and E.L. Rech. 2011a. Expression of functional recombinant human growth hormone in transgenic soybean seeds. Transgenic Research 20(4): 811–826.

Cunha, N.B., A.M. Murad, G.L. Ramos, A.Q. Maranhão, M.M. Brígido, A.G.C. Araújo, C. Lacorte, F.J. Aragão, D.T. Covas, A.M. Fontes, and G.H. Souza. 2011b. Accumulation of functional recombinant human coagulation factor IX in transgenic soybean seeds. Transgenic Research 20(4): 841–855.

Curran, M.E., D.L. Atkinson, A.K. Ewart, C.A. Morris, M.F. Leppert, and M.T. Keating. 1993. The elastin gene is disrupted by a translocation associated with supravalvular aortic stenosis. Cell 73(1): 159–168.

Daamen, W.F., J. Veerkamp, J. Van Hest, and T. Van Kuppevelt. 2007. Elastin as a biomaterial for tissue engineering. Biomaterials 28(30): 4378–4398.

Davies, H.M. 2010. Commercialization of whole-plant systems for biomanufacturing of protein products: Evolution and prospects. Plant Biotechnol. J. 8(8): 845–861.

Davoodi-Semiromi, A., M. Schreiber, S. Nalapalli, D. Verma, N.D. Singh, R.K. Banks, D. Chakrabarti, and H. Daniell. 2010. Chloroplast-derived vaccine antigens confer dual immunity against cholera and malaria by oral or injectable delivery. Plant Biotechnology Journal 8(2): 223–242.

De Block, M., J. Schell, and M. Van Montagu. 1985. Chloroplast transformation by *Agrobacterium tumefaciens*. EMBO J. 4(6): 1367.

De Bolle, M.F., K.M. Butaye, W.J. Coucke, I.J. Goderis, P.F. Wouters, N. van Boxel, W.F. Broekaert, and B.P. Cammue. 2003. Analysis of the influence of promoter elements and a matrix attachment region on the inter-individual variation of transgene expression in populations of *Arabidopsis thaliana*. Plant Sci. 165(1): 169–179.

De Rocher, E.J., T.C. Vargo-Gogola, S.H. Diehn, and P.J. Green. 1998. Direct evidence for rapid degradation of *Bacillus thuringiensis* Toxin mRNA as a Cause of Poor Expression in Plants. Plant Physiol. 117(4): 1445–1461.

de Virgilio, M., F. De Marchis, M. Bellucci, D. Mainieri, M. Rossi, E. Benvenuto, S. Arcioni, and A. Vitale. 2008. The human immunodeficiency virus antigen Nef forms protein bodies in leaves of transgenic tobacco when fused to zeolin. J. Exp. Bot. 59(10): 2815–2829.

Depicker, A., and M. Van Montagu. 1997. Post-transcriptional gene silencing in plants. Curr. Opin. Cell Biol. 9(3): 373–382.

Diamos, A.G., and H.S. Mason. 2018. Chimeric 3'flanking regions strongly enhance gene expression in plants. Plant Biotechnology Journal 16(12): 1971–1982.

Díaz-Pendón, J.A., and S.-W. Ding. 2008. Direct and indirect roles of viral suppressors of RNA silencing in pathogenesis. Annu. Rev. Phytopathol. 46: 303–326.

Doran, P.M. 2006. Foreign protein degradation and instability in plants and plant tissue cultures. Trends Biotechnol. 24(9): 426–432.

Doshi, K.M., N.N. Loukanina, P.L. Polowick, and L.A. Holbrook. 2016. Seed specific expression and analysis of recombinant human adenosine deaminase (hADA) in three host plant species. Transgenic Research 25(5): 629–637.

Dutt, M., S.A. Dhekney, L. Soriano, R. Kandel, and J.W. Grosser. 2014. Temporal and spatial control of gene expression in horticultural crops. Hort. Res. 1: 14047.

Egelkrout, E., V. Rajan, and J.A. Howard. 2012. Overproduction of recombinant proteins in plants. Plant Sci. 184: 83–101.

Emery, D.W. 2011. The use of chromatin insulators to improve the expression and safety of integrating gene transfer vectors. Hum. Gene Ther. 22(6): 761–774.

Esen, A. 1987. A proposed nomenclature for the alcohol-soluble proteins (zeins) of maize (*Zea mays* L.). Journal of Cereal Science 5(2): 117–128.

Espinoza-Sánchez, E.A., J.A. Torres-Castillo, Q. Rascón-Cruz, F. Zavala-García, and S.R. Sinagawa-García. 2016. Production and characterization of fungal β-glucosidase and bacterial cellulases by tobacco chloroplast transformation. Plant Biotechnology Reports 10(2): 61–73.

Feng, W., and S.D. Michaels. 2015. Accessing the inaccessible: The organization, transcription, replication, and repair of heterochromatin in plants. Annu. Rev. Genet. 49: 439–459.

Fischer, R., E. Stoger, S. Schillberg, P. Christou, and R.M. Twyman. 2004. Plant-based production of biopharmaceuticals. Curr. Opin. Plant Biol. 7(2): 152–158.

Floss, D.M., M. Sack, E. Arcalis, J. Stadlmann, H. Quendler, T. Rademacher, E. Stoger, J. Scheller, R. Fischer, and U. Conrad. 2009. Influence of elastin-like peptide fusions on the quantity and quality of a tobacco-derived human immunodeficiency virus-neutralizing antibody. Plant Biotechnol. J. 7(9): 899–913.

Frigerio, L., A. Pastres, A. Prada, and A. Vitale. 2001. Influence of KDEL on the fate of trimeric or assembly-defective phaseolin: Selective use of an alternative route to vacuoles. The Plant Cell 13(5): 1109–1126.

Fukunaga, R., and J.A. Doudna. 2009. dsRNA with 5′ overhangs contributes to endogenous and antiviral RNA silencing pathways in plants. EMBO J. 28(5): 545–555.

Galili, G. 2004. ER-derived compartments are formed by highly regulated processes and have special functions in plants. Plant Physiol. 136(3): 3411–3413.

Gaszner, M., and G. Felsenfeld. 2006. Insulators: Exploiting transcriptional and epigenetic mechanisms. Nat. Rev. Gen. 7(9): 703–713.

Ge, X., K. Trabbic-Carlson, A. Chilkoti, and C.D. Filipe. 2006. Purification of an elastin-like fusion protein by microfiltration. Biotechnol. Bioeng. 95(3): 424–432.

Geli, M.I., M. Torrent, and D. Ludevid. 1994. Two structural domains mediate two sequential events in [gamma]-zein targeting: Protein endoplasmic reticulum retention and protein body formation. The Plant Cell Online 6(12): 1911–1922.

Geyer, B.C., S.P. Fletcher, T.A. Griffin, M.J. Lopker, H. Soreq, and T.S. Mor. 2007. Translational control of recombinant human acetylcholinesterase accumulation in plants. BMC Biotechnol. 7(1): 27.

Gisby, M.F., P. Mellors, P. Madesis, M. Ellin, H. Laverty, S. O'Kane, M.W. Ferguson, and A. Day. 2011. A synthetic gene increases TGFβ3 accumulation by 75-fold in tobacco chloroplasts enabling rapid purification and folding into a biologically active molecule. Plant Biotechnology Journal 9(5): 618–628.

Giddings, G. 2001. Transgenic plants as protein factories. Curr. Opin. Biotechnol. 12(5): 450–454.

Giner, A., L. Lakatos, M. García-Chapa, J.J. López-Moya, and J. Burgyán. 2010. Viral protein inhibits RISC activity by argonaute binding through conserved WG/GW motifs. PLoS Path. 6(7): e1000996.

Girotti, A., J. Reguera, F.J. Arias, M. Alonso, A.M. Testera, and J.C. Rodríguez-Cabello. 2004. Influence of the molecular weight on the inverse temperature transition of a model genetically engineered elastin-like pH-responsive polymer. Macromolecules 37(9): 3396–3400.

González, I., L. Martínez, D.V. Rakitina, M.G. Lewsey, F.A. Atencio, C. Llave, N.O. Kalinina, J.P. Carr, P. Palukaitis, and T. Canto. 2010. Cucumber mosaic virus 2b protein subcellular targets and interactions: Their significance to RNA silencing suppressor activity. Mol. Plant-Microbe Interact. 23(3): 294–303.

Grewal, S.I., and S. Jia. 2007. Heterochromatin revisited. Nat. Rev. Gen. 8(1): 35–46.

Grosse-Holz, F., L. Madeira, M.A. Zahid, M. Songer, J. Kourelis, M. Fesenko, S. Ninck, F. Kaschani, M. Kaiser, and R.A. van der Hoorn. 2018. Three unrelated protease inhibitors enhance accumulation of pharmaceutical recombinant proteins in Nicotiana benthamiana. Plant Biotechnology Journal 16(10): 1797–1810.

Gupta, S., S. Ganguli, and A. Datta. 2014. The Silent Assassins: Informatics of Plant Viral Silencing Suppressors, Agricultural Bioinformatics. Springer, 21–32.

Gustafsson, C., S. Govindarajan, and J. Minshull. 2004. Codon bias and heterologous protein expression. Trends Biotechnol. 22(7): 346–353.

Habibi, P., G.S. Prado, P.B. Pelegrini, K.L. Hefferon, C.R. Soccol, and M.F. Grossi-de-Sa. 2017. Optimization of inside and outside factors to improve recombinant protein yield in plant. Plant Cell, Tissue and Organ Culture (PCTOC), 1–19.

Habibi, P., C.R. Soccol, B.R. O'Keefe, L.R. Krumpe, J. Wilson, L.L.P. de Macedo, M. Faheem, V.O. Dos Santos, G.S. Prado, M.A. Botelho, and S. Lacombe. 2018. Gene-silencing suppressors for high-level production of the HIV-1 entry inhibitor griffithsin in Nicotiana benthamiana. Process Biochemistry 70: 45–54.

Hajibehzad, S.S., H. Honari, J. Nasiri, F.A. Mehrizi, and H. Alizadeh. 2016. High-level transient expression of the N-terminal domain of IpaD from Shigella dysenteriae. *In Vitro* Cell Dev. Biol. 52(3): 293–302.

Hamera, S., X. Song, L. Su, X. Chen, and R. Fang. 2012. Cucumber mosaic virus suppressor 2b binds to AGO4-related small RNAs and impairs AGO4 activities. Plant J. 69(1): 104–115.

Hara-Nishimura, I., R. Matsushima, T. Shimada, and M. Nishimura. 2004. Diversity and formation of endoplasmic reticulum-derived compartments in plants. Are these compartments specific to plant cells? Plant Physiol. 136(3): 3435–3439.

Hefferon, K., P. Kipp, and Y. Moon. 2004. Expression and purification of heterologous proteins in plant tissue using a geminivirus vector system. J. Mol. Microbiol. Biotechnol. 7(3): 109–114.

Hefferon, K. 2013. Plant-derived pharmaceuticals for the developing world. Biotechnol. J. 8(10): 1193–1202.

Hefferon, K. 2014. Plant virus expression vector development: New perspectives. BioMed Res. Int.

Hefferon, K. 2016. Plant virus nanoparticles: New applications and new benefits. Future Virol. 11(8): 591–599.

Hefferon, K.L., and Y. Fan. 2004. Expression of a vaccine protein in a plant cell line using a geminivirus-based replicon system. Vaccine 23(3): 404–410.

Heitzer, M., A. Eckert, M. Fuhrmann, and C. Griesbeck. 2007. Influence of codon bias on the expression of foreign genes in microalgae. Transgenic Microalgae as Green Cell Factories, 46–53.

Hensel, G., D.M. Floss, E. Arcalis, M. Sack, S. Melnik, F. Altmann, T. Rutten, J. Kumlehn, E. Stoger, and U. Conrad. 2015. Transgenic production of an anti HIV antibody in the barley endosperm. PloS one 10(10): e0140476.

Herman, E.M., and B.A. Larkins. 1999. Protein storage bodies and vacuoles. The Plant Cell 11(4): 601–613.

Hernandez-Pinzon, I., N.E. Yelina, F. Schwach, D.J. Studholme, D. Baulcombe, and T. Dalmay. 2007. SDE5, the putative homologue of a human mRNA export factor, is required for transgene silencing and accumulation of trans-acting endogenous siRNA. Plant J. 50(1): 140–148.

Hershberg, R., Petrov, D.A. 2008. Selection on codon bias. Annu. Rev. Genet. 42: 287–299.

Herzog, R.W., T.C. Nichols, J. Su, B. Zhang, A. Sherman, E.P. Merricks, R. Raymer, G.Q. Perrin, M. Häger, B. Wiinberg, and H. Daniell. 2017. Oral tolerance induction in hemophilia B dogs fed with transplastomic lettuce. Molecular Therapy, 25(2).

Hiatt, A., R. Caffferkey, and K. Bowdish. 1989. Production of antibodies in transgenic plants. Nature 342(6245): 76–78.

Hiei, Y., S. Ohta, T. Komari, and T. Kumashiro. 1994. Efficient transformation of rice (*Oryza sativa* L.) mediated by Agrobacterium and sequence analysis of the boundaries of the T-DNA. Plant J. 6(2): 271–282.

Hily, J.-M., S.D. Singer, Y. Yang, and Z. Liu. 2009. A transformation booster sequence (TBS) from Petunia hybrida functions as an enhancer-blocking insulator in *Arabidopsis thaliana*. Plant Cell Rep. 28(7): 1095–1104.

Hirose, T., and M. Sugiura. 1997. Both RNA editing and RNA cleavage are required for translation of tobacco chloroplast ndhD mRNA: A possible regulatory mechanism for the expression of a chloroplast operon consisting of functionally unrelated genes. EMBO J. 16(22): 6804–6811.

Hoekema, A., P. Hirsch, P. Hooykaas, and R. Schilperoort. 1983. A binary plant vector strategy based on separation of vir-and T-region of the *Agrobacterium tumefaciens* Ti-plasmid. Nature 303(5913): 179–180.

Hood, E.E., D.R. Witcher, S. Maddock, T. Meyer, C. Baszczynski, M. Bailey, P. Flynn, J. Register, L. Marshall, and D. Bond. 1997. Commercial production of avidin from transgenic maize:

Characterization of transformant, production, processing, extraction and purification. Mol. Breed. 3(4): 291–306.

Horn, M., S. Woodard, and J. Howard. 2004. Plant molecular farming: systems and products. Plant Cell Rep. 22(10): 711–720.

Hull, A.K., C.J. Criscuolo, V. Mett, H. Groen, W. Steeman, H. Westra, G. Chapman, B. Legutki, L. Baillie, and V. Yusibov. 2005. Human-derived, plant-produced monoclonal antibody for the treatment of anthrax. Vaccine 23(17): 2082–2086.

Huy, N.X., M.S. Yang, and T.G. Kim. 2011. Expression of a cholera toxin B subunit-neutralizing epitope of the porcine epidemic diarrhea virus fusion gene in transgenic lettuce (*Lactuca sativa* L.). Molecular Biotechnology 48(3): 201–209.

Huy, N.X., M.Y. Kim, T.G. Kim, Y.S. Jang, and M.S. Yang. 2016. Immunogenicity of an S1D epitope from porcine epidemic diarrhea virus and cholera toxin B subunit fusion protein transiently expressed in infiltrated Nicotiana benthamiana leaves. Plant Cell, Tissue and Organ Culture (PCTOC) 127(2): 369–380.

Ishida, Y., H. Saito, S. Ohta, Y. Hiei, T. Komari, and T. Kumashiro. 1996. High efficiency transformation of maize (*Zea mays* L.) mediated by *Agrobacterium tumefaciens*. Nat. Biotechnol. 14(6): 745–750.

Ivanov, K.I., K. Eskelin, M. Bašić, S. De, A. Lõhmus, M. Varjosalo, and K. Mäkinen. 2016. Molecular insights into the function of the viral RNA silencing suppressor HCPro. Plant J. 85(1): 30–45.

Jackel, J.N., R.C. Buchmann, U. Singhal, and D.M. Bisaro. 2015. Analysis of geminivirus AL2 and L2 proteins reveals a novel AL2 silencing suppressor activity. J. Virol. 89(6): 3176–3187.

Jamal, A., K. Ko, H.-S. Kim, Y.-K. Choo, H. Joung, and K. Ko. 2009. Role of genetic factors and environmental conditions in recombinant protein production for molecular farming. Biotechnol. Adv. 27(6): 914–923.

Jamous, R.M., K. Boonrod, M.W. Fuellgrabe, M.S. Ali-Shtayeh, G. Krczal, and M. Wassenegger. 2011. The helper component-proteinase of the Zucchini yellow mosaic virus inhibits the Hua Enhancer 1 methyltransferase activity *in vitro*. J. Gen. Virol. 92(9): 2222–2226.

Jin, S., and H. Daniell. 2015. The engineered chloroplast genome just got smarter. Trends Plant Sci. 20(10): 622–640.

Joensuu, J.J., A.J. Conley, M. Lienemann, J.E. Brandle, M.B. Linder, and R. Menassa. 2010. Hydrophobin fusions for high-level transient protein expression and purification in Nicotiana benthamiana. Plant Physiology 152(2): 622–633.

Joseph, M., M.D. Ludevid, M. Torrent, V. Rofidal, M. Tauzin, M. Rossignol, and J.-B. Peltier. 2012. Proteomic characterisation of endoplasmic reticulum-derived protein bodies in tobacco leaves. BMC Plant Biol. 12(1): 36.

Kaldis, A., A. Ahmad, A. Reid, B. McGarvey, J. Brandle, S. Ma, A. Jevnikar, S.E. Kohalmi, and R. Menassa. 2013. High-level production of human interleukin-10 fusions in tobacco cell suspension cultures. Plant Biotechnol. J. 11(5): 535–545.

Kanazawa, A., J.I. Inaba, H. Shimura, S. Otagaki, S. Tsukahara, A. Matsuzawa, B.M. Kim, K. Goto, and C. Masuta. 2011. Virus-mediated efficient induction of epigenetic modifications of endogenous genes with phenotypic changes in plants. Plant J. 65(1): 156–168.

Kempinski, C., Z. Jiang, S. Bell, and J. Chappell. 2015. Metabolic engineering of higher plants and algae for isoprenoid production, Biotechnology of Isoprenoids. Springer, 161–199.

Khan, I., R.M. Twyman, E. Arcalis, and E. Stoger. 2012. Using storage organelles for the accumulation and encapsulation of recombinant proteins. Biotechnol. J. 7(9): 1099–1108.

Kim, T.G., M.Y. Kim, and M.S. Yang. 2010. Cholera toxin B subunit-domain III of dengue virus envelope glycoprotein E fusion protein production in transgenic plants. Protein Expression and Purification 74(2): 236–241.

Kogan, M.J., I. Dalcol, P. Gorostiza, C. Lopez-Iglesias, R. Pons, M. Pons, F. Sanz, and E. Giralt. 2002. Supramolecular properties of the proline-rich γ-zein N-terminal domain. Biophys. J. 83(2): 1194–1204.

Kolotilin, I., A. Kaldis, E.O. Pereira, S. Laberge, and R. Menassa. 2013. Optimization of transplastomic production of hemicellulases in tobacco: Effects of expression cassette configuration and tobacco cultivar used as production platform on recombinant protein yields. Biotechnology for Biofuels 6(1): 65

Komari, T., Y. Takakura, J. Ueki, N. Kato, Y. Ishida, and Y. Hiei. 2006. Binary vectors and super-binary vectors. Agrobacterium Protocols, 15–42.

Kontra, L., T. Csorba, M. Tavazza, A. Lucioli, R. Tavazza, S. Moxon, V. Tisza, A. Medzihradszky, M. Turina, and J. Burgyán. 2016. Distinct effects of p19 RNA silencing suppressor on small RNA mediated pathways in plants. PLoS Path. 12(10): e1005935.

Kumar, S., A. Dhingra, and H. Daniell. 2004. Stable transformation of the cotton plastid genome and maternal inheritance of transgenes. Plant Mol. Biol. 56(2): 203–216.

Kwon, K.C., R. Nityanandam, J.S. New, and H. Daniell. 2013. Oral delivery of bioencapsulated exendin-4 expressed in chloroplasts lowers blood glucose level in mice and stimulates insulin secretion in beta-TC6 cells. Plant Biotechnology Journal 11(1): 77–86.

Lakatos, L., G. Szittya, D. Silhavy, and J. Burgyán. 2004. Molecular mechanism of RNA silencing suppression mediated by p19 protein of tombusviruses. EMBO J. 23(4): 876–884.

Le Mauff, F., G. Mercier, P. Chan, C. Burel, D. Vaudry, M. Bardor, L.P. Vézina, M. Couture, P. Lerouge, and N. Landry. 2015. Biochemical composition of haemagglutinin-based influenza virus-like particle vaccine produced by transient expression in tobacco plants. Plant Biotechnol. J. 13(5): 717–725.

Lee, L.-Y., and S.B. Gelvin. 2008. T-DNA binary vectors and systems. Plant Physiol. 146(2): 325–332.

Lee, S.B., B. Li, S. Jin, and H. Daniell. 2011. Expression and characterization of antimicrobial peptides Retrocyclin-101 and Protegrin-1 in chloroplasts to control viral and bacterial infections. Plant Biotechnology Journal 9(1): 100–115.

Lee, S.M., K. Kang, H. Chung, S.H. Yoo, X.M. Xu, S.-B. Lee, J.-J. Cheong, H. Daniell, and M. Kim. 2006. Plastid transformation in the monocotyledonous cereal crop, rice (*Oryza sativa*) and transmission of transgenes to their progeny. Molecules and Cells 21(3): 401.

Lehtimäki, N., M.M. Koskela, and P. Mulo. 2015. Posttranslational modifications of chloroplast proteins: An emerging field. Plant Physiol. 168(3): 768–775.

Lelivelt, C.L., M.S. McCabe, C.-A. Newell, C. Bastiaan deSnoo, K.M. Van Dun, I. Birch-Machin, J.-C. Gray, K.H. Mills, and J.M. Nugent. 2005. Stable plastid transformation in lettuce (*Lactuca sativa* L.). Plant Mol. Biol. 58(6): 763–774.

Lending, C.R., and B.A. Larkins. 1989. Changes in the zein composition of protein bodies during maize endosperm development. The Plant Cell Online 1(10): 1011–1023.

Levy, A.A. 2016. T-DNA integration: Pol θ controls T-DNA integration. Nature Plants 2: 16170.

Lewis, M.J., D.J. Sweet, and H.R. Pelham. 1990. The ERD2 gene determines the specificity of the luminal ER protein retention system. Cell 61(7): 1359–1363.

Li, B., D.O. Alonso, and V. Daggett. 2001. The molecular basis for the inverse temperature transition of elastin. J. Mol. Biol. 305(3): 581–592.

Li, B., and V. Daggett. 2003. The molecular basis of the temperature-and pH-induced conformational transitions in elastin-based peptides. Biopolymers 68(1): 121–129.

Li, D., J. O'Leary, Y. Huang, N.P. Huner, A.M. Jevnikar, and S. Ma. 2006. Expression of cholera toxin B subunit and the B chain of human insulin as a fusion protein in transgenic tobacco plants. Plant Cell Reports 25(5): 417–424.

Li, L., X. Wang, L. Yang, Y. Fan, X. Zhu, and X. Wang. 2016. Large-scale production of foreign proteins via the novel plant transient expression system in Pisum sativum. Plant Biotechnol. Rep. 10(4): 207–217.

Lin, M., S. Rose-John, J. Grötzinger, U. Conrad, and J. Scheller.2006. Functional expression of a biologically active fragment of soluble gp130 as an ELP-fusion protein in transgenic plants: Purification via inverse transition cycling. Biochem. J. 398(3): 577–583.

Liu, C.-W., C.-C. Lin, J.J. Chen, and M-J. Tseng. 2007. Stable chloroplast transformation in cabbage (*Brassica oleracea* L. var. capitata L.) by particle bombardment. Plant Cell Rep. 26(10): 1733–1744.

Liu, Q., and Q. Xue. 2005. Comparative studies on codon usage pattern of chloroplasts and their host nuclear genes in four plant species. J. Genet. 84(1): 55–62.

Llop-Tous, I., S. Madurga, E. Giralt, P. Marzabal, M. Torrent, and M.D. Ludevid. 2010. Relevant elements of a maize γ-zein domain involved in protein body biogenesis. J. Biol. Chem. 285(46): 35633–35644.

Ludevid, M., M. Torrent, J. Martinez-Izquierdo, P. Puigdomenech, and J. Palau. 1984. Subcellular localization of glutelin-2 in maize (*Zea mays* L.) endosperm. Plant Mol. Biol. 3(4): 227–234.

Ma, J.K.C., E. Barros, R. Bock, P. Christou, P.J. Dale, P.J. Dix, R. Fischer, J. Irwin, R. Mahoney and M. Pezzotti. 2005. Molecular farming for new drugs and vaccines. EMBO Reports 6(7): 593–599.

MacEwan, S.R., W. Hassouneh, and A. Chilkoti. 2014. Non-chromatographic purification of recombinant elastin-like polypeptides and their fusions with peptides and proteins from *Escherichia coli*. J. Vis. Exp. (88).

Maina, C.V., P.D. Riggs, A.G. Grandea, B.E. Slatko, L.S. Moran, J.A. Tagliamonte, L.A. McReynolds. 1988. An *Escherichia coli* vector to express and purify foreign proteins by fusion to and separation from maltose-binding protein. Gene 74(2): 365–373.

Makhzoum, A., R. Benyammi, K. Moustafa, and J. Trémouillaux-Guiller. 2014a. Recent advances on host plants and expression cassettes' structure and function in plant molecular pharming. BioDrugs 28(2): 145–159.

Makhzoum, A., S. Tahir, M.E.O. Locke, J. Trémouillaux-Guiller, and K. Hefferon. 2014b. An *in silico* overview on the usefulness of tags and linkers in plant molecular pharming. Plant Sci. Today 1(4): 201–212.

Marusic, C., F. Novelli, A.M. Salzano, A. Scaloni, E. Benvenuto, C. Pioli and M. Donini. 2016. Production of an active anti-CD20-hIL-2 immunocytokine in Nicotiana benthamiana. Plant Biotechnology Journal 14(1): 240–251.

Matsuo, K., N. Fukuzawa, and T. Matsumura. 2016. A simple agroinfiltration method for transient gene expression in plant leaf discs. J. Biosci. Bioeng. 122(3): 351–356.

Menassa, R., A. Ahmad, and J.J. Joensuu. 2012. Transient expression using agroinfiltration and its applications in molecular farming, Molecular farming in plants: recent advances and future prospects. Springer, 183–198.

Meyer, D.E., and A. Chilkoti. 1999. Purification of recombinant proteins by fusion with thermally-responsive polypeptides. Nat. Biotechnol. 17(11): 1112–1115.

Meyer, D.E., and A. Chilkoti. 2004. Quantification of the effects of chain length and concentration on the thermal behavior of elastin-like polypeptides. Biomacromolecules 5(3): 846–851.

Meyer, P. 2000. Transcriptional transgene silencing and chromatin components. Plant Mol. Biol. 43(2): 221–234.

Meyers, B., A. Zaltsman, B. Lacroix, S.V. Kozlovsky, and A. Krichevsky. 2010. Nuclear and plastid genetic engineering of plants: Comparison of opportunities and challenges. Biotechnol. Adv. 28(6): 747–756.

Meza, T.J., B. Stangeland, I.S. Mercy, M. Skårn, D.A. Nymoen, A. Berg, M.A. Butenko, A.M. Håkelien, C. Haslekås, and L.A. Meza-Zepeda. 2002. Analyses of single-copy Arabidopsis T-DNA-transformed lines show that the presence of vector backbone sequences, short inverted repeats and DNA methylation is not sufficient or necessary for the induction of transgene silencing. Nucleic Acids Res. 30(20): 4556–4566.

Mielenz, J.R., M. Rodriguez, O.A. Thompson, X. Yang, and H. Yin. 2015. Development of Agave as a dedicated biomass source: Production of biofuels from whole plants. Biotechnology for Biofuels 8(1): 79.

Millán, F.S., A. Mingo-Castel, M. Miller and H. Daniell. 2003. A chloroplast transgenic approach to hyper-express and purify Human Serum Albumin, a protein highly susceptible to proteolytic degradation. Plant Biotechnology Journal 1(2): 71–79.

Monreal-Escalante, E., D.O. Govea-Alonso, M. Hernández, J. Cervantes, J.A. Salazar-González, A. Romero-Maldonado, G. Rosas, T. Garate, G. Fragoso, E. Sciutto, and S. Rosales-Mendoza. 2016. Towards the development of an oral vaccine against porcine cysticercosis: Expression of the protective HP6/TSOL18 antigen in transgenic carrots cells. Planta, 243(3): 675–685.

Moustafa, K., A. Makhzoum, and J. Trémouillaux-Guiller. 2016. Molecular farming on rescue of pharma industry for next generations. Crit. Rev. Biotechnol. 36(5): 840–850.

Müntz, K. 1998. Deposition of storage proteins, Protein trafficking in plant cells. Springer, 77–99.

Muskens, M.W., A.P. Vissers, J.N. Mol, and J.M. Kooter. 2000. Role of inverted DNA repeats in transcriptional and post-transcriptional gene silencing. Plant Mol. Biol. 43(2-3): 243–260.

Nandi, S., and G.S. Khush. 2015. Strategies to increase heterologous protein expression in rice grains, recent advancements in gene expression and enabling technologies in crop plants. Springer, 241–262.

Nagaya, S., K. Kawamura, A. Shinmyo, and K. Kato. 2010. The HSP terminator of *Arabidopsis thaliana* increases gene expression in plant cells. Plant and Cell Physiology, 51(2): 328–332.

Nilsson, B., G. Forsberg, T. Moks, M. Hartmanis, and M. Uhlén. 1992. Fusion proteins in biotechnology and structural biology. Curr. Opin. Struct. Biol. 2(4): 569–575.

O'Connell, K.P., and J. Handelsman. 1989. chvA locus may be involved in export of neutral cyclic p-1, 2-linked D-glucan from *Agrobacterium tumefaciens*. Mol. Plant-Microbe Interact 2: 11–16.

Oh, S.-J., J. Jeong, E.-H. Kim, N. Yi, S.I. Yi, I.C. Jang, Y. Kim, S.-C. Suh, B. Nahm, and J.-K. Kim 2005. Matrix attachment region from the chicken lysozyme locus reduces variability in transgene expression and confers copy number-dependence in transgenic rice plants. Plant Cell Rep. 24(3): 145–154.

Ong, C.-T., and V.G. Corces. 2014. CTCF: An architectural protein bridging genome topology and function. Nat. Rev. Gen. 15(4): 234–246.

Ong, E., J.M. Greenwood, N.R. Gilkes, D.G. Kilburn, R.C. Miller, and R.A.J. Warren. 1989. The cellulose-binding domains of cellulases: Tools for biotechnology. Trends Biotechnol. 7(9): 239–243.

Pelham, H.R. 1990. The retention signal for soluble proteins of the endoplasmic reticulum. Trends Biochem. Sci. 15(12): 483–486.

Phan, H.T., and U. Conrad. 2011. Membrane-based inverse transition cycling: An improved means for purifying plant-derived recombinant protein-elastin-like polypeptide fusions. Int. J. Mol. Sci. 12(5): 2808–2821.

Phan, H.T., J. Pohl, D.M. Floss, F. Rabenstein, J. Veits, B.T. Le, H.H. Chu, G. Hause, T. Mettenleiter, and U. Conrad. 2013. ELPylated haemagglutinins produced in tobacco plants induce potentially neutralizing antibodies against H5N1 viruses in mice. Plant Biotechnol. J. 11(5): 582–593.

Phan, H.T., B. Hause, G. Hause, E. Arcalis, E. Stoger, D. Maresch, F. Altmann, J. Joensuu, and U. Conrad. 2014. Influence of elastin-like polypeptide and hydrophobin on recombinant hemagglutinin accumulations in transgenic tobacco plants. PloS one 9(6): e99347.

Pimpl, P., J.P. Taylor, C. Snowden, S. Hillmer, D.G. Robinson, and J. Denecke. 2006. Golgi-mediated vacuolar sorting of the endoplasmic reticulum chaperone BiP may play an active role in quality control within the secretory pathway. The Plant Cell 18(1): 198–211.

Piotrzkowski, N., S. Schillberg, and S. Rasche. 2012. Tackling heterogeneity: A leaf disc-based assay for the high-throughput screening of transient gene expression in tobacco. PloS one 7(9): e45803.

Pompa, A., and A. Vitale. 2006. Retention of a bean phaseolin/maize γ-Zein fusion in the endoplasmic reticulum depends on disulfide bond formation. The Plant Cell 18(10): 2608–2621.

Prat, S., J. Cortadas, P. Puigdomènech, and J. Palau. 1985. Nucleic acid (cDNA) and amino acid sequences of the maize endosperm protein glutelin-2. Nucleic Acids Res. 13(5): 1493–1504.

Pumplin, N., and O. Voinnet. 2013. RNA silencing suppression by plant pathogens: defence, counter-defence and counter-counter-defence. Nature Rev. Microbiol. 11(11): 745–760.

Raab, J.R., and R.T. Kamakaka. 2010. Insulators and promoters: Closer than we think. Nat. Rev. Gen. 11(6): 439–446.

Ram, N., M. Ayala, D. Lorenzo, D. Palenzuela, L. Herrera, V. Doreste, M. Pérez, J.V. Gavilondo, and P. Oramas. 2002. Expression of a single-chain Fv antibody fragment specific for the hepatitis B surface antigen in transgenic tobacco plants. Transgenic Res. 11(1): 61–64.

Rhoades, M.W., B.J. Reinhart, L.P. Lim, C.B. Burge, B. Bartel, and D.P. Bartel. 2002. Prediction of plant microRNA targets. Cell 110(4): 513–520.

Ribeiro, A., F.J. Arias, J. Reguera, M. Alonso, and J.C. Rodríguez-Cabello. 2009. Influence of the amino-acid sequence on the inverse temperature transition of elastin-like polymers. Biophys. J. 97(1): 312–320.

Robert, S., P.V. Jutras, M. Khalf, M.-A. D'Aoust, M.-C. Goulet, F. Sainsbury, and D. Michaud. 2016. Companion protease inhibitors for the *in situ* protection of recombinant proteins in plants. Recombinant Proteins from Plants: Methods and Protocols, 115–126.

Rodríguez, M., N.I. Ramírez, M. Ayala, F. Freyre, L. Pérez, A. Triguero, C. Mateo, G. Selman-Housein, J.V. Gavilondo, and M. Pujol. 2005. Transient expression in tobacco leaves of an aglycosylated recombinant antibody against the epidermal growth factor receptor. Biotechnology and Bioengineering 89(2): 188–194.

Rosales-Mendoza, S., R.E. Soria-Guerra, L. Moreno-Fierros, A.G. Alpuche-Solís, L. Martínez-González, and S.S. Korban. 2010. Expression of an immunogenic F1-V fusion protein in lettuce as a plant-based vaccine against plague. Planta 232(2): 409–416.

Rosenthal, S.H., A.G. Diamos, and H.S. Mason. 2018. An intronless form of the tobacco extensin gene terminator strongly enhances transient gene expression in plant leaves. Plant Molecular Biology 96(4): 429–443.

Roth, B.M., G.J. Pruss, and V.B. Vance. 2004. Plant viral suppressors of RNA silencing. Virus Res. 102(1): 97–108.

Roy, S., A. Tyagi, S. Tiwari, A. Singh, S.V. Sawant, P.K. Singh, and R. Tuli. 2010. Rabies glycoprotein fused with B subunit of cholera toxin expressed in tobacco plants folds into biologically active pentameric protein. Protein Expression and Purification 70(2): 184–190.

Sabalza, M., P. Christou, and T. Capell. 2014. Recombinant plant-derived pharmaceutical proteins: Current technical and economic bottlenecks. Biotechnol. Lett. 36(12): 2367–2379.

Sanford, J.C. 1990. Biolistic plant transformation. Physiol. Plant. 79(1): 206–209.

Scharff, L.B., M. Ehrnthaler, M. Janowski, L.H. Childs, C. Hasse, J. Gremmels, S. Ruf, R. Zoschke, and R. Bock.2017. Shine-Dalgarno sequences play an essential role in the translation of plastid mRNAs in tobacco. The Plant Cell 29(12): 3085–3101.

Scheller, J., D. Henggeler, A. Viviani, and U. Conrad. 2004. Purification of spider silk-elastin from transgenic plants and application for human chondrocyte proliferation. Transgenic Res. 13(1): 51–57.

Semenza, J.C., K.G. Hardwick, N. Dean, and H.R. Pelham. 1990. ERD2, a yeast gene required for the receptor-mediated retrieval of luminal ER proteins from the secretory pathway. Cell 61(7): 1349–1357.

Shewry, P.R., and N.G. Halford. 2002. Cereal seed storage proteins: Structures, properties and role in grain utilization. J. Exp. Bot. 53(370): 947–958.

Shrawat, A.K., D. Becker, and H. Lörz. 2007. *Agrobacterium tumefaciens*-mediated genetic transformation of barley (*Hordeum vulgare* L.). Plant Sci. 172(2): 281–290.

Sidorov, V.A., D. Kasten, S.Z. Pang, P.T. Hajdukiewicz, J.M. Staub, and N.S. Nehra. 1999. Stable chloroplast transformation in potato: Use of green fluorescent protein as a plastid marker. Plant J. 19(2): 209–216.

Sikdar, S., G. Serino, S. Chaudhuri, and P. Maliga. 1998. Plastid transformation in *Arabidopsis thaliana*. Plant Cell Rep. 18(1): 20–24.

Simons, J.P., M. McClenaghan, and A.J. Clark. 1987. Alteration of the quality of milk by expression of sheep β-lactoglobulin in transgenic mice. Nature 328(6130): 530–532.

Singer, S.D., K.D. Cox, and Z. Liu. 2011. Enhancer–promoter interference and its prevention in transgenic plants. Plant Cell Rep. 30(5): 723–731.

Snell, K.D., V. Singh, and S.M. Brumbley. 2015. Production of novel biopolymers in plants: Recent technological advances and future prospects. Curr. Opin. Biotechnol. 32: 68–75.

Sparkes, I.A., J. Runions, A. Kearns, and C. Hawes. 2006. Rapid, transient expression of fluorescent fusion proteins in tobacco plants and generation of stably transformed plants. Nature Protocols 1(4): 2019–2025.

Spiegel, H., S. Schillberg, M. Sack, J. Holzem, J. Nähring, M. Monecke, Y.-C. Liao, and R. Fischer. 1999. Accumulation of antibody fusion proteins in the cytoplasm and ER of plant cells. Plant Sci. 149(1): 63–71.

Streatfield, S.J., J.M. Jilka, E.E. Hood, D.D. Turner, M.R. Bailey, J.M. Mayor, S.L. Woodard, K.K. Beifuss, M.E. Horn, and D.E. Delaney. 2001. Plant-based vaccines: Unique advantages. Vaccine 19(17): 2742–2748.

Su, J., L. Zhu, A. Sherman, X. Wang, S. Lin, A. Kamesh, J.H. Norikane, S.J. Streatfield, R.W. Herzog, and H. Daniell. 2015. Low cost industrial production of coagulation factor IX bioencapsulated in lettuce cells for oral tolerance induction in hemophilia B. Biomaterials 70: 84–93.

Svab, Z., and P. Maliga. 1993. High-frequency plastid transformation in tobacco by selection for a chimeric aadA gene. Proc. Natl. Acad. Sci. 90(3): 913–917.

Tanaka, M., M. Tokuoka, and K. Gomi. 2014. Effects of codon optimization on the mRNA levels of heterologous genes in filamentous fungi. Appl. Microbiol. Biotechnol. 98(9): 3859–3867.

Tavladoraki, P., E. Benvenuto, S. Trinca, D. De Martinis, A. Cattaneo, and P. Galeffi. 1993. Transgenic plants expressing a functional single-chain Fv antibody are specifically protected from virus attack. Nature 366(6454): 469–472.

Torrent, M., B. Llompart, S. Lasserre-Ramassamy, I. Llop-Tous, M. Bastida, P. Marzabal, A. Westerholm-Parvinen, M. Saloheimo, P.B. Heifetz, and M.D. Ludevid. 2009. Eukaryotic protein production in designed storage organelles. BMC Biol. 7(1): 5.

Tremblay, R., M. Feng, R. Menassa, N.P. Huner, A.M. Jevnikar, and S. Ma. 2011. High-yield expression of recombinant soybean agglutinin in plants using transient and stable systems. Transgenic Research 20(2): 345–356.

Tsao, K.-L., B. Debarbieri, H. Michel, and D.S. Waugh. 1996. A versatile plasmid expression vector for the production of biotinylated proteins by site-specific, enzymatic modification in *Escherichia coli.* Gene 169(1): 59–64.

Twyman, R.M., E. Stoger, S. Schillberg, P. Christou, and R. Fischer. 2003. Molecular farming in plants: Host systems and expression technology. Trends Biotechnol. 21(12): 570–578.

Tzfira, T., J. Li, B. Lacroix, and V. Citovsky. 2004. Agrobacterium T-DNA integration: Molecules and models. Trends Genet. 20(8): 375–383.

Ullrich, K.K., M. Hiss, and S.A. Rensing. 2015. Means to optimize protein expression in transgenic plants. Curr. Opin. Biotechnol. 32: 61–67.

Urry, D., M. Long, and E. Gross. 1976. Conformations of the repeat peptides of elastin in solution: An application of proton and carbon-13 magnetic resonance to the determination of polypeptide secondary structur. CRC Crit. Rev. Biochem. 4(1): 1–45.

Urry, D.W., T.M. Parker, M.C. Reid, and D.C. Gowda. 1991. Biocompatibility of the bioelastic materials, poly (GVGVP) and its γ-irradiation cross-linked matrix: Summary of generic biological test results. J. Bioact. Compatible Polym. 6(3): 263–282.

van Kregten, M., S. de Pater, R. Romeijn, R. van Schendel, P.J. Hooykaas, and M. Tijsterman. 2016. T-DNA integration in plants results from polymerase-θ-mediated DNA repair. Nature Plants 2: 16164.

Várallyay, É., E. Oláh, and Z. Havelda. 2013. Independent parallel functions of p19 plant viral suppressor of RNA silencing required for effective suppressor activity. Nucleic Acids Res. 42(1): 599–608.

Verrier, L., F. Taglini, R.R. Barrales, S. Webb, T. Urano, S. Braun, and E.H. Bayne. 2015. Global regulation of heterochromatin spreading by Leo1. Open Biol. 5(5): 150045.

Vézina, L.P., L. Faye, P. Lerouge, M.A. D'Aoust, E. Marquet-Blouin, C. Burel, P.O. Lavoie, M. Bardor, and V. Gomord. 2009. Transient co-expression for fast and high-yield production of antibodies with human-like N-glycans in plants. Plant Biotechnology Journal 7(5): 442–455.

Villani, M.E., B. Morgun, P. Brunetti, C. Marusic, R. Lombardi, I. Pisoni, C. Bacci, A. Desiderio, E. Benvenuto, and M. Donini. 2009. Plant pharming of a full-sized, tumour-targeting antibody using different expression strategies. Plant Biotechnology Journal 7(1): 59–72.

Vitale, A., and J. Denecke. 1999. The endoplasmic reticulum—Gateway of the secretory pathway. The Plant Cell 11(4): 615–628.

Vitale, A., and A. Ceriotti. 2004. Protein quality control mechanisms and protein storage in the endoplasmic reticulum. A conflict of interests? Plant Physiol. 136(3): 3420–3426.

Vitale, A., and E. Pedrazzini. 2005. Recombinant pharmaceuticals from plants: The plant endomembrane system as bioreactor. Mol. Interventions 5(4): 216.

Voinnet, O., S. Rivas, P. Mestre, and D. Baulcombe. 2003. Retracted: An enhanced transient expression system in plants based on suppression of gene silencing by the p19 protein of tomato bushy stunt virus. Plant J. 33(5): 949–956.

Wandelt, C.I., M.R.I. Khan, S. Craig, H.E. Schroeder, D. Spencer, and T.J. Higgins. 1992. Vicilin with carboxy-terminal KDEL is retained in the endoplasmic reticulum and accumulates to high levels in the leaves of transgenic plants. Plant J. 2(2): 181–192.

Wang, J., S.T. Lawry, A.L. Cohen, and S. Jia. 2014. Chromosome boundary elements and regulation of heterochromatin spreading. Cell. Mol. Life Sci. 71(24): 4841–4852.

Wang, J., S.T. Jia, and S. Jia. 2016. New insights into the regulation of heterochromatin. Trends Genet. 32(5): 284–294.

Wang, L., and M.J. Roossinck. 2006. Comparative analysis of expressed sequences reveals a conserved pattern of optimal codon usage in plants. Plant Mol. Biol. 61(4): 699–710.

Wang, Y.C., M. Yu, P.Y. Shih, H.Y. Wu, and E.M. Lai. 2018. Stable pH suppresses defense signaling and is the key to enhance agrobacterium-mediated transient expression in Arabidopsis seedlings. Scientific Reports 8(1): 1–9.

Wei, G.H., L. De Pei, and C.C. Liang. 2005. Chromatin domain boundaries: Insulators and beyond. Cell Res. 15(4): 292–300.

Wu, H.Y., K.H. Liu, Y.C. Wang, J.F. Wu, W.L. Chiu, C.Y. Chen, S.H. Wu J. Sheen, and E.M. Lai. 2014. AGROBEST: An efficient Agrobacterium-mediated transient expression method for versatile gene function analyses in Arabidopsis seedlings. Plant Methods 10(1): 1–16.

Xu, J.-H., and J. Messing. 2008. Organization of the prolamin gene family provides insight into the evolution of the maize genome and gene duplications in grass species. Proc. Natl. Acad. Sci. 105(38): 14330–14335.

Xue, H., Y.-T. Yang, C.-A. Wu, G.-D. Yang, M.-M. Zhang, and C.-C. Zheng. 2005. TM2, a novel strong matrix attachment region isolated from tobacco, increases transgene expression in transgenic rice calli and plants. Theor. Appl. Genet. 110(4): 620–627.

Yadav, N.S., J. Vanderleyden, D.R. Bennett, W.M. Barnes, and M.-D. Chilton. 1982. Short direct repeats flank the T-DNA on a nopaline Ti plasmid. Proc. Natl. Acad. Sci. 79(20): 6322–6326.

Yang, X., Y. Xie, P. Raja, S. Li, J.N. Wolf, Q. Shen, D.M. Bisaro, and X. Zhou. 2011. Suppression of methylation-mediated transcriptional gene silencing by βC1-SAHH protein interaction during geminivirus-betasatellite infection. PLoS Path. 7(10): e1002329.

Yang, Y., S.D. Singer, and Z. Liu. 2011. Evaluation and comparison of the insulation efficiency of three enhancer-blocking insulators in plants. Plant Cell, Tissue and Organ Culture (PCTOC) 105(3): 405–414.

CHAPTER 8

Plant-based Vaccines against Livestock Diseases

Way to Achieve Several Sustainable Development Goals

Mohammad Tahir Waheed,[1,*] *Muhammad Suleman Malik,*[1] *Kiran Saba,*[2] *Iqra Younus,*[1] *Niaz Ahmad,*[3] *Bushra Mirza*[1] and *Andreas Günter Lössl*[4]

Introduction

The term livestock originated between 1650–1660 AD by the combination of two words "live" and "stock". There are however varying definitions existing for the term livestock. Generally, livestock is defined as domesticated animals farmed in an agricultural setting for either food or labour. These include cattle, horses, goat, swine, sheep and fur-bearing animals. It does not include animals kept as pets or farmed birds. However, inclusion of poultry in livestock is not uncommon. Livestock not only provide meat, milk and dairy products but are also a source of wool, animal skin for use in the leather industry, labour and fertilizer.

Livestock farming is done on both a large scale at an industrial level and small scale by local farmers. The trend is gradually changing from small scale farming to a large scale bio-economy-based concept to serve consumers (Von Braun 2010). There has been an increase in the demand of animal-based food in many rapidly growing economies of

[1] Department of Biochemistry, Quaid-i-Azam University, 45320, Islamabad, Pakistan.

[2] Department of Biochemistry, Faculty of Life Sciences, Shaheed Benazir Bhutto Women University, Peshawar 25000, Pakistan .

[3] Agricultural Biotechnology Division, National Institute for Biotechnology & Genetic Engineering (NIBGE), Jhang Road, Faisalabad, 38000, Pakistan.

[4] Department of Applied Plant Sciences and Plant Biotechnology, University of Natural Resources and Applied Life Sciences, Tulln an der Donau, Austria.

* Corresponding author: tahirwaheed@qau.edu.pk

the world which has resulted in the increase in livestock production overtime throughout the globe. This demand and supply chain is maintained mostly by commercializing livestock production at larger scales. However, in the developing world, millions of people in rural areas still keep livestock at small scale. In such resource poor settings, traditional livestock farming for meat and milk is one of the main sources of income and livelihood. The global value of the livestock sector is approximately $1.4 trillion and employs around 1.3 billion people, providing livelihood to 800 million poor small-holders (Robinson et al. 2015).

The global population is expected to increase from 7.4 billion to approximately 9.9 billion by 2050 (Population Reference Bureau 2016). The total demand for animal products in developing countries is expected to be more than double by 2030 due to increase in population. Fulfilling the needs of food for the global population in the next three decades is going to be a challenge. Livestock will play a key role in providing food to an increasing population, this increase in population is expected to largely occur in developing countries. For sustainable production and continued supply, disease control is very important to keep livestock animals healthy and thus ensuring the production of healthy meat and livestock-related products. This not only ensures animal health but also reduces the risk of zoonotic diseases from animals to humans. According to an estimate, thirteen such zoonotic diseases kill 2.2 million people each year throughout the globe (Bryner 2012). Livestock diseases pose a serious threat to the global livestock market, economy, food supply and human health. In particular, this affects people in developing countries more because of the larger increase in population, lesser affordability or non-availability of control measures and poor hygiene conditions. Below, we give an overview of livestock diseases and the losses caused due to these diseases is given.

Livestock Diseases

Infectious diseases cause direct as well as indirect losses to livestock sector. Direct losses occur by causing mortality and reduced livestock productivity. Indirect losses are associated with the costs incurred for controlling diseases, the decreased market value of livestock, loss of trade and food insecurity (Dehove et al. 2012). Livestock diseases cause billions of dollars of loss to the farmers and economies of the countries throughout the world. Many recent livestock disease epidemics, in developing as well as developed countries, have prompted attention again towards the remerging livestock diseases and their management. Additionally, human lives also remain on risk due to zoonotic diseases. Recently, World Organization for Animal Health (2017) has established a single list of notifiable terrestrial and aquatic animal diseases to replace the former lists. This list includes 116 animal diseases, infections and infestations. In the case of livestock, more than 50 notifiable diseases affect the farmed livestock species throughout the world. These diseases can be divided in two main categories; bacterial and viral, depending on the infecting pathogen. Few diseases are caused by fungus, protozoans and tapeworm. Some of the major diseases of livestock are described below. A summarized list of main livestock diseases, available vaccines, deaths and related data is presented in Table 1.

Table 1: List of livestock diseases and vaccines.

Diseases	**Species/Causative Agent**	**Vaccine**	**Industrial Product**	**Additional Information/ Livestock deaths** [(a,b)]	**References**
			Viral Diseases		
Blue Tongue Disease (BTD)	• *Bluetongue virus*	Available	• Bluetongue Vaccine [(c,d,e,)] (OBP) [(f)] • Bluetongue Vaccine Type-10 [(g)] (Colorado Serum Co.)	• Non-contagious • Insects-borne disease • 27 different serotypes • 1,50,169 deaths around the world	(Dungu et al. 2003) (Hofmann et al. 2008) (Jenckel et al. 2015) (Maan et al. 2011)
Bovine Viral Diarrhea (BVD)	• *Bovine viral diarrhea virus 1* (BVDV-1) • *Bovine viral diarrhea virus 2* (BVDV-2)	Available	• Bovi Shield Gold One Shot™ [(c,g,e,)] (Zoetis) • Bovine Virus Diarrhea Vaccine [(g)] (Colorado Serum Co.)	• 2,938 deaths around the world	(Du-Gyeong and Kyoung-Seong 2017)
Epizootic Hemorrhagic Disease (EHD)	• *Epizootic hemorrhagic disease virus* (EHDV)	----------	---------------------	• *Culicoides* sp. act as vectors for transmitting EHDV among susceptible hosts.	(Mullen et al. 1984) (Temizel et al. 2009)
Fog Fever/ Acute Bovine Pulmonary Edema and Emphysema (ABPEE)	• 3-methylindole • 3-methylindoleine	----------	---------------------	• It is disorder of rumen due to udder. • Death occurs in 30% of cases.	(Breeze 1985) (Peek 2005) (Selman et al. 1976)
Foot and Mouth Disease (FMD)	• *Foot and mouth disease virus* (FMDV)	Available	• Aftovax [(c,h)] (Botswana Vaccine Institute/Merial)	• 7 different serotypes. • 88,18,470 deaths around the world	(Grubman and Baxt 2004)
Infectious Bovine Rhinotracheitis (IBR)	• *Bovine herps virus-1* (BHV-1)	Available	• Bovi Shield® IBR [(c,g)] (Zoetis) • Pre-Breed 6 (IBR-LEPTO 5) [(c,g,h)] (Colorado Serum Co.) • Leukopast 3 [(c,h)] (OBP)	• Highly contagious disease • 11,317 deaths around the world	(Nandi et al. 2009) (van Regenmortel et al. 2000)

Rabies	• *Rabies lyssavirus*	Available	• Defensor® 3 [h] (Zoetis)	• 65270 deaths around the world.	(WHO 2008)
Rift Valley Fever (RFV)	• *Rift valley fever virus*	Available	• Inactivated Rift Valley Fever Vaccine [h] (OBP) • Live Rift Valley Fever Vaccine [d,e] (OBP) • RVF Clone 13 [d,e] (OBP)	• RVF is a zoonotic and mosquito borne. • 560 deaths around the world.	(Peters and Linthicum 1994)
Rotaviral Diarrhea	• *Rotavirus*	Available	• Scourguard®4KC [c,h] (Zoetis)	----------------------	(Snodgrass et al. 1990)
Schmallenberg	• *Schmallenberg virus*	Available	• Zulvac SBV [h] (Zoetis)	----------------------	(Beer et al. 2012) (European Medicines Agency 2015)
Bacterial Diseases					
Anthrax	• *Bacillus anthracis*	Available	• Anthrax spore vaccine [i] (Colorado serum Co.)	• Vaccination provide immunity for one year. • 59253 deaths around the world.	(Inglesby et al. 1999)
Black Leg	• *Clostridium chauvoei*	Available	• Ultrabac® 7 [c,h] (Zoetis)	• 105091 deaths around the world	(Mackintosh et al. 2002)
Botulism	• *Clostridium botulinum*	Available	• Botulism (OBP)	• 14419 deaths around the world	(Shapiro et al. 1998) (Sobel 2005)
Bovine Tuberculosis	• *Mycobacterium bovis*	Available	• Mycobacterium bovis Bacille Calmette-Guérin	• BCG vaccination showed disappointing results in livestock. • Animals were slaughtered to prevent infection. • 327618 deaths around the world	(Buddle et al. 1995) (Fine 1989) (Griffin et al. 1999) (Hewinson et al. 2003) (Karlson and Lessel 1970) (Skinner et al. 2001)

Table 1 contd. ...

...Table 1 contd.

Diseases	**Species/Causative Agent**	**Vaccine**	**Industrial Product**	**Additional Information/ Livestock deaths** [(a,b)]	**References**
Bovine Vibriosis	• *Campylobacter fetus* ssp. *fetus* • *Campylobacter fetus* ssp. *venerealis*	Available	• Cattle Master® 4+VL5 [(c,g,h)] (Zoetis) • Campylobacter vibrio [(c,h)] (OBP)	• These species cause infertility and abortion in cattle and sheep	(Skirrow 1994)
Brucellosis	• *Brucella abortus* • *Brucella melitensis* • *Brucella ovis* • *Brucella suis*	Available	• Brucella Rev. 1 [(e,j)] (OBP) • Brusella S19 [(e,j)] (OBP)	• Zoonotic Disease • 557335 deaths around the worlds	(Corbel et al. 1997)
Foot rot/Foul in the Foot	• Baceroides melaninogenicus • *Dichelobacter nodosus* • *Fusobacterium necrophorum*	Available	• Footvax® [(c,h)] (MSD Animal Health)	• Vaccine have low immunization • 1131 deaths around the world	(Bennett et al. 2009) (Berg and Loan 1975)
Johne's Disease	• *Mycobacterium avium* ssp. *paratuberculosis* (MAP)	Available	• Guadair® vaccine [(h)] (Zoetis)	• A chronic infection • Economic losses of livestock occur.	(Nielsen and Toft 2009)
Leptospirosis	• *Leptospira borgpetersenii serovar hardjo* • *Leptospira canicola* • *Leptospira grippotyphosa* • *Leptospira icterohaemorrhagia* • *Leptospira pomona* • *Leptospira tarassovi*	Available	• Cattle Master® 4+VL5 (Zoetis) • Leptoferm-5® [(c,h)] (Zoetis) • Spirovac9® [(h)] (Zoetis)	• Zoonotic disease. • Different serogroups • 12471 deaths around the world	(Bharti et al. 2003) (Sullivan 1974) (Vinetz 2001)
Listeriosis	• *Listeria monocytogenes*	----------	--------------------	• A zoonotic infection • 8387 deaths around the world	(Matto et al. 2017) (Schlech et al. 1983)

Salmonella Dublin Infection/ Salmonellosis/ Intestinal salmonella infection	• *Salmonella abortusovis* • *Salmonella dublin* • *Salmonella* choleraesuis • *Salmonella* gallinarum • *Salmonella typhimurium*	Available	• Salmonella Dublin-Ty Bacterin [c,h] (Colorado Serum Co.)	• It can cause illness in humans. • 756464 deaths around the world	(Fierer 1983) (Schiaffino et al. 1996) (Werner et al. 1979)
Vibrionic dysentery	• *Campylobactor coli* • *Campylobacter jejuni*	Available	• Campyvax4® [c,h] (MSD Animal Health)	• 55478 deaths around the world	(Cole 1949) (Skirrow and Benjamin 1980)
Wooden tongue	• *Actinobacillosis lignieresii*	----------	----------------------	• Careful diagnosis is needed	(Rebhun et al. 1988)
Protozoan borne Diseases					
Coccidiosis	• *Eimeria alabamensis* • *Eimeria auburnensis* • *Eimeria bovis* • *Eimeria cylindrica* • *Eimeria ellipsoidalis* • *Eimeria subspherica* • *Eimeria zuernii*	Available	• Coccivac-D [c,j] (Intervet/Merck Animal Health) • Coccivac-T [c,j] (Intervet/Merck Animal Health)	• 20 different species. • Poultry vaccines are available only. • 2969323 deaths around the world	(Chibunda et al. 1997) Drugs.com
Cryptosporidiosis	• *Cryptosporidium agni* • *Cryptosporidium anserinum* • *Cryptosporidium baileyi* • *Cryptosporidium bovis* • *Cryptosporidium meleagridis* • *Cryptosporidium tyzzeri* • *Cryptosporidium wrairi*	----------	----------------------	• It is water borne and zoonotic disease.	(Barker and Carbonell 1974) (Current et al. 1986) (Levine 1961) (Proctor and Kemp 1974) (Slavin 1955) (Vetterling et al. 1971)
Trypanosomiasis/ sleeping disease/ Nagana	• *Trypanosoma brucei brucei* • *Trypanosoma congolense* • *Trypanosoma simiae* • *Trypanosoma vivax*	----------	----------------------	• It is Zoonotic. • Tsetse flies are vehicle. • 14453 deaths around the world.	(Stephen 1986)

Table 1 contd. ...

...Table 1 contd.

Diseases	**Species/Causative Agent**	**Vaccine**	**Industrial Product**	**Additional Information/ Livestock deaths** [a,b]	**References**
			Tick borne Disease		
Bovine Anemia/ Theileria	• *Theileria annulata* • *Theileria lestoquardi* • *Theileria ovis* • *Theileria. parva* • *Theileria separate* • *Theileria* sp. china 1 • *Theileria* sp. china 2 • *Theileria recondite*	Available	• Tayledoll [j] (Dollvet) • Teylovac™ [j] (Vetal company)	• 78959 deaths around the world.	(Ahmed et al. 2006) (Irvin 1985) (Nagore et al. 2004) (Niu et al. 2009)
Bovine babesiosis/ Red water	• *Babesia argentina* • *Babesia Bigemia* • *Babesia Bovis*	Available	• Anabasan [h] (Limor de Colombia) • Frozen Asiatic Redwater Vaccine for Cattle [j] (OBP) • Frozen African Redwater Vaccine for Cattle [j] (OBP) • Combivac 3 in 1 Live Tick Fever Vaccine [c,j] (State of Queensland)	• Also, Known as tick fever. • 34132 deaths around the world.	(Riek 1968) (Smith et al. 1981)
			Other Diseases		
Bovine Spongiform Encephalopathy (BSE)	• BSE agent	----------	---------------------	• It's an infectious and zoonotic disease. • 315997 deaths around the world.	(Doherr 2003) (Wells et al. 1987)
Bracksen Poisoning	• Bracksin toxins	----------	---------------------	---------------------	(Hirono et al. 1984)
Cold cow syndrome	• Unknown	----------	---------------------	---------------------	The cattle site

(a): Total number of deaths of livestock from 1996 to 2004 available at HANDISTATUS II of OIE (http://web.oie.int/hs2/report.asp), (b): Deaths of following species or animal group are presented (avian, buffaloes, cattle, camelidae, equidae, goats, sheep, and swine), (c): Polyvalent vaccine (vaccine for two or more diseases/two or more species/strains), (d): Live attenuated vaccine, (e): Freeze-dried vaccine, (f): *Onderstepoort Biological Products,* (g): Modified live virus, (h): Inactivated/Killed vaccine, (i): Non-encapsulated live culture, (j) Live vaccine

Bacterial Diseases

The following are most important livestock diseases caused by bacteria that result in a great loss of livestock throughout the world.

Anthrax

Anthrax is caused by *Bacillus anthracis*. It is characterized by dark centred ulcers that develop on the skin of infected animals. Although most mammals are susceptible to infection, anthrax is a more common disease of ruminants and humans. Cattle and sheep with the disease die suddenly. Animals may show some signs such as high fever and muscle tremors. Bleeding from the nose, mouth and anus of carcasses can occur. However, the signs may be absent in many cases. This disease is found in almost all parts of the world including some endemic areas with more frequent outbreaks. Some areas are subjected to sporadic outbreaks due to ingestion of spores in the soil by ruminants. Anthrax is mainly enzootic in many regions of the world, particularly in sub-Saharan Africa, Asia and Central and South America (World Health Organization 2008a). Although, the overall disease burden and its economic impact is not fully known, hundreds and thousands of deaths occur due to epizootics every year (Shadomy et al. 2016). Anthrax remains a threat because of presence of spores and several recent outbreaks in many parts of the world in 2016 have resulted in significant animal and livestock losses. Anthrax is a zoonotic disease and human lives also become at risk due to the outbreaks. It is estimated that 2,000–20,000 human anthrax cases occur annually worldwide (Martin 2010). Due to climatic changes, there is an increasing threat that the persisting spores in soil can result in disease outbreaks showing a seasonal pattern. Disease control measures include the proper disposal of dead animals, use of vaccines for the prevention of the diseases and antibiotics for a cure. The Anthrax vaccine used for livestock health consists of avirulent live spores of *B. anthracis*.

Bovine Tuberculosis (BTB)

Bovine tuberculosis is caused by *Mycobacterium bovis*, a close relative of bacteria that cause tuberculosis in humans and birds. In the early 20th century, it was one of the major diseases of domestic animals in the world. Even today, bovine tuberculosis remains an important disease of livestock and animals and is also a significant zoonotic disease. Although, the disease is found throughout the world, it is more prevalent in Africa and some parts of Asia and America. Many developed countries have either completely eliminated or reduced the bovine TB from the cattle population. However, the infection may persist in wildlife in many countries (Lawes et al. 2016). Cattle are the main hosts for *M. bovis*. Other than cattle, isolation of *M. bovis* has been made from a broad range of animals including buffaloes, sheep, goats, camels, pigs, elephants, etc. Infection can spread by the inhalation of bacteria expelled by the coughing of infected organisms and also by using raw milk from infected cows. The standard control measure is to test the animal and slaughter. Treatment can be done by using antibiotics but is rarely attempted because of the high cost and prolonged time and even it is prohibited in many countries. Preventive measures are used in humans by vaccination but not widely practiced in case of animals. Test and slaughter policies of infected animals have completely or nearly completely eliminated tuberculosis from livestock in many of the developed countries of Europe, North America and Australia. However, the existence and maintenance of *M. bovis* in wild species significantly compromised the disease eradication efforts in many countries such as Ireland, New Zealand, UK and parts of USA (Thoen et al. 2009). BTB, especially in developing countries, can lead to severe economic losses due to chronic disease, livestock deaths and trade restrictions. Additionally, the risk of dispersal of the diseases to human also remains associated.

Brucellosis

Brucellosis is caused by number of bacteria belonging to the family Brucella. Although, specific animal species are infected by specific species of bacteria, most species of Brucella are able to infect more than one animal species. Among different species of Brucella, *Brucella abortus* is the cause in cattle, *B. melitensis* in sheep and goats and *B. suis* in swine. The highest diseases burden lies in the sub-Saharan Africa, Middle East, China, India, Peru, Mexico and Mediterranean region. The cases of Brucellosis are currently on the increase in the countries of central and southwest Asia. Brucellosis is highly contagious diseases of livestock with significant economic impact. It is a zoonotic disease and humans get infected by contact with infected animals or animal products. In some cases, bacteria can spread from person to person. Most cases of human brucellosis take place due to *B. melitensis*. The disease in animals is characterized by reproductive failure or abortions. The animals may recover and will be able to have offspring following the initial abortion. However, the infected animals may continue to shed the bacteria.

Different antibiotics are used for the treatment of brucellosis in animals. The disease is usually prevented in animals by vaccination commonly known as the 'calfhood' vaccination using live vaccines. Vaccination of uninfected animals has helped to eradicate brucellosis from most of the USA (Treanor et al. 2010). There is no vaccine for immunizing humans against brucellosis. Due to the existence of a natural reservoir of *Brucella* species in wild animals, it is expected brucellosis will remain an important public health concern.

Salmonellosis

Salmonellosis is a bacterial disease which is caused by different species of Salmonella: *Salmonella enterica*, and *Salmonella bongori* with many subtypes. Approximately 2,500 different Salmonella serovars exist (World Organization for Animal Health 2008b). Different serovars of Salmonella have a narrow host range. For example, *Salmonella abortusovis* usually infects sheep, *Salmonella choleraesuis* usually infects pigs and *Salmonella dublin* usually infects cattle. There is a strong prevalence of salmonellosis in the cattle industry. Although infection mostly occurs in dairy calves of between 1–10 weeks of age, it is also observed in adult dairy cows and buffalos. Salmonellosis has a serious economic impact on the cattle industry worldwide in terms of livestock mortality, abortion, reduced production, treatment costs and reduced consumer confidence that affects the purchase. Salmonellosis is a zoonotic disease and is a significant public health concern, due to the face that humans and other animals can become infected from the consumption of contaminated drinking water, raw milk and other dairy products, and undercooked meat products. Inactivated/live vaccines are commercially available for livestock protection against some strains of Salmonella.

Haemorrhagic Septicaemia

Haemorrhagic septicaemia (HS) is a contagious bacterial disease that is caused by *Pasteurella multocida.* The disease occurs mainly in cattle and buffalos but is also reported in other livestock such as goats, camels, horses and pigs. The disease has a high mortality rate in infected animals. In South-East Asia, it is regarded as one of the most serious diseases of large. In endemic areas, approximately 2% of apparently healthy cattle and buffalo carry the organism in the lymphatic tissues of the upper respiratory tract. The infection is transmitted by direct contact between animals and by contamination of animal feed or water. The high mortality rate of livestock may lead to serious economic losses. Export of infected animals from endemically infected countries is also restricted as infection can be spread by alive diseased animals in new areas. Treatment is done by antibiotics but is mostly effective if given very early in the diseases. However, multidrug resistance is increasing in some strains of *P. multocida* (Desmolaize et al. 2011) hence vaccination is a more desirable option to prevent the diseases, especially in endemic areas. Although, serotypes of *P. multocida* causing haemorrhagic septicaemia have not been recovered from human infections; however, many serotypes of *P. multocida* have the potential to infect people and hence appropriate precautions are needed when dealing with suspected cases of HS to avoid zoonotic Pasteurellosis.

Contagious Bovine Pleuropneumonia

Contagious bovine pleuropneumonia (CBPP) is a highly contagious disease that is often generally accompanied by the inflammation of pleural tissue. This bacteria is widespread in Africa, Middle East, Southern Europe and many parts of Asia. The causative agent is the bacteria *Mycoplasma mycoides* subsp. mycoides Small Colony-bovine biotype (MmmSC). Droplets released by infected cattle by coughing can spread the disease to susceptible cattle. Infections have also been reported from Asian buffalo, bison and yak. Sheep and goats can also become naturally infected. Wild bovids and camels seem to be resistant, and, so far, do not appear to be important in the transmission of CBPP. Treatment includes the use of antibiotics but the use of antibiotics is mostly not recommended because it may delay recognition of the disease, create chronic carriers and encourage the emergence of resistant MmmSC strains. Currently, attenuated vaccines are used for prevention. However, efficacy of these vaccines is directly related to the virulence of the original strain used in it's production.

Viral Diseases

There are many viral infectious diseases that pose a risk of livestock loss. One of the viral diseases of cattle, rinderpest has been successfully eradicated. Some of viral livestock diseases are zoonotic in nature and increase the risks of infections in human and mortality. A number of viral animal diseases such as foot and mouth disease, West Nile fever, swine fever, Newcastle disease and rabies are covered in other chapters of this book. Here, a brief description of some other common viral diseases of livestock, mostly with zoonotic potential, is given below.

Rift Valley Fever

Rift valley fever is caused by a retrovirus termed as rift valley fever virus (RVFV). The disease is zoonotic, primarily infecting animals but it can also infect humans. RVFV infection can cause severe disease in both animals and humans. The disease can lead to significant economic losses caused due to death and abortion amongst the infected livestock (WHO 2017). The disease is predominant in sub-Saharan Africa and many outbreaks have occurred in African continent, Saudi Arabia and Yemen. Humans are highly susceptible to RVFV and infected with virus in many different ways such as, the bite of infected mosquitoes, contact with infected blood, contact with other body fluids or tissues of infected animals and the consumption of uncooked meat and raw milk from infected animals. People working in slaughter facilities, laboratories or hospitals are at an increased risk of acquiring infections. Attenuated virus vaccine, inactivated virus vaccine and live attenuated mutant vaccines are available for use in animals. However, these vaccines have major safety concerns (Dungu et al. 2013). No vaccine against RVFV is licensed for human use.

Nipah Virus

Nipah virus belongs to the genus Henipavirus in the family Paramyxoviridae. It is an emerging infectious disease that first appeared in 1998 and 1999 in Singapore and Malaysia in domestic pigs. Over 1 million pigs were isolated due to the disease and a total of 105 people died out of 256 infected individuals (Parashar et al. 2005). The virus can cause fatal illness in animals such as pigs, horses, dogs, and cats, sheep can also be infected. The disease is of primary veterinary concern and due its devastating zoonotic potential, it is also a public health risk. Transmission to humans in Malaysia and Singapore has almost always been through direct contact with secretions or excretions from pigs. Outbreaks in Bangladesh have been reported due to involvement of bats without involving an intermediate host by drinking raw palm sap contaminated with bat excrement (Islam et al. 2016).

Japanese Encephalitis

Japanese encephalitis is caused by the Japanese encephalitis virus (JEV). This disease is a mosquito-borne viral disease that affects horses, donkeys, pigs, and humans. Disease is also reported in some other animals such as cattle, however this is rare. JEV has gradually expanded its geographic range within Asia and has spread to parts of the western Pacific region during the last 50 years (Center for Food Security and Public Health, CFSPH 2016). The disease is not directly transferred to humans from infected animals. However, humans can get infected with JEV from the bite of an infected mosquito (vector). Asymptomatic infections have been documented in many domesticated and wild mammals; e.g., cattle, sheep, goats, rabbits, dogs and cats. The morbidity and mortality due to Japanese encephalitis can be high if vaccination is not carried out during epidemics. For instance, 4,000 people died in Japan in the 1924 epidemic and nearly 2,500 in South Korea in 1949. Several live or killed vaccines for protection against JEV are available for use in humans and animals. These vaccines are cross protective against all viral genotypes. However, immunity against genotypes other than the one employed in the vaccine strain may be weaker.

Swine Influenza

Swine influenza is a respiratory disease caused by influenza A viruses in pigs. High morbidity rates, high mortality rates during outbreaks and reduced growth rates in young pigs can cause severe economic losses (Schultz-Cherry et al. 2013). Swine influenza is a zoonotic disease and affects humans, although usually not efficiently transmitted amongst human populations. In 2009–2010, a human pandemic was caused by a virus that appears to have resulted from genetic re-assortment between North American and Eurasian swine influenza viruses (Garten et al. 2009, Smith et al. 2009). This virus now circulates in human populations worldwide as a seasonal influenza virus. In some cases, people have transmitted this virus to herds of pigs, leading to re-assortment with other swine influenza viruses, thus resulting in increased viral diversity. This factor has made the effective vaccination of pigs more difficult. Several vaccines against swine influenza are available in different parts of the world consisting of killed or live strains of virus.

African Swine Fever

African swine fever is a notifiable disease that is caused by African swine fever virus (ASFV), a member of Asfarviridae family. The disease exists in different forms as mild to severe. Most severe form cause almost 100% of death in pigs. The disease may exist in peracute form (highly virulent virus) causing very few symptoms and sudden death (World Organization for Animal Health, OIE 2013). The virus is spread by direct contact with infected pigs and also via contaminated material from the environment. Feeding on infected meat or garbage is one of the major sources of spread of ASF. The disease is predominant in sub-Saharan Africa. It used to exist in some European countries; however, it has been eradicated from most parts of Europe. Nevertheless, the risk of the spread of infection exists internationally through garbage from international airports or seaports. No vaccine is available against ASFV. The control measures include a careful import policy, efficient sterilization of garbage, quarantine and removing the source of the virus.

Current Strategies for Controlling Livestock Diseases

Livestock diseases are controlled in two ways; (1) by using vaccines before the disease to immunize animals (2) by using antibiotics (Fig. 1). Mostly vaccines are used to immunize livestock against various diseases. Currently available vaccines against various livestock diseases mostly consist of either live attenuated/weakened or killed microorganisms. A list of livestock diseases along with causative agent and available vaccines is given in the Table 1. Undoubtedly, these vaccines provide effective protection against respective diseases. By vaccination and other measures, one of the livestock diseases, rinderpest, has

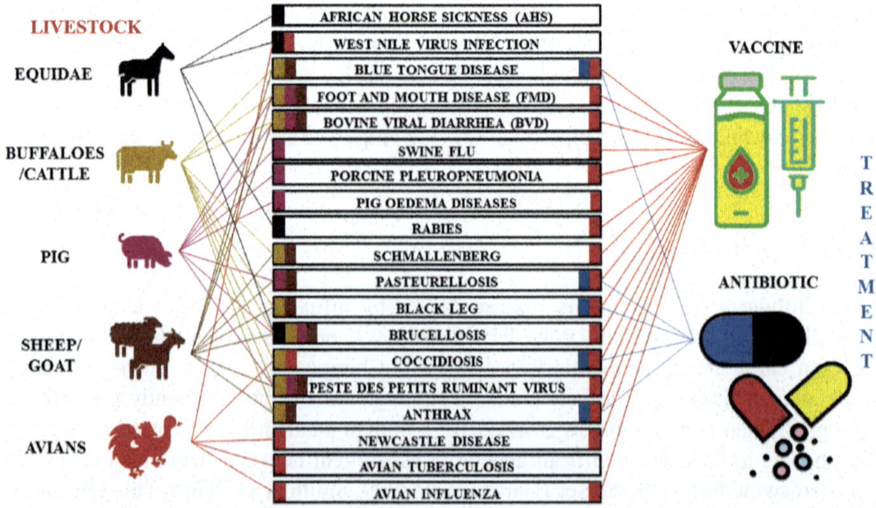

Figure 1: Prevention and treatment strategies of some of livestock diseases.

been eradicated from the world, only the second infectious disease eradicated after small pox (World Organization for Animal Health 2011). Available livestock vaccines have helped many countries to achieve the "disease free" status for some of the major diseases of livestock such as foot and mouth disease (FMD), contagious bovine pleuropneumonia (CBPP), classical swine fever, highly pathogenic avian influenza, Peste des petits ruminants (PPR) and Newcastle disease. However, there are certain shortcomings and disadvantages related to use of these vaccines in livestock. First of all, it is important for animals to be vaccinated to be in good welfare throughout their lives to ensure that they are in a fit state to respond successfully to vaccination (Morton 2007). One of the major issues related with live vaccines is safety. The use of such vaccines can sometimes cause disease either due to infection in immunocompromised hosts or due to genetic reversion. Vaccine virus may mutate (revert) towards more virulence and additionally, may also recombine with other viruses in the host (Lee et al. 2012). Excretion of vaccine virus may affect other animals in the herd. Although this is advantageous as it can lead to herd immunity, it may cause disease in susceptible animals. In some cases, the use of live vaccines may lead to reduced fertility, foetal deformities and abortion (Morton 2007). Some side effects are related to the use of adjuvants also, including adverse allergic reactions such as inflammation and in extreme cases granulomas or sterile abscesses (Spickler and Roth 2003). In case of inactive vaccines, certain problems are also observed. These vaccines may cause allergic reactions and may not stimulate a strong or long lasting immunity. These vaccines have a narrow spectrum of protection and may be of a higher cost than modified live virus-based vaccines. Furthermore, it can be difficult to differentiate between vaccinated and infected animals because both will show positive results when tested. This can cause hurdles in the export of animals since animals should be certified as disease-free before export. In some cases, this has led to restrictions on the use of some vaccines. In case of many viruses, it is very difficult to develop vaccines due to many reasons. Either viruses do not grow well in cultures or require virulent genes for survival. In some cases, inactivated viruses do not effectively induce immune response to the native virus, due to the changes in structure of viral antigens during inactivation process. Rapid mutation of some viruses due to antigenic shift is also a hurdle for developing vaccines against those viral infections. Sometimes, a disease is caused by many serotypes. Vaccines in such cases may not be available for all serotypes. For example, there are seven serotypes of FMD virus and each requires a vaccine specific to it (Institute of Science and Technology, IFST 2013). To control an outbreak related to a novel disease serotype, the rapid development of a new vaccine is necessary. Vaccines are used to prevent the disease before getting infected. Once an animal is infected, a treatment strategy is required to cure the disease.

Livestock diseases are mostly cured by the use of antimicrobials. Antibiotics have been effectively used for curing livestock diseases. However, there are certain issues related to the use of antibiotics in livestock. Antibiotics can remain in animal system and excreted in milk. Consumption of meat and milk of these animals can pass unwanted antibiotics to humans. The widespread use of antibiotics in livestock for controlling the diseases has been associated with the emergence and spread of antimicrobial resistant strains of disease causing microbes through the consumption of meat (Ventola 2015). This could be a serious threat to disease treatment in humans and animals. Additionally, these antibiotics when consumed by using milk or meat can cause allergic reactions in humans. This problem could be more pronounced in developing countries where the information about safety of use of antibiotics in livestock is not disseminated properly. Use of antibiotics also affects the environmental microbiome. It has been estimated that up to 90% of the antibiotics given to livestock are excreted in urine and stool. This in turn is dispersed through fertilizer, groundwater and surface runoff (Bartlett et al. 2013, CDC 2013).

Subunit Vaccines

Subunit vaccines are those vaccines that consist of only a part or subunit of disease-causing agent and not the complete pathogen. Since subunit vaccines do not consists of live microorganisms, these are considered safe for administration. In case of subunit vaccines, it is important to screen many antigens and identify the antigen that elicits a high immune response in the host. Subunit vaccines have number of advantages over live/inactive microorganism-based vaccines. When the vaccine formulation consists of only a specific antigen, it can be relatively easy to differentiate between vaccinated and diseased animals. In those cases, where more than one serotype is involved for causing disease, a domain that is constant in all serotypes, can be used to develop a vaccine that can provide cross-protectivity against different serotypes. Alternatively, antigenic proteins from different serotypes can be combined together to develop multivalent subunit vaccines. In the case of livestock subunit vaccines, a recent development is a new subunit vaccine against FMD which addresses some of the drawbacks of the live vaccines. The subunit vaccine is produced without the requirement for infectious virus. Thus this vaccine is safer. This vaccine is also more stable at ambient temperatures and can tolerate fluctuations in refrigeration temperature, which means that the vaccine is less likely to lose its efficacy during transportation (IFST 2013). However, subunit vaccines are less successful in generating long-lasting immunity and generate specific immunity against the given antigen.

Potential of Plant-Based System for Development of Vaccines against Livestock Disease

In the recent past, a plant-based system has become an attractive expression system for producing valuable products of pharmaceutical importance. The plant-based system has a number of advantages over conventional fermenter-based systems, mainly the safety and high expression of antigens in plants (Bock and Warzecha 2010, Cardi et al. 2010, Lössl and Waheed 2011, Stöger et al. 2014). Plant-based systems are safe production systems, as plants are not host for human or animal pathogens (Shahid and Daniell 2016). In contrast, vaccine antigens obtained from bacterial production system may have pyrogens or endotoxin contamination, which require stringent purification; yet safety issues may still be associated. There is a long list of antigens that have been expressed in plants against humans and animal diseases, either using nuclear transformation or chloroplast-based expression system (reviewed by Waheed et al. 2015, Shahid and Daniell 2016, Shahid et al. 2017). Both of these expression systems have their own advantages and disadvantages. In those cases, where antigenic proteins require glycosylation, expression via nuclear transformation would be the choice. However, nuclear transformation often leads to a low yield of expressed proteins (Daniell et al. 2001). In the case of vaccine antigens, a high level of expression for antigens is desirable. A very high level of expression, 72% of total leaf protein (Ruhlman et al. 2010) and 70% of total soluble protein (Oey et al. 2009) has been achieved. In comparison to other veterinary vaccines, livestock vaccines should be cost-effective. This is particularly important due the fact that livestock in developing countries are mostly kept by small scale farmers in homes. In such systems, the non-affordability of costly vaccines may lead to economic losses. For instance, in sub-Saharan Africa, 75% of the population is engaged in small-scale farming and 80% of these households keep livestock which serve as a critical asset. Plants can be grown at site where the vaccine is needed. In developing countries, if a vaccine is locally produced rather than transported from large distances, it can reduce costs related to the transportation and maintenance of the cooling chain (Waheed et al. 2016). This can increase the affordability of vaccines for small scale local farmers. Plants can be consumed as raw material. Vaccine antigens expressed in those plants that can be mixed with livestock feed may evade the costs related to the administration of vaccines. Another advantage of the plant-based system is the stability of expressed vaccine antigens inside a plant's cellular compartments. Plant material can be stored at room temperatures for longer times without protein degradation. In the case of many expressed proteins in plants and algal chloroplasts, it has been shown that the expressed antigens were immunogenic after up to 20 months (Dreesen et al. 2010, Hayden et al. 2010, Lakshmi et al. 2013). Chloroplast-based expression of antigens has an advantage that multiple antigens can be stacked together as one operon and co-expressed (Bock 2014). This aspect is beneficial for developing multivalent vaccines, against different serotypes of same disease or different diseases. Additionally, this feature of chloroplasts can be utilized for the production of vaccine antigens directly coupled with biological adjuvants such as *Escherichia coli* heat-labile enterotoxin subunit B (LTB) or cholera toxin subunit B (CTB). Adjuvants coupled with antigens can facilitate the transport of the antigen from the gut lumen to the gut-associated lymphoid tissues. Adjuvants can bind to receptors and target the antigens to relevant sites and/or activate antigen-presenting cells (APCs), thus eliciting protective immunity (Granell et al. 2010, Salyaev et al. 2010).

Plants have been used for expression of many vaccine antigens against livestock disease. A list of antigens expressed, method used for transformation and the plant used for expression is summarized in Table 2. It can be seen from Table 2 that various vaccine antigens have been expressed in plants targeting many important livestock disease. These diseases include anthrax, foot and mouth disease, bovine viral diarrhoea, classical swine fever, bluetongue virus, Japanese encephalitis, etc. These vaccine antigens have been expressed using different systems such as nuclear transformation, chloroplast transformation, and the transient expression system and cell suspension culture. A considerably high expression has been achieved in case of many vaccine antigens targeting livestock diseases. These vaccine expression platforms have been very well established and hold strong potential to be used as industrial-scale production systems.

The Go-to-market Potential of Plant-Based Livestock Vaccines

It has been very well documented that plants are an excellent and cost-effective platform for the production of antigen-based vaccines. Plant-based vaccines have are highly immunogenic in animal models. Due to safety concerns, the concept of subunit vaccines will gradually replace conventional whole-cell based vaccines. Various drawbacks are associated with these conventional vaccines and their development platforms as described previously. Despite of safety concerns, live or attenuated vaccines successfully pass through the regulatory framework, get approved and become available in the market; while the alternate effective platforms such as plant-based system lags behind on this path. One reason for this is that the big pharrma companies are reluctant to invest in new platforms for vaccine development when already established profitable systems are in place. The data of pre-clinical trials of plant-based vaccines show promising results in terms of the vaccine antigen's safety, efficacy and stability (Govea-Alonso et al. 2014). In many cases, target animals have been completely immunized against the microorganisms in disease challenge experiments (Shahid and Daniell 2016). However, in the case of humans, clinical trials are necessary for approval from regulatory authorities. Human clinical trials are expensive and this is another point where the lack of investment/funds renders plant-produced vaccines/pharmaceuticals stay stagnant in the course of regulatory approval. Regulations related to the developmemt of veterinary vaccines are not so strict in comparison to regulations related to human vaccine development. This laxity makes the development of novel veterinary vaccines and their platforms more competitive for regulatory approval. In the case of veterinary vaccines in some instances, approvals have been granted even though the efficacy of the vaccine was not appropriately proved in target animals. For example, in case of West Nile Virus (WNV) infection, United States Department of Agriculture (USDA) gave conditional approval to an equine vaccine manufactured by Fort Dodge Animal Health (a division of Wyeth, later purchased by Pfizer in 2009). In this case, although the product's safety was proved, its efficacy against WNV was not. Since the virus was judged as a crisis by USDA, and the horses were particularly susceptible to it, USDA granted the approval for the emergency sale for the use in horses. In contrast, it took more than a decade of research and development and the start of clinical trials two years ago (2020) with the funding from National Institute

Table 2: Plant-based vaccines against livestock diseases.

Disease	Foreign antigen/gene expressed	Transgenic Plants	Transformation Type	References
		Viral Diseases		
African Horse sickness (AHS)	• AHSV serotype 5 VLPs	• Tobacco	• Transient	(Dennis et al. 2017)
Bird Flu/Avian Influenza Virus	• ELPylated HA of H5N1 • Hemagglutinin type 1 (HA1) of H5N1 • HA of Malaysian isolate • H5 of (HPAI) A Virus • M2e Peptide of H5N1 • Recombinant hemagglutinin (rHA0) of HPAIV	• Arabidopsis • Duckweed plant • Tobacco	• Transgenic • Transient	(Farsad et al. 2017) (Firsov et al. 2015) (Kalthoff et al. 2010) (Lee et al. 2015) (Phan et al. 2013) (Pua et al. 2017)
Bluetongue Disease	• VP2, VP3, VP5 and VP7 of BTV serotype 8 • Zera®-VP2ep and Zera®-VP2	• Tobacco	• Transient	(Thuenemann et al. 2013) (van Zyl et al. 2016, 2017)
Bovine Viral Diarrhoea (BVD)	• E0 glycoprotein • E2 glycoprotein • Truncated E2 glycoprotein (tE2) • tE2 fused to APCH gene (APCHtE2) • Erns glycoprotein	• Alfalfa • Astragalus • Panax ginseng • Tobacco	• Transgenic • Transient	(Aguirreburualde et al. 2013) (Gao et al. 2014, 2015) (Nelson et al. 2012) (Santos and Widgorovitz 2005)
Classical Swine Fever	• Glycoprotein 55 (G55) or Envelope Protein 2 (E2)	• Alfalfa • Arabidopsis • Lettuce • Rice • Tobacco	• Cell Suspension Culture • Transgenic • Transplastomic	(Jung et al. 2014) (Legocki et al. 2005) (Park et al. 2016) (Shao et al. 2008)
Crimean-Congo Haemorrhagic Fever (CCHF)	• G1 & G2	• Tobacco	• Transgenic	(Ghiasi et al. 2011)

Foot-and- Mouth Disease (FMD)	• Polypeptide P1 gene • P1-2A3C • VP1 capsid protein • VP1 epitope 135 - 160 • VP1 protein • VP1 protein of serotype O- and Asia 1-type	• Alfalfa • Arabidopsis • Maize • Potato • Quinoa • Rice • Sunn hemp plants • Tobacco • Tomato	• Transgenic • Transient • Transplastomic	(Carillo et al. 1998, 2001) (Habibi-Pirkoohi et al. 2014) (Lentz et al. 2010) (Pan et al. 2008) (Rao et al. 2012) (Santos et al. 2002) (Wang et al. 2012) (Yang et al. 2007) (Zhang et al. 2011)
Gastroenteritis Transmissible Gastroenteritis Virus (TGEV)	• S protein • SIP	• Maize • Tobacco	• Transgenic • Transient	(Lamphear et al. 2004) (Monger et al. 2006)
Infectious bronchitis virus (IBV) Disorder	• S1 glycoprotein	• Potato	• Transgenic	(Zhou et al. 2003)
Infectious Bursal Disease (IBD)	• VP2	• Arabidopsis • Rice • Tobacco	• Transgenic • Transient	(Chen et al. 2012) (Gomez et al. 2013) (Wu et al. 2004, 2007)
Japanese Encephalitis (JE)	• Envelope protein (E) • E Domain III (EDIII) fused to BaMV coat protein (CP)	• Tobacco • Quinoa • Rice	• Transgenic • Transient	(Chen et al. 2017) (Wang et al. 2009)
Newcastle Disease	• Hemagglutining-neuraminidase protein (HN) • Fusion Protein (F) • F and HN • HN ectodomain • F and HN epitope	• Asiatic pennywort • Maize • Potato • Rice • Tobacco	• Transgenic • Transient	(Berinstein et al. 2005) (Gomez et al. 2008, 2009) (Guerrero-Andrade et al. 2006) (Hahn et al. 2007) (Lai et al. 2013) (Shahriari et al. 2015) (Song Lai et al. 2012) (Yang et al. 2007)

Table 2 contd. ...

...Table 2 contd.

Disease	Foreign antigen/gene expressed	Transgenic Plants	Transformation Type	References
Peste des Petits Ruminant Virus (PPRV)	• HN protein	• Peanut plant • Pigeon pea Plant	• Transgenic	(Khandelwal et al. 2011) (Prasad et al. 2004)
Porcine Epidemic Diarrhoea (PED)	• Core Neutralizing epitope COE • CTB-fused COE • LTB-COE • M cell-fused COE	• Lettuce • Potato • Rice • Tobacco	• Transgenic	(Huy et al. 2011, 2012) (Kang et al. 2006) (Kim et al. 2005)
Porcine Reproductive and Respiratory Syndrome (PRRS)	• GP5 • GP5-LTB • M protein • RhoA peptide • Co-expressed GP5, M and N protein	• Banana • Maize • Potato tubers • Tobacco • Woodland Tobacco	• Transgenic • Transient	(Chia et al. 2010, 2011) (Chen and Liu 2011) (Chan et al. 2013) (Hu et al. 2012) (Ortega-Berlanga et al. 2016) (Uribe-Campero et al. 2015)
Rabies	• Protective Antigen (PA) • G protein • G protein fused with CTB • Rabies Glycoprotein fused with ricin toxin-B chain (RGP–RTB)	• Carrot • Lettuce • Maize • Spinach • Tobacco • Tomato	• Transgenic • Transient • Transplastomic	(Loza-Rubio et al. 2012) (Rasouli et al. 2014) (Rojas-Anaya et al. 2009) (Roy et al. 2010) (Sing et al. 2015)
Rota Virus	• VP2 • VP6 • VP7 • VP2, VP6 and VP8* as VLPs • C486 BRV VP8 protein • MucoRice-ARP1	• Alfalfa • Chenopodium Leaves • Potato • Tobacco • Tomato • Rice	• Transgenic • Transient • Transplastomic	(Choi et al. 2005) (Chung et al. 2000) (Dong et al. 2005) (Lentz et al. 2011) (Pera et al. 2015) (Tokuhara 2013) (Wu et al. 2003) (Yu and Langridge 2003) (Zhou et al. 2010)

Swine Flu	• NA gene of H1N1 • NP of H3N2	• Lettuce • Maize	• Transgenic	(Liu et al. 2012) (Nahampun et al. 2015)
West Nile Virus	• VLPs from Norwalk virus • mAbs form West Nile and Ebola viruses • Hu-E16	• Lettuce • Tobacco	• Transient	(Lai et al. 2009, 2012)
Bacterial Diseases				
Anthrax	• Protective antigen (PA) • Domain IV of PA [PA(DIV)] • Deglycosylated rPA vaccine candidate (dPA83) • pp-PA83 and PNGase F • Lethal factor Domain 1 (LFD1)	• Lettuce • Mustard Plant • Tobacco • Tomato	• Transgenic • Transient • Transplastomic	(Aziz et al. 2005) (Chichester et al. 2007) (Gorantala et al. 2011, 2014) (Jones et al. 2017) (Koya et al. 2005) (Mamedov et al. 2016)
Botulism	• BoHC	• Rice	• Transgenic	(Yuki et al. 2012)
Brucellosis	• U-Omp19	• Tobacco	• Transient	(Pasquevich et al. 2011)
***E. coli*-mediated Diarrhoea**	• LT-B • LTK63 • faeG gene	• Carrot • Maize • Potato • Soya been • Tobacco • Tomato	• Transgenic • Transplastomic	(Haq et al. 1995) (Kang et al. 2004) (Loc et al 2014) (Moravec et al. 2007) (Rosales-Mendoza et al. 2007) (Shen et al. 2010) (Tacket et al. 2004)
Infectious plague	• F1 • V • F1-V	• Carrot • Lettuce • Tobacco	• Transgenic • Transplastomic	(Arlen et al. 2008) (Rosales-Mendoza et al. 2010, 2011) (Santi et al. 2006)
Listeriosis	• IFN-a	• Potato	• Transgenic	(Ohya et al. 2005)

Table 2 contd. ...

...Table 2 contd.

Disease	Foreign antigen/gene expressed	Transgenic Plants	Transformation Type	References
Pasteurellosis	• GS60 • Leukotoxin 50	• Alfalfa • Tobacco • White Clover	• Transgenic • Transient	(Lee et al. 2001, 2008)
Pig oedema Diseases	• Stx2EB • Vt2e-B and F18 • Vt2e-B and FedA subunit of F18	• Lettuce • Tobacco	• Transgenic	(Matsui et al. 2011) (Rossi et al. 2013, 2014)
Porcine pleuropneumonia	• ApxIIA antigen • ApxIIA fused with CTB	• Maize	• Transgenic	(Kim et al. 2010) (Shin et al. 2011)
Psiattacosis	• LTB-fused MOMP gene • MOMP gene	• Rice	• Transgenic	(Zhang et al. 2008, 2013)
Tuberculosis	• Ag85B & ESAT-6 antigens fused to ELP • Ag85 and Acr antigens • Ag85B, MPT83, MPT64, ESAT6 • GroES TB • ESAT6 and CFP10 • ESAT-6 fused to LTB • CFP10, ESAT6, dIFN	• Arabidopsis • Carrot • Lettuce • Potato • Tobacco	• Transgenic • Transplastomic	(Floss et al. 2010) (Jose et al. 2014) (Lakshmi et al. 2013) (Pepponi et al. 2014) (Permyakova et al. 2015) (Rigano et al. 2004) (Uvarova et al. 2013) (Zhang et al. 2012)
Protozoic Diseases				
Coccidiosis	• EtMIC2 • EtMIC1 and EtMIC2	• Tobacco	• Transient	(Sathish et al. 2011, 2012)
Parasitic Diseases				
Cysticercosis	• HP6/TSOL18 antigen • Synthetic peptides (KETc1, KETc12, KETc7)	• Carrot • Papaya	• Transgenic	(Hernández et al. 2007) (Monreal-Escalante et al. 2016)
Toxoplasmosis	• GRA4 • SAG1	• Tobacco	• Transient • Transplastomic	(Laguia-Becher et al. 2010) (Yacono et al. 2012)

of Health (NIH) for a WNV vaccine for human use (NIH 2015). In comparison to conventional platforms and methods, the plant-based system has more potential to fulfill the requirements of the regulatory framework in terms of safety and efficacy. Chloroplast-based expression is capable of addressing regulatory concerns of transgene's containment and thus is capable of getting regulatory approval as a safe production platform. Clinical trials of plant-produced livestock vaccines can be directly carried out in target livestock animals to show efficacy and safety. Clinical trials of target animals are comparatively less expensive than human clinical trials. Less expenditure would lead to comparatively lower price of vaccines which can be an attraction for pharmaceutical companies to invest less money at the return of more revenue.

Impact of Livestock Vaccines on Sustainable Development Goals

In continuation to the Millennium Development Goals (MDGs), the United Nations (UN) set Sustainable Development Goals in September 2015. These SDGs were 17 goals set for 2030 for the well-being of people and planet. Maintaining livestock health by vaccine development using innovative platforms is linked to SDGs in many ways. Nine of these goals can be addressed either directly or indirectly by improving and delivering safe livestock vaccination (Fig. 2). These include no poverty, zero hunger, good health and well-being, quality education, gender equality, decent work and economic growth, industry, innovation and infrastructure, reduced inequalities and responsible consumption and production. Livestock vaccination will improve the health of livestock and reduce the risk of zoonotic diseases in humans. Thus, not only will the overall health of people will be at less risk but this will also eliminate the cost of treatment which otherwise would have caused a burden in case of disease occurrence. The livestock vaccination will indirectly provide a positive economic impact on livestock-dependent families living in rural areas and thus in this way contribute to poverty alleviation in developing countries. Improved livestock health will decrease the death rate and will help to feed more people in the future, a necessary goal to meet the increasing food requirements due to the increasing world population, most of which will take place in developing countries and in rural areas. Large number of the population of developing countries live in rural areas where people are mostly involved in agriculture and livestock farming at small scale. Women are actively involved in caring and handling in such rural setups. Sometimes, even the whole setups are run by women in resource poor settings. Affordable and safe livestock vaccines will enable these women to increased earnings and the self sufficiency of female gender in many parts of the developing world. Small farmers can earn more money from keeping livestock healthy by cost-effective vaccination. This in turn would lead to the improvement in household expenditures including child education (Marsh et al. 2016). A positive association of livestock vaccination rates (resulting in fewer cattle deaths) linked with attendance of girls in secondary schools has been found in the recent study of Marsh et al. (2016). Availability of low-cost and preferably multivalent vaccines will enable farmers to vaccinate all livestock animals and against more diseases. This will also lead to significant net income benefits by decreasing livestock mortality, increased milk production and result in more savings by reducing costly antibiotic-based treatment of livestock. More sustainable livelihood will lead to more affordability and aid in

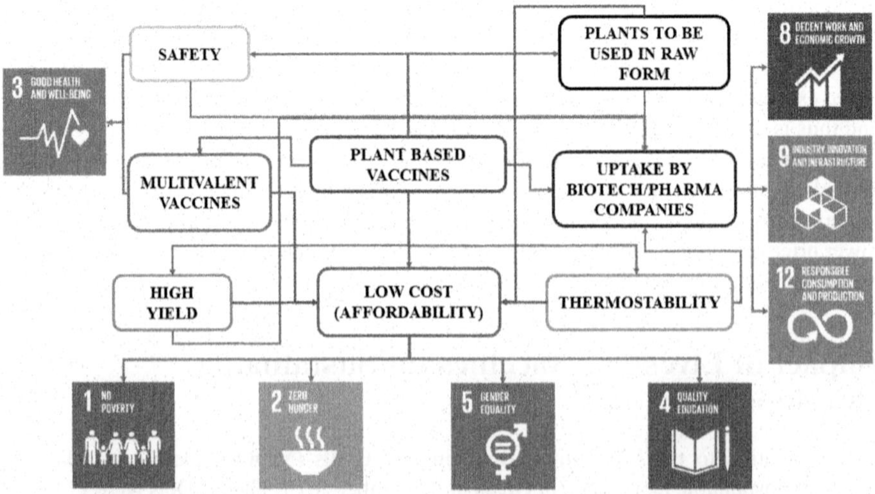

Figure 2: Impact of Plant-based livestock vaccines on Sustainable Development Goals.

reducing inequalities in terms of income and opportunity. Development of vaccines using unconventional platforms such as plant-based system will enable small scale industries to invest in the vaccine sector. Particularly in developing countries, the translation of biotechnology-based innovative platforms will create employment opportunities and give a boost to local industries. Thus, overall, improvement and innovation in livestock vaccination can have a substantial effect on the economy of developing countries.

References

Aguirreburualde, M.S., M.C. Gómez, A. Ostachuk, F. Wolman, G. Albanesi, A. Pecora, A. Odeon, F. Ardila, J.M. Escribano, M.J. Santos, and A. Wigdorovitz. 2013. Efficacy of a BVDV subunit vaccine produced in alfalfa transgenic plants. Vet. Immunol. Immunopathol. 151(3): 315–324.

Ahmed, J.S., J. Luo, L. Schnittger, U. Seitzer, F. Jongejan, and H. Yin. 2006. Phylogenetic position of small ruminant infecting piroplasms. Ann. N. Y. Acad. Sci. 1081(1): 498–504.

Arlen, P.A., M. Singleton, J.J. Adamovicz, Y. Ding, A. Davoodi-Semiromi, and H. Daniell. 2008. Effective plague vaccination via oral delivery of plant cells expressing F1-V antigens in chloroplasts. Infect. Immun. 76(8): 3640–3650.

Aziz, M.A., S. Singh, P.A. Kumar, and R. Bhatnagar. 2002. Expression of protective antigen in transgenic plants: A step towards edible vaccine against anthrax. Biochem. Biophys. Res. Commun. 299(3): 345–351.

Balderas, M.A., C.T. Nguyen, A. Terwilliger, W.A. Keitel, A. Iniguez, R. Torres, F. Palacios, C.W. Goulding, and A.W. Maresso. 2016. Progress toward the development of a NEAT protein vaccine for anthrax disease. Infect. Immun. 84(12): 3408–22.

Barker, I.K., and P.L. Carbonell. 1974. *Cryptosporidium* agni sp. n. from lambs, and *Cryptosporidium* bovis sp. n. from a calf, with observations on the oocyst. Z. Parasitenkd. 44(4): 289–298.

Bartlett, J.G., D.N. Gilbert, and B. Spellberg. 2013. Seven ways to preserve the miracle of antibiotics. Clin. Infect. Dis. 56(10): 1445–1450.

Beer, M., F.J. Conraths, and W.H.M. Van der Poel. 2013. 'Schmallenberg virus'–a novel orthobunyavirus emerging in Europe. Epidemiol. Infect. 141(1): 1–8.

Bennett, G., J. Hickford, R. Sedcole, and H. Zhou. 2009. Dichelobacter nodosus, *Fusobacterium necrophorum* and the epidemiology of footrot. Anaerobe 15(4): 173–176.

Berg, J.N., and R.W. Loan. 1975. *Fusobacterium necrophorum* and *Bacteroides melaninogenicus* as etiologic agents of foot rot in cattle. Am. J. Vet. Res. 36(08): 1115–1122.

Berinstein, A., C. Vazquez-Rovere, S. Asurmendi, E. Gómez, F. Zanetti, O. Zabal, A. Tozzini, D.C. Grand, O. Taboga, G. Calamante, and H. Barrios. 2005. Mucosal and systemic immunization elicited by Newcastle disease virus (NDV) transgenic plants as antigens. Vaccine 23(48): 5583–5589.

Bharti, A.R., J.E. Nally, J.N. Ricaldi, M.A. Matthias, M.M. Diaz, M.A. Lovett, P.N. Levett, R.H. Gilman, M.R. Willig, E. Gotuzzo, and J.M. Vinetz. 2003. Leptospirosis: A zoonotic disease of global importance. Lancet Infect. Dis. 3(12): 757–771.

Bock, R., and H. Warzecha. 2010. Solar-powered factories for new vaccines and antibiotics. Trends Biotechnol. 28: 246–252.

Bock, R. 2014. Genetic engineering of the chloroplast: Novel tools and new applications. Curr. Opin. Biotechnol. 26: 7–13.

Botswana Vaccine Institute, BVI. http://www.bvi.co.bw/. Accessed 30 June 2017.

Breeze, R. 1985. Respiratory disease in adult cattle. Vet. Clin. North. Am. Food. Anim. Pract. 1(2): 311–346.

Bryner, J. 2012. 13 Animal-to-Human Diseases Kill 2.2 Million People Each Year. Live Science. http://www.livescience.com/21426-global-zoonoses-diseases-hotspots.html. Accessed 04 April 2017.

Buddle, B.M., G.W. De Lisle, A. Pfeffer, and F.E. Aldwell. 1995. Immunological responses and protection against *Mycobacterium bovis* in calves vaccinated with a low dose of BCG. Vaccine 13(12): 1123–1130.

Cardi, T., P. Lenzi, and P. Maliga. 2010. Chloroplasts as expression platforms for plant-produced vaccines. Expert Rev. Vaccines 9(8): 893–911.

Carrillo, C., A. Wigdorovitz, J.C. Oliveros, P.I. Zamorano, A.M. Sadir, N. Gomez, J. Salinas, J.M. Escribano, and M.V. Borca. 1998. Protective immune response to foot-and-mouth disease virus with VP1 expressed in transgenic plants. J. Virol. 72(2): 1688–1690.

Carrillo, C., A. Wigdorovitz, K. Trono, M.J. Dus Santos, S. Castanon, A.M. Sadir, R. Ordas, J.M. Escribano, and M.V. Borca. 2001. Induction of a virus-specific antibody response to foot and mouth disease virus using the structural protein VP1 expressed in transgenic potato plants. Viral Immunol. 14(1): 49–57.

Center for Disease Control and Prevention, CDC. 2013. Antibiotic resistance threats in the United States. http://www.cdc.gov/drugresistance/threat-report-2013. Accessed 25 December, 28, 2017.

Center for Food Security and Public Health, CFSPH. 2016. Japanese Encephalitis fact sheet. http://www.cfsph.iastate.edu/Factsheets/pdfs/japanese_encephalitis.pdf. Accessed on 25 December 2017.

Chan, H.T., M.Y. Chia, V.F. Pang, C.R. Jeng, Y.Y. Do, and P.L. Huang. 2013. Oral immunogenicity of porcine reproductive and respiratory syndrome virus antigen expressed in transgenic banana. Plant Biotechnol. J. 11(3): 315–324.

Chen, T.H., T.H. Chen, C.C. Hu, J.T. Liao, C.W. Lee, J.W. Liao, M.Y. Lin, H.J. Liu, M.Y. Wang, N.S. Lin, and Y.H. Hsu. 2012. Induction of protective immunity in chickens immunized with plant-made chimeric bamboo mosaic virus particles expressing very virulent infectious bursal disease virus antigen. Virus Res. 166(1): 109–115.

Chen, T.H., C.C. Hu, J.T. Liao, Y.L. Lee, Y.W. Huang, N.S. Lin, Y.L. Lin, and Y.H. Hsu. 2017. Production of Japanese encephalitis virus antigens in plants using bamboo mosaic virus-based Vector. Front. Microbiol. 8.

Chen, X., and J. Liu. 2011. Generation and immunogenicity of transgenic potato expressing the GP5 protein of porcine reproductive and respiratory syndrome virus. J. Virol. Methods 173(1): 153–158.

Chia, M.Y., S.H. Hsiao, H.T. Chan, Y.Y. Do, P.L. Huang, H.W. Chang, Y.C. Tsai, C.M. Lin, V.F. Pang, and C.R. Jeng. 2010. Immunogenicity of recombinant GP5 protein of porcine reproductive and respiratory syndrome virus expressed in tobacco plant. Vet. Immunol. Immunopathol. 135(3): 234–242.

Chia, M.Y., S.H. Hsiao, H.T. Chan, Y.Y. Do, P.L. Huang, H.W. Chang, Y.C. Tsai, C.M. Lin, V.F. Pang, and C.R. Jeng. 2011. Evaluation of the immunogenicity of a transgenic tobacco plant expressing the recombinant fusion protein of GP5 of porcine reproductive and respiratory syndrome virus and B subunit of *Escherichia coli* heat-labile enterotoxin in pigs. Vet. Immunol. Immunopathol. 140(3): 215–225.

Chibunda, R.T., A.P. Muhairwa, D.M. Kambarage, M.M. Mtambo, L.J. Kusiluka, and R.R. Kazwala. 1997. Eimeriosis in dairy cattle farms in Morogoro municipality of Tanzania. Prev. Vet. Med. 31(3-4): 191–197.

Chichester, J.A., K. Musiychuk, P. de la Rosa, A. Horsey, N. Stevenson, N. Ugulava, S. Rabindran, G.A. Palmer, V. Mett, and V. Yusibov. 2007. Immunogenicity of a subunit vaccine against *Bacillus anthracis*. Vaccine 25(16): 3111–3114.

Choi, N.W., M.K. Estes, and W.H. Langridge. 2005. Synthesis and assembly of a cholera toxin B subunit-rotavirus VP7 fusion protein in transgenic potato. Mol. Biotechnol. 31(3): 193.

Chung, I.S., C.H. Kim, K.I. Kim, S.H. Hong, J.H. Park, J.K. Kim, and W.Y. Kim. 2000. Production of recombinant rotavirus VP6 from a suspension culture of transgenic tomato (*Lycopersicon esculentum* Mill.) cells. Biotechnol. Lett. 22(4): 251–255.

Cole, C.R. 1949. Vibrionic dysentery of swine. Ohio State Univ. Conf. Vet. Proc. 18: 96.

Colorado Serum Company. http://www.thepeakofquality.com/colorado-serum-company/. Accessed 30 June 2017.

Corbel, M.J. 1997. Brucellosis: An overview. Emerg. Infect. Dis. 3(2): 213.

Current, W.L., S.J. Upton, and T.B. Haynes. 1986. The life cycle of *Cryptosporidium* baileyi n. sp. (Apicomplexa, Cryptosporidiidae) infecting chickens. J. Protozool. 33(2): 289–296.

Daniell, H., S.J. Streatfield, and K. Wycoff. 2001. Medical molecular farming: Production of antibodies, biopharmaceuticals and edible vaccines in plants. Trends Plant Sci. 6: 219–226.

Dehove, A., J. Commault, M. Petitclerc, M. Teissier, and J. Macé. 2012. Economic analysis and costing of animal health: A literature review of methods and importance. Rev. Sci. Tech. 31(2): 605–617.

Dennis, S.J., A.E. Meyers, A.J. Guthrie, I.I. Hitzeroth, and E.P. Rybicki. 2017. Immunogenicity of plant-produced African horse sickness virus-like particles: Implications for a novel vaccine. Plant Biotech. J. 2017.

Desmolaize, B., S. Rose, R. Warrass, and S. Douthwaite. 2011. A novel Erm monomethyltransferase in antibiotic-resistant isolates of *Mannheimia haemolytica* and *Pasteurella multocida*. Mol. Microbiol. 80(1): 184–94.

Doherr, M.G. 2003. Bovine spongiform encephalopathy (BSE)–infectious, contagious, zoonotic or production disease? Acta. Vet. Scand. Suppl. 44(1): S33.

Dollvet. http://www.dollvet.com.tr/. Accessed on 30 June 2017.

Dong, J.L., B.G. Liang, Y.S. Jin, W.J. Zhang, and T. Wang. 2005. Oral immunization with pBsVP6-transgenic alfalfa protects mice against rotavirus infection. Virology 339(2): 153–163.

Dreesen, I.A., G. Charpin-El Hamri, and M. Fussenegger. 2010. Heat-stable oral alga-based vaccine protects mice from *Staphylococcus* aureus infection. J. Biotechnol. 14: 273–280.

Drugs.com. https://www.drugs.com/. Accessed on 30 June 2017.

Du-Gyeong, H., and C. Kyoung-Seong. 2017. Non-cytopathic bovine viral diarrhea virus 2 induce autophagy in MDBK cell. J. Immunol. May 198(S1): 214–216.

Dungu, B., C. Potgieter, B. Von Teichman, and T. Smit. 2003. Vaccination in the control of bluetongue in endemic regions: The South African experience. Dev. Biol. (Basel) 119: 463–472.

Dungu, B., M. Donadeu, and M. Bouloy. 2013. Vaccination for the control of Rift Valley fever in enzootic and epizootic situations. Dev. Biol. (Basel) 135: 61–72.

European Medicines Agency, EMA. http://www.ema.europa.eu/ema/index.jsp?curl=pages/medicines/veterinary/medicines/002781/vet_med_000315.jsp&mid=WC0b01ac058008d7a8. Accessed on 30 June 2017.

Farsad, A.S., S. Malekzadeh-Shafaroudi, N. Moshtaghi, F. Fotouhi, and S. Zibaee. 2017. Transient Expression of HA1 Antigen of H5N1 Influenza Virus in Tobacco (*Nicotiana tabacum* L.) via Agro-infiltration. J. Agr. Sci. Tech. 19(2): 439–451.

Fierer, J. 1983. Invasive Salmonella dublin infections associated with drinking raw milk. West. J. Med. 138(5): 665–669.

Fine, P.E. 1989. The BCG story: Lessons from the past and implications for the future. Rev. Infect. Dis. 11(S2): 353–359.

Firsov, A., I. Tarasenko, T. Mitiouchkina, N. Ismailova, L. Shaloiko, A. Vainstein, and S. Dolgov. 2015. High-yield expression of M2e peptide of avian influenza virus H5N1 in transgenic duckweed plants. Mol. Biotechnol. 57(7): 653–661.

Floss, D.M., M. Mockey, G. Zanello, D. Brosson, M. Diogon, R. Frutos, T. Bruel, V. Rodrigues, E. Garzon, C. Chevaleyre, and M. Berri. 2010. Expression and immunogenicity of the mycobacterial Ag85B/ESAT-6 antigens produced in transgenic plants by elastin-like peptide fusion strategy. Biomed. Res. Int. 2010.

Gao, Y., X. Zhao, P. Zang, Q. Liu, G. Wei, and L. Zhang. 2014. Generation of the bovine viral diarrhea virus E0 Protein in transgenic astragalus and its immunogenicity in Sika Deer. Evid. Based Complement. Alternat. Med.

Gao, Y., X. Zhao, C. Sun, P. Zang, H. Yang, R. Li, and L. Zhang. 2015. A transgenic ginseng vaccine for bovine viral diarrhea. Virol. J. 12(1): 73.

Garcia, M.M., M.D. Eaglesome, and C. Rigby. 1983. Campylobacters important in veterinary medicine. Vet. Bull. 53(9): 793–818.

Garten, R.J., C.T. Davis, C.A. Russell, B. Shu, S. Lindstrom, A. Balish, W.M. Sessions, X. Xu, E. Skepner, V. Deyde, and M. Okomo-Adhiambo. 2009. Antigenic and genetic characteristics of swine-origin 2009 A(H1N1) influenza viruses circulating in humans. Science 325(5937): 197–201.

Ghiasi, S.M., A.H. Salmanian, S. Chinikar, and S. Zakeri. 2011. Mice orally immunized with a transgenic plant expressing the glycoprotein of Crimean-Congo hemorrhagic fever virus. Clin. Vaccine Immunol. 18(12): 2031–2037.

Gómez, E., S.C. Zoth, E. Carrillo, M.E. Roux, and A. Berinstein. 2008. Mucosal immunity induced by orally administered transgenic plants. Immunobiology 213(8): 671–675.

Gómez, E., S.C. Zoth, S. Asurmendi, C.V. Rovere, and A. Berinstein. 2009. Expression of Hemagglutinin-Neuraminidase glycoprotein of Newcastle Disease Virus in agroinfiltrated Nicotiana benthamiana plants. J. Biotechnol. 144(4): 337–340.

Gómez, E., M.S. Lucero, S.C. Zoth, J.M. Carballeda, M.J. Gravisaco, and A. Berinstein. 2013. Transient expression of VP2 in Nicotiana benthamiana and its use as a plant-based vaccine against infectious bursal disease virus. Vaccine 31(23): 2623–2627.

Gorantala, J., S. Grover, A. Rahi, P. Chaudhary, R. Rajwanshi, N.B. Sarin, and R. Bhatnagar. 2014. Generation of protective immune response against anthrax by oral immunization with protective antigen plant-based vaccine. J. Biotechnol. 176: 1–10.

Govea-Alonso, D., E. Rybicki, and S. Rosales-Mendoza. 2014. Plant-Based vaccines as a global vaccination approach: Current perspectives. pp: 265–280. *In*: Rosales-Mendoza, S. (ed). Genetically Engineered Plants as a Source of Vaccines Against Wide Spread Diseases. Springer, New York, NY.

Granell, A., A. Fernández-del-Carmen, and D. Orzáez. 2010. In planta production of plant-derived and non-plant-derived adjuvants. Expert Rev. Vaccines 9: 843–858.

Griffin, J.F.T., C.G. Mackintosh, L. Slobbe, A.J. Thomson, and G.S. Buchan. 1999. Vaccine protocols to optimise the protective efficacy of BCG. Tuber Lung Dis. 79(3): 135–143.

Grubman, M.J., and B. Baxt. 2004, Foot-and-mouth disease. Clin. Microbiol. Rev. 17(2): 465–493.

Guerrero-Andrade, O., E. Loza-Rubio, T. Olivera-Flores, T. Fehérvári-Bone, and M.A. Gómez-Lim. 2006. Expression of the Newcastle disease virus fusion protein in transgenic maize and immunological studies. Transgenic Res. 15(4): 455–463.

Habibi-Pirkoohi, M., S. Malekzadeh-Shafaroudi, H. Marashi, N. Moshtaghi, M. Nassiri, and S. Zibaee. 2014. Transient expression of foot and mouth disease virus (FMDV) coat protein in tobacco (Nicotiana tabacom) via agroinfiltration. Iran J. Biotechnol. 12(3): 28–34.

Hahn, B.S., I.S. Jeon, Y.J. Jung, J.B. Kim, J.S. Park, S.H. Ha, K.H. Kim, H.M. Kim, J.S. Yang, and Y.H. Kim. 2007. Expression of hemagglutinin-neuraminidase protein of Newcastle disease virus in transgenic tobacco. Plant Biotechnol. Rep. 1(2): 85–92.

Haq, T.A., H.S. Mason, J.D. Clements, and C.J. Arntzen. 1995. Oral immunization with a recombinant bacterial antigen produced in transgenic plants. Science 268(5211): 714.

Hayden, C.A., E.M. Egelkrout, A.M. Moscoso, C. Enrique, T.K. Keener, R. Jimenez-Flores, C.W. Jaffrey, and J.A. Howard. 2012. Production of highly concentrated, heat-stable hepatitis B surface antigen in maize. Plant Biotechnol. J. 10(8): 979–984.

Hernández, M., J.L. Cabrera-Ponce, G. Fragoso, F. López-Casillas, A. Guevara-García, G. Rosas, C. León-Ramírez, P. Juárez, G. Sánchez-García, J. Cervantes, and G. Acero. 2007. A new highly effective anticysticercosis vaccine expressed in transgenic papaya. Vaccine 25(21): 4252–4260.

Hewinson, R.G., H.M. Vordermeier, and B.M. Buddle. 2003. Use of the bovine model of tuberculosis for the development of improved vaccines and diagnostics. Tuberculosis (Edinb) 83(1): 119–130.

Hirono, I., Y. Kono, K. Takahashi, K. Yamada, H. Niwa, M. Ojika, H. Kigoshi, K. Niiyama, and Y. Uosaki. 1984. Reproduction of acute bracken poisoning in a calf with ptaquiloside, a bracken constituent. Vet. Rec. 115(15): 375–378.

Hofmann, M.A., S. Renzullo, M. Mader, V. Chaignat, G. Worwa, and B. Thuer. 2008. Genetic characterization of toggenburg orbivirus, a new bluetongue virus, from goats, Switzerland. Emerg. Infect. Dis. 14(12): 1855–1861.

Hu, J., Y. Ni, B.A. Dryman, X.J. Meng, and C. Zhang. 2012. Immunogenicity study of plant-made oral subunit vaccine against porcine reproductive and respiratory syndrome virus (PRRSV). Vaccine 30(12): 2068–2074.

Huy, N.X., M.S. Yang, and T.G. Kim. 2011. Expression of a cholera toxin B subunit-neutralizing epitope of the porcine epidemic diarrhea virus fusion gene in transgenic lettuce (*Lactuca sativa* L.). Mol. Biotechnol. 48(3): 201–209.

Huy, N.X., S.H. Kim, M.S. Yang, and T.G. Kim. 2012. Immunogenicity of a neutralizing epitope from porcine epidemic diarrhea virus: M cell targeting ligand fusion protein expressed in transgenic rice calli. Plant Cell Rep. 31(10): 1933–1942.

Inglesby, T.V., D.A. Henderson, J.G. Bartlett, M.S. Ascher, E. Eitzen, A.M. Friedlander, J. Hauer, J. McDade, M.T. Osterholm, T. O'toole, and G. Parker. 1999. Anthrax as a biological weapon: Medical and public health management. JAMA 281(18): 1735–1745.

Institute of Food Science Technology, IFST. 2013. Livestock Vaccines. Avaiable at. http://www.ifst.org/sites/default/files/Livestock%20Vaccines-PN-433_0.pdf. Accessed 25 December, 2017.

Irvin, A.D. 1985. Immunity in theileriosis. Parasitol. Today 1(5): 124–128.

Islam, M.S., H.M. Sazzad, S.M. Satter, S. Sultana, M.J. Hossain, M. Hasan, M. Rahman, S. Campbell, D.L. Cannon, U. Ströher, and P. Daszak. 2016. Nipah virus transmission from bats to humans associated with drinking traditional liquor made from date palm sap, Bangladesh, 2011–2014. Emerg. Infect. Dis. 22(4): 664–670.

Jenckel, M., E. Bréard, C. Schulz, C. Sailleau, C. Viarouge, B. Hoffmann, D. Höper, M. Beer, and S. Zientara. 2015. Complete coding genome sequence of putative novel bluetongue virus serotype 27. Genome Announc. 3(2): e00016-15.

Jones, R.M., M. Burke, D. Dubose, J.A. Chichester, S. Manceva, A. Horsey, S.J. Streatfield, J. Breit, and V. Yusibov. 2017. Stability and pre-formulation development of a plant-produced anthrax vaccine candidate. Vaccine 35(41): 5463–5470.

Jose, S., S. Ignacimuthu, M. Ramakrishnan, K. Srinivasan, G. Thomas, P. Kannan, and S. Narayanan. 2014. Expression of GroES TB antigen in tobacco and potato. Plant Cell Tissue Organ Cult. 119(1): 157–169.

Jung, M., Y.J. Shin, J. Kim, S.B. Cha, W.J. Lee, M.K. Shin, S.W. Shin, M.S. Yang, Y.S. Jang, T.H. Kwon, and H.S. Yoo. 2014. Induction of immune responses in mice and pigs by oral administration of classical swine fever virus E2 protein expressed in rice calli. Arch. Virol. 159(12): 3219–3230.

Kalthoff, D., A. Giritch, K. Geisler, U. Bettmann, V. Klimyuk, H.R. Hehnen, Y. Gleba, and M. Beer. 2010. Immunization with plant-expressed hemagglutinin protects chickens from lethal highly pathogenic avian influenza virus H5N1 challenge infection. J. Virol. 84(22): 12002–12010.

Kang, T.J., S.C. Han, M.Y. Kim, Y.S. Kim, and M.S. Yang. 2004. Expression of non-toxic mutant of *Escherichia coli* heat-labile enterotoxin in tobacco chloroplasts. Protein Expr. Purif. 38(1): 123–128.

Kang, T.J., S.C. Han, M.S. Yang, and Y.S. Jang. 2006. Expression of synthetic neutralizing epitope of porcine epidemic diarrhea virus fused with synthetic B subunit of *Escherichia coli* heat-labile enterotoxin in tobacco plants. Protein Expr. Purif. 46(1): 16–22.

Karlson, A.G., and E.F. Lessel. 1970. *Mycobacterium bovis* nom. nov. Int. J. Syst. Evol. Microbiol. 20(3): 273–282.

Khandelwal, A., G.J. Renukaradhya, M. Rajasekhar, G.L. Sita, and M.S. Shaila. 2011. Immune responses to hemagglutinin-neuraminidase protein of peste des petits ruminants virus expressed in transgenic peanut plants in sheep. Vet. Immunol. Immunopathol. 140(3): 291–296.

Kim, H.A., H.S. Yoo, M.S. Yang, S.Y. Kwon, J.S. Kim, and P.S. Choi. 2010. The development of transgenic maize expressing *Actinobacillus pleuropneumoniae* ApxIIA gene using Agrobacterium. Plant Biotechnol. J. 37(3): 313–318.

Kim, Y.S., T.J. Kang, Y.S. Jang, and M.S. Yang. 2005. Expression of neutralizing epitope of porcine epidemic diarrhea virus in potato plants. Plant Cell Tissue Organ Cult. 82(2): 125–130.

Koya, V., M. Moayeri, S.H. Leppla, and H. Daniell. 2005. Plant-based vaccine: mice immunized with chloroplast-derived anthrax protective antigen survive anthrax lethal toxin challenge. Infect. Immun. 73(12): 8266–8274.

Laguía-Becher, M., V. Martín, M. Kraemer, M. Corigliano, M.L. Yacono, A. Goldman, and M. Clemente. 2010. Effect of codon optimization and subcellular targeting on Toxoplasma gondii antigen SAG1 expression in tobacco leaves to use in subcutaneous and oral immunization in mice. BMC Biotechnol. 10(1): 52.

Lai, H., J. He, M. Engle, M.S. Diamond, and Q. Chen. 2012. Robust production of virus-like particles and monoclonal antibodies with geminiviral replicon vectors in lettuce. Plant Biotechnol. J. 10(1): 95–104.

Lai, H., M. Engle, A. Fuchs, T. Keller, S. Johnson, S. Gorlatov, M.S. Diamond, and Q. Chen. 2010. Monoclonal antibody produced in plants efficiently treats West Nile virus infection in mice. Proc. Natl. Acad. Sci. USA 107(6): 2419–2424.

Lakshmi, P.S., D. Verma, X. Yang, B. Lloyd, and H. Daniell. 2013. Low cost tuberculosis vaccine antigens in capsules: Expression in chloroplasts, bio-encapsulation, stability and functional evaluation *in vitro*. PLoS ONE 8: e54708.

Lamphear, B.J., J.M. Jilka, L. Kesl, M. Welter, J.A. Howard, and S.J. Streatfield. 2004. A corn-based delivery system for animal vaccines: An oral transmissible gastroenteritis virus vaccine boosts lactogenic immunity in swine. Vaccine 22(19): 2420–2424.

Lawes, J.R., K.A. Harris, A. Brouwer, J.M. Broughan, N.H. Smith, and P.A. Upton. 2016. Bovine TB surveillance in Great Britain in 2014. Vet. Rec. 178(13): 310–315.

Lee, G., Y.J. Na, B.G. Yang, J.P. Choi, Y.B. Seo, C.P. Hong, C.H. Yun, D.H. Kim, E.J. Sohn, J.H. Kim, and Y.C. Sung. 2015. Oral immunization of haemaggulutinin H5 expressed in plant endoplasmic reticulum with adjuvant saponin protects mice against highly pathogenic avian influenza A virus infection. Plant Biotechnol. J. 13(1): 62–72.

Lee, R.W., J. Strommer, D. Hodgins, P.E. Shewen, Y. Niu, and R.Y. Lo. 2001. Towards development of an edible vaccine against bovine pneumonic pasteurellosis using transgenic white clover expressing amannheimia haemolytica A1 Leukotoxin 50 Fusion Protein. Infect. Immun. 69(9): 5786–5793.

Lee, R.W., M. Cornelisse, A. Ziauddin, P.J. Slack, D.C. Hodgins, J.N. Strommer, P.E. Shewen, and R.Y. Lo. 2008. Expression of a modified Mannheimia haemolytica GS60 outer membrane lipoprotein in transgenic alfalfa for the development of an edible vaccine against bovine pneumonic pasteurellosis. J. Biotechnol. 135(2): 224–231.

Lee, S.W., P.F. Markham, M.J. Coppo, A.R. Legione, J.F. Markham, A.H. Noormohammadi, G.F. Browning, N. Ficorilli, C.A. Hartley, and J.M. Devlin. 2012. Attenuated vaccines can recombine to form virulent field viruses. Science 337(6091): 188–188.

Legocki, A.B., K. Miedzinska, M. Czaplińska, A. Płucieniczak, and H. Wędrychowicz. 2005. Immunoprotective properties of transgenic plants expressing E2 glycoprotein from CSFV and cysteine protease from Fasciola hepatica. Vaccine 23(15): 1844–1846.

Lentz, E.M., M.E. Segretin, M.M. Morgenfeld, S.A. Wirth, M.J. Santos, M.V. Mozgovoj, A. Wigdorovitz, and F.F. Bravo-Almonacid. 2010. High expression level of a foot and mouth disease virus epitope in tobacco transplastomic plants. Planta 231(2): 387.

Lentz, E.M., M.V. Mozgovoj, D. Bellido, M.D. Santos, A. Wigdorovitz, and F.F. Bravo-Almonacid. 2011. VP8* antigen produced in tobacco transplastomic plants confers protection against bovine rotavirus infection in a suckling mouse model. J. Biotechnol. 156(2): 100–107.

Levine, N.D. 1961. Protozoan parasites of domestic animals and of man. Publisher, Burgess Pub. Co.

Limor de Colombia. http://www.limorcolombia.com/. Accessed on 30 June 2017.

Liu, C.W., J.J. Chen, C.C. Kang, C.H. Wu, and J.C. Yiu. 2012. Transgenic lettuce (*Lactuca sativa* L.) expressing H1N1 influenza surface antigen (neuraminidase). Sci. Hortic. 139: 8–13.

Loc, N.H., D.T. Long, T.G. Kim, and M.S. Yang. 2014. Expression of *Escherichia coli* Heat-labile Enterotoxin B Subunit in Transgenic Tomato. Czech. J. Genet. Plant Breed. 50(1): 26–31.

Lössl, A.G., and M.T. Waheed. 2011. Chloroplast-derived vaccines against human diseases: Achievements, challenges and scopes. Plant Biotechnol. J. 9(5): 527–539.

Loza-Rubio, E., E. Rojas-Anaya, J. López, M.T. Olivera-Flores, M. Gómez-Lim, and G. Tapia-Pérez. 2012. Induction of a protective immune response to rabies virus in sheep after oral immunization with transgenic maize, expressing the rabies virus glycoprotein. Vaccine 30(37): 5551–5556.

Maan, S., N.S. Maan, K. Nomikou, C. Batten, F. Antony, M.N. Belaganahalli, A.M. Samy, A.A. Reda, S.A. Al-Rashid, M. El Batel, and C.A. Oura. 2011. Novel bluetongue virus serotype from Kuwait. Emerg. Infect. Dis. 17(5): 886.

Mackintosh, C., J.C. Haigh, and F. Griffin. 2002. Bacterial diseases of farmed deer and bison. Rev. Sci. Tech. 21(2): 249–263.

Mamedov, T., J.A. Chichester, R.M. Jones, A. Ghosh, M.V. Coffin, K. Herschbach, A.I. Prokhnevsky, S.J. Streatfield, and V. Yusibov. 2016. Production of functionally active and immunogenic non-glycosylated protective antigen from *Bacillus anthracis* in *Nicotiana benthamiana* by Co-Expression with Peptide-N-Glycosidase F (PNGase F) of *Flavobacterium meningosepticum*. PloS one 11(4): e0153956.

Marsh, T.L., J. Yoder, T. Deboch, T.F. McElwain, and G.H. Palmer. 2016. Livestock vaccinations translate into increased human capital and school attendance by girls. Sci. Adv. 2(12): e1601410.

Martin, G.J., and A.M. Friedlander. 2010. *Bacillus anthracis* (anthrax). pp. 2715–25. *In*: Mandell, G.L., J.E. Bennett, and R. Dolin (eds.). Mandell, Douglas, and Bennett's principles and practice of infectious diseases, 7th ed. Philadelphia, Churchill Livingstone.

Matsui, T., E. Takita, T. Sato, M. Aizawa, M. Ki, Y. Kadoyama, K. Hirano, S. Kinjo, H. Asao, K. Kawamoto, and H. Kariya. 2011. Production of double repeated B subunit of Shiga toxin 2e at high levels in transgenic lettuce plants as vaccine material for porcine edema disease. Transgenic Res. 20(4): 735–748.

Matto, C., G. Varela, M.I. Mota, R. Gianneechini, and R. Rivero. 2017. Rhombencephalitis caused by Listeria monocytogenes in a pastured bull. J. Vet. Diagn. Invest. 29(2): 228–231.

Monger, W., J.M. Alamillo, I. Sola, Y. Perrin, M. Bestagno, O.R. Burrone, P. Sabella, J. Plana-Duran, L. Enjuanes, J.A. Garcia, and G.P. Lomonossoff. 2006. An antibody derivative expressed from viral vectors passively immunizes pigs against transmissible gastroenteritis virus infection when supplied orally in crude plant extracts. Plant Biotechnol. J. 4(6): 623–631.

Monreal-Escalante, E., D.O. Govea-Alonso, M. Hernández, J. Cervantes, J.A. Salazar-González, A. Romero-Maldonado, G. Rosas, T. Garate, G. Fragoso, E. Sciutto, and S. Rosales-Mendoza. 2016. Towards the development of an oral vaccine against porcine cysticercosis: Expression of the protective HP6/TSOL18 antigen in transgenic carrots cells. Planta 243(3): 675–685.

Moravec, T., M.A. Schmidt, E.M. Herman, and T. Woodford-Thomas. 2007. Production of *Escherichia coli* heat labile toxin (LT) B subunit in soybean seed and analysis of its immunogenicity as an oral vaccine. Vaccine 25(9): 1647–1657.

Morton, D.B. 2007. Vaccines and animal welfare. Rev. Sci. Tech. Off. Int. Epiz. 26(1): 157.

MSD Animal Health. http://www.msd-animal-health.co.nz/product_types/vaccines/livestock.aspx. Accessed 30 June 2017.

Mullen, G.R., M.E. Hayes, and K.E. Nusbaum. 1984. Potential vectors of bluetongue and epizootic hemorrhagic disease viruses of cattle and white-tailed deer in Alabama. Prog. Clin. Biol. Res. 178: 201–206.

Nagore, D., J. Garcıa-Sanmartın, A.L. Garcıa-Pérez, R.A. Juste, and A. Hurtado. 2004. Identification, genetic diversity and prevalence of *Theileria* and *Babesia* species in a sheep population from Northern Spain. Int. J. Parasitol. 34(9): 1059–1067.

Nahampun, H.N., B. Bosworth, J. Cunnick, M. Mogler, and K. Wang. 2015. Expression of H3N2 nucleoprotein in maize seeds and immunogenicity in mice. Plant Cell Rep. 34(6): 969–980.

Nandi, S., M. Kumar, M. Manohar, and R.S. Chauhan. 2009. Bovine herpes virus infections in cattle. Anim. Health Res. Rev. 10: 85–98.

National Institute of Health, NIH. 2015. NIH-funded vaccine for West Nile virus enters human clinical trials. https://www.nih.gov/news-events/news-releases/nih-funded-vaccine-west-nile-virus-enters-human-clinical-trials. Accessed 14 May 2017.

Nelson, G., P. Marconi, O. Periolo, J. La Torre, and M.A. Alvarez. 2012. Immunocompetent truncated E2 glycoprotein of bovine viral diarrhea virus (BVDV) expressed in Nicotiana tabacum plants: A candidate antigen for new generation of veterinary vaccines. Vaccine 30(30): 4499–4504.

Nielsen, S.S., and N. Toft. 2009. A review of prevalences of paratuberculosis in farmed animals in Europe. Prev. Vet. Med. 88(1): 1–14.

Niu, Q., J. Luo, G. Guan, M. Ma, Z. Liu, A. Liu, Z. Dang, J. Gao, Q. Ren, Y. Li, and J. Liu. 2009. Detection and differentiation of ovine Theileria and Babesia by reverse line blotting in China. Parasitol. Res. 104(6): 1417–1423.

Oey, M., M. Lohse, B. Kreikemeyer, and R. Bock. 2009. Exhaustion of the chloroplast protein synthesis capacity by massive expression of a highly stable protein antibiotic. Plant J. 57: 436–445.

Ohya, K., T. Matsumura, N. Itchoda, K. Ohashi, M. Onuma, and C. Sugimoto. 2005. Ability of orally administered IFN-α-containing transgenic potato extracts to inhibit Listeria monocytogenes infection. J. Interferon Cytokine Res. 25(8): 459–466.

Onderstepoort Biological Products, OBP. https://www.obpvaccines.co.za/products. Accessed 30 June 2017.

Ortega-Berlanga, B., K. Musiychuk, Y. Shoji, J.A. Chichester, V. Yusibov, O. Patiño-Rodríguez, D.E. Noyola, and A.G. Alpuche-Solís. 2016. Engineering and expression of a RhoA peptide against respiratory syncytial virus infection in plants. Planta 243(2): 451–458.

Pan, L., Y. Zhang, Y. Wang, B. Wang, W. Wang, Y. Fang, S. Jiang, J. Lv, W Wang, Y. Sun, and Q. Xie. 2008. Foliar extracts from transgenic tomato plants expressing the structural polyprotein, P1-2A, and protease, 3C, from foot-and-mouth disease virus elicit a protective response in guinea pigs. Vet. Immunol. Immunopathol. 121(1): 83–90.

Parashar, U.D., L.M. Sunn, F. Ong, A.W. Mounts, M.T. Arif, T.G. Ksiazek, M.A. Kamaluddin, A.N. Mustafa, H. Kaur, L.M. Ding, and G. Othman. 2000. Case-control study of risk factors for human infection with a new zoonotic paramyxovirus, Nipah virus, during a 1998–1999 outbreak of severe encephalitis in Malaysia. J. Infect. Dis. 181(5): 1755–9.

Park, M., K. Min, S. Gu, S. Park, N.H. Kim, H. Lee, H. Kim, Y. Lee, S. Lim, H. Jeong, D. An, J. Song, I. Hwang, and E. Sohn. 2016. Transgenic Plants Producing Green-vaccine for CSFV (classical swine fever virus) Lead on Plant Biotechnology-based Product on Market. Gene, Genome and New Technology for Plant Breeding pp. 328.

Pasquevich, K.A., A.E. Ibañez, L.M. Coria, C.G. Samartino, S.M. Estein, A. Zwerdling, P. Barrionuevo, F.S. Oliveira, C. Seither, H. Warzecha, and S.C. Oliveira. 2011. An oral vaccine based on U-Omp19 induces protection against *B. abortus* mucosal challenge by inducing an adaptive IL-17 immune response in mice. PLoS One 6(1): e16203.

Peek, S.F. 2005. Respiratory emergencies in cattle. Vet. Clin. North Am. Food. Anim. Pract. 2(3): 697–710.

Pepponi, I., G.R. Diogo, E. Stylianou, C.J. Dolleweerd, P.M. Drake, M.J. Paul, L. Sibley, J.K. Ma, and R. Reljic. 2014. Plant-derived recombinant immune complexes as self-adjuvanting TB immunogens for mucosal boosting of BCG. Plant Biotechnol. J. 12(7): 840–850.

Peters, C.J., and K.J. Linthicum. 1994. Rift valley fever. Handbook of Zoonoses B: Viral. 125–138.

Phan, H.T., J. Pohl, D.M. Floss, F. Rabenstein, J. Veits, B.T. Le, H.H. Chu, G. Hause, T. Mettenleiter, and U. Conrad. 2013. ELPylated haemagglutinins produced in tobacco plants induce potentially neutralizing antibodies against H5N1 viruses in mice. Plant Biotechnol. J. 11(5): 582–593.

Population Reference Bureau, PRB. 2016. Population Reference Bureau: 2016 World Population Data Sheet. http://www.prb.org/pdf16/prb-wpds2016-web-2016.pdf. Accessed 04 April, 2017.

Prasad, V., V.V. Satyavathi, K.M. Valli, A. Khandelwal, M.S. Shaila, and G.L. Sita. 2004. Expression of biologically active Hemagglutinin-neuraminidase protein of Peste des petits ruminants virus in transgenic pigeonpea [*Cajanus cajan* (L.) Millsp.]. Plant Sci. 166(1): 199–205.

Proctor, S.J., and R.L. Kemp. 1974. *Cryptosporidium anserinum* sp. n.(Sporozoa) in a domestic goose *Anser anser* L., from Iowa. J. Protozool. 21(5): 664–666.

Pua, T.L., X.Y. Chan, H.S. Loh, A.R. Omar, V. Yusibov, K. Musiychuk, A.C. Hall, M.V. Coffin, Y. Shoji, J.A. Chichester, and H. Bi. 2017. Purification and immunogenicity of hemagglutinin from highly pathogenic avian influenza virus H5N1 expressed in Nicotiana benthamiana. Hum. Vaccin. Immunothe. 13(2): 306–313.

Rao, J.P., P. Agrawal, R. Mohammad, S.K. Rao, G.R. Reddy, H.J. Dechamma, and V.V. Suryanarayana. 2012. Expression of VP1 protein of serotype A and O of foot-and-mouth disease virus in transgenic sunnhemp plants and its immunogenicity for guinea pigs. Acta. Virol. 56(2): 91.

Rasouli, R., H. Honari, H. Alizadeh, M. Gorjian, M. Jalali, and M. Aalayi. 2014. Expression of protective antigen of bacillus anthracis in Iranian variety of lettuce plastid (*Lactuca sativa* L.). Middle-East J. Sci. Res. 21: 1855–1861.

Rebhun, W.C., J.M. King, and R.B. Hillman. 1988. Atypical actinobacillosis granulomas in cattle. Cornell. Vet. 78(2): 125–130.

Riek, R.F. 1968. Babesiosis. pp. 219. *In*: Weinman, D., and M. Ristic (eds.). Infectious Blood Diseases of Man and Animals. Vol. II. Academic Press, New York.

Rigano, M.M., M.L. Alvarez, J. Pinkhasov, Y. Jin, F. Sala, C.J. Arntzen, and A.M. Walmsley. 2004. Production of a fusion protein consisting of the enterotoxigenic *Escherichia coli* heat-labile toxin B subunit and a tuberculosis antigen in *Arabidopsis thaliana*. Plant Cell Rep. 22(7): 502–508.

Robinson, T., W. Wint, G. Conchedda, G. Cinardi, T.P.V. Boeckel, M. Macleod, B. Bett, D. Grace, and M. Gilbert. 2015. The global livestock sector: Trends, drivers and implications for society, health and the environment. Presented at the Annual Conference of the British Society of Animal Science (BSAS), Chester, UK, 14–15 April 2015. Nairobi, Kenya: ILRI.

Rosales-Mendoza, S., R.E. Soria-Guerra, M.T. de Jesús Olivera-Flores, R. López-Revilla, G.R. Argüello-Astorga, J.F. Jiménez-Bremont, R.F. García-de la Cruz, J.P. Loyola-Rodríguez, and A.G. Alpuche-Solís. 2007. Expression of *Escherichia coli* heat-labile enterotoxin b subunit (LTB) in carrot (*Daucus carota* L.). Plant Cell Rep. 26(7): 969–976.

Rosales-Mendoza, S., R.E. Soria-Guerra, L. Moreno-Fierros, A.G. Alpuche-Solís, L. Martínez-González, and S.S. Korban. 2010. Expression of an immunogenic F1-V fusion protein in lettuce as a plant-based vaccine against plague. Planta 232(2): 409–416.

Rosales-Mendoza, S., R.E. Soria-Guerra, L. Moreno-Fierros, Y. Han, A.G. Alpuche-Solís, and S.S. Korban. 2011. Transgenic carrot tap roots expressing an immunogenic F1–V fusion protein from *Yersinia pestis* are immunogenic in mice. J. Plant Physiol. 168(2): 174–180.

Rossi, L., A. Di Giancamillo, S. Reggi, C. Domeneghini, A. Baldi, V. Sala, V. Dell'Orto, A. Coddens, E. Cox, and C. Fogher. 2013. Expression of verocytotoxic *Escherichia coli* antigens in tobacco seeds and evaluation of gut immunity after oral administration in mouse model. J. Vet. Sci. 14(3): 263–270.

Rossi, L., V. Dell'Orto, S. Vagni, V. Sala, S. Reggi, and A. Baldi. 2014. Protective effect of oral administration of transgenic tobacco seeds against verocytotoxic *Escherichia coli* strain in piglets. Vet. Res. Commun. 38(1): 39–49.

Roy, S., A. Tyagi, S. Tiwari, A. Singh, S.V. Sawant, P.K. Singh, and R. Tuli. 2010. Rabies glycoprotein fused with B subunit of cholera toxin expressed in tobacco plants folds into biologically active pentameric protein. Protein Expr. Purif. 70(2): 184–190.

Ruhlman, T., D. Verma, N. Samson, and H. Daniell. 2010. The role of heterologous chloroplast sequence elements in transgene integration and expression. Plant Physiol. 152: 2088–2104.

Salyaev, R.K., M.M. Rigano, and N.I. Rekoslavskaya. 2010. Development of plant-based mucosal vaccines against widespread infectious diseases. Expert Rev. Vaccines 9: 937–946.

Santi, L., A. Giritch, C.J. Roy, S. Marillonnet, V. Klimyuk, Y. Gleba, R. Webb, C.J. Arntzen, and H.S. Mason. 2006. Protection conferred by recombinant Yersinia pestis antigens produced by a rapid and highly scalable plant expression system. Proc. Natl. Acad. Sci. USA 103(4): 861–866.

Santos, M.J., A. Wigdorovitz, K. Trono, R.D. Ríos, P.M. Franzone, F. Gil, J. Moreno, C. Carrillo, J.M. Escribano, and M.V. Borca. 2002. A novel methodology to develop a foot and mouth disease virus (FMDV) peptide-based vaccine in transgenic plants. Vaccine 20(7): 1141–1147.

Santos, M.J., and A. Wigdorovitz. 2005. Transgenic plants for the production of veterinary vaccines. Immunol. Cell Biol. 83(3): 229–238.

Santos, M.J., C. Carrillo, F. Ardila, R.D. Ríos, P. Franzone, M.E. Piccone, A. Wigdorovitz, and M.V. Borca. 2005. Development of transgenic alfalfa plants containing the foot and mouth disease virus structural polyprotein gene P1 and its utilization as an experimental immunogen. Vaccine 23(15): 1838–1843.

Sathish, K., R. Sriraman, B.M. Subramanian, N.H. Rao, B. Kasa, J. Donikeni, M.L. Narasu, and V.A. Srinivasan. 2012. Plant expressed coccidial antigens as potential vaccine candidates in protecting chicken against coccidiosis. Vaccine 30(30): 4460–4464.

Sathish, K., R. Sriraman, B.M. Subramanian, N.H. Rao, K. Balaji, M.L. Narasu, and V.A. Srinivasan. 2011. Plant expressed EtMIC2 is an effective immunogen in conferring protection against chicken coccidiosis. Vaccine 29(49): 9201–9208.

Schiaffino, A., C.R. Beuzon, S. Uzzau, G. Leori, P. Cappuccinelli, J. Casadesús, and S. Rubino. 1996. Strain typing with IS200 fingerprints in *Salmonella abortusovis*. App. Environ. Microbiol. 62(7): 2375–2380.

Schlech III, W.F., P.M. Lavigne, R.A. Bortolussi, A.C. Allen, E.V. Haldane, A.J. Wort, A.W. Hightower, S.E. Johnson, S.H. King, E.S. Nicholls, and C.V. Broome. 1983. Epidemic listeriosis—evidence for transmission by food. N. Engl. J. Med. 308(4): 203–206.

Schultz-Cherry, S., C.W. Olsen, and B.C. Easterday. 2013. History of swine influenza. Curr. Top. Microbiol. Immunol. 370: 21–28.

Selman, I.E., A. Wiseman, R.G. Breeze, and H.M. Pirie. 1976. Fog fever in cattle: Various theories on its aetiology. Vect. Rec. 99(10): 181.

Shadomy, S., A. El Idrissi, E. Raizman, M. Bruni, E. Palamara, C. Pittiglio, and J. Lubroth. 2016. Anthrax outbreaks: A warning for improved prevention, control and heightened awareness. Food and Agricultural Organization of the United Nations. Available at. http://www.fao.org/3/a-i6124e.pdf. Accessed 06 May 2017.

Shahid, N., and H. Daniell. 2016. Plant-based oral vaccines against zoonotic and non-zoonotic diseases. Plant Biotechnol. J. 14: 2079–2099.

Shahid, N., A.Q. Rao, P.E. Kristen, M.A. Ali, B. Tabassum, S. Umar, S. Tahir, A. Latif, A. Ahad, A.A. Shahid, and T. Husnain. 2017. A concise review of poultry vaccination and future implementation of plant-based vaccines. Worlds Poult. Sci. J. 73(3): 471–482.

Shahriari, A.G., A. Bagheri, M.R. Bassami, S. Malekzadeh Shafaroudi, and A.R. Afsharifar. 2015. Cloning and expression of fusion (F) and haemagglutinin-neuraminidase (HN) epitopes in hairy roots of tobacco (Nicotiana tabaccum) as a step toward developing a candidate recombinant vaccine against Newcastle disease. J. Cell. Mol. Res. 7(1): 11–18.

Shao, H.B., D.M. He, K.X. Qian, G.F. Shen, and Z.L. Su. 2008. The expression of classical swine fever virus structural protein E2 gene in tobacco chloroplasts for applying chloroplasts as bioreactors. C. R. Biol. 331(3): 179–184.

Shapiro, R.L., C. Hatheway, and D.L. Swerdlow. 1998. Botulism in the United States: A clinical and epidemiologic review. Ann. Intern. Med. 129(3): 221–228.

Shen, H., B. Qian, W. Chen, Z. Liu, L. Yang, D. Zhang, and W. Liang. 2010. Immunogenicity of recombinant F4 (K88) fimbrial adhesin FaeG expressed in tobacco chloroplast. Acta. Biochim. Biophys. Sin. (Shanghai) 42(8): 558–567.

Shin, M.K., M.H. Jung, W.J. Lee, P.S. Choi, Y.S. Jang, and H.S. Yoo. 2011. Generation of transgenic corn-derived *Actinobacillus pleuropneumoniae* ApxIIA fused with the cholera toxin B subunit as a vaccine candidate. J. Vet. Sci. 12(4): 401–403.

Singh, A., S. Srivastava, A. Chouksey, B.S. Panwar, P.C. Verma, S. Roy, P.K. Singh, G. Saxena, and R. Tuli. 2015. Expression of Rabies Glycoprotein and Ricin Toxin B Chain (RGP–RTB) fusion protein in tomato hairy roots: A step towards oral vaccination for rabies. Mol. Biotechnol. 57(4): 359–370.

Skinner, M.A., D.N. Wedlock, and B.M. Buddle. 2001. Vaccination of animals against *Mycobacterium bovis*. Rev. Sci. Tech. 20(1): 112–132.

Skirrow, M.B., and J. Benjamin. 1980. Differentiation of enteropathogenic Campylobacter. J. Clin. Pathol. 33(11): 1122.

Skirrow, M.B. 1994. Diseases due to Campylobacter, Helicobacter and related bacteria. J. Comp. Pathol. 111(2): 113–149.

Slavin, D. 1955. *Cryptosporidium meleagridis* (sp. nov.). J. Comp. Pathol. 65: 262IN20–266IN23.

Smith, G.J., D. Vijaykrishna, J. Bahl, S.J. Lycett, M. Worobey, O.G. Pybus, S.K. Ma, C.L. Cheung, J. Raghwani, S. Bhatt, J.S. Peiris, Y. Guan, and A. Rambaut. 2009. Origins and evolutionary genomics of the 2009 swine-origin H1N1 influenza A epidemic. Nature 459(7250): 1122–1125.

Smith, R.D., M.A. James, M. Ristic, and M. Aikawa. 1981. Bovine babesiosis: Protection of cattle with culture-derived soluble Babesia bovis antigen. Science 212(4492): 335–338.

Snodgrass, D.R., T. Fitzgerald, I. Campbell, F.M. Scott, G.F. Browning, D.L. Miller, A.J. Herring, and H.B. Greenberg. 1990. Rotavirus serotypes 6 and 10 predominate in cattle. J. Clin. Microbiol. 28(3): 504–507.

Sobel, J. 2005. Botulism. Clin. Infect. Dis. 41(8): 1167–1173.

Song Lai, K., K. Yusoff, and M. Mahmood. 2012. Heterologous expression of hemagglutinin-neuraminidase protein from Newcastle disease virus strain AF2240 in Centella asiatica. Acta. Biol. Crac. Ser. Bot. 54(1): 142–147.

Spickler, A.R., and J.A. Roth. 2003. Adjuvants in veterinary vaccines: modes of action and adverse effects. J. Vet. Intern. Med. 17(3): 273–281.

State of Queensland. https://www.business.qld.gov.au/industries/farms-fishing-forestry/agriculture/livestock/cattle/tick-fever-vaccines. Accessed 30 June 2017.

Stöger, E., R. Fischer, M. Moloney, and J.K.C. Ma. 2014. Plant molecular pharming for the treatment of chronic and infectious diseases. Annu. Rev. Plant Biol. 65: 743–768.

Sullivan, N.D. 1974. Leptospirosis in animals and man. Aust. Vet. J. 50(5): 216–223.

Tacket, C.O., M.F. Pasetti, R. Edelman, J.A. Howard, and S. Streatfield. 2004. Immunogenicity of recombinant LT-B delivered orally to humans in transgenic corn. Vaccine 22(31): 4385–4389.

Temizel, E.M., K. Yesilbag, C. Batten, S. Senturk, N.S. Maan, P.P.C. Mertens, and H. Batmaz. 2009. Epizootic hemorrhagic disease in cattle, Western Turkey. Emerg. Infect. Dis. 15(2): 317.

The Cattle Site. http://www.thecattlesite.com/. Accessed 30 June 2017.

Thoen, C.O., P. LoBue, D.A. Enarson, J.B. Kaneene, and I.N. de Kantor. 2009. Tuberculosis: A re-emerging disease of animals and humans. Vet. Ital. 45(1): 135–181.

Thuenemann, E.C., A.E. Meyers, J. Verwey, E.P. Rybicki, and G.P. Lomonossoff. 2013. A method for rapid production of heteromultimeric protein complexes in plants: Assembly of protective bluetongue virus-like particles. Plant Biotechnol. J. 11(7): 839–846.

Tokuhara, D., B. Álvarez, M. Mejima, T. Hiroiwa, Y. Takahashi, S. Kurokawa, M. Kuroda, M. Oyama, H. Kozuka-Hata, T. Nochi, and H. Sagara. 2013. Rice-based oral antibody fragment prophylaxis and therapy against rotavirus infection. J. Clin. Invest. 123(9): 3829–3838.

Treanor, J.J., J.S. Johnson, R.L. Wallen, S. Cilles, P.H. Crowley, J.J. Cox, D.S. Maehr, P.J. White, and G.E. Plumb. 2010. Vaccination strategies for managing brucellosis in Yellowstone bison. Vaccine 28(Suppl 5): F64–F72.

Uribe-Campero, L., A. Monroy-García, A.L. Durán-Meza, M.V. Villagrana-Escareño, J. Ruíz-García, J. Hernández, H.G. Núñez-Palenius, and M.A. Gómez-Lim. 2015. Plant-based porcine reproductive and respiratory syndrome virus VLPs induce an immune response in mice. Res. Vet. Sci. 102: 59–66.

Uvarova, E.A., P.A. Belavin, N.V. Permyakova, A.A. Zagorskaya, O.V. Nosareva, A.A. Kakimzhanova, and E.V. Deineko. 2013. Oral immunogenicity of plant-made *Mycobacterium tuberculosis* ESAT6 and CFP10. Biomed. Res. Int. pp. 1–8.

van Zyl, A.R., A.E. Meyers, and E.P. Rybicki. 2016. Transient Bluetongue virus serotype 8 capsid protein expression in *Nicotiana benthamiana*. Biotechnol. Rep. (Amst) 9: 15–24.

van Zyl, A.R., A.E. Meyers, and E.P. Rybicki. 2017. Development of plant-produced protein body vaccine candidates for bluetongue virus. BMC Biotechnol. 17(1): 47.

Ventola, C.L. 2015. The antibiotic resistance crisis: part 1: Causes and threats. Pharm. Ther. 40(4): 277.

Vetal company. http://www.vetal.com.tr/. Accessed on 30 June 2017.

Vetterling, J.M., H.R. Jervis, T.G. Merrill, and H. Sprinz. 1971 *Cryptosporidium wrairi* sp. n. from the guinea pig *Cavia porcellus*, with an emendation of the genus. J. Protozool. 18(2): 243–247.

Vinetz, J.M. 2001. Leptospirosis. Curr. Opin. Infect. Dis. 14: 527–538.

Von Braun, J. 2010. The role of livestock production for a growing world population. Lohmann Information 45(2): 3–9. Available at. http://www.lohmann-information.com/content/l_i_45_artikel10.pdf. Accessed 03 April 2017.

Waheed, M.T., H. Ismail, J. Gottschamel, B. Mirza, and A.G. Lössl. 2015. Plastids: The green frontiers for vaccine production. Front. Plant. Sci. 6: 1005.

Waheed, M.T., M. Sameeullah, F.A. Khan, T. Syed, M. Ilahi, J. Gottschamel, and A.G. Lössl. 2016. Need of cost-effective vaccines in developing countries: What plant biotechnology can offer? SpringerPlus 5(1): 65.

Wang, Y., H. Deng, X. Zhang, H. Xiao, Y. Jiang, Y. Song, L. Fang, S. Xiao, Y. Zhen, and H. Chen. 2009. Generation and immunogenicity of Japanese encephalitis virus envelope protein expressed in transgenic rice. Biochem. Biophys. Res. Commun. 380(2): 292–297.

Wang, Y., Q. Shen, Y. Jiang, Y. Song, L. Fang, S. Xiao, and H. Chen. 2012. Immunogenicity of foot-and-mouth disease virus structural polyprotein P1 expressed in transgenic rice. J. Virol. Methods 181(1): 12–17.

Wells, G.A., A.C. Scott, C.T. Johnson, R.F. Gunning, R.D. Hancock, M. Jeffrey, M. Dawson, and R. Bradley. 1987. A novel progressive spongiform encephalopathy in cattle. Vet. Rec. 21(18): 419–420.

Werner, S.B., G.L. Humphrey, and I. Kamei. 1979. Association between raw milk and human *Salmonella dublin* infection. Br. Med. J. 2(6184): 238–241.

World Health Organization, WHO. 2008. WHO expert consultation on rabies. Fact sheet no. 99. Geneva; December 2008.

World Health Organization, WHO. 2008a. Anthrax in humans and animals, 4th ed. Geneva.

World Health Organization, WHO. 2015. Global Tuberculosis Report. Available at. http://apps.who.int/iris/bitstream/10665/191102/1/9789241565059_eng.pdf?ua=10. Accessed 13 May 2017.

World Health Organization, WHO. 2017. Rift Valley Fever. Available at. http://www.who.int/mediacentre/factsheets/fs207/en/. Accessed 30 July 2017.

World Organization for Animal Health. 2008b. Salmonellosis. Available at. http://www.oie.int/fileadmin/Home/eng/Health_standards/tahm/2008/pdf/2.09.09_SALMONELLOSIS.pdf. Accessed 23 May 2017.

World Organization for Animal Health. 2011. Joint FAO/OIE Committee on Global Rinderpest Eradication. Available at. http://www.oie.int/fileadmin/Home/eng/Media_Center/docs/pdf/Final_Report_May2011.pdf. Accessed on 10 April 2017.

World Organization for Animal Health. 2013. African Swine Fever. Available at. https://www.oie.int/fileadmin/Home/eng/Animal_Health_in_the_World/docs/pdf/Disease_cards/AFRICAN_SWINE_FEVER.pdf. Accessed 27 July 2017.

World Organization for Animal Health. 2017. OIE-Listed diseases, infections and infestations in force in 2017. http://www.oie.int/en/animal-health-in-the-world/oie-listed-diseases-2017/. Accessed 06 May 2017.

Wu, H., N.K. Singh, R.D. Locy, K. Scissum-Gunn, and J.J. Giambrone. 2004. Immunization of chickens with VP2 protein of infectious bursal disease virus expressed in *Arabidopsis thaliana*. Avian Dis. 48(3): 663–668.

Wu, J., L. Yu, L. Li, J. Hu, J. Zhou, and X. Zhou. 2007. Oral immunization with transgenic rice seeds expressing VP2 protein of infectious bursal disease virus induces protective immune responses in chickens. Plant Biotechnol. J. 5(5): 570–578.

Wu, Y.Z., J.T. Li, Z.R. Mou, L. Fei, B. Ni, M. Geng, Z.C. Jia, W. Zhou, L.Y. Zou, and Y. Tang. 2003. Oral immunization with rotavirus VP7 expressed in transgenic potatoes induced high titers of mucosal neutralizing IgA. Virology 313(2): 337–342.

Yácono, M., I. Farran, M.L. Becher, V. Sander, V.R. Sanchez, V. Martín, J. Veramendi, and M. Clemente. 2012. A chloroplast-derived Toxoplasma gondii GRA4 antigen used as an oral vaccine protects against toxoplasmosis in mice. Plant Biotechnol. J. 10(9): 1136–1144.

Yang, C.D., J.T. Liao, C.Y. Lai, M.H. Jong, C.M. Liang, Y.L. Lin, N.S. Lin, Y.H. Hsu, and S.M. Liang. 2007. Induction of protective immunity in swine by recombinant bamboo mosaic virus expressing foot-and-mouth disease virus epitopes. BMC Biotechnol. 7(1): 62.

Yang, Z.Q., Q.Q. Liu, Z.M. Pan, H.X. Yu, and X.A. Jiao. 2007. Expression of the fusion glycoprotein of newcasstle disease virus in transgenic rice and its immunogenicity in mice. Vaccine 25(4): 591–598.

Yu, J., and W. Langridge. 2003. Expression of rotavirus capsid protein VP6 in transgenic potato and its oral immunogenicity in mice. Transgenic Res. 12(2): 163–169.

Yuki, Y., M. Mejima, S. Kurokawa, T. Hiroiwa, I.G. Kong, M. Kuroda, Y. Takahashi, T. Nochi, D. Tokuhara, T. Kohda, and S. Kozaki. 2012. RNAi suppression of rice endogenous storage proteins enhances the production of rice-based Botulinum neutrotoxin type A vaccine. Vaccine 30(28): 4160–4166.

Zhang, S.Z., G.L. Zhang, T.Z. Rong, P.A. Li, Z.H. Peng, and Y.G. Zhang. 2011. Transformation of two VP1 genes of O-and Asia 1-type foot-and-mouth disease virus into maize. Agr. Sci. *China* Journal 10(5): 661–667.

Zhang, X., Z. Yuan, X. Guo, J. Li, Z. Li, and Q. Wang. 2008. Expression of Chlamydophila psittaci MOMP heat-labile toxin B subunit fusion gene in transgenic rice. Biologicals 36(5): 296–302.

Zhang, X.X., H. Yu, X.H. Wang, X.Z. Li, Y.P. Zhu, H.X. Li, S.J. Luo, and Z.G. Yuan. 2013. Protective efficacy against Chlamydophila psittaci by oral immunization based on transgenic rice expressing MOMP in mice. Vaccine 31(4): 698–703.

Zhang, Y., S. Chen, J. Li, Y. Liu, Y. Hu, and H. Cai. 2012. Oral immunogenicity of potato-derived antigens to *Mycobacterium tuberculosis* in mice. Acta. Biochim. Biophys. Sin. (Shanghai) 44(10): 823–830.

Zhou, B., Y. Zhang, X. Wang, J. Dong, B. Wang, C. Han, J. Yu, and D. Li. 2010. Oral administration of plant-based rotavirus VP6 induces antigen-specific IgAs, IgGs and passive protection in mice. Vaccine 28(37): 6021–6027.

Zhou, J.Y., J.X. Wu, L.Q. Cheng, X.J. Zheng, H. Gong, S.B. Shang, and E.M. Zhou. 2003. Expression of immunogenic S1 glycoprotein of infectious bronchitis virus in transgenic potatoes. J. Virol. 77(16): 9090–9093.

Zoetis Australia. https://www.zoetis.com.au/product-class/index.aspx. Accessed 30 June 2017.

Zoetis United States. https://www.zoetisus.com/products/index.aspx. Accessed 30 June 2017.

CHAPTER 9

Chloroplast Biotechnology Tools for Industrial and Clinical Applications

Kwang-Chul Kwon

Introduction

The emergence of photosynthetic eukaryotes dates back more than 1000 million years via endosymbiosis of the current cyanobacterial ancestor with a eukaryotic host, and the conversion of the endosymbiont to photosynthetic organelle, chloroplast, gave rise to green algae and plants. Over the past 30 years, chloroplast genomes have experienced far more dramatic changes than ever due to the human-made modifications to them. After gene transfer mechanism between *Agrobacterium* and plants was discovered in 1977 (Schell and Van Montagu 1977), the first transgenic plant transformed by *Agrobacterium* carrying a disarmed Ti plasmid vector was created in 1983 (Zambryski et al. 1983). Since then, the first trial for the production of a biopharmaceutical molecule in plants was carried out with human growth hormone (hGH) (Barta et al. 1986), but the study confirmed only the expression of functional hGH pre-mRNA in tobacco and sunflower callus tissues. In 1989, Hiatt et al. transformed tobacco with the γ- or κ-chain cDNAs derived from mouse hybridoma and successfully created transgenic tobacco expressing functional antibodies assembled through a sexual cross between the individual transgenic tobacco (Hiatt et al. 1989). The early plastid transformation studies date back about 30 years ago. In 1987, Daniell and McFadden found that the EDTA-treated etioplasts can uptake plasmids (Daniell and McFadden 1987). In the next year, Boynton et al. were successful in the stable transformation of the chloroplast genome of *Chlamydomonas reinhardtii* using a gene gun (Boynton et al. 1988b). Two years later, plant plastid was also successfully transformed using the gene gun (Svab et al. 1990). The first biopharmaceutical expressed in chloroplasts was

MicroSynbiotiX Ltd. 11011 N Torrey Pines Rd Ste. #135, La Jolla, CA 92037.
Emails: kwang-chul.kwon@microsynbiotix.com; kwangchul.kwon@gmail.com

human somatotropin. The biologically active and disulfide-bonded form of the human somatotropin in tobacco chloroplasts was accumulated up to a 300–fold higher level than that of a similar gene expressed in the nucleus (Staub et al. 2000). In the following year, Daniell et al. created transplastomic tobacco expressing cholera toxin subunit B (CTB) of *Vibrio cholerae*. The expressed monomeric CTB in the chloroplast showed the proper formation of the disulfide bonds and functional oligomeric structures (Daniell et al. 2001a); the CTB expression level was 410–fold higher than that of B subunit of enterotoxigenic *Escherichia coli* (LTB) expressed in the nuclear genome.

These initial technical successes have sparked chloroplast engineering to be extended to various fields such as diseases control for humans and animals, production of biopharmaceuticals and industrial proteins, and improvements of agronomic traits. There have been a large number of biomolecules expressed in chloroplasts, such as vaccine antigens, autoantigens, enzymes, antibodies, immunotoxins, hormones, bioactive peptides, blood clotting factors, biomaterials, industrial enzymes, biofuels, secondary/medicinal metabolites, and dsRNAs (Kwon and Daniell 2016, Adem et al. 2017, Zhang et al. 2017a, Daniell et al. 2005, Bock 2015, Fuentes et al. 2018, Specht and Mayfield 2014, Taunt et al. 2018, Shahid and Daniell 2016, Jin and Daniell 2015, Bock 2007, Maliga and Bock 2011, Zhang et al. 2017b, Scranton et al. 2015, Rasala and Mayfield 2015). The main attributes of using chloroplasts as an expression platform are that the organelle can provide a high level of expression, maternal inheritance, and polycistronic expression. Further details about these advantageous characteristics will be described below.

In addition, plants and algae have several advantages over conventional platforms in terms of biopharmaceutical productions. Biomass production can be achieved in a cheaper and safer way. For example, mammalian cell culture systems for the production of biopharmaceuticals requires expensive substrates and supplements. The culture media is composed of up to 100 ingredients such as hormones, growth factors, vitamins, amino acids, and hydrolysates to support cell culture (McGillicuddy et al. 2018). Furthermore, the animal-derived components can be a potential source of contamination of human pathogens. In contrast, plants (Su et al. 2015b), plant suspension cells such as tobacco and carrot cells (Tekoah et al. 2015, Paul and Ma 2011), and algal cells (Fields et al. 2018) can be cultivated with a simple nutrient composition in a bioreactor or a hydroponic system (Fig. 1). Furthermore, the current main expression platforms, such as mammalian cells, yeast and bacteria, and their downstream processes and delivery of the drugs into patients have several disadvantages. For example, a new cGMP mammalian cell culture production unit costs ~ 77 – 500 million Euros (€) and it takes several years to be constructed and operational (Spök et al. 2008). Moreover, these facilities and their manufactured products have a risk of contamination with toxins or pathogens (Ma et al. 2003). The protein drugs produced using the conventional fermentor systems should be extracted and purified, which is a very expensive process. For example, protein A which is used for the purification of antibody drugs cost millions of dollars per kilogram (Morrow 2007). The purified proteins are then formulated for an injectable form which needs a cold chain for both transport and storage, in addition, these drugs have short shelf lives. In contrast, plants and algae don't serve as hosts for animal pathogens. Their cell walls can provide the protection to biomolecules expressed in the cells from the harsh environment such as stomach acid (Dreesen et al. 2010), so edible plant and algal

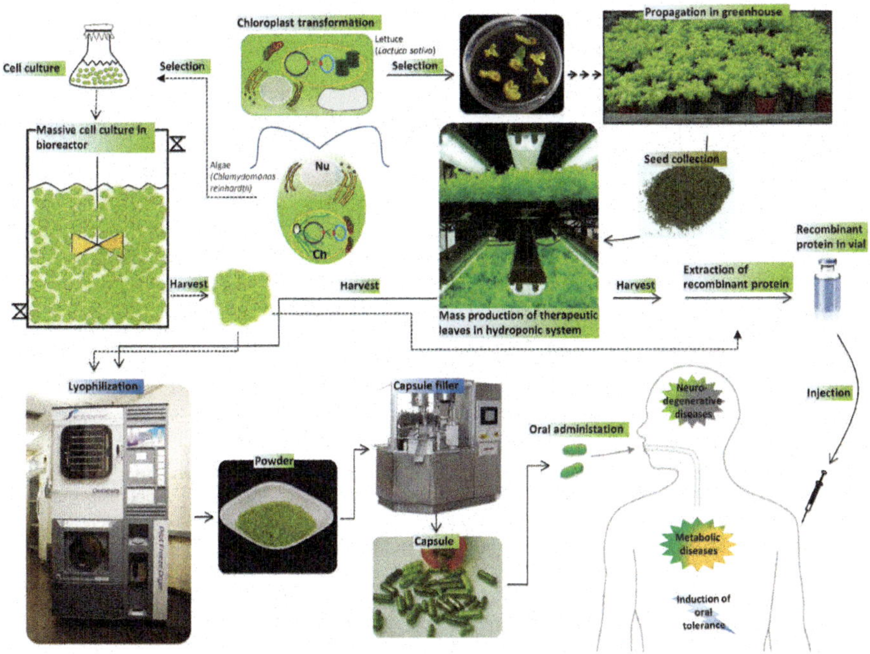

Figure 1. Schematic diagram of chloroplast engineering to express recombinant protein drugs. Selected homoplasmic lines from transformed cells are grown in a bioreactor or in a hydroponic cGMP facility. Harvested cells are subject to freeze-dry for oral delivery or extraction of the expressed recombinant proteins for injection. The lyophilized and powdered cells formulated in capsules or extracted proteins in vial are used to prevent or treat human diseases by oral administration or by injection, respectively. Figure adapted from Fig. 1 in Kwon et al. (2015).

cells can be used as a vehicle for the oral delivery of therapeutic proteins (Tran et al. 2013, Gregory et al. 2012, Herzog et al. 2017, Kwon et al. 2018a, Daniell et al. 2018, Chan et al. 2016, Su et al. 2015a, 2015b, Shil et al. 2014, Shenoy et al. 2014, Sherman et al. 2014, Kohli et al. 2014, Kwon et al. 2013, Davoodi-Semiromi et al. 2010, Verma et al. 2010, Boyhan and Daniell 2011, Dreesen et al. 2010). Further, freeze-dried plant cells can be stored for several years at ambient temperatures (Su et al. 2015b), which can eliminate cold chains. Moreover, the lyophilized and powdered plant cell materials formulated in a capsule can be delivered via oral administration to humans or animals for therapeutic or prophylactic purposes. Also, these harvested cells can be subject to extraction of the expressed recombinant proteins for injection (Fig. 1).

Along with the many merits of plant chloroplasts as a platform for the production of biomolecules, there have been continuous technological advancements to strengthen the merits and to extend chloroplast engineering to the platform available for the commercial production of biopharmaceuticals (Table 1). The molecular tools for chloroplast engineering have been evolved to increase yield, efficiency, and safety to make this technology a viable and profitable system. This chapter will highlight the technical advances which have been achieved over the last 30 years and consider some of the current issues that need to be addressed in the future.

Table 1: Chloroplast engineering tools. Tools applied to engineer chloroplasts are summarized and more details of each subject are described in the main text.

Subject	Molecular tool	Effect and/or Method
Improvement of the level of transgene expression	Promoter and 5′UTR	• Improves transcriptional and translational activity using strong and efficient 5′UTR region
	Codon optimization	• Improves translation efficiency using codon usage hierarchy of host chloroplast
	Downstream sequence	• Improves translational initiation by adding 5′-coding region fragment in between promoter/5′UTR and a transgene
	Avoidance of auto-repression	• Achieves high level of transgene expression under the control of *psbA*/5′UTR in *psbA*-deficient genetic background
	Linker	• Improves stability of fusion proteins by insertion of a linker sequence between genes
	N-end rule	• Improves stability of protein by replacing second amino acid with a more stable amino acid in chloroplast
	Integration site of expression cassette into chloroplast genome	• Increases copy number of transgene by copy correction • Improves transcription of transgene by insertion of the transgene into transcriptional active regions
Avoidance of toxic effect of transgenes on *E. coli* host	Codon reassignment	• Secures scale-up of transformation vector plasmid carrying toxic genes to *E. coli* by codon reassignment
Inducible expression system	Transcriptional switch	• Avoids toxic effect of transgene to chloroplast using transcriptional factors responsive to chemical inducers
	Translational switch	• Avoids toxic effect of transgene to chloroplast using RNA sensors responsive to chemical inducers
Multiple gene expression	Polycistronic expression with following cleavage	• Converts polycistronic transcript to translatable monocistronic mRNAs by cleavage at intercistronic expression element (IEE)
	Polycistronic expression without further process	• Expresses polycistronic transcript using bacterial type operon • Translates polycistronic transcript directly by introduction of ribosome binding sequence at intercistronic spacer regions
Avoidance of the use of antibiotic selection marker	Screening by restoration of photosynthesis or amino acid synthesis	• Restores photosynthesis or amino acid synthesis ability using photosynthesis-deficient or auxotrophic mutants as an expression platform
	Screening by detoxication or metabolization of toxic or non-metabolizable compounds	• Uses a gene which is feedback-insensitive to end product • Uses a gene which can convert toxic or non-metabolizable compounds to non-toxic or metabolizable ones

Table 1 contd. ...

...Table 1 contd.

Subject	Molecular tool	Effect and/or Method
Excision of antibiotic selection marker	Direct repeats	• Excises selection marker flanked with direct repeats by loop-out recombination event • Excises selection marker placed outside the flanking sequences of expression cassette by two-step process: formation of transient cointegrates and following direct repeat-induced loop-out event
	Recombinase system	• Uses phage recombinases and their cognate recognition sites to excise antibiotic selection marker
Alternative chloroplast transformation methods to particle bombardment	Glass bead-mediated transformation	• Agitates cell-wall deficient cells with glass beads and pores created by glass beads allow cells to take up DNA
	Polyethylene glycol (PEG)-mediated transformation	• Incubates cell-wall deficient protoplasts with PEG to allow cells to take up DNA
	Electroporation-mediated transformation	• Allows microalgal chloroplast whose transformation is not available with particle bombardment due to cell size limitation to receive DNA
Biomass increase	Fermentor	• Increases biomass of algae using fermentor which can automatically control feeding nutrients and acid for optimal pH adjustment
	Hydroponics	• Increases biomass of plant leaf materials in cGMP hydroponic system which is optimized with temperature, light cycle and nutrition for growth of a specific plant

Advantages of Plant Chloroplast as an Expression Platform

Although there has been a long track record of biopharmaceutical productions using nuclear transformation, the low level of expression is still a hurdle to extend the nuclear transformation to the industrial-scale production system, which is largely attributed to various positional effects: (1) integration of transgenes into transcriptionally silent regions, (2) integration of transgenes into a region where the transgene is susceptible to DNA methylation, (3) post transcriptional silencing which is associated with the existence of multiple transgene copies within a cell. In addition, there is no direct correlation between the expression level and the copy number of integrated transgenes (Gelvin 2003, Hobbs et al. 1990, Peach and Velten 1991). Furthermore, the control of the escape of engineered genes into the ecosystem via pollen and the inadvertent contamination of food/feed chains by transgenic seeds is a challenging part of nuclear transformation. Although the traditional nuclear transformations have struggled to increase the expression level of transgenes, a recent advanced system stretched from transient expression based on viral vectors, called the deconstructed viral vector system, has shown outperformance in terms of the expression level over the traditional nuclear transformation, showing the expression level of genes of interest up to 30 – 80% of total soluble proteins (Gleba et al. 2004, Marillonnet et al. 2004, Mardanova et al. 2017). In the system, whole plants or detached mature leaves are infiltrated by *Agrobacterium* carrying T-DNA binary vector system. Deconstructed viral amplicon integrated into T-DNA is inserted into host nuclear DNA chromosome and then the released amplicon from the chromosome produces a recombinant protein of interest in the plant cells (Gleba et al. 2007). This system clearly addresses several issues of the earlier version of full virus strategy for transient expression; overcoming of the unstability of the viral vector by integration of the viral amplicon into the host chromosome, insertion of larger than 1 kb fragments up to 2.3 kb, achievement of systemic infection and movement through agroinfiltration, and the increase of the level of expression of heterologous genes through amplicon resulting from copy number increase of the transgene and movement from cell-to-cell (Gleba et al. 2007). Apparently, this system has several advantages over the traditional nuclear transformations such as a high yield of recombinant proteins, rapid scale-up, and fast production. However, this system requires an intensive upstream processing to make sure that each plant is infected with engineered *Agrobacterium* cells via vacuum infiltration. In addition, this system is not appropriate when using plants as an oral delivery vehicle of recombinant therapeutic proteins to human or animals due to the safety issue of viral components used in the transformation process.

In contrast to the conventional nuclear transformation, chloroplast transformation can provide a high level of transgene expression. Approximately, 100 – 200 copies of the genome are contained in a chloroplast (Shaver et al. 2006) and a mature plant cell has around 100 chloroplasts. So the conversion of wild type chloroplast genomes to all the transformed genomes, homoplasmic status, through a selection process can make the transgenes highly expressed, up to > 70% of total soluble protein (Oey et al. 2009, Ruhlman et al. 2010). Moreover, copy correction mechanism happening between two identical inverted repeats (IR) can further increase the copy number of transgenes if the integration of transgenes is targeted into the IR region (Dhingra and Daniell 2006, Boynton et al. 1988a). Any location of chloroplast genome can be targeted for the integration of transgenes, however, in terms of high expression, transcriptionally active regions are preferred (Daniell et al. 2016). The specific integration of the expression cassette into the chloroplast genome via double homologous recombination is another big advantage so no positional effect theoretically can be expected. However, it was found that there was a variation in the expression level of a transgene in chloroplast transformants. As shown in a study by Dreesen et al. the difference in the expression level of CTB-D2 in the chloroplast of *C. reinhardtii* between individual clones was approximately 15–fold (Dreesen et al. 2010). However, the expression difference found in the transplastomic clones is considered relatively lower when compared to nuclear transformants in which, due to the positional effect, the variation was much higher than chloroplast transformants, up to 50–fold (Hobbs et al. 1990). Moreover, the screening process of nuclear transformation requires extensive time and labor to find the lines whose expressions are stable and highest. Furthermore, the highly amplified transgenes through chloroplast transformation can be contained via maternal inheritance which eventually prevents the possible accidental escape of transgenes to the environment via pollen. Also, the polycistronic expression property allows multiple transgenes to be expressed by a single transformation event, so it doesn't require sequential or co-transformation of multiple plasmids, which imposes the need of multiple selection markers as performed under nuclear transformations (Ye 2000, Ma et al. 1995, Kebeish et al. 2007).

When using algal chloroplasts as an expression platform for recombinant proteins, additional features are gained besides the advantages from the use of plant chloroplasts. Algal biomass production is very fast as described below and it can be easily achieved by using enclosed reactors in which algae grow in defined optimal conditions, which allows the biomass production to be unaffected by seasonal variations or soil conditions without any possible transgene escape into the environment (Specht and Mayfield 2014).

In contrast to terrestrial plants, algae have very few chloroplasts, e.g., *Chlamydomonas reinhardtii* has only one chloroplast and it has around 80 copies of chloroplast genomes (Gallaher et al. 2018). As the number of copies of the chloroplast genome per an algal cell is much less than that of plants, the expected levels of transgene expression are relatively low. In fact, the overall level of expression of recombinant proteins in *C. reinhardtii* chloroplasts is less than 1% (Manuell et al. 2007). However, as seen in a recent study, the yield of the algal cell can be dramatically increased using a mixotrophic fed-batch fermentor in a very short period of time. The cell density reached about 24 g/L (dry weight) after 168 h; the previous highest cell density was 9 g/L (Fields et al. 2018). This study represents a practical approach to increase the overall protein yield by the optimization of growth conditions when the protein level is restricted by a lesser number of chloroplast and genome copy numbers.

In addition, algal chloroplasts, like plant chloroplasts, are also able to make posttranslational modifications on proteins, such as phosphorylation, disulfide bond formation, multimeric protein assembly and lipidation (Lehtimäki et al. 2015). Further, chaperones (Schroda 2004) in chloroplasts allow for complex proteins to be assembled in functional forms.

Chloroplast Engineering Tools

Modification of chloroplast requires a series of procedures: vector design to express heterologous genes using various DNA elements operable in chloroplasts; preparation of plasmids for bombardment by amplifying the plasmids in *E. coli*; delivery of the plasmids into chloroplasts using particle delivery system; integration of the expression cassettes into a specific location of the chloroplast genome induced by homologous recombination between the identical sequences located on both the delivered plasmids and the host chloroplast genomes; and selection of transformants and subsequent subculture under a selection pressure to achieve homoplasmic lines. After selection of the homoplasmic lines, biomass increase of the lines is followed using a defined facility or bioreactors.

To highly express transgenes, expression cassettes constructed with endogenous strong promoters and their 5′UTRs or the 5′UTR from phage T7*g10* (Daniell et al. 2016, Adem et al. 2017, Maliga 2002, Doron et al. 2016) are usually used. Also, the use of inducible promoters is very desirable for transgenes if their expression affects chloroplast function. (Bock 2015, Emadpour et al. 2015). Once a transformation vector is constructed, the plasmids are delivered into chloroplasts, in general, by biolistic particle delivery system. The plasmids are coated onto microparticles such as gold or tungsten, which are then shot by pressured helium gas to cells or tissues placed in the vacuum-created bombardment chamber. To increase transformation efficiency with less cell damage, the kind and size of the particles, the distance between the tissue/algal cells and the rupture disk, and helium gas pressure should be empirically determined.

The expression cassettes of plasmids are then integrated into a specific site of plastome via double homologous recombination between the flanking sequences in the vector and the identical sequences in the endogenous plastome (Maliga 2002, Jin and Daniell 2015) so the flanking sequences need to be defined. The flanking sequences can be chosen and defined using the GenBank at the National Center for Biotechnology Information (NCBI) where the chloroplast genome sequences of more than 800 are now available (Daniell et al. 2016). Also, depending on the integration location of expression cassettes in plastome such as single copy or inverted repeat regions, the expression level of transgenes could be affected (Herz et al. 2005, Krichevsky et al. 2010). The initial integration of the transgene just occurred to a few copies of the chloroplast genome, which need to go through multiple rounds of shoot regeneration to reach homoplasmy, meaning that all the plastomes carry the foreign DNA in response to the long-term selective pressure. Antibiotic-resistant genes are most widely used in the selection process (Esland et al. 2018). The aminoglycoside 3′-adenylyltransferase (*aadA*) gene among the selection markers has been exclusively used in chloroplast transformations due to its high enzymatic activity and specificity of the AadA protein to spectinomycin. Although there have been some alternative antibiotics that have developed as a selection marker: neomycin phosphotransferase, *nptII*, and aminoglycoside phosphotransferase, *aphA-6*, both confer resistance to kanamycin (Carrer et al. 1993, Huang et al. 2002); chloramphenicol acetyltransferase, *cat*, which confers resistance to chloramphenicol (Li et al. 2011), however, the efficiency of these selection markers is much less than that of *aadA* so their use is not prevalent. A new selection marker has been recently developed, which was a bacterial-origin bifunctional aminoglycoside acetyltransferase/phosphotransferase gene (*ac(6')-Ie/aph(2")-Ia*) and tested its selectability by transforming tobacco chloroplasts using aminoglycoside antibiotics, and tobramycin. The new selection system showed a similar efficiency to *aadA* with no spontaneous resistance mutations (Tabatabaei et al. 2017).

After the confirmation of homoplasmic lines by PCR and Southern blot, the selected lines are subject to biomass increase on demand. The biomass increase can be done using a bioreactor, greenhouse or hydroponic system. Biomass increase of unicellular algae can be routinely achieved using bioreactors. A recent study for the biomass increase of transplastomic algae showed that the dramatic increase of cell mass was achieved using a fed-batch fermentor with a simple nutritional composition (Fields et al. 2018). For genetically modified plants, the biomass increase is generally performed in a confined facility. An automated greenhouse system is widely used for biomass production (Su et al. 2015b, Kwon et al. 2017) but the plant materials grown in microbe-infested soil are prohibited from the clinical use. The plant materials intended to be used in clinics should be cultivated under a Current Good Manufacturing Practice (cGMP) facility (Herzog et al. 2017). Transplastomic lettuce expressing a human blood clotting factor was cultivated in cGMP hydroponic system at the Fraunhofer USA Center, and the amount harvested per 1000 ft^2 per annum comes to ~ 870 kg of fresh lettuce leaves expressing factor FIX, which is an amount that can provide 24,000–36,000 doses for 20–kg (weight) pediatric patients (Su et al. 2015b).

In the next section, key technical achievements made in chloroplast engineering will be covered in more detail. These chloroplast engineering tools are briefly summarized in Table 1.

Improvement of the Level of Transgene Expression

Promoter and 5′UTR

The achievement of the high expression level of transgenes is one of the top main interests across all the expression platforms because it can eventually reduce production costs, increase potency, and provide flexibility in dosage regimen.

Obviously, the high copy number of plastome in the chloroplast is a remarkable advantage for the high level of expression. Nevertheless, the studies on various promoters/5′UTR and their synthetic derivatives have been continuously and extensively carried out, which include promoters for *psbA, 16S rRNA, rbcL, psbD, psaA, atpA, atpB* and their corresponding 5′UTR and gene 10 leader sequence of T7 phage (Gimpel and Mayfield 2013, Blowers et al. 1990, Yang et al. 2013, Barnes et al. 2005, Marín-Navarro et al. 2007, Kuroda and Maliga 2001a, Ishikura et al. 1999, Rasala et al. 2011, Eibl et al. 1999, Anthonisen et al. 2001, Ruhlman et al. 2010, Ye et al. 2001, Bock 2014, Staub and Maliga 1994, Zou et al. 2003, Drechsel and Bock 2011, Kuroda and Maliga 2001b, Herz et al. 2005). Among these studies, the super-high expression levels of transgenes were up to 70% (Oey et al. 2009, Ruhlman et al. 2010). In 2009, there was a report that the expression level of phage-derived synthetic antibacterial lysin gene in tobacco chloroplasts was accumulated up to ~ 74% under the control of synthetic promoter: 16S ribosomal promoter was fused to the gene 10 leader sequence from phage T7 (P*rrn*:T7*g10*) but the transplastomic lines suffered from growth retardation due to the exhaustion of the protein synthesis capacity for the production of endogenous plastid-encoded proteins (Oey et al. 2009). Another super-high expression was reported that the native sequence of the fusion gene of cholera nontoxic B-subunit and human proinsulin (CTB:Pins) was expressed up to 72% of total leaf proteins in tobacco chloroplasts under the control of endogenous *psbA* promoter and 5′UTR without showing any deleterious phenotype (Ruhlman et al. 2010). In Bally et al.'s study, hydroxyphenyl pyruvate dioxygenase (HPPD) and GFP were expressed in tobacco chloroplasts in the range of 30% to 40% using endogenous *psbA* promoter/5′UTR or corn *16S rRNA* plastid promoter/T7*g10* (Bally et al. 2009). There are many other examples of the high expression of foreign genes in the chloroplast. CTB fused human proinsulin via three furin cleavage sites (CTB-PFx3) was expressed up to 47% and 53% total leaf protein (TLP) in tobacco and lettuce chloroplasts, respectively (Boyhan and Daniell 2011). Tobacco chloroplasts transformed with *cry2Aa2* operon expressed the cry2Aa2 up to 45.3% TSP in mature leaves (De Cosa et al. 2001) and anthrax protective antigen (PA) accumulated up to 29.6% (Ruhlman et al. 2010). Human insulin-like growth factor-1 (hIGF-1) and interferon-α2b were expressed up to 32% and 20% total soluble protein (TSP), respectively, in tobacco chloroplasts (Daniell et al. 2009, Arlen et al. 2007). However, the expression of foreign genes is not always high in the chloroplast. In many cases, the expression of the native sequences of foreign genes faced a very low expression in chloroplasts, less than 0.2%, such as human blood clotting factor VIII heavy chain, factor IX, and poliovirus capsid protein 1 (VP1) (Kwon et al. 2016, Verma et al. 2010). The strong promoters and 5′UTRs are necessary enough for the high expression of some transgenes in chloroplasts but other genes seem to need additional expression signals. Firstly, the codon preference of the chloroplast is different from that of the organism where the transgenes are originated so chloroplast ribosomes could stall at rare codons residing in the foreign genes, resulting in the drop of translational efficiency. Secondly, the incompatibility between 5′UTR and 5′-coding region of transgene interrupts the proper secondary structure necessary for the translation initiation complex so the incompatibility reduces or abolishes the expression of transgenes. Lastly, a unique chloroplast gene regulation system is not well compatible with transgenes.

Codon Optimization

Codon-adjustment of transgenes is necessary when the codon preference of the heterologous genes is different from that of platform chloroplasts since non-preferably used codons cause stalls of ribosomes and result in the decrease of the translational efficiency of the transgene. The level of expression of codon-optimized human transforming growth factor-β3 (TGF-β3) was expressed 75–fold higher than that of native sequence (Gisby et al. 2011). Likewise, the expression of GFP was increased ~ 80-fold after codon optimization (Franklin et al. 2002). In a recent study, a codon-optimization table was created for the universal expression of transgenes in plant chloroplasts, which was developed based on the codon usage analysis of only *psbA* genes, the most highly expressed gene in plant chloroplasts, from 133 plant species, instead of using codon usage percentage from combined analysis of all the chloroplast genes. The codon usage hierarchy was applied to foreign gene expressions such as human clotting factor FVIII and capsid protein of poliovirus VP1 and the enhanced expression levels of the synthetic genes were found 8 and 125–fold higher in increase than those of native sequences, respectively (Kwon et al. 2016). In the same vein, a codon usage table was generated by using highly expressing genes in chloroplasts such as *psbA* and *psbD* as reference genes and codon-optimized VP28 was expressed up to 21% TSP in *Chlamydomonas* (Surzycki et al. 2009). However, codon-optimization doesn't always increase the expression of transgenes in chloroplasts (Lenzi et al. 2008, Nakamura et al. 2016, Ye et al. 2001, Wang et al. 2015, Ruhlman et al. 2010, Daniell et al. 2009, Surzycki et al. 2009).

Downstream Sequence

Klein et al. (1994) reported that the transcriptional activity of 81 bp-long *rbcL* basic promoter (positioned –18 to +63, +1 as the site of initiation of transcription) enhanced transcription of the bacterial *uidA* gene, when fusing the 5′-coding region of the *rbcL* gene to the *uidA* gene, indicating that core promoters of chloroplast genes are not enough to drive heterologous genes efficiently (Klein et al. 1994). In 2003, Kasai et al. applied the Kelin's finding to other promoters to assess the impact of the addition of the 5′-coding region to corresponding promoters on the transcription of the reporter gene, *uidA*. In the study, the fusion of a series of *rbcL* 5′-coding region to *rbcL* promoter increased the level of the *uidA* mRNA with the increase of the length of the *rbcL* coding regions, while the fusion of a series of *psbA* 5′-coding regions to P*psbA* showed the increase when the length of the coding region was above 60 bp, and P*atpA* series showed loss of the *uidA* mRNA level as the corresponding 5′-coding region was added (Kasai et al. 2003). The importance of downstream sequences of the translation initiation was further demonstrated by Kuroda and Maliga (Kuroda and Maliga 2001b). The neomycin phosphotransferase (*nptII*) reporter gene was constructed in a way that both wild type and silent mutant versions of 14 N-terminal amino acid sequences (DS, sequence downstream of initiation codon) from each *rbcL* and *atpB* gene were fused to the *nptII* gene and the chimeric genes were driven by P*rrn*/*rbcL* 5′UTR or *atpB* 5′UTR. The *nptII* mRNA levels between the wild type and mutant sequences for each gene was not dramatically affected. However, the silent mutant sequence of *rbcL* segment downstream of AUG dramatically reduced the accumulation of NPTII protein by 35–fold over wild type sequence, while the *atpB* mutant segment just reduced the NPTII level by 2–fold (Kuroda and Maliga 2001b). The expression of the codon-optimized *bar* gene in chloroplasts was enhanced by the P*rrn*/*atpB* 5′UTR + 14 codons of 5′-coding region of *atpB*, up to > 7% TSP (Lutz et al. 2001). Also, the expression of *aadA-gfp* fusion construct was driven by P*rrn*/5′*atpB* UTR+DS or P*rrn*/5′*rbcL* UTR+DS and the fusion proteins were accumulated 8% or 18% TSP, respectively (Khan and Maliga 1999).

The expression levels of both native and codon-optimized 5-enolpyruvylshikimate-3-phosphate synthase (EPSPS) gene sequences, the target enzyme of glyphosate herbicide, were 0.001% and 0.002% TSP, respectively, under the control of P*rrn*/*rbcL* 5′UTR. When the gene was driven by P*rrn*/T7*g10*, the expression levels were 0.2% and 0.3% TSP for native and synthetic sequences, respectively (Ye et al. 2001). Even the worst cases were reported in Lenzy et al.'s study (Lenzi et al. 2008). The various forms of human papillomavirus (HPV) L1 capsid gene constructs were tested for expression in tobacco chloroplasts and there were no L1 proteins detected in the western blots when the gene was driven by three different expression signals without sequence downstream to start codon: plastid *16S rRNA* promoter fused to either of three different 5′UTRs such as *rbcL*, *psbA* or T7*g10*. Similarly, the human immunodeficiency virus *tat* coding region fused directly to the tobacco chloroplast *rbcL*, *psbA* or T7*g10* 5′UTR with no additional N-terminal fusion showed no detectable translation of the transgene (Nakamura et al. 2016). The increase of the expression level of the proteins, EPSPS, L1, and Tat proteins, was achieved by the addition of downstream sequences to the N-terminus of the transgenes. The level of expression of codon-optimized EPSPS reached more than 10% TSP once the first 14 codons of *gfp* were added to the synthase gene (Ye et al. 2001). Also, when the first 14 codons of *atpB* or *rbcL* tobacco genes were fused to HPV L1, the virus gene was translatable and the highest %TLP was 1.5% (Lenzi et al. 2008). The level of the HIV Tat protein was significantly detectable once the codon-optimized synthetic *tat* gene was fused with more than 10 codons of *psbA* 5′-coding segment (Nakamura et al. 2016).

Although there are many similarities of gene expression mechanisms between chloroplasts and *Escherichia coli* due to the prokaryotic origin of chloroplasts, some characteristic features found in *E. coli* don't work in plastids in the same way. The efficiency of phage T7*g10* leader sequence as a translational enhancer was first reported in *E. coli*. The fusion of T7*g10* leader sequence to *recA* promoter enhanced translational efficiency of transgenes dramatically, up to more than 340–fold higher than those of the reference sequences which contained consensus ribosome binding sequence (Olins et al. 1988). In addition to the interaction between the Shine-Dalgarno (SD) sequence upstream of start codon, as seen in T7*g10* sequence, and the anti-Shine-Dalgarno sequence (ASD) located in the 3′ end of 16S RNA, *E. coli* translation is further facilitated by another interaction between the downstream box (DB) sequence and the *16S rRNA* penultimate stem. Likewise, for plant plastids, the fusion of T7*g10* leader sequence to P*rrn* or P*psbA* also led to the dramatic increase of protein expression but when the N-terminal fusion tag, downstream sequence (DS), was added to the transgenes (Kuroda and Maliga 2001a, Ye et al. 2001, Herz et al. 2005). However, the increasing complementarity between the downstream sequence and plastid *16S rRNA* penultimate stem reduced the translation efficiency by 100–fold (Kuroda and Maliga 2001b). Rather, not perfect complementarity between the sequences led to the significant increase of expression.

Avoidance of Auto-repression

In contrast to terrestrial plants, there is a limitation in using *psbA* promoter/5′UTR for the expression of foreign genes in *C. reinhardtii*. High level of transgene expression under the control of *psbA* promoter/5′UTR was only achieved in a *psbA*-deficient genetic background (Minai et al. 2006, Manuell et al. 2007) due to *psbA*/D1-dependent auto-repression, which is attributed to the control by epistasy of synthesis (CES) (Choquet and Wollman 2002) so unassembled and accumulated D1 protein negatively regulates *psbA* expression by binding to the *psbA* regulatory elements. As an alternative approach for high-level expression of transgenes in photosynthetic-competent microalgae, Rasala et al. evaluated synthetic expression signals for the high-level expression of transgenes in *C. reinhardtii*, so *16S rRNA* strong promoter was fused to *atpA* 5′UTR and also to *psbA* 5′UTR. The *16S/atpA* promoter/5′UTR drove reporter gene, luciferase (luxCt), highly in the wild type background, and the expression was comparable to the level driven by *16S rRNA* promoter/*psbA* 5′UTR in the D1-deficient cells. Although, in the expression study of a therapeutic, 14FN3 domain of fibronectin, *16S/atpA* promoter/5′UTR-14FN3 construct (wild type background) expressed 14FN3 domain two-fold less than that of *psbA* promoter/*psbA* 5′UTR (D1-deficient background), this new synthetic expression element is undoubtedly a useful molecular tool which has a higher capacity of increasing the expression of heterologous proteins in photosynthetically competent algae (Rasala et al. 2011).

Linker

Another factor which needs to be considered when expressing a gene of interest in a fusion form such as a fusion to an adjuvant or a carrier protein is the presence of a linker. The expression level of human blood clotting factor IX (FIX) in tobacco chloroplasts was affected by the absence of the linker sequence, furin cleavage site, between the FIX and CTB. The expression level of the CTB-FIX fusion with the furin cleavage site reached up to 3.8% TSP in the tobacco chloroplasts, while the same fusion construct without the cleavage site was expressed only 0.19% TSP (Verma et al. 2010). For the *aadA-gfp* fusion construct, a 16-mer (ELVEGKLELVEGLKVA) linker between the genes was favored against an 11-mer (ELAVEGKLEVA) for chloroplast transformation (Khan and Maliga 1999).

N-end rule

N-terminal amino acid determines the half-life of a protein, and the rule applies to both prokaryotes and eukaryotes. For eukaryotes, N-terminal amino acid is recognized by N-degrons and then followed by ubiquitination for degradation (Tasaki et al. 2012). But for prokaryotes, the first Met amino acid is formylated and then removed so the second residue becomes the N-terminal amino acid and governs the stability of proteins (Tobias et al. 1991). To investigate whether the N-end rule operates in plastids, Apel et al. did a systemic test on the stability of GFP so 20 different GFP mutant sequences were created by positioning 20 amino acids in the second place of GFP since the N-terminal amino acid is also removed in plastids like bacteria. The twenty created transplastomic tobacco lines were analyzed for the stability of GFP. The protein accumulated highest was the lines with the penultimate amino acids of Glu, Met, and Val, while the most destabilized lines were found as Cys and His. In addition, the study also found that the stability of plastid proteins are largely influenced by the N-terminal part (Apel et al. 2010).

Integration site of Expression Cassette into Chloroplast Genome

In contrast to the general notion that the chloroplast expression system has no positional effects such as gene silence and repression. However, the expression level of cholera toxin B (CTB) fused to the fibronectin-binding domain D2 derived from the *Staphylococcus aureus* showed significant variations among individual transformants up to 15 times in both *rbcL* 5′UTR-CTB-D2 and *atpA* 5′UTR-CTB-D2 transplastomic *Chlamydomonas* lines (Dreesen et al. 2010). But it is not clear whether the variation is attributed to some negative effects of the possible mutations of nuclear genes caused by bombardment or not. The integration site of the expression cassette into the chloroplast genome affects expression levels. In the Herz et al.'s study, the expression level of *uidA* integrated into large single copy region (LSC) such as *trnS*/*ORF74* was two-fold less than the same gene integrated into inverted repeat region (IR) such as *trnN*/*trnR* or *ORF131*/*rps12*, which was assumed due to the copy correction between inverted repeat regions (Herz et al. 2005). Furthermore, even an expression cassette is integrated into IR, the expression level can be severely affected by whether the integration region is transcriptionally active or not. When *lux* operon, a bacterial light-emission enzymatic system, was inserted into *trnI*/*trnA* region of tobacco plastome, the photon emission was 25 times higher than when inserted into *rps12*/*trnV* region, due to the higher read-through transcriptional activity in the *trnI*/*trnA* locus. Also, it is evident that the transcriptionally active spacer region, *trnI*/*trnA*, has been preferentially used for transgene expression as reported in a recent review, 71 out of 114 transplastomic plants used the site for the integration of transgenes (Daniell et al. 2016).

Avoidance of Toxic Effect of Transgenes on *E. coli* Host

Codon Reassignment

The delivery of the transformation vector into chloroplasts via the biolistic particle delivery system requires a certain amount of plasmids. The plasmids are usually scaled up by culturing *E. coli* after transformation. However, the prokaryotic nature of chloroplast promoters and 5′UTRs used in the vector constructs allows transgenes to be expressed in *E. coli* and some of the translated products of transgenes such as metabolic enzymes, antimicrobial peptides, and integral membrane proteins could be toxic to the host. To avoid the toxic effect of the transgenes, a new system was developed, which was based on the codon reassignment of the tryptophan codon (UGG) to the stop codon (UGA), because the UGA stop codon is not used by any of the 69 genes in the chloroplast of *Chlamydomonas*. The Trp codon (UGG) of toxic genes is replaced with the UGA so the modified transgene can't be expressed in both *E. coli* and the chloroplast of *Chlamydomonas*, however, the gene can be translated through co-introduction of a modified plastid *trnW* gene which can read through the UGA (Young and Purton 2016).

Inducible Expression System

Transcriptional Switch

The most well-known inducible expression system in the chloroplast is *Cyc6* promoter, RNA stability factor Nac2 and *psbD* 5′UTR (Schwarz et al. 2007, Rochaix et al. 2014, Surzycki et al. 2007). *Cyc6* promoter is copper-responsive so it turns on when copper is absent and off when present. The chloroplast protein, Nac2, encoded by the nucleus is involved in the stable accumulation of the *psbD* mRNA through binding to the 5′UTR of the gene which encodes the D2 reaction center protein of photosystem II. So, the transgene expression in chloroplasts can be tightly regulated by placing the transgenes under the control of *psbD* 5′UTR in the *nac2* mutant background whose nucleus is transformed with the construct of *Cyc6* promoter-*Nac2*. In the presence of copper, the expression of the transgene is turned off but the removal of the copper turns on the transgene expression.

Earlier than 2007, the first trans-activation system was developed by Lössl et al. (Lössl et al. 2005) for the production of the polyester polyhydroxybutyric acid (PHB), otherwise, the constitutive expression of PHB in chloroplasts was toxic to the host. In the system, heterologous *alc* regulon derived from fungus, *Aspergillus nidulans*, and T7 RNA polymerase (RNAP) from T7 phage were adopted and tobacco nucleus was transformed with two expression cassettes: one was to constitutively express alcR transcription factor under the control of 35S promoter and the other one was able to inducibly express T7 RNAP under the control of *alcA* promoter. Upon ethanol application, the alcR activated by the ethanol binds to the *alcA* promoter and then the expressed T7 RNAP fused to a transit peptide is targeted to chloroplasts. Once the T7 RNAP moves in chloroplasts, which turns on *phb* operon by binding to T7 gene 10 promoter/5′UTR. The constitutive expression of *phb* operon caused it to stunt plant growth in a very early stage and pollen sterility (Lossl et al. 2003). However, the ethanol-inducible PHB expression system allowed for plants to normally grow and set seeds by inducing PHB production in late stage.

E. coli lac regulation system was also investigated in the chloroplast of *Chlamydomonas* (Kato et al. 2007). In the study, the inducible system consisted of *lac* operator (*lacO*) sequence, which was embedded into *rbcL* or *16S rRNA* promoter, and was incorporated into the chloroplast genome. For the *rbcL* promoter embedded with *lacO*, the repression of transcription of a reporter gene, *uidA*, was not complete but the repression was maintained at a level of 10% over that of the transcript driven by intact *rbcL* promoter. Induction of the reporter gene by IPTG occurred rapidly at 5 mM and the transcript level reached up to 95% of that of transcript driven by intact *rbcL* promoter. For the modified *16S rRNA* promoter embedded with *lacO* sequence, the repression was too strong to fully induce the transcription of the reporter gene due presumably to the formation and binding of homotetrameric *lac* repressor to two operator sequences, which consequently impaired the induction so the level of induced transcript was dramatically reduced compared to that of transcript driven by intact *16S rRNA* promoter. Considering the expensive IPTG, this system currently doesn't look practical for the use of mass production of biomolecules but the system can be applied for the fine-tuning of the metabolic pathway in the production of high-value biomolecules.

Translational Switch

As seen in the Lössl et al.'s (2005) study, the inducible system needs an additional nuclear transformation, which needs to be selected by another selection marker, and the system has basal leakiness. To regulate the inducible system more tightly without nuclear transformation, riboswitches, natural RNA sensors, were investigated by Verhounig et al. in tobacco chloroplasts (Verhounig et al. 2010). The conformational change of secondary structures in riboswitches by metabolite binding can turn the switches on or off. Riboswitches which were derived from bacteria were modified and the resultant six riboswitches were tested in tobacco chloroplasts. Among them, synthetic theophylline riboswitch, translational "on" switch, only worked, and the strong induction of *gfp* reporter gene by the theophylline was found at a concentration of 2.5 mM. In addition, theophylline-independent accumulation of *gfp* mRNA indicates that the synthetic theophylline riboswitch serves as a translational switch.

However, the level of GFP induced by the synthetic theophylline riboswitch was very low. In 2015, the same research group developed a new riboswitch system to increase the expression level, called RNA amplification-enhanced riboswitch (RAmpER) (Emadpour et al. 2015). The new inducible vector system introduced RNA amplification step by placing T7 RNA polymerase gene under the control of the synthetic theophylline riboswitch, and the expression of the reporter gene was designed to be driven by T7 RNA polymerase promoter/*g10* 5′UTR. When the expression level of GFP was compared between the new transplastomic line and the previous one, the new riboswitch system enhanced GFP expression at least more than 50 times higher than the previous riboswitch system.

Multiple Gene Expression

Multiple gene expression is necessary when introducing multiple agronomic traits, engineering metabolic pathway or producing multiprotein complex. Like prokaryotes, most of the chloroplast genes are organized in operons and expressed in polycistronic mRNAs. As seen in recent studies of RNA-sequencing data of *Chlamydomonas* chloroplast, 84 out of the 109 genes are clustered in 22 operonic units (Cavaiuolo et al. 2017), while Gallaher et al. (Gallaher et al. 2018) confirmed the presence of 16 polycistronic transcripts. Like eubacteria in which polycistronic transcripts are directly translated, some polycistronic transcripts of chloroplast can be translated to multiple proteins without further processes, such as *psbE* (Willey and Gray 1989) and *psaA/B* (Meng et al. 1988) operons. In the studies, the pea chloroplast *psbE* and *psbF* genes are co-transcribed with another two genes and translated without post-transcriptional cleavage. Likewise, tobacco *psaA* and *psaB* genes are co-transcribed with *rps14* and the single transcript is translated. However, most of the chloroplast genes transcribed in polycistronic mRNAs are processed to mature monocistronic translatable mRNAs, otherwise, the unprocessed transcripts are rendered to be untranslatable or undergo inefficient translation.

Polycistronic Expression with following Cleavage

According to Zhou et al.'s study, operon-type transcription shows that the post-transcriptional process of polycistronic mRNAs is mediated by intercistronic expression element (IEE) at which cleavage happens and cleaved monocistronic transcripts become translatable. In addition, the generation of proper 5′UTR of next cistron is crucial for the gene's mRNA stability and translation. The aim of the study was to find the minimum length of IEE sequence which can process two genes tied bicistronically to mature translatable monocistronic mRNAs. Their finding showed that a minimum 50-nt IEE derived from the intercistronic spacer between *psbT* and *psbH* was enough to produce two translatable monocistronic mRNAs, kanamycin resistance gene (*nptII*) and yellow fluorescent protein (*yfp*) (Zhou et al. 2007).

In the following application of the IEE by the same research group, to improve the production of tocochromanols (tocopherols and tocotrienols), three key genes for the production of the metabolite were constructed in an operonic expression cassette in which two identical IEE sequences were placed in between the genes and the expression cassette was then driven by a single promoter, tobacco *16S rRNA* promoter (Lu et al. 2013). For comparison, bacterial-type operonic expression cassette was also constructed, so that the same three genes were connected via two endogenous intergenic spacer regions such as between *psbH* and *petB* and between *rps2* and *atpI*. In terms of metabolite production, tobacco transplastomic lines expressing the IEE built-in synthetic operon increased the tocochromanol 5 times higher than wild type, while the bacterial type operon increased 1.7-fold in the tocopherol accumulation over control. The synthetic operon including IEE was also tested in tomato, which showed that the transplastomic tomatoes produced stable monocistronic mRNAs for the three genes like tobacco lines, and the level of tocochromanols in the tomato leaves accumulated up to 10 times higher than in wild type.

The IEE sequence was also efficiently used when transferring the whole biosynthetic pathway of artemisinic acid from *Artemisia annua* to tobacco (Fuentes et al. 2016). In the study, the integration of four core genes into the chloroplast genome was done using two bicistronic operons in which the genes in each operon was separated by the IEE and the operons were driven by two different promoters. Artemisinin is a natural compound and the plant secondary metabolite is the only available cure for malaria, which has saved millions of lives worldwide since its discovery. Tu Youyou, who discovered and proved the anti-malaria efficacy of artemisinin, was one of the Nobel laureates in Physiology or Medicine in 2015 (Liu and Liu 2016).

To establish the entire biosynthetic pathway in the tobacco, the Fuentes et al. also transformed the nucleus of transplastomic tobacco whose chloroplasts were already transformed with the four core genes of the artemisinic acid pathway with a set of accessory genes to maximize the flux through the pathway via combinatorial supertransformation approach, called as COSTREL (for COmbinatorial Supertransformation of Transplastomic REcipient Lines). Five additional accessory genes were individually constructed for nuclear transformation and the mixed plasmids were co-bombarded onto the transplastomic background line: 199 COSTREL lines were screened to find optimum combination of genes for the right balanced enzyme activities, and the highest producer was found to be 77–fold higher in artemisinic acid content than its background transplastomic line.

Only recently, there was a report on a study of multiple gene expression in algae. Macedo-Osorio et al. (Macedo-Osorio et al. 2018) tested five putative intercistronic regions and found that two regions, such as *psbN-psbH* (569 bp) and *tscA-chlN* (650 bp), have IEE. Bicistronic transcript, *aphA-6-gfp,* was cleaved at the intercistronic sequence and the resultant two monocistronic transcripts were properly translated.

Polycistronic Expression with no Further Process

In a previous study on the correlation between the Shine-Dalgarno (SD) sequence recognition and translation initiation efficiency (Drechsel and Bock 2011), the efficiently recognized internal SD sequences in *E. coli* didn't work as efficiently as in chloroplasts, instead, the 5′-most SD was predominantly utilized, and which could explain why the post-transcriptional cleavage of polycistrons mediated via IEE are more prevalently in chloroplasts. However, many studies showed that the plant chloroplast has the full ability to express bacterial type operon (De Cosa et al. 2001, Krichevsky et al. 2010, Ruiz et al. 2003). *Cry2Aa2* operon from *Bacillus thuringiensis* (Bt) was integrated into tobacco chloroplast genomes and the last gene of the three gene operon was expressed up to 45.3% of the total protein in mature leaves and the level of operonic expression of the *Cry2Aa2* was shown 126-fold higher than that of single gene expression. In another study, *lux* operon (luxCDABEG, a bacterial light-emission enzymatic system) was amplified using *Photobacterium leiognathi* genomic DNA as a template and tobacco chloroplasts transformed with the amplified fragments emitted autoluminescence (Krichevsky et al. 2010). This bacterial type operonic expression system was also applied for phytoremediation. To detoxify mercury (Hg), the native bacterial *merAB* bicistronic operon, composed of mercuric ion reductase (*merA*) and organomercurial lyase (*merB*), was introduced into tobacco chloroplast genomes, which protected the transplastomic plants from the toxic heavy metal. Also, mRNA transcript analysis showed that most of the detected mRNAs were bicistronic transcripts (Ruiz et al. 2003).

Furthermore, multiple artificial polycistronic expression cassettes whose intercistronic spacer regions contain ribosome binding sequence (GGAGG) produced polycistronic transcripts predominantly in plastids. This clearly demonstrates that the chloroplast is able to transcribe polycistronic mRNAs and directly translate them like bacteria. Genes cloned into operonic expression cassettes, such as *aadA-orf1-orf2-HSA*, *aadA-TPS1*, *aadA-CTB*, and *CTB-GFP*, were properly translated and the genes' products were detected by anti-orf2, -HSA, -TPS1 and -CTB antibodies, respectively (Quesada-Vargas et al. 2005).

In 2004, Kumar et al. (Kumar et al. 2004a) created transplastomic carrot expressing bicistronic transcript which was composed of *aadA* and betaine aldehyde dehydrogenase (*BADH*) gene under the control of a single promoter, *16S rRNA*, and the intercistronic space consisted of *psbA* 3′UTR downstream to *aadA* and the phage T7 gene 10 leader sequence upstream to the 5′ end of *BADH* gene. BADH metabolizes the toxic betaine aldehyde to glycine betaine which is well known as an osmoprotectant and found in crops such as sugar beet, cotton, and other high salt and drought-tolerant plants (Rhodes and Hanson 1993, Nishimura et al. 2001). The BADH proteins were detected in all the tissues of transplastomic lines and the lines showed the high-salt resistance up to 400 mM, which demonstrates the bacterial operon expression is fully workable in chloroplasts.

Further application of the bacterial type operonic expression was made to enhance isoprenoid metabolites by the integration of the entire cytosolic mevalonate pathway (MEV), which consisted of six genes plus a selection marker, *aadA* (Kumar et al.

2012). The expression of seven genes was controlled by a single promoter, *16S rRNA*, the transplastomic line created by a single transformation event showed improved accumulation of isoprenoid metabolites such as mevalonate, carotenoids, squalene, sterols, and triacylglycerols over untransformed wild type.

Based on the confirmation of the improved accumulation of isoprenoid metabolites, the cytosolic MEV pathway was combined with artemisinic acid (AA) pathway to lead isopentenyl-diphosphate (IPP) generated by the introduced MEV pathway to AA biosynthesis (Saxena et al. 2014). This combination created a new energy flux route so the exogenous sugar can enter into the MEV pathway and the consequent IPP along with the IPP produced by plastid MEP pathway can be used by the introduced AA biosynthesis pathway. In the study, five genes for AA synthesis and an *aadA* gene as a selection marker were cloned into a bacterial type operonic expression cassette under the control of P*psbA*/5′UTR and another six genes for MEV pathway under the control of *16S rRNA* promoter were constructed in the other bacterial type operonic cassette. The 5′-end of each gene was engineered with a ribosome binding site. The two bacterial type expression cassettes were constructed into pMEV-Arte vector (23 kb) and the vector was delivered into tobacco chloroplasts. This combination of two pathways facilitated the biosynthesis of artemisinic acid in tobacco but the plant suffered from stunt growth, resulting in the difficulty in biomass increase. Although there was no data for the quantitative and qualitative studies on transcripts of the 11 transgenes due to the shortage of leaf materials, the detection of artemisinic acid using 13^{C}-NMR and ESI-MS indicates again that the bacterial type multigene expression is fully operable and functional in chloroplasts.

Avoidance of the use of Antibiotic Selection Marker

Antibiotics-based selection is the most widely used method when screening transplastomic plants. However, besides the metabolic burden, the use of antibiotic-resistant genes has long posed public concerns. For example, the *aadA* gene is the most widely used antibiotic-resistant selection marker in chloroplast engineering, and the target antibiotics (streptomycin and spectinomycin) were used in clinics to treat tuberculosis and gonorrhea (Goldstein et al. 2005). Not only could the horizontal transfer of antibiotic resistance genes between bacteria, but the probability of the gene transfer from chloroplasts to bacteria could be enhanced by the presence of several thousand copies of the antibiotic resistance gene in a cell of transplastomic plants so the released genetic material from dead plant cells can be taken up by bacteria in the gastrointestinal tract or in soil (Heinemann and Traavik 2004, Pontiroli et al. 2007). The removal of the antibiotic selection marker or use of non-antibiotic based selection system is required prior to commercialization.

Screening by Restoration of Photosynthesis or Amino Acid Synthesis

Chlamydomonas can be grown heterotrophically so photosynthesis-deficient mutant can be served as an expression platform, for example, growth in TAP medium in the dark. Ishikura et al. (Ishikura et al. 1999) created *rbcL*-deleted nonphotosynthetic *Chlamydomonas* mutant (DEVL) in which the *rbcL* gene was replaced by an *aadA* expression cassette. The successful chloroplast transformation of the mutant line with a vector containing both transgene and intact wild type *rbcL* gene can render transformants to be survived in the photoautosynthetic conditions. The same strategy, but with a different knock-out mutant of the photosynthesis gene, was also investigated for the selection of chloroplast transformant as below.

TN72 (cell wall deficient strain cw15 mt+), the *psbH* knock-out mutant by the insertion of *aadA* cassette, was developed. Since the *psbH* gene is one of the component genes for photosystem II, the disruption of the gene makes the strain not survive in the normal light conditions (Bateman and Purton 2000, Wannathong et al. 2016). The transformation vectors for the TN72, such as pAsapI, pSRSapI, pPSapI, pCSapI (Wannathong et al. 2016, Charoonnart et al. 2019), pCD, pWUCA2, pSty and pSde (Young and Purton 2016), were equipped with the intact wild type *psbH* gene so the transformants can survive under light by restoration of their photosynthesis. Another *psbH::aadA* mt+ mutant *Chlamydomonas* but with the cell wall, CC4388 (O'Connor et al. 1998), can be also available at *Chlamydomonas* collection center.

Chen and Melis (Chen and Melis 2013) used CC2653 *Chlamydomonas* line with point-mutated *rbcL* for chloroplast transformation of alcohol dehydrogenase (*adh1*). The integration of the transgene and rescue of the mutated *rbcL* was achieved by the double homologous recombination between the *psaB* and *rbcL* (pointed mutated) gene of the plastome and the *psaB* and functional *rbcL* gene in a vector.

Auxotrophic mutant was also used as a recipient cell for chloroplast transformation. *N*-acetyl ornithine aminotransferase is encoded by *Arg9* and the enzyme is targeted into plastids where arginine biosynthesis happens. In the study of Remacle et al. (Remacle et al. 2009), arginine auxotrophic mutant, *arg9-2*, was transformed with *aadA* cassette to be integrated into the intergenic region between *psaA-3* and *tranL2* of the plastome of *Chlamydomonas.* And the same expression cassette used for the *aadA* transformation was used to transform a functional *Arabidopsis Arg9* gene. After the transformation, the mutant recipient cells were screened for the restoration of arginine prototrophy. The selected chloroplast transformants showed a similar growth rate and enzymatic activity to the nuclear transformants with the same *Arg9* gene. The functional relocation of *Arg9* into the chloroplast genome of the mutant recipient strain demonstrates that *Arg9* can serve as a selection marker.

Screening by Detoxification or Metabolization of Toxic or Non-metabolizable Compounds

In contrast to the complementing system of a knock-out mutant deficient with a gene of a biosynthetic pathway, feedback-insensitive anthranilate synthase α-subunit gene (*ASA2*) was investigated as a possible selection marker for chloroplast transformation. Anthranilate synthase (AS) is the first step enzyme of the tryptophan biosynthesis and the enzyme is inhibited by the end product, Trp. The α-subunit of the enzyme that is composed of two large α-subunits and two small β-subunits is allosterically inhibited by Trp. But naturally occurring feedback-insensitive *ASA2* gene, isolated from a tobacco suspension-culture cell line, is resistant to toxic Trp analogue 5-methyltryptophan (5MT) (Song et al. 1998). The suitability of *ASA2* as a selection marker was investigated by Barone et al. using the indole analogue 4-methylindole (4MI) or the tryptophan analogue 7-methyl-DL-tryptophan (7MT) as selection agents. The chloroplast transformation study showed that five and six individual shoots out of bombarded 130 and 140 leaves showed resistance to 7MT and 4MI, respectively. And the frequency of plastid transformed lines from regenerated shoots was similar between 7MT, 4MI, and spectinomycin (Barone et al. 2009).

Genes that can convert toxic or non-metabolizable compounds to non-toxic or metabolizable ones were also tested as a selection marker. Betaine aldehyde dehydrogenase (*BADH*) genes from both spinach and sugar beet which can convert toxic betaine aldehyde (BA) to non-toxic glycine betaine were investigated. According to an early study, tobacco nuclear transformants expressing transit-peptide fused BADH of either spinach or beet showed the reduction of betaine aldehyde and concurrent increase of glycine betaine in the transgenic tobacco, resulting from the activity of the transgene BADH products. The usability of *BADH* gene as a selection marker was then evaluated, and which showed that the leaf disks of the transgenic tobacco expressing BADH regenerated shoots, while the shoot regeneration from wild type leaf disks was severely inhibited (Rathinasabapathi et al. 1994). As a selection marker, the spinach *BADH* gene was also tested for the selection of tobacco chloroplast transformants (Daniell et al. 2001b), but *BADH* gene as a primary selection marker wasn't supported by another group (Whitney and Sharwood 2008). Considering Rathinasabapathi et al.'s study, the beet BADH showed higher enzymatic activity over the spinach one as a selection marker so the further investigation of the beet BADH will be helpful to determine whether BADH gene can be suitable as a selection marker for chloroplast transformation.

Also, there have been studies on the usability of D-amino acid detoxifying enzyme as a selection marker. Gisby et al. developed a dual marker system using D-amino acid oxidase (DAAO) derived from *Schizosaccharomyces pombe*. In the study, transplastomic plants transformed with DAAO were tolerant to D-alanine but sensitive to D-valine. However, wild type represents the opposite pattern (Gisby et al. 2012). D-amino acids are cheap and relatively nontoxic to animals and microbes so the dual marker system can be very useful for the outside use. For example, transgenic crops in germination and seedling stages can be positively selected from weeds by spraying D-alanine, and unwanted spread of the transgenic crops then be prevented by spraying D-Valine.

Another study showed the possible use of D-serine ammonia lyase (*dsdA*) originated from *E. coli* as a selection marker. Li et al. created transplastomic tobacco expressing codon-optimized *dsdA*. Tolerance test to D-serine was performed with seeds harvested from lines regenerated from the primary selection using spectinomycin. The tolerance level to D-serine was quantitatively measured and the effective concentrations of D-serine for seed germination, shoot regeneration and spraying were found at 10, 30 and 75 mM, respectively (Li et al. 2016).

As we can expect, not all selection markers which work for plants always work for algae. For example, D-alanine and BADH are not suited for *Chlamydomonas* due to the absence of a system able to uptake the D-amino acid and the presence of an endogenous BADH enzyme (Nishimura et al. 2007, Wood and Joel Duff 2009, Esland et al. 2018).

Phosphorus (P) is an essential element for cellular activity, which involves the synthesis of nucleic acids and phospholipids, and cell signaling mediated by phosphorylation. Most organisms use phosphate (PO_4^{3-}, Pi) as P source but a few prokaryotes can assimilate phosphite (PO_3^{3-}, Phi) using phosphite oxidoreductase (ptxD) by which Phi is converted to Pi. There have been many studies to investigate the usability of *ptxD* gene as a positive selection marker for plant nuclear transformations such as *Arabidopsis*, maize, and cotton (Nahampun et al. 2016, Pandeya et al. 2017, López-Arredondo and Herrera-Estrella 2013). But so far, there has been no published report that *ptxD* can be used as a selection marker for chloroplast transformation. A report done by Sandoval-Vargas et al. (Sandoval-Vargas et al. 2018) showed the possibility of *ptxD* as a selection marker. In the study, the transplastomic *Chlamydomonas* expressing *ptxD* showed normal growth on solid TAPhi (0.1 mM phosphite) media while wild type didn't. In addition, the growth rate in liquid culture of TAPhi (0.1 mM phosphite) showed that the transplastomic lines reached stationary phase at around day 8 but untransformed wild type cells failed to grow in the liquid culture. Since those viability tests were done with *ptxD* transplastomic lines regenerated from spectinomycin selection, the suitability of *ptxD* as a selection marker needs to be further determined.

Excision of Antibiotic Selection Marker

Direct Repeats

It was well known that the excision of a transgene or a DNA region by homologous recombination between two direct repeats within the chloroplast genome. The integrated *aadA* gene into the chloroplast genome but flanked by 216 bp direct repeats was looped out (Cerutti et al. 1995) and the homologous recombination between *atpA* 5′ regions in plastome and in transformation vector induced the deletion of a 2kb-long DNA fragment (Künstner et al. 1995). More detailed research on the excision efficiency of the antibiotic selection marker by direct repeats was carried out by Fisher et al. A series of transformation plasmids containing *aadA* expression cassette with the 4 different sizes of flank sequence including 100-, 230-, 483- and 832-bp were constructed. From the spectinomycin sensitivity test with transplastomic plants, excision of *aadA* didn't happen with 100- and 230-bp flanking sequences, but 483- and 832-bp flanking sequences produced *Chlamydomonas* sensitive to spectinomycin (Fischer et al. 1996). Gene excision using direct repeats was also investigated in plants. Iamtham and Day designed chloroplast transformation vector containing three genes including *uidA, aadA* and *bar*, and two different kinds of direct repeats such as 174-bp rrnHv and 418-bp 3′-NtpsA. Two copies of 174-bp repeat were used to flank the *uidA-aadA*, and each of the three genes was terminated by 418-bp 3′-NtpsA. The order of the repeats and genes in the construction was like this: rrnHvA-*uidA*-NtpsA1-*aadA*-NtpsA2-rrnHvB-*bar*-NtpsA3. In the study, 174-bp rrnHv repeats created the excision of *uidA* and *aadA* at T0 generation, whereas excision of *aadA* and *bar* by three copies of 418-bp NtpsA was made at T2 generation. However, single gene excision by 418-bp direct repeats wasn't detected in the three-gene construct, and the excision happened at the very low frequency when the efficiency of the single gene excision was tested using a two-gene construct. From the study, it was found that the 418-bp direct repeats weren't efficient (Iamtham and Day 2000) so the next study tested 649-bp direct repeats (DR1 and DR2) corresponding to the 5′ *atpB* regulatory region of tobacco for the excision of 5.4-kbp long three genes (*rbcL-aadA-uidA*) (Kode et al. 2006). The integration of the construct (DR1-*rbcL-aadA-uidA*-DR2) into the chloroplast genome resulted in the replacement of the endogenous *rbcL* with the trans-*rbcL*. The shoot regenerations under antibiotic-free culture media created Δ*rbcL* mutant lines as a result of the excision of the three genes via homologous recombination between the direct repeats. This study shows that the more precise deletion of the *aadA* gene can be achieved by homology-based excision. Dufourmantel et al. created transplastomic plants of tobacco and soybean which were herbicide-tolerant but marker-free using two 210-bp 3′ *rbcL* repeats (Dufourmantel et al. 2007). Also, 84-bp direct repeats derived from *rRNA* operon promoter sequences were tested in tobacco but it was shown that the excision rate was relatively rare (Lutz and Maliga 2008). Overall, the size of the direct repeat is the major factor influencing excision by homologous recombination. The longer the repeat is the higher excision frequency is but it needs to further study the upper limit for the length of the repeat, by which precise excision can happen. Also, a study that investigates what sequence composition can be more amenable to the excision by homologous recombination (Mudd et al. 2014). A very recent study showed the successful removal of *aadA* marker in transplastomic lettuce using direct repeats. The chloroplast transformation vector consisted of two expression cassettes: for selection, *aadA* expression cassette was used but it was flanked by the 649-bp direct repeats of the 5′ *atpB* regulatory region mentioned above; for expression of precursor human insulin-like growth factor-1 (*Pro-IGF-1*), P*psbA*/5′UTR-driven expression cassette was used. Both expression cassettes were surrounded by flanking sequences (*16S rRNA-trnI* and *trnA-23S rRNA*) for the integration of the expression cassettes in the chloroplast genome (Park et al. 2019). There is a noteworthy finding in the report that the loop-out excision event of the marker can happen as early as at T0. The marker removal from the edible plant will help shorten the course of regulatory approval for the clinical use of the transplastomic lettuce.

The marker excision system aforementioned used direct repeats which were derived from other organisms and flanked the selection marker expression cassette so the repeat sequences are different from the flanking sequences that were used to surround both expression cassettes. In contrast, an approach tried by Klaus et al. was that, without the use of additional direct repeat sequences, the marker gene is placed outside flanking sequences of the expression cassette of a transgene so the single homologous recombination between left or right flanking sequence and their cognate endogenous sequences created the plastome having the cointegrated marker gene. But in the following recombination events, the marker gene was looped out via homologous recombination between two copies of left or right flank repeats (Klaus et al. 2004). Although this system has clear advantages over *Cre/lox* or *Int/attP* and *attB* systems (refer to next section), such as no requirement for retransformation and no gene deletion, however, there are some disadvantages. Due to the simultaneous events of plastome transformation and marker excision, the achievement of the maker-free homoplasmic line seemed to be difficult if the excision happens earlier than the removal of all the wild type plastomes. Also, the cointegration by a single flanking sequence-mediated homologous recombination seemed rarely happened because the detection of the cointegrate molecules by Southern blot was failed, however, the presence of the molecules was confirmed by PCR. In addition, the system used photosynthesis-deficient or plastid-encoded RNA polymerase mutants as recipients such as Δ*petA* or Δ*rpoA* so the transformation vectors carry the respective wild type genes as well as *aphA-6* gene for kanamycin resistance. Most of all, there was a high probability that the integration of the gene of interest (GOI) could have been happened by double homologous recombination between the flanking sequences in vector and the corresponding endogenous sequences in plastome directly by skipping single homologous recombination and cointegration status. It looked impossible to tell whether the integration of GOI was made by double or single homologous recombination by this system. Therefore, a large number of the marker would never be integrated into plastomes, instead, the maker was just simply excluded when the GOI was integrated by double homologous recombination.

Recombinase System

Another strategy for marker excision was investigated using phage recombinases and their cognate recognition sites such as Cre/*lox* and Int/*attB*-*attP*. In 2001, two papers which used the *Cre/lox* system to remove selection markers were published back-to-back in The Plant Journal (Hajdukiewicz et al. 2001, Corneille et al. 2001). GFP reporter gene disrupted by insertion of the *aadA* gene which was flanked by *lox* sequences was integrated into the chloroplast genome of tobacco and the homoplasmic transplastomic plants were selected via spectinomycin selection pressure. The transplastomic plants were then retransformed to introduce *Cre* recombinase gene into the nucleus by *Agrobacterium* with *nptII* as a selection marker. The *Cre* gene product was designed to target to chloroplast and the progeny seeds were selected on kanamycin. The excision event was very rapid so the germinated shoot on the kanamycin showed predominant GFP fluorescence, indicating the excision of the *aadA* gene. Further, the transgenes used in the transformation such as *Cre* and *nptII* were washed out by self-fertilization. Although all the transplastomic lines excised the transgene *aadA*, some independent lines showed off-target homologous recombinations by the two copies of P*rrn* sequences, resulting in the removal of *aadA*, *gfp* and endogenous plastid *trnV* genes. Another off-target excision removed *lox*, *Prrn* and *aadA* by homologous recombination between the *lox* placed in the downstream of 3′ region of the *aadA* gene and the repeated sequence of a 5-bp-log TATTA which assumed to acts as a recombinational hot spot placed in the upstream of the *Prrn-lox-aadA* DNA sequence (Hajdukiewicz et al. 2001). In the Corneille et al.'s study, the same strategy was used to remove a negative selection marker, cytosine deaminase (*codA*) gene, from plastome. Since the *codA* converts 5-fluorocytosine (5FC) to toxic 5-fluorouracil (5FU), the removal of *codA* gene flanked by two copies of *lox* sequence by the introduction of transit-peptide fused *Cre* chimeric gene into nucleus can be identified by phenotype showing resistance to 5-fluorocytosine (5FC). Likewise, this study also found a deletion of the large plastid genome. But such deletion wasn't detected when the *Cre* gene was introduced via pollen which was derived from an independent tobacco line transformed with the *Cre* gene (Corneille et al. 2001).

Also, a phiC31 phage derived recombinase, integrase (Int), was employed to excise marker gene (Kittiwongwattana et al. 2007). The integrase recognizes non-homologous sequences such as *attB* and *attP* so the system can avoid the undesired recombinations between *att* and pseudo-*att* sequences in the chloroplast genome. However, the study observed the enhanced homologous recombination via the P*rrn* direct repeats when they are adjacent to *attB* and *attP* sites.

Tissue-culture based plastid genome manipulation using recombinases is a long-haul work. In 2012, Tungsuchat-Huang and Maliga (Tungsuchat-Huang and Maliga 2012) developed *in planta* excision system by which the excision of plastid aurea *bar* gene (bar^{au}) was achieved with greenhouse-grown plants. *Agrobacterium* cells transformed with binary vectors carrying *Cre* or *Int* genes designed to be targeted to chloroplasts were injected at the axillary bud site of transplastomic tobacco whose chloroplast genomes are integrated with aurea *bar* marker, flanked with *attP* and *attB* or *loxP* sites. After the *Agrobacterium* injection, the new apical meristem was forcibly differentiated by cutting off the plants above the injection site. To detect marker excision, green pigmentation was used as an indicator because the removal of the marker changed the leaf color from golden-yellow to green. The differentiated shoots showed bar^{au}-free plastids in 30–40% of the injected plants and 7% of the plants passed down the bar^{au}-free plastids to the next generation. However, to extend this *in planta* excision system to other plants, visual maker system for the detection of excision event needs to be developed.

Alternative Chloroplast Transformation Methods to Particle Bombardment

Glass bead-mediated Transformation

All the studies described above used particle bombardment method, which requires expensive equipment and consumables, for example, Biolistic PDS-1000/He Particle Delivery System. As an alternative method, a very simple and economical chloroplast transformation approach for *Chlamydomonas* has been developed (Kindle et al. 1991, Economou et al. 2014). The method requires cell-wall deficient cells, which can be achieved by mutation or enzymatic removal of the cell walls. The cell wall-deficient cells are subject to agitation with glass beads for DNA uptake. Although transformation efficiency is less than particle bombardment, considering that specific integration of expression cassette into the predetermined locus of the chloroplast genome by homologous recombination, a small number of transformants should not be a problem.

Polyethylene glycol (PEG)-mediated Transformation

Another easier and cheaper method for plant chloroplast transformation can be achieved using polyethylene glycol (PEG). To introduce DNA, the cell wall of plant cells needs to be removed and the consequently isolated protoplasts are then treated with PEG. The PEG-treated protoplasts are embedded in alginate film because which greatly enhances the viability and subsequent regeneration of protoplasts. It is important to maintain that the film layer should be thin to facilitate the exchange of metabolites and the absorption of nutrients. However, this method has some drawbacks. Fresh protoplasts need to be prepared upon transformation, and a plant species which can be regenerated from protoplasts should be used (Kofer et al. 1998, Koop et al. 1996, Huang et al. 2002)

Electroporation-mediated Transformation

Microparticle-based gene delivery system has been widely used in chloroplast transformation but its use is restricted by cell size, in particular, which limits gene delivery to microalgal chloroplasts such as *Nannochloropsis* ap. and *Chlorella* sp. (the cell size of both is ~ 2 μm). The smallest size of particles for bombardment is 0.6 μm (BioRad) or 0.55 μm (Seashell technology) so bombardment with these sizes would kill most of the small-size microalgae. Even though there is a much smaller size of gold particles available from Cytodiagnostics Inc. in the range from 0.05 μm to 0.4 μm, but the nanoparticles are hard to get enough speed to penetrate cell walls. In a recent paper, Gan et al. successfully established an electroporation-mediated chloroplast transformation of *Nannochloropsis oceanica.* The stable integration of transgenes such as *gfp* and *ble* (zeocin resistance gene) was confirmed by detection of GFP fluorescence signals under fluorescence and laser confocal microscopy, and by PCR and Southern blot for *ble* gene (Gan et al. 2018).

Biomass Increase

Fermentor

In addition to the molecular tools, the level of the yield of transgenes can be further enhanced by optimizing growth conditions. Algae are versatile in the utilization of carbon source so they can be cultivated mixotrophically or heterotrophically as well as autotrophically. Braun-Galleani et al. (Braun-Galleani et al. 2015) studied the effect of different cultivation conditions on protein yield, the protein level of the vivid Verde Fluorescent Protein (VFP) expressed in the chloroplast of *Chlamydomonas* was the highest at 25°C in heterotrophic cultivation, but, at 30°C, the level was highest when cultivated in mixotrophic condition. However, whatever the growth temperature was, the mixotrophic growth conditions showed the highest value of cell density up to 4.06 at OD_{750}. It is also noteworthy that the trends observed from the *Chlamydomonas*-VFP weren't the same as when the different transgene was expressed so the optimal growth conditions need to be empirically decided for each transgene.

In another recent study, the mixotrophic fed-batch fermentor system dramatically improved the biomass of transplastomic *Chlamydomonas* expressing GFP, driven by P*psbD*/5'UTR, up to 24g/L (dry weight, DW), comparing to the previous highest yield, 9g/L (DW) and conventional batch fermentor, 2.3 g/L (DW) (Fields et al. 2018). The supply of concentrated feed of acetic acid and nutrition was automatically controlled by a pH-auxostat method by which the feed flows into fermenter only when the pH rises above 7.0. However, in terms of GFP content per total soluble protein (TSP), the conventional batch fermentor expressed 2 times higher on average than the fed-batch, which seemed to be caused by the less accessibility of the light-regulated promoter to light due to high cell density, resulting in non-optimal activity of the *psbD* promoter, and by high-salt stress. Although some issues were revealed from the study, the issues could be addressable enough by mitigating salt stress and using light-independent protomers. Considering copy number difference of chloroplast genome between phototorophic and mixotrophic conditions, ~ 40 and ~ 100 copies, respectively (Lau et al. 2000), the use of mixotrophic cultivation for biomass increase should be favored.

Hydroponics

There were also several studies of plant biomass increase using hydroponic cGMP cultivation system (Su et al. 2015b, Herzog et al. 2017, Kwon et al. 2018b). A commercial-scale production of transplastomic lettuce expressing clotting factor was achieved using the cGMP hydroponic system at the Fraunhofer USA Center and estimated productivity was ~ 870 kg fresh or 43.5 kg dry weight of lettuce leaf materials per 1000 ft^2 per annum yielding 24,000 – 36,000 doses for 20-kg pediatric patients (Su et al. 2015b). Moreover, the concentration of CTB-FIX expressed in transplastomic lettuce increased gradually as the lettuce grew in the hydroponic cultivation system and peaked 1.81 mg/g DW at day 101 (Herzog et al. 2017). Most of all, the optimized hydroponic system enhanced expression levels of CTB-Factor XIII heavy and single chains. For the single chain (SC) protein, the expression level of 26 ug/g DW from the transplastomic lettuce cultivated in greenhouse was increased up to 615 ug/g DW when grown in the hydroponic system. For heavy chain (HC), the expression level was improved 11.2-fold higher by the hydroponic system (3622 μg/g DW) than greenhouse system (324 μg/g DW) (Kwon et al. 2018b).

Conclusions and Future Perspectives

Over the past 30 years, many studies have been conducted to engineer chloroplasts and their consequent technologies have been employed for molecular farming, metabolic engineering, and crop improvement. The traction for those efforts is largely attributed to the unique characteristics of chloroplast: high copy number of the chloroplast genome, maternal inheritance, multiple gene expression, and almost no positional effects. However, it is also noteworthy that many technical issues still need to be addressed. As seen in this chapter, lots of basic and biotechnological studies have been carried out using photosynthetic chloroplasts of two popular model organisms such as tobacco and *C. reinhardtii*, which has helped broaden the knowledge of plastid science a lot and promoted successful stable plastid transformations of many other species: for plants, tomato (Ruf et al. 2001), potato (Nguyen et al. 2005, Sidorov et al. 1999), *Arabidopsis* (Sikdar et al. 1998), *Lesquerella* (Skarjinskaia et al. 2003), oilseed rape (Hou et al. 2003), petunia ((Zubkot et al. 2004), cotton (Kumar et al. 2004b), soybean (Dufourmantel et al. 2004), lettuce (Lelivelt et al. 2005), *Scoparia dulcis* L. (Muralikrishna et al. 2016) and rice (Lee et al. 2006, Wang et al. 2018); and, for algae, red algae *Porphyridium* sp. (Lapidot 2002), Diatom *Phaeodactylum tricornutum* (Xie et al. 2014), and *Nannochloropsis oceanica* (Gan et al. 2018). However there haven't yet been reliable plastid transformation protocols for major food crops such as wheat and maize, and for industrially important algae such as *Chlorella* and *Euglena*. Although there was a report for chloroplast transformation of *Euglena gracilis*, stable integration of transgene failed and the delivered transgene was maintained as an extrachromosomal element (Doetsch et al. 2001). As expected, the addition of traits to crops or industrially valued algae should improve yield or increase nutritional or therapeutic values, which will benefit farmers and further enhance consumers' health. The development of plastid transformation protocols for the intractable species needs to be persistently carried out. Furthermore, to improve the accumulation of the biomolecules in the crops, the molecular biology tools which can make non-photosynthetic plastids such as tubers, fruits, and seeds transformable should also be developed in the near future.

Even though the maternal inheritance is one of the touting features of chloroplast transformation over nuclear transformation, by which chloroplast transformation can avoid transgene transfer to the ecosystem, however, there are concerns on the possible transfer of transgenes contained in chloroplasts to relative weedy plants or bacteria in the gut or soil. Paternal transmission of plastids was also observed in tobacco (Avni and Edelman 1991, Medgyesy et al. 1986) and the chloroplast introgression from conventional oilseed rape (*Brassica napus*) into wild *Brassica rapa* can happen at an extremely rare rate if they grow in sympatry (Scott and Wilkinson 1999). Also, the released gene from decaying transgenic plants could be possibly horizontally transferred to soil bacteria (Pontiroli et al. 2007). So, a reliable method that can remove antibiotic selection markers from chloroplasts needs to be secured. In terms of easiness and simplicity, the direct repeats method can be applied by choosing an accurate release point of selection pressure. In addition, inadvertent contamination of the food chain by transgene products should be avoided. The constitutive expression system is a non-starter for solving the contamination issue. However, if the transplastomic line is equipped with a tightly regulated inducible system, this accidental contamination can be fully controlled. There has been a great advancement in the inducible system workable in the chloroplast (Emadpour et al. 2015), but the background leakiness of transgene still needs to be further addressed.

Plants have been used as a therapeutic purpose for a long time, due to their therapeutically active secondary metabolite. Sixteen plant natural products and derivatives had been approved for clinical use by the U.S. FDA in the period of 2001–2010 (Kinghorn et al. 2011). Also, there is a report that 122 plant natural compounds which are structurally defined are used as drugs globally (Fabricant and Farnsworth 2001). About 200,000 secondary metabolites have been identified in plants (Dixon and Strack 2003). Algae also have great potential as a source of natural products (Alves et al. 2018) and more than 3,000 natural products have been identified (Leal et al. 2013). The first algae-derived natural product, iota-carrageenan (Carragelose®), is already in the market, which is an anti-viral polymer and used to prevent respiratory viruses' infection (Ludwig et al. 2013, Calado et al. 2018). Considering that more than half of the 50,000 medicinal plants are still wildcrafted for supply of medicinal materials, which could endanger the sustainability of those medicinal plants (Atanasov et al. 2015), and that the very low collection yield of therapeutic metabolites from natural source, metabolic pathway engineering for the valued compounds in chloroplasts is a very compelling subject because a large number of biosynthetic pathways actively operates in chloroplasts. There have been several studies to produce medicinal compounds in chloroplasts by the introduction of whole or core biosynthetic pathway (Saxena et al. 2014, Kumar et al. 2012, Fuentes et al. 2016), but their yield was too low to be used commercially. To overcome the low productivity, a new combinatorial expression system was developed, called COSTREL (combinatorial supertransformation of transplastomic recipient lines), which improved artemisinic acid yield 77-fold higher than that of the transplastomic line transformed with core biosynthetic pathway (Fuentes et al. 2016). As an ongoing project, the SynPLASTome 2.0 project initiated by The University of Tennessee Institute of Agriculture (https://ag.tennessee.edu/racheff/Pages/default.aspx) could be another approach for metabolic engineering in the near future by which artificially synthesized complete chloroplast genome could provide a high yield of biologically active compounds.

Plant chloroplast engineering has also been extensively studied as the production and delivery platform using edible plants for therapeutic purpose. The therapeutic effects of orally delivered bioencapsulated protein drugs have been proved in many studies (Kwon and Daniell 2015, 2016, Jin and Daniell 2015). As seen in a very recent study, hemophilia drugs were orally delivered with lyophilized lettuce containing FIX and the anti-coagulation symptom was markedly improved (Herzog et al. 2017). However, for clinical applications, the oral delivery system needs to develop appropriate oral formulations. The concept of the system is based on the fact that the protein drugs expressed in plant cells can be protected by cell walls from the acid in stomach but the cell walls are degraded in intestines by commensal bacteria and the proteins are released and taken up by epithelial cells. Considering that lots of cells are broken in the preparation of powders from lyophilized leave materials and the existence of plasmodesmata of plant cell walls, the powdered materials become vulnerable to the stomach acid. Another protection layer, like acid-resistant coating (enteric coating) or capsule, could be applied for oral formulation. In addition, there are variations in composition and number of commensal bacteria in intestines between individuals and between different ethnic groups. To achieve consistent therapeutic efficacy, the complete degradation of cell walls is required at the intestines, which could be achievable by delivering cell-wall degrading enzymes together.

A few of the cornerstone studies made at the dawn of chloroplast-based molecular farming have led to a lot of breakthrough achievements for the last three decades. In the next coming decades, we hope to see more translations of the conceptual studies to commercially available products. In conclusion, the continued research and investment on chloroplast engineering should enhance the productivity, efficiency, and efficacy of transplastomic functional plants and algae, and which will eventually help the sustainability of industries and global health.

Acknowledgments

I would like to seek consideration from those colleagues whose work could not be covered here because of word limitation.

References

Adem, M., D. Beyene, and T. Feyissa. 2017. Recent achievements obtained by chloroplast transformation. Plant Methods 13: 30.

Alves, C., J. Silva, S. Pinteus, H. Gaspar, M.C. Alpoim, L.M. Botana, and R. Pedrosa. 2018. From marine origin to therapeutics: The antitumor potential of marine algae-derived compounds. Front. Pharmacol. 9: 777.

Anthonisen, I.L., M.L. Salvador, and U. Klein. 2001. Specific sequence elements in the 5′ untranslated regions of rbcL and atpB gene mRNAs stabilize transcripts in the chloroplast of *Chlamydomonas reinhardtii*. RNA 7: 1024–1033.

Apel, W., W.X. Schulze, and R. Bock. 2010. Identification of protein stability determinants in chloroplasts. Plant J. 63: 636–650.

Arlen, P.A., R. Falconer, S. Cherukumilli, A. Cole, A.M. Cole, K.K. Oishi, and H. Daniell. 2007. Field production and functional evaluation of chloroplast-derived interferon-alpha2b. Plant Biotechnol. J. 5: 511–525.

Atanasov, A.G., B. Waltenberger , E.-M. Pferschy-Wenzig, T. Linder, C. Wawrosch, P. Uhrin, V. Temml, L. Wang, S. Schwaiger, E.H. Heiss, J.M. Rollinger, D. Schuster, J.M. Breuss, V. Bochkov, M.D. Mihovilovic, B. Kopp, R. Bauer, V.M. Dirsch, and H. Stuppner. 2015. Discovery and resupply of pharmacologically active plant-derived natural products: A review. Biotechnol. Adv. 33: 1582–1614.

Avni, A., and M. Edelman. 1991. Direct selection for paternal inheritance of chloroplasts in sexual progeny of Nicotiana. Mol. Gen. Genet. 225: 273–277.

Bally, J., M. Nadai, M. Vitel, A. Rolland, R. Dumain, and M. Dubald. 2009. Plant physiological adaptations to the massive foreign protein synthesis occurring in recombinant chloroplasts. Plant Physiol. 150: 1474–1481.

Barnes, D., S. Franklin, J. Schultz, R. Henry, E. Brown, A. Coragliotti, and S.P. Mayfield. 2005. Contribution of 5′- and 3′-untranslated regions of plastid mRNAs to the expression of *Chlamydomonas reinhardtii* chloroplast genes. Mol. Genet. Genomics 274: 625–636.

Barone, P., X.-H. Zhang, and J.M. Widholm. 2009. Tobacco plastid transformation using the feedback-insensitive anthranilate synthase [α]-subunit of tobacco (ASA2) as a new selectable marker. J. Exp. Bot. 60: 3195–3202.

Barta, A., K. Sommergruber, D. Thompson, K. Hartmuth, M.A. Matzke, and A.J.M. Matzke. 1986. The expression of a nopaline synthase? human growth hormone chimaeric gene in transformed tobacco and sunflower callus tissue. Plant Mol. Biol. 6: 347–357.

Bateman, J.M., and S. Purton. 2000. Tools for chloroplast transformation in Chlamydomonas: Expression vectors and a new dominant selectable marker. Mol. Gen. Genet. 263: 404–410.

Blowers, A.D., G.S. Ellmore, U. Klein, and L. Bogorad. 1990. Transcriptional analysis of endogenous and foreign genes in chloroplast transformants of Chlamydomonas. Plant Cell 2: 1059–1070.

Bock, R. 2007. Plastid biotechnology: Prospects for herbicide and insect resistance, metabolic engineering and molecular farming. Curr. Opin. Biotechnol. 18: 100–106.

Bock, R. 2014. Engineering chloroplasts for high-level foreign protein expression. Methods Mol. Biol. 1132: 93–106.

Bock, R. 2015. Engineering plastid genomes: Methods, tools, and applications in basic research and biotechnology. Annu. Rev. Plant Biol. 66: 211–241.

Boyhan, D., and H. Daniell. 2011. Low-cost production of proinsulin in tobacco and lettuce chloroplasts for injectable or oral delivery of functional insulin and C-peptide. Plant Biotechnol. J. 9: 585–598.

Boynton, J.E., N.W. Gillham, E.H. Harris, J.P. Hosler, A.M. Johnson, A.R. Jones, B.L. Randolph-Anderson, D. Robertson, T.M. Klein, and K.B. Shark. 1988. Chloroplast transformation in Chlamydomonas with high velocity microprojectiles. Science 240: 1534–1538.

Braun-Galleani, S., F. Baganz, and S. Purton. 2015. Improving recombinant protein production in the *Chlamydomonas reinhardtii* chloroplast using vivid Verde Fluorescent Protein as a reporter. Biotechnol. J. 10: 1289–1297.

Calado, R., M.C. Leal, H. Gaspar, S. Santos, A. Marques, M.L. Nunes, and H. Vieira. 2018. How to succeed in marketing marine natural products for nutraceutical, pharmaceutical and cosmeceutical markets. Grand Challenges in Marine Biotechnology: 317–403.

Carrer, H., T.N. Hockenberry, Z. Svab, and P. Maliga. 1993. Kanamycin resistance as a selectable marker for plastid transformation in tobacco. Mol. Gen. Genet. 241: 49–56.

Cavaiuolo, M., R. Kuras, F.-A. Wollman, Y. Choquet, and O. Vallon. 2017. Small RNA profiling in Chlamydomonas: Insights into chloroplast RNA metabolism. Nucleic Acids Res. 45: 10783–10799.

Cerutti, H., A.M. Johnson, J.E. Boynton, and N.W. Gillham. 1995. Inhibition of chloroplast DNA recombination and repair by dominant negative mutants of *Escherichia coli* RecA. Mol. Cell. Biol. 15: 3003–3011.

Chan, H.-T., Xiao, Y., Weldon, W.C., Oberste, S.M., Chumakov, K., and Daniell, H. 2016. Cold chain and virus-free chloroplast-made booster vaccine to confer immunity against different poliovirus serotypes. Plant Biotechnol. J. 14: 2190–2200.

Charoonnart, P., N. Worakajit, J.A.Z. Zedler, M. Meetam, C. Robinson, and V. Saksmerprome. 2019. Generation of microalga *Chlamydomonas reinhardtii* expressing shrimp antiviral dsRNA without supplementation of antibiotics. Scientific Reports 9.

Chen, H.-C., and A. Melis. 2013. Marker-free genetic engineering of the chloroplast in the green microalga *Chlamydomonas reinhardtii*. Plant Biotechnol. J. 11: 818–828.

Choquet, Y., and F.-A. Wollman. 2002. Translational regulations as specific traits of chloroplast gene expression. FEBS Lett. 529: 39–42.

Corneille, S., K. Lutz, Z. Svab, and P. Maliga. 2001. Efficient elimination of selectable marker genes from the plastid genome by the CRE-lox site-specific recombination system. Plant J. 27: 171–178.

Daniell, H., and B.A. McFadden. 1987. Uptake and expression of bacterial and cyanobacterial genes by isolated cucumber etioplasts. Proc. Natl. Acad. Sci. U. S. A. 84: 6349–6353.

Daniell, H., S.B. Lee, T. Panchal, and P.O. Wiebe. 2001a. Expression of the native cholera toxin B subunit gene and assembly as functional oligomers in transgenic tobacco chloroplasts. J. Mol. Biol. 311: 1001–1009.

Daniell, H., B. Muthukumar, and S.B. Lee. 2001b. Marker free transgenic plants: Engineering the chloroplast genome without the use of antibiotic selection. Curr. Genet. 39: 109–116.

Daniell, H., S. Kumar, and N. Dufourmantel. 2005. Breakthrough in chloroplast genetic engineering of agronomically important crops. Trends Biotechnol. 23: 238–245.

Daniell, H., G. Ruiz, B. Denes, L. Sandberg, and W. Langridge. 2009. Optimization of codon composition and regulatory elements for expression of human insulin like growth factor-1 in transgenic chloroplasts and evaluation of structural identity and function. BMC Biotechnol. 9: 33.

Daniell, H., C.-S. Lin, M. Yu, and W.-J. Chang. 2016. Chloroplast genomes: Diversity, evolution, and applications in genetic engineering. Genome Biol. 17: 134.

Daniell, H., V. Rai, and Y. Xiao. 2019. Cold chain and virus-free oral polio booster vaccine made in lettuce chloroplasts confers protection against all three poliovirus serotypes. Plant Biotechnol. J. 17: 1357–1368.

Davoodi-Semiromi, A., M. Schreiber, S. Nalapalli, D. Verma, N.D. Singh, R.K. Banks, D. Chakrabarti, and H. Daniell. 2010. Chloroplast-derived vaccine antigens confer dual immunity against cholera and malaria by oral or injectable delivery. Plant Biotechnol. J. 8: 223–242.

De Cosa, B., W. Moar, S.B. Lee, M. Miller, and H. Daniell. 2001. Overexpression of the Bt cry2Aa2 operon in chloroplasts leads to formation of insecticidal crystals. Nat. Biotechnol. 19: 71–74.

Dhingra, A., and H. Daniell. 2006. Chloroplast genetic engineering via organogenesis or somatic embryogenesis. Methods Mol. Biol. 323: 245–262.

Dixon, R.A., and D. Strack. 2003. Phytochemistry meets genome analysis, and beyond....... Phytochemistry 62: 815–816.

Doetsch, N.A., M.R. Favreau, N. Kuscuoglu, M.D. Thompson, and R.B. Hallick. 2001. Chloroplast transformation in *Euglena gracilis*: Splicing of a group III twintron transcribed from a transgenic psbK operon. Curr. Genet. 39: 49–60.

Doron, L., N. Segal, and M. Shapira. 2016. Transgene expression in microalgae—from tools to applications. Front. Plant Sci. 7: 234.

Drechsel, O., and R. Bock. 2011. Selection of Shine-Dalgarno sequences in plastids. Nucleic Acids Res. 39: 1427–1438.

Dreesen, I.A.J., G. Charpin-El Hamri, and M. Fussenegger. 2010. Heat-stable oral alga-based vaccine protects mice from *Staphylococcus aureus* infection. J. Biotechnol. 145: 273–280.

Dufourmantel, N., B. Pelissier, F. Garçon, G. Peltier, J.-M. Ferullo, and G. Tissot. 2004. Generation of fertile transplastomic soybean. Plant Mol. Biol. 55: 479–489.

Dufourmantel, N., M. Dubald, M. Matringe, H. Canard, F. Garcon, C. Job, E. Kay, J.-P. Wisniewski, J.-M. Ferullo, B. Pelissier, A. Sailland, and G. Tissot. 2007. Generation and characterization of soybean and marker-free tobacco plastid transformants over-expressing a bacterial 4-hydroxyphenylpyruvate dioxygenase which provides strong herbicide tolerance. Plant Biotechnol. J. 5: 118–133.

Economou, C., T. Wannathong, J. Szaub, and S. Purton. 2014. A simple, low-cost method for chloroplast transformation of the green alga *Chlamydomonas reinhardtii*. Methods Mol. Biol. 1132: 401–411.

Eibl, C., Z. Zou, A. Beck, M. Kim, J. Mullet, and H.-U. Koop. 1999. *In vivo* analysis of plastid psbA, rbcL and rpl32 UTR elements by chloroplast transformation: Tobacco plastid gene expression is controlled by modulation of transcript levels and translation efficiency. Plant J. 19: 333–345.

Emadpour, M., D. Karcher, and R. Bock. 2015. Boosting riboswitch efficiency by RNA amplification. Nucleic Acids Res. 43: e66.

Esland, L., M. Larrea-Alvarez, and S. Purton. 2018. Selectable markers and reporter genes for engineering the chloroplast of *Chlamydomonas reinhardtii*. Biology 7:46.

Fabricant, D.S., and N.R. Farnsworth. 2001. The value of plants used in traditional medicine for drug discovery. Environ. Health Perspect. 109 Suppl 1: 69–75.

Fields, F.J., J.T. Ostrand, and S.P. Mayfield. 2018. Fed-batch mixotrophic cultivation of *Chlamydomonas reinhardtii* for high-density cultures. Algal Research 33: 109–117.

Fischer, N., O. Stampacchia, K. Redding, and J.D. Rochaix. 1996. Selectable marker recycling in the chloroplast. Mol. Gen. Genet. 251: 373–380.

Franklin, S., B. Ngo, E. Efuet, and S.P. Mayfield. 2002. Development of a GFP reporter gene for *Chlamydomonas reinhardtii* chloroplast. Plant J. 30: 733–744.

Fuentes, P., F. Zhou, A. Erban, D. Karcher, J. Kopka, and R. Bock. 2016. A new synthetic biology approach allows transfer of an entire metabolic pathway from a medicinal plant to a biomass crop. Elife 5.

Fuentes, P., T. Armarego-Marriott, and R. Bock. 2018. Plastid transformation and its application in metabolic engineering. Curr. Opin. Biotechnol. 49: 10–15.

Gallaher, S.D., S.T. Fitz-Gibbon, D. Strenkert, S.O. Purvine, M. Pellegrini, and S.S. Merchant. 2018. High-throughput sequencing of the chloroplast and mitochondrion of *Chlamydomonas reinhardtii* to generate improved *de novo* assemblies, analyze expression patterns and transcript speciation, and evaluate diversity among laboratory strains and wild isolates. Plant J. 93: 545–565.

Gan, Q., J. Jiang, X. Han, S. Wang, and Y. Lu. 2018. Engineering the chloroplast genome of oleaginous marine microalga nannochloropsis oceanica. Front. Plant Sci. 9: 439.

Gelvin, S.B. 2003. *Agrobacterium*-Mediated plant transformation: The Biology behind the "Gene-Jockeying" Tool. Microbiol. Mol. Biol. Rev. 67: 16–37.

Gimpel, J.A., and S.P. Mayfield. 2013. Analysis of heterologous regulatory and coding regions in algal chloroplasts. Appl. Microbiol. Biotechnol. 97: 4499–4510.

Gisby, M.F., P. Mellors, P. Madesis, M. Ellin, H. Laverty, S. O'Kane, M.W.J. Ferguson, and A. Day. 2011. A synthetic gene increases TGFβ3 accumulation by 75-fold in tobacco chloroplasts enabling rapid purification and folding into a biologically active molecule. Plant Biotechnol. J. 9: 618–628.

Gisby, M.F., E.A. Mudd, and A. Day. 2012. Growth of transplastomic cells expressing D-amino acid oxidase in chloroplasts is tolerant to D-alanine and inhibited by D-valine. Plant Physiol. 160: 2219–2226.

Gleba, Y., S. Marillonnet, and V. Klimyuk. 2004. Engineering viral expression vectors for plants: The 'full virus' and the 'deconstructed virus' strategies. Curr. Opin. Plant Biol. 7: 182–188.

Gleba, Y., V. Klimyuk, and S. Marillonnet. 2007. Viral vectors for the expression of proteins in plants. Curr. Opin. Biotechnol. 18: 134–141.

Goldstein, D.A., B. Tinland, L.A. Gilbertson, J.M. Staub, G.A. Bannon, R.E. Goodman, R.L. McCoy, and A. Silvanovich. 2005. Human safety and genetically modified plants: A review of antibiotic resistance markers and future transformation selection technologies. J. Appl. Microbiol. 99: 7–23.

Gregory, J.A., F. Li, L.M. Tomosada, C.J. Cox, A.B. Topol, J.M. Vinetz, J.M., and S. Mayfield. 2012. Algae-produced Pfs25 elicits antibodies that inhibit malaria transmission. PLoS One 7: e37179.

Hajdukiewicz, P.T.J., L. Gilbertson, and J.M. Staub. 2001. Multiple pathways for Cre/lox-mediated recombination in plastids. Plant J. 27: 161–170.

Heinemann, J.A., and T. Traavik. 2004. Problems in monitoring horizontal gene transfer in field trials of transgenic plants. Nat. Biotechnol. 22: 1105–1109.

Herzog, R.W., T.C. Nichols, J. Su, B. Zhang, A. Sherman, E.P. Merricks, R. Raymer, G.Q. Perrin, M. Häger, B. Wiinberg, and H. Daniell. 2017. Oral tolerance induction in hemophilia B dogs fed with transplastomic lettuce. Mol. Ther. 25: 512–522.

Herz, S., M. Füssl, S. Steiger, and H.-U. Koop. 2005. Development of novel types of plastid transformation vectors and evaluation of factors controlling expression. Transgenic Res. 14: 969–982.

Hiatt, A., R. Cafferkey, and K. Bowdish. 1989. Production of antibodies in transgenic plants. Nature 342: 76–78.

Hobbs, S.L., P. Kpodar, and C.M. DeLong. 1990. The effect of T-DNA copy number, position and methylation on reporter gene expression in tobacco transformants. Plant Mol. Biol. 15: 851–864.

Hou, B.-K., Y.-H. Zhou, L.-H. Wan, Z.-L. Zhang, G.-F. Shen, Z.-H. Chen, and Z.-M. Hu. 2003. Chloroplast transformation in oilseed rape. Transgenic Res. 12: 111–114.

Huang, F.-C., S.M.J. Klaus, S. Herz, Z. Zou, H.-U. Koop, and T.J. Golds. 2002. Efficient plastid transformation in tobacco using the aphA-6 gene and kanamycin selection. Mol. Genet. Genomics 268: 19–27.

Iamtham, S., and A. Day. 2000. Removal of antibiotic resistance genes from transgenic tobacco plastids. Nat. Biotechnol. 18: 1172–1176.

Ishikura, K., Y. Takaoka, K. Kato, M. Sekine, K. Yoshida, and A. Shinmyo. 1999. Expression of a foreign gene in *Chlamydomonas reinhardtii* chloroplast. J. Biosci. Bioeng. 87: 307–314.

Jin, S., and H. Daniell. 2015. The engineered chloroplast genome just got smarter. Trends Plant Sci. 20: 622–640.

Kasai, S., S. Yoshimura, K. Ishikura, Y. Takaoka, K. Kobayashi, K. Kato, and A. Shinmyo. 2003. Effect of coding regions on chloroplast gene expression in *Chlamydomonas reinhardtii*. J. Biosci. Bioeng. 95: 276–282.

Kato, K., T. Marui, S. Kasai, and A. Shinmyo. 2007. Artificial control of transgene expression in *Chlamydomonas reinhardtii* chloroplast using the lac regulation system from *Escherichia coli*. J. Biosci. Bioeng. 104: 207–213.

Kebeish, R., M. Niessen, K. Thiruveedhi, R. Bari, H.-J. Hirsch, R. Rosenkranz, N. Stäbler, B. Schönfeld, F. Kreuzaler, and C. Peterhänsel. 2007. Chloroplastic photorespiratory bypass increases photosynthesis and biomass production in *Arabidopsis thaliana*. Nat. Biotechnol. 25: 593–599.

Khan, M.S., and P. Maliga. 1999. Fluorescent antibiotic resistance marker for tracking plastid transformation in higher plants. Nat. Biotechnol. 17: 910–915.

Kindle, K.L., K.L. Richards, and D.B. Stern. 1991. Engineering the chloroplast genome: Techniques and capabilities for chloroplast transformation in *Chlamydomonas reinhardtii*. Proc. Natl. Acad. Sci. U. S. A. 88: 1721–1725.

Kinghorn, A.D., L. Pan, J.N. Fletcher, and H. Chai. 2011. The relevance of higher plants in lead compound discovery programs. J. Nat. Prod. 74: 1539–1555.

Kittiwongwattana, C., K. Lutz, M. Clark, and P. Maliga. 2007. Plastid marker gene excision by the phiC31 phage site-specific recombinase. Plant Mol. Biol. 64: 137–143.

Klaus, S.M.J., F.-C. Huang, T.J. Golds, and H.-U. Koop. 2004. Generation of marker-free plastid transformants using a transiently cointegrated selection gene. Nat. Biotechnol. 22: 225–229.

Klein, U., M.L. Salvador, and L. Bogorad. 1994. Activity of the Chlamydomonas chloroplast rbcL gene promoter is enhanced by a remote sequence element. Proc. Natl. Acad. Sci. U. S. A. 91: 10819–10823.

Kode, V., E.A. Mudd, S. Iamtham, and A. Day. 2006. Isolation of precise plastid deletion mutants by homology-based excision: A resource for site-directed mutagenesis, multi-gene changes and high-throughput plastid transformation. The Plant Journal 46: 901–909.

Kofer, W., C. Eibl, K. Steinmüller, and H.-U. Koop. 1998. PEG-mediated plastid transformation in higher plants. *In Vitro* Cellular & Developmental Biology - Plant 34: 303–309.

Kohli, N., D.R. Westerveld, A.C. Ayache, A. Verma, P. Shil, T. Prasad, P. Zhu S.L. Chan, Q. Li, and H. Daniell. 2014. Oral delivery of bioencapsulated proteins across blood–brain and blood–retinal barriers. Mol. Ther. 22: 535–546.

Koop, H.U., K. Steinmüller, H. Wagner, C. Rössler, C. Eibl, and L. Sacher. 1996. Integration of foreign sequences into the tobacco plastome via polyethylene glycol-mediated protoplast transformation. Planta 199: 193–201.

Krichevsky, A., B. Meyers, A. Vainstein, P. Maliga, and V. Citovsky. 2010. Autoluminescent plants. PLoS One 5: e15461.

Kumar, S., A. Dhingra, and H. Daniell. 2004a. Plastid-expressed betaine aldehyde dehydrogenase gene in carrot cultured cells, roots, and leaves confers enhanced salt tolerance. Plant Physiol. 136: 2843–2854.

Kumar, S., A. Dhingra, and H. Daniell. 2004b. Stable transformation of the cotton plastid genome and maternal inheritance of transgenes. Plant Mol. Biol. 56: 203–216.

Kumar, S., F.M. Hahn, E. Baidoo, T.S. Kahlon, D.F. Wood, C.M. McMahan, K. Cornish, J.D. Keasling, H. Daniell, and M.C. Whalen. 2012. Remodeling the isoprenoid pathway in tobacco by expressing the cytoplasmic mevalonate pathway in chloroplasts. Metab. Eng. 14: 19–28.

Künstner, P., A. Guardiola, Y. Takahashi, and J.D. Rochaix. 1995. A mutant strain of *Chlamydomonas reinhardtii* lacking the chloroplast photosystem II psbI gene grows photoautotrophically. J. Biol. Chem. 270: 9651–9654.

Kuroda, H., and P. Maliga. 2001a. Complementarity of the 16S rRNA penultimate stem with sequences downstream of the AUG destabilizes the plastid mRNAs. Nucleic Acids Res. 29: 970–975.

Kuroda, H., and P. Maliga. 2001b. Sequences downstream of the translation initiation codon are important determinants of translation efficiency in chloroplasts. Plant Physiol. 125: 430–436.

Kwon, K.-C., R. Nityanandam, J.S. New, and H. Daniell. 2013. Oral delivery of bioencapsulated exendin-4 expressed in chloroplasts lowers blood glucose level in mice and stimulates insulin secretion in beta-TC 6 cells. Plant Biotechnol. J. 11: 77–86.

Kwon, K.-C., and H. Daniell. 2015. Low-cost oral delivery of protein drugs bioencapsulated in plant cells. Plant Biotechnol. J. 13: 1017–1022.

Kwon, K.-C., H.-T. Chan, I.R. León, R. Williams-Carrier, A. Barkan, and H. Daniell. 2016. Codon optimization to enhance expression yields insights into chloroplast translation. Plant Physiol. 172: 62–77.

Kwon, K.-C., and H. Daniell. 2016. Oral delivery of protein drugs bioencapsulated in plant cells. Mol. Ther. 24: 1342–1350.

Kwon, K.-C., A. Sherman, W.-J. Chang, A. Kamesh, M. Biswas, R.W. Herzog, and H. Daniell. 2018a. Expression and assembly of largest foreign protein in chloroplasts: Oral delivery of human FVIII made in lettuce chloroplasts robustly suppresses inhibitor formation in haemophilia A mice. Plant Biotechnol. J. 16: 1148–1160.

Kwon, K.-C., A. Sherman, W.-J. Chang, A. Kamesh, M. Biswas, R.W. Herzog, and H. Daniell. 2018b. Expression and assembly of largest foreign protein in chloroplasts: Oral delivery of human FVIII made in lettuce chloroplasts robustly suppresses inhibitor formation in haemophilia A mice. Plant Biotechnol. J. 16: 1148–1160.

Lapidot, M. 2002. Stable Chloroplast transformation of the unicellular red alga *Porphyridium* species. Plant Physiol. 129: 7–12.

Lau, K.W., J. Ren, and M. Wu. 2000. Redox modulation of chloroplast DNA replication in *Chlamydomonas reinhardtii*. Antioxid. Redox Signal. 2: 529–535.

Leal, M.C., M.H.G. Munro, J.W. Blunt, J. Puga, B. Jesus, R. Calado, R. Rosa, and C. Madeira. 2013. Biogeography and biodiscovery hotspots of macroalgal marine natural products. Nat. Prod. Rep. 30: 1380–1390.

Lee, S.M., K. Kang, H. Chung, S.H. Yoo, X.M. Xu, S.-B. Lee, J.-J. Cheong, H. Daniell, and M. Kim. 2006. Plastid transformation in the monocotyledonous cereal crop, rice (*Oryza sativa*) and transmission of transgenes to their progeny. Mol. Cells 21: 401–410.

Lehtimäki, N., M.M. Koskela, and P. Mulo. 2015. Posttranslational modifications of chloroplast proteins: An emerging field. Plant Physiol. 168: 768–775.

Lelivelt, C.L.C., M.S. McCabe, C.A. Newell, C.B. deSnoo, K.M.P. van Dun, I. Birch-Machin, J.C. Gray, K.H.G. Mills, and J.M. Nugent. 2005. Stable plastid transformation in lettuce (*Lactuca sativa* L.). Plant Mol. Biol. 58: 763–774.

Lenzi, P., N. Scotti, F. Alagna, M.L. Tornesello, A. Pompa, A. Vitale, A. De Stradis, L. Monti, S. Grillo, F.M. Buonaguro, P. Maliga, and T. Cardi. 2008. Translational fusion of chloroplast-expressed human papillomavirus type 16 L1 capsid protein enhances antigen accumulation in transplastomic tobacco. Transgenic Res. 17: 1091–1102.

Liu, W., and Y. Liu. 2016. Youyou Tu: Significance of winning the 2015 Nobel Prize in Physiology or Medicine. Cardiovasc Diagn Ther 6: 1–2.

Li, W., S. Ruf, and R. Bock. 2011. Chloramphenicol acetyltransferase as selectable marker for plastid transformation. Plant Mol. Biol. 76: 443–451.

Li, Y., R. Wang, Z. Hu, H. Li, S. Lu, J. Zhang, Y. Lin, and F. Zhou. 2016. Expression of a codon-optimized dsda gene in tobacco plastids and rice nucleus confers D-serine tolerance. front. Plant Sci. 7: 640.

López-Arredondo, D.L., and L. Herrera-Estrella. 2013. A novel dominant selectable system for the selection of transgenic plants under *in vitro* and greenhouse conditions based on phosphite metabolism. Plant Biotechnol. J. 11: 516–525.

Lössl, A., C. Eibl, H. Harloff, C. Jung, and H.U. Koop. 2003. Polyester in transplastomic tobacco: Significant contents of polyhydroxybutyrate are associated with growth reduction. Plant Cell Rep. 21: 891–899.

Lössl, A., K. Bohmert, H. Harloff, C. Eibl, S. Mühlbauer, and H.-U. Koop. 2005. Inducible trans-activation of plastid transgenes: Expression of the *R. eutrophaphb* Operon in Transplastomic Tobacco. Plant Cell Physiol. 46: 1462–1471.

Ludwig, M., E. Enzenhofer, S. Schneider, M. Rauch, A. Bodenteich, K. Neumann, E. Prieschl-Grassauer, A. Grassauer, T. Lion, and C.A. Mueller. 2013. Efficacy of a Carrageenan nasal spray in patients with common cold: A randomized controlled trial. Respir. Res. 14: 124.

Lutz, K.A., J.E. Knapp, and P. Maliga. 2001. Expression of bar in the plastid genome confers herbicide resistance. Plant Physiol. 125: 1585–1590.

Lutz, K.A., and P. Maliga. 2008. Plastid genomes in a regenerating tobacco shoot derive from a small number of copies selected through a stochastic process. Plant J. 56: 975–983.

Lu, Y., H. Rijzaani, D. Karcher, S. Ruf, and R. Bock. 2013. Efficient metabolic pathway engineering in transgenic tobacco and tomato plastids with synthetic multigene operons. Proc. Natl. Acad. Sci. U. S. A. 110: E623–32.

Macedo-Osorio, K.S., V.H. Pérez-España, C. Garibay-Orijel, D. Guzmán-Zapata, N.V. Durán-Figueroa, and J.A. Badillo-Corona. 2018. Intercistronic expression elements (IEE) from the chloroplast of *Chlamydomonas reinhardtii* can be used for the expression of foreign genes in synthetic operons. Plant Mol. Biol. 98: 303–317.

Ma, J.K., A. Hiatt, M. Hein, N.D. Vine, F. Wang, P. Stabila, C. van Dolleweerd, K. Mostov, and T. Lehner. 1995. Generation and assembly of secretory antibodies in plants. Science 268: 716–719.

Ma, J.K.-C., P.M.W. Drake, and P. Christou. 2003. The production of recombinant pharmaceutical proteins in plants. Nat. Rev. Genet. 4: 794–805.

Maliga, P. 2002. Engineering the plastid genome of higher plants. Curr. Opin. Plant Biol. 5: 164–172.

Maliga, P., and R. Bock. 2011. Plastid biotechnology: Food, fuel, and medicine for the 21st century. Plant Physiol. 155: 1501–1510.

Manuell, A.L., M.V. Beligni, J.H. Elder, D.T. Siefker, M. Tran, A. Weber, T.L. McDonald, and S.P. Mayfield. 2007. Robust expression of a bioactive mammalian protein in Chlamydomonas chloroplast. Plant Biotechnol. J. 5: 402–412.

Mardanova, E.S., E.A. Blokhina, L.M. Tsybalova, H. Peyret, G.P. Lomonossoff, and N.V. Ravin. 2017. Efficient transient expression of recombinant proteins in plants by the novel pEff vector based on the genome of potato virus X. Front Plant Sci. 8: 247.

Marín-Navarro, J., A.L. Manuell, J. Wu, and P.S. Mayfield. 2007. Chloroplast translation regulation. Photosynth. Res. 94: 359–374.

Marillonnet, S., A. Giritch, M. Gils, R. Kandzia, V. Klimyuk, and Y. Gleba. 2004. *In planta* engineering of viral RNA replicons: Efficient assembly by recombination of DNA modules delivered by *Agrobacterium*. Proc. Natl. Acad. Sci. U. S. A. 101: 6852–6857.

McGillicuddy, N., P. Floris, S. Albrecht, and J. Bones. 2018. Examining the sources of variability in cell culture media used for biopharmaceutical production. Biotechnol. Lett. 40: 5–21.

Medgyesy, P., A. Páy, and L. Márton. 1986. Transmission of paternal chloroplasts in Nicotiana. Mol. Gen. Genet. 204: 195–198.

Meng, B.Y., M. Tanaka, T. Wakasugi, M. Ohme, K. Shinozaki, and M. Sugiura. 1988. Cotranscription of the genes encoding two P700 chlorophyll a apoproteins with the gene for ribosomal protein CS14: Determination of the transcriptional initiation site by *in vitro* capping. Curr. Genet. 14: 395–400.

Minai, L., K. Wostrikoff, F.-A. Wollman, and Y. Choquet. 2006. Chloroplast biogenesis of photosystem II cores involves a series of assembly-controlled steps that regulate translation. Plant Cell 18: 159–175.

Morrow, K.J., Jr. 2007. Improving protein production processes. Gen Eng News 27: 50–41.

Mudd, E.A., P. Madesis, E.M. Avila, and A. Day. 2014. Excision of plastid marker genes using directly repeated DNA sequences. Methods Mol. Biol. 1132: 107–123.

Muralikrishna, N., K. Srinivas, K.B. Kumar, and A. Sadanandam. 2016. Stable plastid transformation in *Scoparia dulcis* L. Physiol. Mol. Biol. Plants 22: 575–581.

Nahampun, H.N., D. López-Arredondo, X. Xu, L. Herrera-Estrella, and K. Wang. 2016. Assessment of ptxD gene as an alternative selectable marker for *Agrobacterium*-mediated maize transformation. Plant Cell Rep. 35: 1121–1132.

Nakamura, M., Y. Hibi, T. Okamoto, and M. Sugiura. 2016. Cooperation between the chloroplast psbA 5′-untranslated region and coding region is important for translational initiation: The chloroplast translation machinery cannot read a human viral gene coding region. Plant J. 85: 772–780.

Nguyen, T.T., G. Nugent, T. Cardi, and P.J. Dix. 2005. Generation of homoplasmic plastid transformants of a commercial cultivar of potato (*Solanum tuberosum* L.). Plant Sci. 168: 1495–1500.

Nishimura, N., J. Zhang, M. Abo, A. Okubo, and S. Yamazaki. 2001. Application of capillary electrophoresis to the simultaneous determination of betaines in plants. Anal. Sci. 17: 103–106.

Nishimura, K., Y. Tomoda, Y. Nakamoto, T. Kawada, Y. Ishii, and Y. Nagata. 2007. Alanine racemase from the green alga *Chlamydomonas reinhardtii*. Amino Acids 32: 59–62.

O'Connor, H.E., S.V. Ruffle, A.J. Cain, Z. Deak, I. Vass, J.H.A. Nugent, and S. Purton. 1998. The 9-kDa phosphoprotein of photosystem: II. Generation and characterisation of Chlamydomonas mutants lacking PSII-H and a site-directed mutant lacking the phosphorylation site. Biochim. Biophys. Acta. 1364: 63–72.

Oey, M., M. Lohse, B. Kreikemeyer, and R. Bock. 2009. Exhaustion of the chloroplast protein synthesis capacity by massive expression of a highly stable protein antibiotic. Plant J. 57: 436–445.

Olins, P.O., C.S. Devine, S.H. Rangwala, and K.S. Kavka. 1988. The T7 phage gene 10 leader RNA, a ribosome-binding site that dramatically enhances the expression of foreign genes in *Escherichia coli*. Gene 73: 227–235.

Pandeya, D., L.M. Campbell, E. Nunes, D.L. Lopez-Arredondo, M.R. Janga, L. Herrera-Estrella, and K.S. Rathore. 2017. ptxD gene in combination with phosphite serves as a highly effective selection system to generate transgenic cotton (*Gossypium hirsutum* L.). Plant Mol. Biol. 95: 567–577.

Park, J., G. Yan, K.-C. Kwon, M. Liu, P.A. Gonnella, S. Yang, and H. Daniell 2019. Oral delivery of novel human IGF-1 bioencapsulated in lettuce cells promotes musculoskeletal cell proliferation, differentiation and diabetic fracture healing. Biomaterials 233: 119591.

Paul, M., and J.K.-C. Ma. 2011. Plant-made pharmaceuticals: Leading products and production platforms. Biotechnol. Appl. Biochem. 58: 58–67.

Peach, C., and J. Velten. 1991. Transgene expression variability (position effect) of CAT and GUS reporter genes driven by linked divergent T-DNA promoters. Plant Mol. Biol. 17: 49–60.

Pontiroli, A., P. Simonet, A. Frostegard, T.M. Vogel, and J.-M. Monier. 2007. Fate of transgenic plant DNA in the environment. Environ. Biosafety Res. 6: 15–35.

Quesada-Vargas, T., O.N. Ruiz, and H. Daniell. 2005. Characterization of heterologous multigene operons in transgenic chloroplasts: Transcription, processing, and translation. Plant Physiol. 138: 1746–1762.
Rasala, B.A., M. Muto, J. Sullivan, and S.P. Mayfield. 2011. Improved heterologous protein expression in the chloroplast of *Chlamydomonas reinhardtii* through promoter and 5′ untranslated region optimization. Plant Biotechnol. J. 9: 674–683.
Rasala, B.A., and S.P. Mayfield. 2015. Photosynthetic biomanufacturing in green algae; production of recombinant proteins for industrial, nutritional, and medical uses. Photosynth. Res. 123: 227–239.
Rathinasabapathi, B., K. McCue, D. Gage, and A. Hanson. 1994. Metabolic engineering of glycine betaine synthesis: Plant betaine aldehyde dehydrogenases lacking typical transit peptides are targeted to tobacco chloroplasts where they confer betaine aldehyde resistance. Planta 193: 155–162.
Remacle, C., S. Cline, L. Boutaffala, S. Gabilly, V. Larosa, M.R. Barbieri, N. Coosemans, and P.P. Hamel. 2009. The ARG9 gene encodes the plastid-resident N-acetyl ornithine aminotransferase in the green alga *Chlamydomonas reinhardtii*. Eukaryot. Cell 8: 1460–1463.
Rhodes, D., and A.D. Hanson. 1993. Quaternary ammonium and tertiary sulfonium compounds in higher plants. Annu. Rev. Plant Physiol. Plant Mol. Biol. 44: 357–384.
Rochaix, J.-D., R. Surzycki, and S. Ramundo. 2014. Tools for regulated gene expression in the chloroplast of Chlamydomonas. pp. 413–424. *In:* Maliga, P. (ed.). Chloroplast Biotechnology: Methods and Protocols. Humana Press: Totowa, New York, USA.
Ruf, S., M. Hermann, I.J. Berger, H. Carrer, and R. Bock. 2001. Stable genetic transformation of tomato plastids and expression of a foreign protein in fruit. Nat. Biotechnol. 19: 870–875.
Ruhlman, T., D. Verma, N. Samson, and H. Daniell. 2010. The role of heterologous chloroplast sequence elements in transgene integration and expression. Plant Physiol. 152: 2088–2104.
Ruiz, O.N., H.S. Hussein, N. Terry, and H. Daniell. 2003. Phytoremediation of organomercurial compounds via chloroplast genetic engineering. Plant Physiol. 132: 1344–1352.
Sandoval-Vargas, J.M., K.S. Macedo-Osorio, N.V. Durán-Figueroa, C. Garibay-Orijel, and J.A. Badillo-Corona. 2018. Chloroplast engineering of *Chlamydomonas reinhardtii* to use phosphite as phosphorus source. Algal Research 33: 291–297.
Saxena, B., M. Subramaniyan, K. Malhotra, N.S. Bhavesh, S.D. Potlakayala, and S. Kumar. 2014. Metabolic engineering of chloroplasts for artemisinic acid biosynthesis and impact on plant growth. J. Biosci. 39: 33–41.
Schell, J., and M. Van Montagu. 1977. The Ti-Plasmid of *Agrobacterium Tumefaciens*, a natural vector for the introduction of NIF genes in plants? Basic Life Sci. 9: 159–179.
Schroda, M. 2004. The Chlamydomonas genome reveals its secrets: Chaperone genes and the potential roles of their gene products in the chloroplast. Photosynth. Res. 82: 221–240.
Schwarz, C., I. Elles, J. Kortmann, M. Piotrowski, and J. Nickelsen. 2007. Synthesis of the D2 protein of photosystem II in Chlamydomonas is controlled by a high molecular mass complex containing the RNA stabilization factor Nac2 and the translational activator RBP40. Plant Cell 19: 3627–3639.
Scott, S.E., and M.J. Wilkinson. 1999. Low probability of chloroplast movement from oilseed rape (*Brassica napus*) into wild Brassica rapa. Nat. Biotechnol. 17: 390–392.
Scranton, M.A., J.T. Ostrand, F.J. Fields, and S.P. Mayfield. 2015. Chlamydomonas as a model for biofuels and bio-products production. Plant J. 82: 523–531.
Shahid, N., and H. Daniell. 2016. Plant-based oral vaccines against zoonotic and non-zoonotic diseases. Plant Biotechnology Journal 14: 2079–2099.
Shaver, J.M., D.J. Oldenburg, and A.J. Bendich. 2006. Changes in chloroplast DNA during development in tobacco, Medicago truncatula, pea, and maize. Planta 224: 72–82.
Shenoy, V., K.-C. Kwon, A. Rathinasabapathy, S. Lin, G. Jin, C. Song, P. Shil, A. Nair, Y. Qi, Q. Li, J. Francis, M.J. Katovich, H. Daniell, and M.K. Raizada. 2014. Oral delivery of angiotensin-converting enzyme 2 and angiotensin-(1–7) bioencapsulated in plant cells attenuates pulmonary hypertension. Hypertension 64: 1248–1259.
Sherman, A., J. Su, S. Lin, X. Wang, R.W. Herzog, and H. Daniell. 2014. Suppression of inhibitor formation against FVIII in a murine model of hemophilia A by oral delivery of antigens bioencapsulated in plant cells. Blood 124: 1659–1668.
Shil, P.K., K.-C. Kwon, P. Zhu, A. Verma, H. Daniell, and Q. Li. 2014. Oral delivery of ACE2/Ang-(1–7) bioencapsulated in plant cells protects against experimental uveitis and autoimmune uveoretinitis. Mol. Ther. 22: 2069–2082.

Sidorov, V.A., D. Kasten, S.-Z. Pang, P.T.J. Hajdukiewicz, J.M. Staub, and N.S. Nehra. 1999. Stable chloroplast transformation in potato: Use of green fluorescent protein as a plastid marker. Plant J. 19: 209–216.

Sikdar, S.R., G. Serino, S. Chaudhuri, and P. Maliga. 1998. Plastid transformation in *Arabidopsis thaliana*. Plant Cell Reports 18: 20–24.

Skarjinskaia, M., Z. Svab, and P. Maliga. 2003. Plastid transformation in *Lesquerella fendleri*, an oilseed Brassicacea. Transgenic Res. 12: 115–122.

Song, H.-S., J.E. Brotherton, R.A. Gonzales, and J.M. Widholm. 1998. Tissue culture-specific expression of a naturally occurring tobacco feedback-insensitive anthranilate synthase. Plant Physiol. 117: 533–543.

Specht, E.A., and S.P. Mayfield. 2014. Algae-based oral recombinant vaccines. Front. Microbiol. 5: 60.

Spök, A., S. Karner, A.J. Stein, and E. Rodríguez-Cerezo. 2008. Plant molecular farming: Opportunities and challenges. JRC Scientific and Technical Reports. 3: 153–172.

Staub, J.M., and P. Maliga. 1994. Translation of psbA mRNA is regulated by light via the 5'-untranslated region in tobacco plastids. Plant J. 6: 547–553.

Staub, J.M., B. Garcia, J. Graves, P.T. Hajdukiewicz, P. Hunter, N. Nehra, V. Paradkar, M. Schlittler, J.A. Carroll, L. Spatola, D. Ward, G. Ye, and D.A. Russell. 2000. High-yield production of a human therapeutic protein in tobacco chloroplasts. Nat. Biotechnol. 18: 333–338.

Su, J., A. Sherman, P.A. Doerfler, B.J. Byrne, R.W. Herzog, and H. Daniell. 2015a. Oral delivery of Acid Alpha Glucosidase epitopes expressed in plant chloroplasts suppresses antibody formation in treatment of Pompe mice. Plant Biotechnol. J. 13: 1023–1032.

Su, J., L. Zhu, A. Sherman, X. Wang, S. Lin, A. Kamesh, J.H. Norikane, S.J. Streatfield, R.W. Herzog, and H. Daniell. 2015b. Low cost industrial production of coagulation factor IX bioencapsulated in lettuce cells for oral tolerance induction in hemophilia B. Biomaterials 70: 84–93.

Surzycki, R., L. Cournac, G. Peltier, and J.-D. Rochaix. 2007. Potential for hydrogen production with inducible chloroplast gene expression in Chlamydomonas. Proc. Natl. Acad. Sci. U. S. A. 104: 17548–17553.

Surzycki, R., K. Greenham, K. Kitayama, F. Dibal, R. Wagner, J.-D. Rochaix, T. Ajam, and S. Surzycki. 2009. Factors effecting expression of vaccines in microalgae. Biologicals 37: 133–138.

Svab, Z., P. Hajdukiewicz, and P. Maliga. 1990. Stable transformation of plastids in higher plants. Proc. Natl. Acad. Sci. U. S. A. 87: 8526–8530.

Tabatabaei, I., S. Ruf, and R. Bock. 2017. A bifunctional aminoglycoside acetyltransferase/ phosphotransferase conferring tobramycin resistance provides an efficient selectable marker for plastid transformation. Plant Mol. Biol. 93: 269–281.

Tasaki, T., S.M. Sriram, K.S. Park, and Y.T. Kwon. 2012. The N-end rule pathway. Annu. Rev. Biochem. 81: 261–289.

Taunt, H.N., L. Stoffels, and S. Purton. 2018. Green biologics: The algal chloroplast as a platform for making biopharmaceuticals. Bioengineered 9: 48–54.

Tekoah, Y., A. Shulman, T. Kizhner, I. Ruderfer, L. Fux, Y. Nataf, D. Bartfeld, T. Ariel, S. Gingis-Velitski, U. Hanania, and Y. Shaaltiel. 2015. Large-scale production of pharmaceutical proteins in plant cell culture—the protalix experience. Plant Biotechnol. J. 13: 1199–1208.

Tobias, J.W., T.E. Shrader, G. Rocap, and A. Varshavsky. 1991. The N-end rule in bacteria. Science 254: 1374–1377.

Tran, M., C. Van, D.J. Barrera, P.L. Pettersson, C.D. Peinado, J. Bui, and S.P. Mayfield. 2013. Production of unique immunotoxin cancer therapeutics in algal chloroplasts. Proc. Natl. Acad. Sci. U. S. A. 110: E15–22.

Tungsuchat-Huang, T., and P. Maliga. 2012. Visual marker and *Agrobacterium*-delivered recombinase enable the manipulation of the plastid genome in greenhouse-grown tobacco plants. Plant J. 70: 717–725.

Verhounig, A., D. Karcher, and R. Bock. 2010. Inducible gene expression from the plastid genome by a synthetic riboswitch. Proc. Natl. Acad. Sci. U. S. A. 107: 6204–6209.

Verma, D., B. Moghimi, P.A. LoDuca, H.D. Singh, B.E. Hoffman, R.W. Herzog, and H. Daniell. 2010. Oral delivery of bioencapsulated coagulation factor IX prevents inhibitor formation and fatal anaphylaxis in hemophilia B mice. Proc. Natl. Acad. Sci. U. S. A. 107: 7101–7106.

Wang, Y.-P., Z.-Y. Wei, X.-F. Zhong, C.-J. Lin, Y.-H. Cai, J. Ma, Y.-Y. Zhang, Y.-Z. Liu, and S.-C. Xing. 2015. Stable expression of basic fibroblast growth factor in chloroplasts of tobacco. Int. J. Mol. Sci. 17.

Wang, Y., Z. Wei, and S. Xing. 2018. Stable plastid transformation of rice, a monocot cereal crop. Biochem. Biophys. Res. Commun. 503: 2376–2379.

Wannathong, T., J.C. Waterhouse, R.E.B. Young, C.K. Economou, and S. Purton. 2016. New tools for chloroplast genetic engineering allow the synthesis of human growth hormone in the green alga *Chlamydomonas reinhardtii*. Appl. Microbiol. Biotechnol. 100: 5467–5477.

Whitney, S.M., and R.E. Sharwood. 2008. Construction of a tobacco master line to improve Rubisco engineering in chloroplasts. J. Exp. Bot. 59: 1909–1921.

Willey, D.L., and J.C. Gray. 1989. Two small open reading frames are co-transcribed with the pea chloroplast genes for the polypeptides of cytochrome b-559. Curr. Genet. 15: 213–220.

Wood, A.J., and R. Joel Duff. 2009. The aldehyde dehydrogenase (ALDH) gene superfamily of the moss Physcomitrella patens and the algae *Chlamydomonas reinhardtii* and *Ostreococcus tauri*. The Bryologist 112: 1–11.

Xie, W.-H., C.-C. Zhu, N.-S. Zhang, D.-W. Li, W.-D. Yang, J.-S. Liu, R. Sathishkumar, and H.-Y. Li. 2014. Construction of novel chloroplast expression vector and development of an efficient transformation system for the diatom *Phaeodactylum tricornutum*. Mar. Biotechnol. 16: 538–546.

Yang, H., B.N. Gray, B.A. Ahner, and M.R. Hanson. 2013. Bacteriophage 5′ untranslated regions for control of plastid transgene expression. Planta 237: 517–527.

Ye, G.N., P.T. Hajdukiewicz, D. Broyles, D. Rodriguez, C.W. Xu, N. Nehra, and J.M. Staub. 2001. Plastid-expressed 5-enolpyruvylshikimate-3-phosphate synthase genes provide high level glyphosate tolerance in tobacco. Plant J. 25: 261–270.

Ye, X. 2000. Engineering the provitamin A (-Carotene) biosynthetic pathway into (Carotenoid-Free) Rice Endosperm. Science 287: 303–305.

Young, R.E.B., and S. Purton. 2016. Codon reassignment to facilitate genetic engineering and biocontainment in the chloroplast of *Chlamydomonas reinhardtii*. Plant Biotechnol. J. 14: 1251–1260.

Zambryski, P., H. Joos, C. Genetello, J. Leemans, M.V. Montagu, and J. Schell. 1983. Ti plasmid vector for the introduction of DNA into plant cells without alteration of their normal regeneration capacity. EMBO J. 2: 2143–2150.

Zhang, B., B. Shanmugaraj, and H. Daniell. 2017a. Expression and functional evaluation of biopharmaceuticals made in plant chloroplasts. Curr. Opin. Chem. Biol. 38: 17–23.

Zhang, J., S.A. Khan, D.G. Heckel, and R. Bock. 2017b. Next-generation insect-resistant plants: RNAi-mediated crop protection. Trends Biotechnol. 35: 871–882.

Zhou, F., D. Karcher, and R. Bock. 2007. Identification of a plastid intercistronic expression element (IEE) facilitating the expression of stable translatable monocistronic mRNAs from operons. Plant J. 52: 961–972.

Zou, Z., C. Eibl, and H.-U. Koop. 2003. The stem-loop region of the tobacco psbA 5′UTR is an important determinant of mRNA stability and translation efficiency. Mol. Genet. Genomics 269: 340–349.

Zubkot, M.K., E.I. Zubkot, K. van Zuilen, P. Meyer, and A. Day. 2004. Stable transformation of petunia plastids. Transgenic Res. 13: 523–530.

CHAPTER 10

Plant Virus-Based Expression Vectors and their Applications in Foreign Protein/Antigen Expression

Srividhya Venkataraman, Erum Shoaeb and *Kathleen Hefferon**

Introduction

Plant viruses have long been engineered for a variety of applications. Douglas and Young (1998) first proposed the idea of manipulating the capsids of viruses, which were found to be suitable in size, symmetrical structure and with a loading capacity for drugs or heterogeneous proteins. These attributes made viral capsids an ideal choice for therapeutic and diagnostic applications.

Biopharmaceutical products generated in plants have proven to be a reliable emerging technology for delivering low-cost, effectual prophylactics and therapeutics for developing countries, to combat threats of infectious disease pandemics as well as for the administration of personalized medicine. Plants serve as efficient biofactories for producing pharmaceutically important proteins and other metabolites. Plants possess some inherent advantages for the mass production of these molecules as their propagation is simplified by their dependence on sunlight, they are bereft of human pathogens and can be grown on a large scale. Among these plant-based systems, plant viral expression vectors serve as highly prominent biofactories capable of synthesizing commercially significant biomolecules (Hefferon 2017). The major advantage of the use of plant virus expression vectors is the production of immense levels of pharmaceutical proteins (Kopertekh and Schiemann 2017) within a brief period, while surmounting concerns of genetically modified organisms engendered by genetically engineered transgenic plants.

Cell and System Biology, University of Toronto, Toronto, Canada.
* Corresponding author: Kathleen.hefferon@alumni.utoronto.ca

Plant virus nanoparticles provide an advantage over synthetic nanomaterials as they are natural, biocompatible and biodegradable (Lefeuvre et al. 2019). For gene delivery into mammalian cells, initially VLPs of animal origin were used (Schaffer et al. 2008) but now plant virus like particles are gaining in popularity due to their simplicity and strength. Furthermore, VNPs derived from plant viruses are predominantly beneficial, as there is little likelihood that they can induce adverse effects when introduced to animal systems and are nonpathogenic to humans (Steinmetz 2010). Plant virus capsid proteins are known to protect genetic material and have shown to be more efficient in packaging variable lengths of nucleic acid compared to those of animal viruses (Azizgolshani et al. 2013). Plant viruses therefore are potentially able to provide a platform for multiple medical applications such as vaccination, drug delivery and tumor targeting.

Studies on the safety and immunogenicity of plant virus-like nanoparticles (VLPs) have attracted the interest of many researchers around the world (Lebel et al. 2015). Plant virus genomes contain genes for coat proteins (CP), and it is the modification of the CP that allows for the expression of desired antigens to elicit an effective immune response (Fausther-Bovendo and Kobinger 2014). The fusion of the sequence of a foreign antigen at the N-terminal or C-terminal region of the CP gene construct is the most common method of modification for the expression of the desired epitope on the surface of the VLP (Babin et al. 2013). Empty spherical nanoparticles of TMV CP were developed as an immunogenic particle platform in 2011 (Atabekov et al. 2011), that could be modified with the introduction of RNA as needed.

The following chapter describes the use of plant viruses as platforms for the expression of vaccines and other therapeutic proteins. A description of plant virus nanoparticles and their uses to combat cancer is also discussed here. This chapter provides examples of three well known plant viruses (Tobacco mosaic virus, Potato virus X and Cowpea mosaic virus) and their uses for these applications. The chapter concludes with a discussion of the future prospects of plant virus applications in medicine.

Biopharming using Plant virus Expression Vectors

Genetically engineered plant viruses have been used thus far to produce prophylactics such as vaccines and monoclonal antibodies as well as therapeutics. Principally, (+)-sense RNA viruses and ssDNA viruses have been used as plant virus expression vectors (Gleba et al. 2007, Klimyuk et al. 2014, Kagale et al. 2012). Plant virus expression vectors are efficient tools for studying cell biology, engineering plant genomes, molecular farming and functional genomics research (Ruiz-Ramon et al. 2019, Baltes et al. 2014, Mardanova et al. 2017, Yamamoto et al. 2018, Hefferon 2017, Porta and Lomonossoff 2002, Liu and Lomonossoff 2002). Engineered plant virus genomes have been generated through strategies such as deconstructed/modular systems, gene insertion, gene substitution and peptide display fusion (Hefferon 2017, Salazar-González et al. 2015, Gleba et al. 2007, Lico et al. 2008) to obtain a stable and robust expression of prophylactic/therapeutic foreign genes as biopharmaceuticals. Mostly these expression vectors have been derived from tobamoviruses, potexviruses, comoviruses, geminiviruses and tobraviruses (Peyret and Lomonossoff 2015).

Initially, first generation viral expression vectors were made wherein the entire viral genome was retained with the foreign gene of interest incorporated into the CP ORF as a fusion protein with the CP or through a robust subgenomic promoter inserted into the viral genome. The exposed epitopes of the capsid protein could be replaced by those of the vaccines such that the vaccine epitopes containing the repetitive protein structures can be directly presented to the immune system of the human/animal host to elicit effectual immune responses. However, the use of the full virus genome in these vectors offers undue constraints with respect to the size of the vaccine protein under consideration.

Compared to the above first generation viral vectors, the second generation vectors are the relatively new "deconstructed vectors" containing the vaccine antigen/therapeutic gene of interest along with the minimal coding sequence of the virus essential for its replication (Yusibov et al. 2013, Gleba et al. 2004, Gleba et al. 2005). Thus, by getting rid of the genes required for the virus movement and assembly, more of the foreign gene can be accommodated. However, this necessitates the delivery of these expression vectors by techniques such as transformation of *Agrobacterium* with the viral expression vector followed by vacuum infiltration of the transformed bacteria into the leaves of the plants (Leuzinger et al. 2013). The transformed plants can now produce the pharmaceutical antigen/protein systemically across the entire plant thus generating large amounts of the foreign protein(s) within a short time period. The most preferred plant-based platform for recombinant protein production is transient expression by the *Agrobacterium*-mediated delivery of viral or non-viral expression vectors. This method has several advantages including the high yield of the foreign antigen, the enhanced speed of the procedure and the diminished concern for the escape of the transgenes. Among the plant virus-based expression systems, those composed of the deconstructed viral vectors derived from BeYDV (Huang et al. 2010), CPMV (Sainsbury et al. 2008), PVX (Giritch et al. 2006) and TMV (Gleba et al. 2004) have proved to be most promising in terms of obtaining high yields of the foreign proteins.

Recently, third generation viral vectors have been engineered wherein the plant viruses are used as nanoparticle payloads to direct and deliver antivirals into cancerous tumors (Shukla et al. 2013). The rod-shaped potexviruses such as PVX containing the vaccine antigen on their outer surfaces have been used for such tumor-targeted delivery. The icosahedral structure of the Cowpea Mosaic Virus has been exploited to carry molecular cargos such as drugs into the internal cavities of the tumors (Wen et al. 2012, Lico et al. 2013). In recent times, plant virus-based nanoparticles have been used to express recombinant foreign proteins in plants in a rapid, simplified and scalable manner (Röder et al. 2018, Wen and Steinmetz 2016). Plant virus nanoparticles have the advantage of self-assembling capability with perfect symmetry and polyvalency. Also, they are stable and monodisperse under a wide range of conditions and are inherently safe while being biocompatible and non-infectious in mammals. This makes them superior to synthetic nanoparticles and enabling them to be used for the production of a wide range of biopharmaceuticals.

Tobamovirus Expression Vectors

TMV was the first candidate virus used to build the deconstructed viral vector (Hefferon 2017, Gleba et al. 2004, Gleba et al. 2005). TMV belongs to the genus Tobamovirus (Ishibashi and Ishikawa 2016, Knapp and Lewandowski 2001, Scholthof 2004) and contains a plus-sense genomic RNA encapsidated within the viral capsid that forms rigid, rod-shaped helical virions 300–310 nm in length and 18 nm in diameter. The genome encodes 2 replication proteins of 126 and 183 kDa, a movement protein of 30 kDa and a capsid protein of 17.5 kDa. While the former two proteins (replication proteins) are translated from the genomic RNA, the latter two proteins are translated from 3' terminal subgenomic RNAs. A crucial advantage of TMV as an expression system is its ability to multiply and accumulate large quantities of the viral capsid protein in infected plants. TMV has been used as an effective vector for the production of several valuable biopharmaceuticals using strategies such as expression of the foreign gene under the control of the TMV CP promoter while eliminating the CP gene. Instead, encapsidation is achieved by insertion of the CP ORF from another tobamovirus into the TMV genome.

Plants were transfected by the German biotechnology company, Icon Genetics using these TMV vectors carrying the foreign antigen introduced by Agro-infiltration into plants wherein the movement of the viral vector is disabled. This process was labeled as Magnifection (Gleba et al. 2005). Several reports of the use of TMV to express whole proteins or their epitopes have emerged in recent times (Noris et al. 2011, Huang et al. 2010, Musiychuk et al. 2007, Mett et al. 2008). The Influenza M2e epitope, the Human Papilloma Virus (HPV) E7 antigen and Norwalk VLPs have all been produced in deconstructed TMV vectors (Noris et al. 2011, Massa et al. 2007, Santi et al. 2008). Vaccines against influenza, cholera and plague have also been developed using TMV-based "launch vectors" (Petukhova et al. 2014, Hamorsky et al. 2013, Chichester et al. 2009, Santi et al. 2006).

TMV-based foreign antigen expression is augmented to a great extent by the co-expression of TBSV P19, an RNA silencing suppressor (Voinnet et al. 2003) along with that of the foreign protein. In another invention, a TMV RNA-Based Overexpression (TRBO) vector was constructed with the ORF of the foreign gene closer to the TMV RNA 3' end while deleting the capsid protein ORF entirely and being unable to spread systemically. This enhanced the protein antigen expression significantly (Lindbo 2007a). Fraunhofer USA has generated a TMV-based potent vaccine against the pandemic causing H1N1 Influenza Virus, which is now at the stage of clinical trials (Shoji et al. 2013, Cummings et al. 2013, Neuhaus et al. 2013, Vezina et al. 2011). In another study, the Influenza virus protein hemagglutinin (HA) was expressed in *Nicotiana benthamiana* leaves with maximum, stable expression occurring at 20°C in a temperature-specific manner (Matsuda et al. 2017). Transport of foreign gene products was also influenced by light intensity (Patil and Fauquet 2015). In another report, Sun hemp mosaic virus, a tobamovirus infecting legumes was used for controlled, inducible expression of the foreign gene ("SHEC" vector), while being unable to form virions due to lack of the CP and showing poor replication capabilities unless in the presence of the P19 RNA-silencing suppressor (Liu and Kearney 2010).

TMV naturally has adjuvant properties which makes it useful for acting as a delivery vehicle for immune therapeutics. Strong CD8+ T cell stimulation was observed when TMV particles were taken up by dendritic cells, thus inducing immune activation (Kemnade et al. 2014, Liu et al. 2013). Similarly, TMV particles were used as adjuvant as well as an epitope display system against *Francisella tularensis*, a facultative intracellular pathogen. This was shown to be protective against a high dosage of *F. tularensis* Live Vaccine Strain LVS (Banik et al. 2015) when injected into mice. Another deconstructed TMV vector was used to generate a HPV vaccine with the L1 protein displayed on the surface of VLPs which could serve as a safe, low-cost vaccine for developing countries (Ortega-Berlanga et al. 2015, Zahin et al. 2016).

Deconstructed MagnICON TMV vectors containing the scFv subunit and full-length idiotype IgG molecules expressed as heavy and light chains against non-Hodgkin's lymphoma (NHL) (McCormick et al. 2008) were shown to be assembled into complete immunoglobulins in plants (McCormick 2011, Bendandi et al. 2010). This vaccine has passed through Phase I clinical trials successfully and is safe with few adverse side-effects. Also, this vaccine can be easily and quickly produced in less than 3 months (Tuse 2011, Tuse et al. 2015, Klimyuk et al. 2012).

TMV has been engineered against tumors by incorporating the tumor homing peptide cRGD into the virus surface, which enables them to be internalized into tumor cells. Also, anti-cancer drugs such as doxorubicin, when conjugated to TMV were shown to be internalized by cancer cells and subsequently TMV expression vectors when co-expressed with genes determining the plant cell cycle progression yielded enhanced virus production and the corresponding augmentation of foreign protein expression. These TMV vectors either in their full or deconstructed forms have been used for transient expression of foreign antigens of which systems such as GENEWARE_ (Shivprasad et al. 1999), magnICON_ (Gleba et al. 2007), TRBO (Lindbo 2007a) and TMV launch vector (Musiychuk et al. 2007) are important examples.

Co-delivery of expression constructs coding for the TMV-scFv-TM43-E10 and the cell cycle checkpoint genes At-CycD2/At-CDC27a from Arabidopsis thaliana, augmented the levels of the scFv-TM43-E10 antibody fragment in N. benthamiana (Kopertekh and Schiemann 2019). Earlier studies by Marillonnet et al. 2005 showed that deletion of cryptic splice sites and inclusion of introns in the TMV sequence resulted in a 712-fold enhancement in viral replication efficiency. Also, ectopic expression of TBSV p19 silencing suppressor increased heterologous foreign gene expression by 100–fold (Lindbo 2007b). Alternatively, suppression of plant genes involved in the antiviral response such as the 33K subunit of the photosystem II oxygen-evolving complex augmented TMV accumulation by 10–fold (Abbink et al. 2002).

Shi et al. 2019 developed a novel TMV-based expression vector that enables rapid and simple cloning of foreign ORFs using the Gibson assembly reaction (Gibson et al. 2009). The Gibson assembly reaction has proved to be highly efficient for insertion of several DNA fragments into a plasmid within a single step provided that the fragments possess appropriate overlapping ends. This reaction therefore enables the insertion of molecular tags and protein domains. Using this technique, two anti-fungal proteins (AFPs), Aspergillus giganteus AFP and the Penicillium digitatum AfpB were produced by transient expression using this TMV-based vector in N. benthamiana which resulted in high yields of the proteins when targeted to the apoplastic space of plant cells (Shi et al. 2019). This process also simplified the downstream purification of these proteins. These AFPs were found to be effective against target pathogens and could protect tomato plants from grey mold infection by Botrytris cinerea (that causes great economic losses), showing that plants could be used as appropriate biofactories for commercial production of these AFPs. Also, in addition to vacuum and syringe infiltration of Agrobacterium harboring this expression vector, the plants were effectively inoculated by spraying the leaves with Agrobacterial cultures similar to that reported by Hahn et al. 2015 for the MagnICON system, thus simplifying the infection process.

Furthermore, the AFPs were stable for up to 2 months at room temperature when the N. benthamiana leaves were stored as tissue prior to purification. Thus, the AFP production step can be decoupled from the purification step, enabling the leaves to be easily stored ahead of purification. This is a safe and economically viable procedure. Since there is no CP produced within this TMV vector, there are no virus particles and therefore no virus spread into the environment. Additionally, the AFP genes are not integrated into the genome of the plant ruling out any undesirable inherited traits. The application of the green AFPs reported here works as a viable alternative to chemical fungicides which makes this procedure environment friendly. In another reported study, Kagale et al. 2012 used Gateway cloning to insert recombinant ORFs into TMV-based vectors containing epitope molecular tags and fluorescent proteins.

The recombinant human plasminogen activator (rhPA) was recently (Ma et al. 2019) overexpressed transiently in tobacco leaves using a TMV-based vector. At 7 days post-inoculation, rhPA formed about 0.6% of the total soluble protein in fresh leaf biomass and was found to be enzymatically active in converting inactive plasminogen into its active form, plasmin. This TMV-based expression system can therefore be a rapid, economically viable alternative for large-scale production of rhPA.

Potexvirus Expression Vectors

The Potexvirus, Potato Virus X (PVX) has been widely engineered as an expression vector for generating biopharmaceuticals. Potato virus X (PVX), the type member of Potexviruses, contains a 6.4 kb RNA genome and a protein coat composed of ~ 1270 capsid protein subunits that are arranged helically to form flexible, filamentous virions (Huisman et al. 1988, Orman et al. 1990, Skryabin et al. 1988). The C-termini of all the CP subunits are located within the interior of the PVX particle while their N-termini are exposed on the surface of the virus particle. The latter provides an ideal site for antigen/epitope presentation of foreign proteins to the immune system (Kendall et al. 2013, Sober et al. 1988, Baratova et al. 1992a, b, Nemykh et al. 2008, Parker et al. 2002). Several modes of expression of foreign antigens in the form of full-length proteins, CP fusion proteins, foreign epitopes exhibited on the external surface of the assembled PVX particle and of late, PVX nanoparticles have been engineered to combat various diseases. Shukla et al. 2015, have shown that PVX nanoparticles can inhibit tumor progression using animal models. Further, PVX has been engineered to generate N and M proteins of Severe Acute Respiratory Syndrome Coronavirus (SARS-CoV) in plants (Demurtas et al. 2016), which can be used to detect the emergence of antibodies specific to these SARS virus proteins in SARS-infected patients.

Other important developments regarding the use of PVX vectors for the production of human vaccines include the H1N1 Influenza Virus, whose extracellular domain epitope (M2e) of the virus matrix protein 2 has been fused to a robust mucosal adjuvant, the bacterial flagellin to augment its immunogenicity (Mardanova et al. 2015). Experiments on animal models showed that this vaccine was successful in protecting mice against the H1N1 influenza infection. Using *N. benthamiana* plants, it was demonstrated that this PVX vector could produce the fusion protein up to 30% of the total soluble protein and as much as 1 mg/g fresh leaf tissue. PVX has also been engineered to produce fusion protein containing the HVR1 epitope of the Hepatitis C Virus (HCV). By parenteral administration, mice were shown to successfully exhibit a strong IgG immune response. Chronic HCV patient sera also exhibited positive reaction with the PVX-HVR1 epitope (Uhde-Holzem et al. 2010). In another report, epitopes from the HCV core antigen were fused to Hepatitis B virus surface antigen (HBsAg) and these chimeric constructs were expressed using the CP promoter of an engineered PVX vector (Mohammadzadeh et al. 2016).

Of late, PVX-based nanoparticles have been designed for anti-tumor immunotherapy. Due to its rod-shaped structure, PVX nanofilaments have been generated to carry Trastuzumab (Herceptin monoclonal antibodies) as fusion proteins into patients having breast cancer (Esfandiari et al. 2016). This was demonstrated to cause apoptosis in breast cancer cell lines. HPV16 E7 oncoprotein mutant was expressed in PVX as a fusion with lichenase that exhibited robust CTL immune response (Plchova et al. 2011, Venuti et al. 2015) and blocked tumor growth. The capsid protein of PVX by virtue of the presence of external lysine residues, can be functionalized to carry conjugates. In another study, PVX was shown to accumulate inside solid tumors, which renders it suitable for tumor-specific drug delivery and imaging (Lico et al. 2015).

Another Potexvirus, the Papaya mosaic virus (PapMV) has been employed to express the envelope protein epitope of HCV which exhibited a protracted humoral immune response in mice (Denis et al. 2007). The influenza virus M2e epitope has been fused to the external surface of PapMV nanoparticles (Denis et al. 2008) that acts as a carrier for a potential influenza vaccine. PapMV also possesses adjuvant/immunostimulatory properties and can additionally act as ideal nanoparticles. These nanoparticles can induce strong α-IFN-dependent response and when injected into tumors, can decrease the progression of melanomas and extend better survival rates in animal studies (Venuti et al. 2015, Lico et al. 2015, Lebel et al. 2016, Lacasse et al. 2008, Hanafi et al. 2010). PapMV is inherently and rapidly taken in by antigen-presenting cells through endocytosis which results in strong CD8+ T cell immune response. Demonstrably, PapMV nanoparticles when administered systemically to mice, followed by the injection of B16-OVA cells after 6 hours showed a decreased number of tumor nodules in mice (Lebel et al. 2016). The immune protection strictly required the entire assembled PapMV nanoparticle itself, and the PapMV capsid monomers or naked viral RNA could not induce any immune response.

The potexvirus PVX has been engineered to generate a movement-deficient deconstructed vector capable of coding for TM43-E10 diagnostic antibodies against the Salmonella Typhimurium OmpD protein (Kopertekha et al. 2018). This PVX vector had the added advantage of encoding the Poa semilatent virus gb silencing suppressor to enhance the recombinant antibody production. Also, unlike the deleted PVX Vector generated by Giritch et al. 2006, this study used a fusion of the TGB1 and capsid protein subgenomic promoter to drive the expression of the foreign gene while being debilitated in terms of virus movement between cells as well as systemic movement through the plant. N. benthamiana plants were used to successfully produce both the scFv-TM43-E10 and scFv-Fc-TM43-E10 antibody derivatives using this PVX vector. These antibodies displayed similar antigen-recognition specificity as that of its counterparts generated in microbial or mammalian cellular systems and could successfully recognize the S. Typhimurium OmpD antigen. This PVX expression vector expressing diagnostic antibodies has emerged to be of great importance in animal husbandry in light of the pressing need for increased surveillance of bacterial/Salmonella-based food contamination amidst the emergence of drug resistant strains.

Several RNA silencing suppressors such as the gb from Barley stripe mosaic virus (BSMV) (Gao et al. 2013), HcPro from Potato virus Y (PVY) (Vezina et al. 2009), p25 from PVX (Lombardi et al. 2009) and p19 from either Artichoke mottled crinkle virus (AMCV) (Villani et al. 2009) or Tomato bushy stunt virus (TBSV) (Garabagi et al. 2012), have been produced to augment foreign gene expression. Among these, p19 is the most powerful silencing inhibitor.

The surface properties of plant virus nanoparticles can be genetically modified or chemically conjugated to make them serve as vehicles for drug-delivery as well as for electronic devices and tissue scaffolds (Lico et al. 2015, Wen and Steinmetz 2016, Lomonossoff and Evans 2011, Culver et al. 2015, Steele et al. 2017, Montague et al. 2011). Viral nanoparticles that are fluorescently tagged can be used in biomedical imaging (Brunel et al. 2010, Lee et al. 2013, Lewis et al. 2006, Manchester and Singh 2006, Shukla et al. 2014, Wen et al. 2012, Niehl et al. 2016). Several investigations have shown that PVX particles conjugated to the molecular contrast agent, Alexa Fluor 647 can effectively penetrate and accumulate in solid tumors (Shukla et al. 2013, Shukla et al. 2014, Shukla et al. 2016).

One strategy to tag the CP could be to directly fuse the amino acid or peptide to the N-terminus of the CP. Nonetheless, fusion of longer peptides may result in steric hindrance, thus requiring the obligate presence of wild type CP, resulting in the generation of hybrid virus particles composed of both the wild type as well as the CP fusion proteins. For this purpose, the FMDV 2A sequence is inserted between the transgene and the CP gene sequence obtaining the CP fusion protein, the free target protein and the wild type CP that is generated due to a ribosomal skip in the course of translation (the overcoat method) (Cruz et al. 1996, Donnelly et al. 2001a, b). Recently, a PVX-mCherry overcoat strategy for live tumor imaging in mouse models was reported (Shukla et al. 2014).

While fluorescent proteins provide insights into several cellular mechanisms and molecular interactions, they are large with complex structures, displaying pH and oxygen dependence (Gawthorne et al. 2012, Shaner et al. 2007). The iLOV polypeptide shares similarity to the GFP in terms of spectral characteristics (Christie 2007). However, it is monomeric, with a lower molecular weight, is more photostable and can undergo reversible photobleaching which makes it ideal for tumor imaging purposes (Christie et al. 2012, Mukherjee et al. 2013, Khrenova et al. 2015, Chapman et al. 2008) and for antiviral drug screening (Dang et al. 2015). The iLOV fluorescent protein can be targeted to the surface of the PVX particles by fusing it to the N-terminus of the PVX capsid protein and has many advantages over the other available fluorescent proteins and delivers a new platform for tumor imaging.

High expression levels of the human interferon gamma (IFN-gamma), a medically important therapeutic protein was obtained using a Bamboo Mosaic Potexvirus (BaMV)-based expression vector (Jiang et al. 2019) in N. benthamiana plants. This vector was engineered to delete the capsid protein of the BaMV that resulted in augmented IFN-gamma synthesis. Subsequently, upon deletion of the N-terminal signal peptide of the IFN-gamma, now labeled as mIFN-gamma, a 7-fold higher IFN protein yield was obtained. Further enhancement of mIFN-gamma expression (up to 40%) was produced by replacing the gene for the BaMV movement polypeptide with the P19 silencing suppressor gene. Also, when the endoplasmic reticulum (ER) retention signal was fused with the mIFN-gamma, this increased the accretion of the mIFN-gamma in its biologically active dimeric form to 87% compared to its non-active monomeric equivalent. Finally, the construct harboring all of the above modifications (designated pKB19mIFN-gamma-ER), yielded, up to 119 ± 0.8 μg/g fresh weight, equal to 2.5% of the total soluble protein (TSP). This study proved that BaMV-based expression vectors can be used to drive the large-scale production of therapeutic proteins in N. benthamiana.

Comovirus Expression Vectors

Cowpea Mosaic Virus, a Comovirus is an icosahedral virus with a diameter of 30 nm whose protein coat is composed of 60 large and 60 small capsid subunits. CPMV is bipartite with the genome split into RNA-1 and RNA-2, with the latter being the main target for genetic engineering towards generating biopharmaceuticals. CPMV has been extensively used for presentation of foreign antigens both as full-length proteins as well as fusion proteins that can be proteolytically cleaved to release the biopharmaceutical protein. CPMV is also used in building biosensors and magnetic clusters (Sainsbury et al. 2008, Sainsbury et al. 2009a, b, Lebedev et al. 2016, Jaafar et al. 2014).

CPMV VLPs displaying Influenza Virus HA antigens (Mardanova et al. 2016) on their surface have been generated by Medicago, Inc. (Durham, NC, USA) and these VLPs have been tested in animal models wherein they are afforded protection against lethal doses of the virus. Subsequently, they are being tested in humans and are at the stage of Phase 2 clinical trials. Medicago's CPMV system generates Influenza vaccine rapidly within 3 weeks of the discovery of the virus strain sequence along with an easily resilient upscaling capability.

The pEAQ system derived from the deconstructed CPMV vector is used in foreign protein expression while dispensing with the necessity for viral replication (Montague et al. 2011, Meshcheriakova et al. 2014). This relieves the target cells from supporting the virus life cycle and therefore from any negative impact on foreign antigen accretion. For this purpose, the gene for the foreign antigen is inserted into RNA-2 between the 5' leader and the 3' UTR (untranslated region) while also deleting an in-frame AUG initiation codon present upstream of the major translation initiation site. This modified CPMV vector has been shown to considerably augment antigen production (Peyret and Lomonossoff 2013).

The new Cowpea Mosaic Virus Hyper-Translational expression system called pCPMV-HT, affords enhanced protein translation efficiency (Sainsbury et al. 2010, Vardakou et al. 2012) and has been used to produce vaccines against the Bluetongue Virus, Dengue Virus, HIV and Influenza Virus (Thuenemann et al. 2013a, Brillault et al. 2017). This is in contrast to the TMV-based vector system which needs RNA replication and cell-to-cell movement that leads to an increased likelihood of genetic instability with the increasing size of the recombinant insert. The non-replicating pEAQ system overcomes this problem while at the same time generating tremendously high protein levels through the engineered 5' UTR.

Vimentin molecules present on the HeLa cell surface have been shown to bind to CPMV particles and CPMV can be endocytosed by antigen-presenting cells triggering antibody response (Koudelka et al. 2009, Steinmetz et al. 2011, Yildiz et al. 2013). Like PapMV, CPMV have been shown to stimulate immune system-based anti-cancer effects. CPMV capsid proteins can self-assemble in the absence of the viral genomic RNA to form empty virus-like particles (eVLPs), which when injected into tumors demonstrably potentiate anti-tumor immunity by stimulating the quiescent neutrophils within the target site. As this immune reaction is restricted to the vicinity of the injected areas, adverse side-effects are precluded (Yildiz et al. 2013). In another report, mouse models were used to show that CPMV nanoparticles could act as efficient immune stimulants against melanoma, lung and other cancers wherein they activated cell-mediated anti-tumor immunity (Lizotte et al. 2016). In the same study, it was also shown that CPMV nanoparticles augmented proinflammatory cytokine levels when provided to cultures of bone marrow cells. When CPMV nanoparticles were injected on a weekly basis, it was proven to diminish tumor burden in ovarian, breast and colon tumors. CPMV can therefore be used as payload carrier for the transport of anti-cancer drugs into the heart of the tumors.

pEAG-HT, a CPMV expression vector has be employed to express the rat epidermal growth factor-related protein ErbB2 extracellular domain in tobacco plants (Mati´c et al. 2016). Aberrant expression of ErbB2 can potentiate cancer development. Tobacco plant extracts expressing the CPMV-based ErbB2, elicited strong immune reaction and antitumor effects when injected into the tumors.

The African horse sickness virus (AHSV) is prevalent in much of southern Africa and Europe where it has an enormous impact on the equine industry. Therefore there is a compelling need to generate effective, safe and economical DNA vaccines. Recently (Dennis et al. 2018), a cowpea mosaic virus-based HyperTrans (CPMV-HT) and the associated pEAQ plant expression vector system has been used to express four capsid proteins of the African horse sickness virus (AHSV, a dsRNA orbivirus) through Agrobacterium-mediated transient expression that led to the formation of VLPs that mimic the structure of native virions and synthesis of the AHSV serotype five candidate vaccine. This system proved to be simple and rapid along with the favorable scalability and economic viability. Antiserum generated against this vaccine in guinea pigs was able to neutralize the live AHSV in cell-based *in vitro* assays.

VLPs are advantageous in being safe to administer, are highly immunogenic by virtue of being nonreplicating protein assemblies wherein the virus epitopes are displayed on the surface of the particles in ordered repetitive arrays (Noad and Roy 2003). VLP-based vaccines pose no risk of genetic reversion to virulence nor of reassortment of dsRNA segments with wild type virus strains as they are bereft of the viral RNA genome as well as non-structural proteins (Lomonossoff and D'Aoust 2016, Rybicki 2010, 2014, Steele et al. 2017, Topp et al. 2016). In another study, Medicago Inc. demonstrated the transient expression of VLPs of the rotavirus in N. benthamiana by the expression and assembly of the four rotavirus capsid proteins that led to formation of these VLPs (D'Aoust et al. 2013).

The pEAQ system has been mainly used for transient expression by agro-infiltration of leaves. However, it has been shown to useful for protein synthesis in stably transformed plants and cell cultures as well. The TBSV p19 gen silencing suppressor was seen to augment protein expression levels in these plants. Enhanced levels of proteins such as DsRed, GFP, human anti-HIV 2G12 antibody, HBV core antigen (HBcAg) (Sainsbury and Lomonossoff 2008, Sainsbury et al. 2010) and Influenza hemagglutinin (HA) (D'Aoust et al. 2009) have been expressed using this system. Presently, industrial levels of Influenza HA are being produced in the pEAQ vector which are now being considered seriously for clinical trials and commercial production (D'Aoust et al. 2010).

The Bovine Papilloma Virus Type 1 (BPV-1) L1 major capsid protein was synthesized in N. benthamiana using the pEAQ-HT system with tissue accumulation levels of 224 mg/kg fresh weight and a recovered yield of 183 mg/kg FWT (Love et al. 2012). In this instance, it was demonstrated that the L1 protein self-assembled into VLPs and induced a robust immune reaction specific to the BPV-1 in rabbits.

The pEAQ expression system has been successfully employed to generate the Bluetongue virus (BTV) VLP. The BTV codes for proteins VP3 and VP7 that together form CLPs or core-like particles. Then the BTV VP2 and VP5 polypeptides assemble on the surface of the core to make the VLPs which have been shown to elicit neutralizing immune response in sheep (Thuenemann et al. 2013a). The pEAQ vector was shown to be most appropriate for the synthesis of the VP3/7 CLPs and the VP3/7/2/5 VLPs. Also, the pEAQ system was successful in modulating the expression of all of these four proteins such that it provided the correct ratio of expression to enable efficient VLP formation.

Apart from being useful to generate animal virus VLPs, the pEAQ system has also been shown to drive the synthesis of CPMV empty VLPs (eVLPs) free of the RNA genome in plants. Formation of these eVLPs required the synthesis of VP60 capsid protein precursor as well as the 24 K protease encoded by the CPMV RNA-1 wherein the latter processed the VP60 into the mature CP S and L proteins (Saunders et al. 2009). The CPMV eVLPs were shown to be identical in structure to the wild type particles except for the absence of the RNA genome. High yields of the CPMV eVLPs was reported by Montague et al. 2011 in which both the VP60 and 24 K protease genes were placed within the same T-DNA construct. These eVLPs show great promise in being able to be loaded with molecular cargo, while at the same time this non-replicating system enables modifications of the CPMV capsid proteins for display on its surface or for targeting of these particles. Sainsbury et al. 2011 showed efficient mineralization of the CPMV eVLPs with cobalt at both internal and external regions of the eVLPs.

The Influenza virus M2e surface epitope was fused to the HBcAg immunodominant e1 loop which yielded nearly 15–50 mg/kg of the recombinant VLPs using the pEAQ system (Thuenemann et al. 2013b). Matic et al. 2011 reportedly generated chimeric VLPs in N. benthamiana wherein the Influenza A M2e antigen was expressed on the carrier Human papillomavirus type 16 (HPV) L1 protein. The yield of the VLPs was up to 120 mg/kg FWT and the VLPs could be recognized by monoclonal antibodies specific to both HPV and Influenza viruses. Another Influenza virus protein, the hemagglutinin (HA) is a viable candidate for generating strain-specific vaccines unlike the M2e which shows low variability. Further, the pEAQ was used for transient expression of avian Influenza A H7N7 HA protein to obtain a yield of 200 mg/kg FWT.

In addition to viral proteins, enzymes such as human gastric lipase (hGL) have been produced in high titers using the pEAQ expression system. Up to 0.5g/kg FWT of hGL has been produced with a maximum enzyme activity of 193 U/g FWT (Vardakou et al. 2012) through transient expression in N. benthamiana. This recombinant hGL has molecular properties similar to the native human-derived protein in terms of being stable at 40C and at low pH, has higher affinity for short-chain lipids than for long ones in addition to being recalcitrant to pepsin digestion. The pEAQ vector has also been used to produce recombinant OsChia4a, a rice chitinase in N. benthamiana wherein its anti-fungal activity was studied in addition to the molecular control of its gene by the plant hormone jasmonic acid (Miyamoto et al. 2012).

The pEAQ vectors have also been used for impacting entire biochemical synthesis pathways through co-production of multiple enzymes. The genes for two sesquiterpene synthases, amorpha-3,11-diene synthase (ADS) and *epi*-cedrol synthase (ECS) from *Artemisia annua*, have been expressed using the pEAQ vector in N. benthamiana in which rich yields of 90 and 96 mg/kg FWT of the two enzymes respectively were obtained (Kanagarajan et al. 2012a). These enzymes were proved to be biologically active in converting farnesyl diphosphate into amorpha-4,11-diene and *epi*-cedrol respectively.

These reaction products are precursor intermediates in the biochemical pathways leading to the synthesis of the antimalarial drug, artemisinin, proving that these plant-produced enzymes can successfully function in planta with authentic reaction specificity.

Two more enzymes, the oxidosqualene cyclase (OSC) and the cytochrome P450 (CYP450) from the monocot Avena sativa (oat) have been expressed using the pEAQ vectors (Sainsbury et al. 2012) and they showed an authentic reaction specificity in planta. The enzyme OSC generated the product ß-Amyrin and upon co-expression of both enzymes, a new compound was synthesized. Thus, the pEAQ system can be used not just to obtain significant amounts of several enzymes, but also to track their modes of activities while creating novel reaction products.

Furthermore, proteins have been over-expressed in cell cultures using the pEAQ vectors. Sun et al. 2011 have generated recombinant human serum albumin in transgenic tobacco Bright Yellow-2 (BY-2) cell suspension cultures wherein protein yields of 22.1 mg/l of growth medium (or 0.7% total soluble protein) were obtained. However, the pEAQ system is more suited for transient expression of recombinant proteins/enzymes through agroinfiltration of leaves even though they have been successful for protein expression in hairy root cultures and cell suspension cultures as well.

Conclusions

Traditional vaccines have been successful against many, but not all infectious diseases. For diseases such as hepatitis C, HIV/AIDS, malaria, tuberculosis and cancer, vaccines have been less effective. Recently, researchers have explored non-infectious plant virus-like nanoparticles (VNPs) for the activation of the immune response in humans. Besides their size, VNPs have highly repetitive molecular structures and therefore a tendency to initiate an immune response. At least three vaccines composed of VLPs are now used in humans; hepatitis B vaccine, human papilloma vaccine, and a vaccine against hepatitis E (Riedmann 2012). Recent studies have shown the use of plants, such as tobacco, spinach, lettuce, and tomato for low-cost vaccine production at a commercial scale (Narayanan and Han 2018). Plant viruses can be used to carry targeted antigenic epitopes to induce an immune response against pathogens (Lebel et al. 2015).

In recent years, the efficiency of several nanoparticles has also been examined based on their drug delivery mechanisms to accurately and specifically target the desired tissues. These were tested for their stability, carrying capacity, and release rate of the target drug. Nanoparticles have also been analyzed for their impacts on both the biological and non-biological environments involved in the process (Iravani 2011, Ju-Nam and Lead 2008). Empty coat proteins assembled into VLPs can encapsulate a drug depending on the interior space available. These nanostructured molecules are capable of holding large drug payloads, which are useful for drug delivery and various other medical applications (Roder et al.2019). Spherical particles are more suitable to carry large loads while rod shaped particles are more helpful to target molecular receptors (Steinmetz 2010). A pH-dependent gating mechanism has been introduced for loading the cargo into VLPs. The drug molecule can also be covalently bonded, orthogonally or through hydrazine ligation to the capsid protein for drug delivery, instead of being loaded inside the capsid (Alemzadeh et al. 2017).

As outlined in this chapter, biopharmaceuticals based upon plant virus expression vectors and virus nanoparticles have many potential applications in medicine. Expression vectors are rapid and facile to use, and do not require refrigeration or sophisticated instrumentation. Plant virus nanoparticles are biodegradable, cheap, and can carry drug payloads to destroy tumor cells. The time has come for plant made biopharmaceuticals to make an impact on the commercial stage.

References

Abbink, T.E., J.R. Peart, T.N. Mos, D.C. Baulcombe, J.F. Bol, and H.J. Linthorst. 2002. Silencing of a gene encoding a protein component of the oxygen-evolving complex of photosystem II enhances virus replication in plants. Virology 295: 307–319.

Alemzadeh, E., K. Izadpanah, and F. Ahmadi. 2017. Generation of recombinant protein shells of Johnson grass chlorotic stripe mosaic virus in tobacco plants and their use as drug carrier. J. Virol. Methods 248: 148–153. doi: 10.1016/j.jviromet.2017.07.003.

Atabekov, J., N. Nikitin, M. Arkhipenko, S. Chirkov, and O. Karpova. 2011. Thermal transition of native tobacco mosaic virus and RNA-free viral proteins into spherical nanoparticles. Journal of General Virology 92(2): 453–456.

Azizgolshani, O., R.F. Garmann, R. Cadena-Nava, C.M. Knobler, and W.M. Gelbart. 2013. Reconstituted plant viral capsids can release genes to mammalian cells. Virology 441(1): 12–7. doi: 10.1016/j.virol.2013.03.001.

Babin, C., N. Majeau, and D. Leclerc. 2013. Engineering of papaya mosaic virus (PapMV) nanoparticles with a CTL epitope derived from influenza NP. Journal of Nanobiotechnology 11(1): 10.

Baltes, N.J., J. Gil-Humanes, T. Cermak, P.A. Atkins, and D.F. Voytas. 2014. DNA replicons for plant genome engineering. Plant Cell. 26: 151–63.

Baratova, L.A., N.I. Grebenshchikov, and E.N. Dobrov. 1992a. The organization of potato virus X coat proteins in virus particles studied by tritium planigraphy and model building. Virology 188(1): 175–180.

Baratova, L.A., N.I. Grebenshchikov, and A.V. Shishkov. 1992b. The topography of the surface of potato virus X: Tritium planigraphy and immunology analysis. Journal of General Virology 73, part 2, pp. 229–235.

Bendandi, M., S. Marillonnet, R. Kandzia, F. Thieme, A. Nickstadt, S. Herz, R. Fröde, S. Inogés, A. Lòpez-Dìaz de Cerio, and E. Soria. 2010. Rapid, high-yield production in plants of individualized idiotype vaccines for non-Hodgkin's lymphoma. Ann. Oncol. 21: 2420–2427.

Betti, C., and D. Lico, Maf. 2012. Potato virus X movement in Nicotiana benthamiana: New details revealed by chimeric coat protein variants. Molecular Plant Pathology 13(2): 198–203.

Brillault, L., P.V. Jutras, N. Dashti, E.C. Thuenemann, G. Morgan, G.P. Lomonossoff, M.J. Landsberg, and F. Sainsbury. 2017. Engineering recombinant virus-like nanoparticles from plants for cellular delivery. ACS Nano. 11: 3476–3484.

Bruckman, M.A., A.E. Czapar, A. VanMeter, L.N. Randolph, and N.F. Steinmetz. 2016. Tobacco mosaic virus-based protein nanoparticles and nanorods for chemotherapy delivery targeting breast cancer. J. Control. Release 231: 103–113.

Brunel, F.M., J.D. Lewis, and G. Destito. 2010. Hydrazone ligation strategy to assemble multifunctional viral nanoparticles for cell imaging and tumor targeting. Nano Letters 10(3): 1093–1097.

Chapman, S., C. Faulkner, and E. Kaiserli. 2008. Te photoreversible fuorescent protein iLOV outperforms GFP as a reporter of plant virus infection. Proceedings of the National Acadamy of Sciences of the United States of America 105(50): 20038–20043.

Christie, J.M. 2007. Phototropin blue-light receptors. Annual Review of Plant Biology 58: 21–45.

Christie, J.M., K. Hitomi, and A.S. Arvai. 2012. Structural tuning of the fuorescent protein iLOV for improved photostability. The Journal of Biological Chemistry 287(26): 22295–22304.

Cruz, S.S., S. Chapman, A.G. Roberts, I.M. Roberts, D.A.M. Prior, and K.J. Oparka. 1996. Assembly and movement of a plant virus carrying a green fuorescent protein overcoat. Proceedings of the National Academy of Sciences of the United States of America 93(13): 6286–6290.

Culver, J.N., A.D. Brown, F. Zang, M. Gnerlich, K. Gerasopoulos, and R. Ghodssi. 2015. Plant virus directed fabrication of nanoscale materials and devices. Virology 479-480: 200–212.

Dang, X., S. Chalkias, and I.J. Koralnik. 2015. JC virus-iLOV fuorescent strains enable the detection of early and late viral protein expression. Journal of Virological Methods 223: 25–29.

D'Aoust, M.-A., M. Couture, F. Ors, S. Trépanier, P.-O. Lavoie, M. Dargis, L.-P. Vézina, and N. Landry. 2009. Recombinant Influenza Virus-like particles (VLPs) produced in transgenic plants expressing hemagglutinin. E.P. Office, ed, EP 2 238 253 B1.

D'Aoust, M.-A., M.M.J. Couture, N. Charland, S. Trépanier, N. Landry, F. Ors, and L.-P. Vézina. 2010. The production of hemagglutinin-based virus-like particles in plants: A rapid, efficient and safe response to pandemic influenza. Plant Biotechnol J. 8: 607–619.

D'Aoust, M.-A., N. Landry, P.-O. Lavoie, M. Arai, N. Asahara, D.L.R. Mutepfa, I.I. Hitzeroth, and E.P. Rybicki. 2013. Rotavirus-like particle production in plants. Google Patents EP2847324A1.

Demurtas, O.C., S. Massa, E. Illiano, D. De Martinis, P.K. Chan, P. Di Bonito, and R. Franconi. 2016. Antigen production in plant to tackle infectious diseases flare up: The case of SARS. Front. Plant. Sci. 7: 54.

Denis, J., N. Majeau, E. Acosta-Ramirez, C. Savard, M.C. Bedard, S. Simard, K. Lecours, M. Bolduc, C. Pare, and B. Willems. 2007. Immunogenicity of papaya mosaic virus-like particles fused to a hepatitis C virus epitope: Evidence for the critical function of multimerization. Virology 363: 59–68.

Denis, J.E., Y. Acosta-Ramirez, M.E. Zhao, I. Hamelin, M. Koukavica, Y. Baz, C. Abed, C. Savard, C. Pare, and C. Lopez Macias. 2008. Development of a universal influenza A vaccine based on the M2e peptide fused to the papaya mosaic virus (PapMV) vaccine platform. Vaccine 26: 3395–3403.

Dennis, S.J., A.E. Meyers, A.J. Guthrie, I.I. Hitzeroth, and E.P. Rybicki. 2018. Immunogenicity of plant-produced African horse sickness virus-like particles: Implications for a novel vaccine. Plant Biotechnology Journal 16: 442–450.

Donnelly, M.L.L., L.E. Hughes, and G. Luke. 2001a. Te 'cleavage' activities of foot-and-mouth disease virus 2A site-directed mutants and naturally occurring '2A-like' sequences. Journal of General Virology, vol. 82, part 5, pp. 1027–1041.

Donnelly, M.L.L., G. Luke, and A. Mehrotra. 2001b. Analysis of the aphthovirus 2A/2B polyprotein 'cleavage' mechanism indicates not a proteolytic reaction, but a novel translational efect: A putative ribosomal 'skip'. Journal of General Virology, vol. 82, part 5, pp. 1013–1025.

Esfandiari, N., M.K. Arzanani, M. Soleimani, M. Kohi-Habibi, and W.E. Svendsen. 2016. A new application of plant virus nanoparticles as drug delivery in breast cancer. Tumour. Biol. 37: 1229–1236.

Fausther-Bovendo, H., and G.P. Kobinger. 2014. Pre-existing immunity against Ad vectors: Humoral, cellular, and innate response, what's important? Human Vaccines & Immunotherapeutics 10(10): 2875–2884.

Gao, S.-J., M.B. Damaj, J.-W. Park, G. Beyene, M.T. Buenrostro-Nava, J. Molina, X. Wang, J.J. Ciomperlik, S.A. Manabayeva, V.Y. Alvarado, K.S. Rathore, H.B. Scholthof, and T.E. Mirkov. 2013. Enhanced transgene expression in sugarcane by Coexpression of virus-encoded RNA silencing suppressors. PLoS One 8(6): e66046, doi:http://dx.doi.org/10.1371/journal.pone.0066046.

Garabagi, F., E. Gilbert, A. Loos, M.D. McLean, and J.C. Hall. 2012. Utility of the P19 suppressor of gene-silencing protein for production of therapeutic antibodies in Nicotiana expression hosts, Plant Biotechnol. J. 10: 1118–1128, doi: http://dx.doi.org/10.1111/j.1467-7652.2012.00742.x.

Gawthorne, J.A., L.E. Reddick, and S.N. Akpunarlieva. 2012. Express your LOV: An engineered favoprotein as a reporter for protein expression and purifcation. PLoS ONE 7(12), Article ID e52962.

Gibson, D.G., L. Young, R.-Y. Chuang, J.C. Venter, C.A. III. Hutchison, and H.O. Smith. 2009. Enzymatic assembly of DNA molecules up to several hundred kilobases. Nat. Methods 6: 343.

Giritch, A., S. Marillonnet, C. Engler, G. van Eldik, J. Botterman, V. Klimyuk, and Y. Gleba. 2006. rapid high-yield expression of full-size IgG antibodies in plants coinfected with noncompeting viral vectors, Proc. Natl. Acad. Sci. U. S. A. 103.2006. 14701–14706, doi:http://dx.doi.org/10.1073/pnas.0606631103.

Gleba, Y., S. Marillonnet, and V. Klimyuk. 2004. Engineering viral expression vectors for plants: The 'full virus' and the 'deconstructed virus' strategies. Curr. Opin. Plant Biol. 7: 182–188, doi:http://dx.doi.org/10.1016/j.pbi.2004.01.003.

Gleba, Y., V. Klimyuk, and S. Marillonnet. 2007. Viral vectors for the expression of proteins in plants. Curr. Opin. Biotechnol. 18: 134–41.

Goff, S.P., and P. Berg. 1976. Construction of hybrid viruses containing SV40 and lambda phage DNA segments and their propagation in cultured monkey cells. Cell 9(4 pt 2): 695–705. doi: 10.1016/0092-8674(76)90133-1.

Hahn, S., A. Giritch, D. Bartels, L. Bortesi, and Y. Gleba. 2015. A novel and fully scalable Agrobacterium spray-based process for manufacturing cellulases and other cost-sensitive proteins in plants. Plant Biotechnol. J. 13: 708–716.

Hanafi, L.A., M. Bolduc, M.E. Gagné, F. Dufour, Y. Langelier, M.R. Boulassel, J.P. Routy, D. Leclerc, and R. Lapointe. 2010. Two distinct chimeric potexviruses share antigenic cross-presentation properties of MHC class I epitopes. Vaccine 28: 5617–5626.

Hefferon, K. 2017. Plant virus expression vectors: A powerhouse for global health. Biomedicines 5: 44.

Huang, Z., W. Phoolcharoen, H. Lai, K. Piensook, G. Cardineau, L. Zeitlin, K.J. Whaley, C.J. Arntzen, H.S. Mason, and Q. Chen. 2010. High-level rapid production of full-size monoclonal antibodies in plants by a single-vector DNA replicon system. Biotechnol. Bioeng. 106: 9–17, doi:http://dx.doi.org/10.1002/ bit.22652.

Huisman, M.J., H.J. Linthorst, J.F. Bol, and J.C. Cornelissen. 1988. Te complete nucleotide sequence of potato virus X and its homologies at the amino acid level with various plus-stranded RNA viruses. Journal of General Virology, vol. 69, part 8, pp. 1789–1798.

Ishibashi, K., and M. Ishikawa. 2016. Replication of tobamovirus RNA. Annu. Rev. Phytopathol. 54: 55–78.

Iravani, S. 2011. Green Synthesis of Metal Nanoparticles using Plants. Vol. 13.

Jaafar, M., A.A. Aljabali, I. Berlanga, R. Mas-Ballesté, P. Saxena, S. Warren, G.P. Lomonossoff, D.J. Evans, and P.J. de Pablo. 2014. Structural insights into magnetic clusters grown inside virus capsids. ACS Appl. Mater. Interfaces 6: 20936–20942.

Jiang, M.C., C.C. Hu, N.S. Lin, and Y.H. Hsu. 2019. Production of human IFNγ protein in *Nicotiana benthamiana* plant through an enhanced expression system based on *Bamboo mosaic* virus. Viruses. 2019 Jun 3;11(6). pii: E509. doi: 10.3390/v11060509.

Ju-Nam, Y., and J.R. Lead. 2008. Manufactured nanoparticles: An overview of their chemistry, interactions and potential environmental implications. Sci. Total Environ. 400(1-3): 396–414. doi: 10.1016/j.scitotenv.2008.06.042.

Kagale, S., S. Uzuhashi, M. Wigness, T. Bender, W. Yang, M.H. Borhan, and R. Rozwadowski. 2012. TMV-Gate vectors: Gateway compatible tobacco mosaic virus based expression vectors for functional analysis of proteins. Sci. Rep. 2: 874.

Kanagarajan, S., S. Muthusamy, A. Gliszczyńska, A. Lundgren, and P. Brodelius. 2012a. Functional expression and characterization of sesquiterpene synthases from *Artemisia annua* L. using transient expression system in Nicotiana benthamiana. Plant Cell Rep. 31: 1309–1319.

Kanagarajan, S., C. Tolf, A. Lundgren, J. Waldenström, and P.E. Brodelius. 2012b. Transient expression of hemagglutinin antigen from low pathogenic avian influenza A (H7N7) in Nicotiana benthamiana. PLoS ONE 7: e33010.

Kendall, A., W. Bian, and A. Maris. 2013. A common structure for the potexviruses. Virology 436(1): 173–178.

Khrenova, M.G., A.V. Nemukhin, and T. Domratcheva. 2015. Teoretical characterization of the favin-based fuorescent protein iLOV and its Q489K mutant. Te Journal of Physical Chemistry B 119(16): 5176–5183.

Klimyuk, V., G. Pogue, S. Herz, J. Butler, and H. Haydon. 2012. Production of recombinant antigens and antibodies in Nicotiana benthamiana using "Magnifection" technology: GMP-compliant facilities for small- and large-scale manufacturing. Curr. Top. Microbiol. Immunol. 375: 127–154.

Knapp, E., and D.J. Lewandowski. 2001. Tobacco mosaic virus, not just a single component virus anymore. Mol. Plant Pathol. 2: 117–123.

Koudelka, K.J., G. Destito, E.M. Plummer, S.A. Trauger, G. Siuzdak, and M. Manchester. 2009. Endothelial targeting of cowpea mosaic virus (CPMV) via surface vimentin. PLoS Pathog. 5: e1000417.

Lacasse, P., J. Denis, R. Lapointe, D. Leclerc, and A. Lamarre. 2008. Novel plant virus-based vaccine induces protective cytotoxic T-lymphocyte-mediated antiviral immunity through dendritic cell maturation. J. Virol. 82: 785–794.

Lebedev, N., I. Griva, W.J. Dressick, J. Phelps, J.E. Johnson, Y. Meshcheriakova, C. Lico, Q. Chen, and L. Santi. 2008. Viral vectors for production of recombinant proteins in plants. J. Cell Physiol. 216: 366–77.

Lebel, M.E., K. Chartrand, D. Leclerc, and A. Lamarre. 2015. Plant viruses as nanoparticle-based vaccines and adjuvants. Vaccines (Basel) 3(3): 620–37. doi: 10.3390/vaccines3030620.

Lebel, M.È., K. Chartrand, E. Tarrab, P. Savard, D. Leclerc, and A. Lamarre. 2016. Potentiating cancer immunotherapy using papaya mosaic virus-derived nanoparticles. Nano. Lett. 16: 1826–1832.

Lee, K.L., L.C. Hubbard, S. Hern, I. Yildiz, M. Gratzl, and N.F. Steinmetz. 2013. Shape matters: Te difusion rates of TMV rods and CPMV icosahedrons in a spheroid model of extracellular matrix are distinct. Biomaterials Science 1(6): 581–588.

Lefeuvre, P., D.P. Martin, S.F. Elena, D.N. Shepherd, P. Roumagnac, and A. Varsani. 2019. Evolution and ecology of plant viruses. Nature Reviews Microbiology. doi: 10.1038/s41579-019-0232-3.

Lewis, J.D., G. Destito, and A. Zijlstra . 2006. Viral nanoparticles as tools for intravital vascular imaging. Nature Medicine 12(3): 354–360.

Lico, C., F. Capuano, and G. Renzone. 2006. Peptide display on Potato virus X: Molecular features of the coat protein-fused peptide afecting cell-to-cell and phloem movement of chimeric virus particles. Journal of General Virology, vol. 87, part 10, pp. 3103–3112.

Lico, C., E. Benvenuto, and S. Baschieri. 2015. The two-faced potato virus X: From plant pathogen to smart nanoparticle. Front. Plant. Sci. 6: 1009.

Lindbo, J.A. 2007a. TRBO: A high-efficiency tobacco mosaic virus RNA-based overexpression vector. Plant Physiology, December 2007, 145: 1232–1240.

Lindbo, J.A. 2007b. High-efficiency protein expression in plants from agroinfection-compatible Tobacco mosaic virusexpression vectors. BMC Biotechnol. 7: 52.

Liu, L., and G.P. Lomonossoff. 2002. Agroinfection as a rapid method for propagating Cowpea mosaic virus-based constructs. J. Virol. Methods 105: 343–8.

Lizotte, P.H., A.M. Wen, M.R. Sheen, J. Fields, P. Rojanasopondist, N.F. Steinmetz, and S. Fiering. 2016. *In situ* vaccination with cowpea mosaic virus nanoparticles suppresses metastatic cancer. Nat. Nanotechnol. 11: 295–303.

Lombardi, R., P. Circelli, M.E. Villani, G. Burianil, L. Nardi, V. Coppola, L. Bianco, E. Benvenuto, M. Donini, and C. Marusic. 2009. High-level HIV-1 Nef transient expression in Nicotiana benthamiana using the P19 gene silencing suppressor protein of Artichoke Mottled Crinckle Virus, BMC Biotechnol. 9: 96, doi:http://dx. doi.org/10.1186/1472-6750-9-96.

Lomonossof, G.P., and D.J. Evans. 2011. Applications of plant viruses in bionanotechnology. In Plant Viral Vectors, vol. 375 of Current Topics in Microbiology and Immunology, pp. 61–87, Springer, Berlin, Germany.

Lomonossoff, G.P., and C.M. Soto. 2016. A virus-based nanoplasmonic structure as a surface-enhanced Raman biosensor. Biosens. Bioelectron. 77: 306–314.

Lomonossoff, G.P., and M.-A. D'Aoust. 2016. Plant-produced biopharmaceuticals: A case of technical developments driving clinical deployment. Science 353: 1237–1240.

Love, A., S. Chapman, S. Matić, E. Noris, G. Lomonossoff, and M. Taliansky. 2012. In planta production of a candidate vaccine against bovine papillomavirus type 1. Planta 236: 1305–1313.

Ma, J., L. Wu, X. Ding, Z. Li, and S. Wang. 2019. Transient expression of bioactive recombinant human plasminogen activator in tobacco leaf]. [Article in Chinese]. Nan. Fang Yi Ke Da Xue Xue Bao. 2019 May 30; 39(5): 515–522.

Manchester, M., and P. Singh. 2006. Virus-based nanoparticles (VNPs): Platform technologies for diagnostic imaging. Advanced Drug Delivery Reviews 58(14): 1505–1522.

Mardanova, E.S., R.Y. Kotlyarov, V.V. Kuprianov, L.A. Stepanova, L.M. Tsybalova, G.P. Lomonosoff, and N.V. Ravin. 2015. Rapid high-yield expression of a candidate influenza vaccine based on the ectodomain of M2 protein linked to flagellin in plants using viral vectors. BMC Biotechnol. 15: 42.

Mardanova, E.S., R.Y. Kotlyarov, V.V. Kuprianov, L.A. Stepanova, L.M. Tsybalova, G.P. Lomonossoff, and N.V. Ravin. 2016. High immunogenicity of plant-produced candidate influenza vaccine based on the M2e peptide fused to flagellin. Bioengineered 7: 28–32.

Mardanova, E.S., E.A. Blokhina, L.M. Tsybalova, H. Peyret, G.P. Lomonossoff, and N.V. Ravin. 2017. Efficient transient expression of recombinant proteins in plants by the novel pEff vector based on the genome of potato virus X. Front Plant Sci. 8: 1–8.

Marillonnet, S., C. Thoeringer, R. Kandzia, V. Klimyuk, and Y. Gleba. 2005. Systemic *Agrobacterium tumefaciens*-mediated transfection of viral replicons for efficient transient expression in plants. Nat. Biotechnol. 23: 718–723.

Matić, S., R. Rinaldi, V. Masenga, and E. Noris. 2011. Efficient production of chimeric Human papillomavirus 16 L1 protein bearing the M2e influenza epitope in Nicotiana benthamiana plants. BMC Biotechnol 11: 106.

Mati´c, S., E. Quaglino, L. Arata, F. Riccardo, M. Pegoraro, M. Vallino, F. Cavallo, and E. Noris. 2016. The rat ErbB2 tyrosine kinase receptor produced in plants is immunogenic in mice and confers protective immunity against ErbB2+ mammary cancer. Plant. Biotechnol. J. 14: 153–159.

McCormick, A.A., S. Reddy, S.J. Reinl, T.I. Cameron, D.K. Czerwinkski, F. Vojdani, K.M. Hanley, S.J. Garger, E.L. White, and J. Novak. 2008. Plant-produced idiotype vaccines for the treatment of non-Hodgkin's lymphoma: Safety and immunogenicity in a phase I clinical study. Proc. Natl. Acad. Sci. USA 105: 10131–10136.

McCormick, A.A. 2011. Tobacco derived cancer vaccines for non-Hodgkin's lymphoma: Perspectives and progress. Hum. Vaccin. 7: 305–312.

Meshcheriakova, Y.A., P. Saxena, and G.P. Lomonossoff. 2014. Fine-tuning levels of heterologous gene expression in plants by orthogonal variation of the untranslated regions of a nonreplicating transient expression system. Plant. Biotechnol. J. 12: 718–727.

Miyamoto, K., T. Shimizu, F. Lin, F. Sainsbury, E. Thuenemann, G.P. Lomonossoff, H. Nojiri, H. Yamane, and K. Okada. 2012. Identification of an E-box motif responsible for the expression of jasmonic acid-induced chitinase gene OsChia4a in rice. J. Plant Physiol. 169: 621–627.

Mohammadzadeh, S., F. Roohvand, A. Memarnejadian, A. Jafari, S. Ajdary, A.H. Salmanian, and P. Ehsani. 2016. Co-expression of hepatitis C virus polytope-HBsAg and p19-silencing suppressor protein in tobacco leaves. Pharm. Biol. 54: 465–473.

Montague, N.P., E.C. Thuenemann, P. Saxena K. Saunders P. Lenzi, and G.P. Lomonossoff. 2011. Recent advances of Cowpea mosaic virus-based particle technology. Hum. Vaccin. 7: 383–390.

Mukherjee, A., J. Walker, K.B. Weyant, and C.M. Schroeder. 2013. Characterization of favin-based fuorescent proteins: An emerging class of fuorescent reporters. PLoS ONE 8(5), Article ID e64753.

Narayanan, K.B., and S.S. Han. 2018. Recombinant helical plant virus-based nanoparticles for vaccination and immunotherapy. Virus Genes 54(5): 623–637. doi: 10.1007/s11262-018-1583-y.

Nemykh, M.A., A.V. Efmov, and V.K. Novikov. 2008. One more probable structural transition in potato virus X virions and a revised model of the virus coat protein structure. Virology 373(1): 61–71.

Niehl, A., F. Appaix, and S. Bosca. 2016. Fluorescent Tobacco mosaic ´ virus-derived bio-nanoparticles for intravital two-photon imaging. Frontiers in Plant Science, vol. 6, article 1244.

Noad, R., and P. Roy. 2003. Virus-like particles as immunogens. Trends Microbiol. 11: 438–444.

Orman, B.E., R.M. Celnik, A.M. Mandel, H.N. Torres, and A.N. Mentaberry. 1990. Complete cDNA sequence of a South American isolate of potato virus X. Virus Research 16(3): 293–305.

Ortega-Berlanga, B., K. Musiychuk, Y. Shoji, J.A. Chichester, V. Yusibov, O. Patiño-Rodríguez, D.E. Noyola, and A.G. Alpuche-Solís. 2015. Engineering and expression of a RhoA peptide against respiratory syncytial virus infection in plants. Planta 243: 451–458.

Parker, L., A. Kendall, and G. Stubbs. 2002. Surface features of potato virus X from fber difraction. Virology 300(2): 291– 295.

Peyret, H., and G.P. Lomonossoff. 2013. The pEAQ vector series: The easy and quick way to produce recombinant proteins in plants. Plant Molecular Biology. March 2013.

Peyret, H., and G.P. Lomonossoff. 2015. When plant virology met Agrobacterium: The rise of the deconstructed clones. Plant Biotechnol J. 13: 1121–35.

Plchova, H., T. Moravec, H. Hoffmeisterova, J. Folwarczna, and N. Cerovska. 2011. Expression of Human papillomavirus 16 E7ggg oncoprotein on N- and C-terminus of Potato virus X coat protein in bacterial and plant cells. Protein Expr. Purif. 77: 146–152.

Porta, C., and G.P. Lomonossoff. 2002. Viruses as vectors for the expression of foreign sequences in plants. Biotechnol. Genet. Eng. Rev. 19: 245–92.

Riedmann, E.M. 2012. Chinese biotech partnership brings first hepatitis E vaccine to the market. Human Vaccines & Immunotherapeutics 8(12): 1743–44.

Röder, J., C. Dickens, R. Fischer, and U. Commandeur. 2018. Systemic Infection of Nicotiana benthamiana with Potato virus X nanoparticles presenting a fluorescent iLOV polypeptide fused directly to the coat protein. Hindawi BioMed Research International Volume 2018, Article ID 9328671, 12 pages.

Roder, J., C. Dickmeis, and U. Commandeur. 2019. Small, smaller, nano: New applications for potato virus X in nanotechnology. Front Plant Sci. 10: 158. doi: 10.3389/fpls.2019.00158.

Ruiz-Ramon, F., R.N. Sempere, E. Mendez-Lopez, M. Sanchez-Pina, and M.A. Aranda. 2019. Second generation of pepino mosaic virus vectors: Improved stability in tomato and a wide range of reporter genes. Plant Methods 15:58.

Rybicki, E.P. 2010. Plant-made vaccines for humans and animals. Plant Biotechnol. J. 8: 620–637.

Rybicki, E.P. 2014. Plant-based vaccines against viruses. Virol. J. 11: 205.

Sainsbury, F., and G.P. Lomonossoff. 2008. Extremely high-level and rapid transient protein production in plants without the use of viral replication. Plant Physiol. 148: 1212–1218.

Sainsbury, F., P.O. Lavoie, M.A. D'Aoust, L.P. Vezina, and G.P. Lomonossoff. 2008. Expression of multiple proteins using full-length and deleted versions of Cowpea mosaic virus RNA-2. Plant Biotechnol. J. 6: 82–92, doi:http://dx. doi.org/10.1111/j.1467-7652.2007.00303.x.

Sainsbury, F., L. Liu, and G.P. Lomonossoff. 2009a. Cowpea mosaic virus-based systems for the expression of antigens and antibodies in plants. Methods Mol. Biol. 483: 25–39.

Sainsbury, F., E.C. Thuenemann, and G.P. Lomonossoff. 2009b. pEAQ: Versatile expression vectors for easy and quick transient expression of heterologous proteins in plants. Plant Biotechnol. J. 7: 682–693.

Sainsbury, F., M. Sack, J. Stadlmann, H. Quendle, R. Fischer, and G.P. Lomonossoff. 2010. Rapid transient production in plants by replicating and non-replicating vectors yields high quality functional anti-HIV antibody. PLoS ONE 5: e13976.

Sainsbury, F., K. Saunders, A.A.A. Aljabali, D.J. Evans, and G.P. Lomonossoff. 2011. Peptide-controlled access to the interior surface of empty virus nanoparticles. ChemBioChem. 12: 2435–2440.

Sainsbury, F., P. Saxena, K. Geisler, A. Osbourn, and G.P. Lomonossoff. 2012. Chapter nine: Using a virus-derived system to manipulate plant natural product biosynthetic pathways. pp. 185–202. *In*: Hopwood, D.A. (ed.). Methods in Enzymology, vol. 517. Academic Press, Waltham, Massachusetts, USA.

Salazar-González, J.A., B. Bañuelos-Hernández, and S. Rosales-Mendoza. 2015. Current status of viral expression systems in plants and perspectives for oral vaccines development. Plant Mol. Biol. 87: 203–17.

Saunders, K., F. Sainsbury, and G.P. Lomonossoff. 2009. Efficient generation of cowpea mosaic virus empty virus-like particles by the proteolytic processing of precursors in insect cells and plants. Virology 393: 329–337.

Schaffer, D.V., J.T. Koerber, and K.I. Lim. 2008. Molecular engineering of viral gene delivery vehicles. Annu. Rev. Biomed. Eng. 10: 169–94. doi: 10.1146/annurev.bioeng.10.061807.160514.

Scholthof, K.G. 2004. Tobacco mosaic virus: A model system for plant biology. Annu. Rev. Phytopathol. 42: 13–34.

Shaner, N.C., G.H. Patterson, and M.W. Davidson. 2007. Advances in fuorescent protein technology. Journal of Cell Science, vol. 120, part 24, pp. 4247–4260.

Shukla, S., A.L. Ablack, A.M. Wen, K.L. Lee, J.D. Lewis, and N.F. Steinmetz. 2013. Increased tumor homing and tissue penetration of the flamentous plant viral nanoparticle potato virus X. Molecular Pharmaceutics 10(1): 33–42.

Shukla, S., C. Dickmeis, A.S. Nagarajan, R. Fischer, U. Commandeur, and N.F. Steinmetz. 2014. Molecular farming of fluorescent virus-based nanoparticles for optical imaging in plants, human cells and mouse models. Biomaterials Science 2(5): 784–797.

Shukla, S., N.A. DiFranco, A.M. Wen, U. Commandeur, and N.F. Steinmetz. 2015. To target or not to target: Active vs. passive tumor homing of filamentous nanoparticles based on potato virus X. Cell Mol. Bioeng. 8: 433–444.

Shukla, S., R.D. Dorand, and J.T. Myers. 2016. Multiple administrations of viral nanoparticles alter *in vivo* behavior—insights from intravital microscopy. ACS Biomaterials Science and Engineering 2(5): 829–837.

Skryabin, K.G., A.S. Kraev, and S.Y. Morozov. 1988. Te nucleotide sequence of potato virus x RNA. Nucleic Acids Research 16(22): 10929–10930.

Sober, J., L. Jarvekulg, I. Toots, J. Radavsky, R. Villems, and M. Saarma. 1988. Antigenic characterization of potato virus X with monoclonal antibodies. Journal of General Virology 69(8): 1799–1807.

Steele, J.F., H. Peyret, K. Saunders, R. Castells-Graells, J. Marsian, Y. Meshcheriakova, and G.P. Lomonossoff. 2017. Synthetic plant virology for nanobiotechnology and nanomedicine. WIRES Nanomed. Nanobiotechnol. 9: e1447.

Steinmetz, N.F. 2010. Viral nanoparticles as platforms for next-generation therapeutics and imaging devices. Nanomedicine: Nanotechnology, Biology and Medicine 6(5): 634–641. doi: https://doi.org/10.1016/j.nano.2010.04.005.

Steinmetz, N.F., C.-F. Cho, A. Ablack, J.D. Lewis, M. Manchester. 2011. Cowpea mosaic virus nanoparticles target surface vimentin on cancer cells. Nanomedicine 6: 351–364.

Sun, Q.-Y., L.-W. Ding, G.P. Lomonossoff, Y.-B. Sun, M. Luo, C.-Q. Li, L. Jiang, and Z.-F. Xu. 2011. Improved expression and purification of recombinant human serum albumin from transgenic tobacco suspension culture. J. Biotechnol. 155: 164–172.

Thuenemann, E.C., A.E. Meyers, J. Verwey, E.P. Rybicki, and G.P. Lomonossoff. 2013a. A method for rapid production of heteromultimeric protein complexes in plants: Assembly of protective bluetongue virus-like particles. Plant Biotechnol. J. 11: 839–846.

Thuenemann, E.C., P. Lenzi, A.J. Love, M. Taliansky, M. Bécares, S. Zuñiga, L. Enjuanes, G.G. Zahmanova, I.N. Minkov, S. Matic, E. Noris, A. Meyers, A. Hattingh, E.P. Rybicki, O.I. Kiselev, N.V. Ravin, M.A. Eldarov, K.G. Skryabin, and G.P. Lomonossoff. 2013b. The use of transient expression systems for the rapid production of virus-like particles in plants. Current Pharm. Des., PMID: 23394559.

Topp, E., R. Irwin, T. McAllister, M. Lessard, J.J. Joensuu, I. Kolotilin, and U. Conrad. 2016. The case for plant-made veterinary immunotherapeutics. Biotechnol. Adv. 34: 597–604.

Trevor, D., and M. Young. 1998. Host-guest encapsulation of materials by assembled virus protein cages. Nature 393(6681): 152–155. doi: 10.1038/30211.

Tusé, D. 2011. Safety of plant-made pharmaceuticals: Product development and regulatory considerations based on case studies of two autologous human cancer vaccines. Hum. Vaccin. 7: 322–330.

Tusé, D., N. Ku, N.M. Bendandi, C. Becerra, R. Collins, Jr., N. Langford, S.I. Sancho, A. López-Díaz de Cerio, F. Pastor, and R. Kandzia. 2015. Clinical Safety and Immunogenicity of Tumor-Targeted, Plant-Made Id-KLH Conjugate Vaccines for Follicular Lymphoma. Biomed. Res. Int., 648143.

Uhde-Holzem, K., R. Fischer, and U. Commandeur. 2007. Genetic stability of recombinant potato virus X virus vectors presenting foreign epitopes. Archives of Virology 152(4): 805– 811.

Uhde-Holzem, K., V. Schlösser, S. Viazov, R. Fischer, and U. Commandeur. 2010. Immunogenic properties of chimeric potato virus X particles displaying the hepatitis C virus hypervariable region I peptide R9. J. Virol. Methods 166: 12–20.

Vardakou, M., F. Sainsbury, N. Rigby, F. Mulholland, and G.P. Lomonossoff. 2012. Expression of active recombinant human gastric lipase in Nicotiana benthamiana using the CPMV-HT transient expression system. Protein Expr. Purif. 81: 69–74.

Venuti, A., G. Curzio, L. Mariani, and F. Paolini. 2015. Immunotherapy of HPV-associated cancer: DNA/plant-derived vaccines and new orthotopic mouse models. Cancer Immunol. Immunother. 64: 1329–1338.

Vezina, L.P., L. Faye, P. Lerouge, M.A. D'Aoust, E. Marquet-Blouin, C. Burel, P.O. Lavoie, M. Bardor, and V. Gomord. 2009. Transient co-expression for fast and high-yield production of antibodies with human-like N-glycans in plants, Plant Biotechnol. J. 7: 442–455, doi:http://dx.doi.org/10.1111/j.1467- 7652.2009.00414.x.

Villani, M.E., B. Morgun, P. Brunetti, C. Marusic, R. Lombardi, I. Pisoni, C. Bacci, A. Desiderio, E. Benvenuto, and M. Donini. 2009. Plant pharming of a full-sized, tumour- targeting antibody using different expression strategies. Plant Biotechnol. J. 7: 59–72, doi:http://dx.doi.org/10.1111/j.1467-7652.2008.00371.x.

Wen, A.M., S. Shukla, and P. Saxena. 2012. Interior engineering of a viral nanoparticle and its tumor homing properties. Biomacromolecules 13(12): 3990–4001.

Wen, A.M., and N.F. Steinmetz. 2016. Design of virus-based nanomaterials for medicine, biotechnology, and energy. Chemical Society Reviews 45(15): 4074–4126.

Xiaoqing Shi, X., T. Teresa Cordero, T., S. Sandra Garrigues, J.F. Jose F. Marcos, J. Antonio Daros, and M. Coca. 2019. Efficient production of antifungal proteins in plants using a new transient expression vector derived from tobacco mosaic virus. Plant Biotechnology Journal 17: 1069–1080.

Yamamoto, T., K. Hoshikawa, K. Ezura, R. Okazawa, S. Fujita, and M. Takaoka. 2018. Improvement of the transient expression system for production of recombinant proteins in plants. Sci. Rep. 8: 1–10.

Yildiz, I., K.L. Lee, K. Chen, S. Shukla, and N.F. Steinmetz. 2013. Infusion of imaging and therapeutic molecules into the plant virus-based carrier cowpea mosaic virus: Cargo-loading and delivery. J. Control. Release 172: 568–578.

Zahin, M., J. Joh, S. Khanal, A. Husk, H. Mason, H. Warzecha, S.J. Ghim, D.M. Miller, N. Matoba, and A.B Jenson. 2016. Scalable production of HPV16 L1 protein and VLPs from tobacco leaves. PLoS ONE 2016, 11, e0160995.

CHAPTER 11

Plant Molecular Pharming of Biologics to Combat HIV

Goabaone Gaobotse,[1,*] *Jocelyne Trémouillaux-Guiller,*[2,*] *Srividhya Venkataraman,*[4] *Mohammed Kamil Sherif,*[3] *Abdullah Makhzoum*[1] and *Kathleen Hefferon*[4]

Introduction

The HIV Pandemic

In 1981, Acquired Immunodeficiency Syndrome (AIDS) was first recognized in the United States of America (USA) as a disease characterized primarily by a reduced CD4+T cell count. AIDS is characterized by an immune deficiency that opens the door to opportunistic pathologies leading to death in the absence of antiretroviral treatments (ARTs). Two years after the discovery of AIDS, a retrovirus was identified as the causative agent of this infectious pathology. Subsequently, this virus was named the Human Immunodeficiency Virus (HIV), due to the strong depletion caused by this on the defense cells of infected hosts.

The HIV virus, a Lentivirus of the family Retroviridae (Ataie Kachoie et al. 2018), has a genome of two similar single-stranded, linear RNA molecules and several viral proteins packaged within a capsid. Two types of HIV (HIV-1 and HIV-2) have been established based on genetic characteristics and various viral antigens. HIV-1, and its subtypes, is currently the major group responsible for the global AIDS pandemic (Haddox et al. 2018). Twenty-five million deaths have been attributed to this virus since

[1] Department of Biological Sciences & Biotechnology, Botswana International University of Science & Technology, Botswana.
[2] Faculty of Pharmaceutical Sciences, University FranÇois Rabelais, Tours, France.
[3] Central Medical Centre, Palapye, Botswana.
[4] Department of Microbiology, Cornell University, USA.
* Corresponding authors: gaobotseg@biust.ac.bw; jocelyne.tremouillauxguiller@gmail.com

the beginning of the AIDS pandemic (Lythgo 2004). In general, antiretroviral drugs have reduced the incidence of HIV across the globe while prolonging the lifespan of HIV/AIDS infected people (Caputo et al. 2009).

At the end of 2018, the World Health Organization (WHO) estimated more than 32 million people were dead of AIDS and, more than 37.9 million people were living with the HIV-1 virus worldwide, of which approximately, 25.8 million of them are mainly located in Southern Africa. Globally, the mortality rate from AIDS is approximately 1.6 million people per year. Despite the tremendous efforts made in the last several decades to develop an immunization processes against HIV, no AIDS vaccine is currently in sight (Hoelscher et al. 2018). Fortunately, antiviral therapies can treat people infected by the HIV virus. Among the 37.9 million people infected with HIV, 23.3 million received antiretroviral therapy (ART). Antiretroviral drugs have reduced the incidence of HIV across the globe and particularly in sub-Saharan Africa and Asia (Lythgo 2004) while prolonging the lifespan of HIV/AIDS infected people (Caputo et al. 2009).

In addition to public health education programs, reducing viral loads to undetectable levels with ARTs to prevent viral transmission is the primary means to control the pandemic. While there is currently no widely effective cure for HIV infection or a vaccine to prevent acquisition of the virus, ARTS have had a discernible impact on the quality of life and longevity of HIV infected individuals. These are also preventing transmission of HIV to vulnerable partners and have reduced the incidence of new infections (Table 1). Appropriate support for the widespread implementation of ART in remote locations or in politically destabilized areas are public health challenges. Although considerable progress has been made in increasing the number of infected individuals receiving treatment, the need for antiretrovirals and effective vaccines that can be produced where they are needed without a requirement for sophisticated laboratory skills or equipment will ensure a reduced viral load and prevent transmission. Plant biotechnology has considerable potential to produce biologics to combat HIV that may be uniquely suited to resource-poor environments where HIV transmission is prevalent. For example: the microbicides, as the antiviral lectin griffithsin, that can potentially prevent HIV infection in blocking viral entry into human cells (Hoelscher et al. 2018).

The total eradication of the AIDS pandemic requires the production of prophylactic anti-HIV vaccines. Over the past few decades, many HIV-1 vaccine candidates based on conventional approaches have been tested. No result is convincing to date. In 2015, six AIDS vaccines reached Phase II b clinical trials, among of them the only vaccine that had modest success came from clinical trial RV144, in which a recombinant canary pox vaccine (ALVAC) combined to a recombinant gp120 vaccine (AIDSVAX) was tested. The other vaccine candidates were not capable of providing protection against HIV acquisition (Li et al. 2017).

Main Hurdles for a Prophylactic HIV-1 Vaccine

Although HIV-1 is one of the most well-characterized viruses, there is no efficient vaccine against this pathogen so far (Rudometov et al. 2019). Why does a safe and efficient vaccine remain elusive today? To develop a vaccine with broad coverage, the major hurdle resides in the extraordinary genetic diversity of HIV-1 subtypes (Asbach et al. 2018) and, more specifically, in the higher variability of the HIV-1 envelope (Env) glycoprotein, the sole target for neutralizing antibodies (Nabs) (van Schooten and van

Table 1: Examples of Research Institutes/Companies with HIV vaccine/antiretroviral clinical trials currently underway.

Company/Research Institute	Vaccine and Antiretroviral drug	Notes
The International AIDS Vaccine Initiative (IAVI) Scripps Research Institute	Antiretroviral drugs, vaccine candidate eOD-GT8 60mer to elicit broadly neutralizing antibodies	Basic research on HIV/AIDS, sponsored by Gates Foundation and USAID
UC San Diego AIDS Research Institute	All manner of HIV/AIDS research	Regional resource for HIV/AIDS research, supported by the National Institute of Health
Lifespan/Tufts/Brown Center for AIDS Research	All manner of HIV/AIDS research	Joint research effort between Tufts and Brown Universities and their affiliated Hospitals and Centers
Bristol-Myers Squibb	Antiretroviral therapy	Focuses on patients with chronic viral diseases
AIDS Research Alliance	Vaccine	NIH-sponsored HIV Vaccine Trials Network (HVTN)
Delaney AIDS Research Enterprise (DARE), University of California, San Francisco	Antiretroviral therapy for patients to control any residual virus after ART is interrupted	International group of more than 30 researchers and doctors, funded by the National Institutes of Health
GeoVax Labs, Inc.	Innovative human vaccines, focusing on those that prevent and fight HIV infections	Vaccines developed are specific for HIV-1 Clades B and C
ViiV Healthcare	Antiretroviral drugs to block mother-to-infant HIV transmission	Launched by Pfizer and GlaxoSmithKline
amfAR (American Foundation for AIDS Research)	Antiretroviral drugs to block mother-to-infant HIV transmission	Observational database to monitor disease course and treatment outcomes

Source: https://www.healthcareglobal.com/top-10/top-10-research-companies-forefront-hivaids-eradication

Gils 2018). Globally, there are more than a dozen HIV-1 subtypes and hundreds of circulating HIV-1 recombinant forms (CRFs) (Stephenson et al. 2015). The descendants of the HIV ancestor appeared 100 years ago and have evolved so rapidly that Env today has as little as a 65% protein identity (Haddox et al. 2018). Due to its high mutation rate, 1–10 mutations/genome/replication per cycle, or because the reverse transcriptase makes some random errors, the conformational flexibility and the extensive glycans coverage, the HIV-1 virus has developed a unique arsenal of tricks to evade neutralizing antibodies (Rathore et al. 2018, Dorgham et al. 2019).

Until now, the discovery of a prophylactic vaccine is hampered by the rapid mutations of the HIV-1 envelope proteins. Indeed, the fast evolution of the HIV's envelope (Env) protein (gp 120 subunit) has serious consequences for the anti-HIV immunity in eroding most neutralizing antibodies (Haddox et al. 2018). New approaches and ideas need to be tested to develop an effective HIV-1 vaccine, for example, the novel vaccine strategy targeting the viral protease cleavage sites to disrupt virus maturation is a feasible and promising approach (Li et al. 2017). Another type of strategy is building upon the possible efficacy to combine two HIV vaccines, as described by the recent RV144 clinical trial reported above (Burton et al. 2012). Polyvalent Env vaccines are one approach to overcome the viral diversity and involve the use of heterologous HIV-1 Env mixtures to expose developing immune responses to diverse Env conformations (Pankrac et al. 2018).

History of Plant Molecular Pharming

Genetically engineered (GE) plants were initially developed in the early 1980s, and the first commercially available engineered crops were new varieties of corn and soybean that were resistant to insect pests. These were first available in the market in the USA in 1996. Subsequently, transgenic plants have been commercially available in many countries including Canada, Brazil, South Africa, India, China, and Australia. The first generation of transgenic plants was designed to resist pathogens and tolerate herbicides with the intention of improving agricultural produce. Today these crops have been adopted more widely in developing countries than in industrialized parts of the world. Given the anticipated impact of climate change, the second generation of transgenic plants were generated to tolerate drought, high temperatures and flooding. Furthermore, second generation GE crops are biofortified to increase their nutrient density to accommodate food shortages and the growing global population. Extensive research and development in the field of plant-made pharmaceuticals (PMPs) is taking place concurrently to apply similar approaches to the production of biologics.

Plant molecular pharming (PMP) emerged just over three decades ago. Initially considered as an academic endeavor, it is now becoming a novel protein production system with applications for manufacturing pharmaceuticals (Shih and Doran 2009). In 1986, the potential for plants to express human genes was first demonstrated with the successful transcription of the human growth hormone in transgenic plant cells (Barta et al. 1986). A few years later in 1990, recombinant human serum albumin was expressed in engineered tobacco and potato plants (Sijmons et al. 1990). The increasing demand for biologics such as enzymes, antibodies and vaccines has generated interest plant-based manufacturing approaches. This has led to the development of several specialized startup and biotechnology companies worldwide. In 2012, Elelyso™ (i.e., Taliglucerase alfa) was the first PMP approved for human use in the USA. This enzyme was prescribed for the treatment of the type 1 Gaucher's disorder, a rare genetic lysosomal disorder. Plant-derived antibodies are appearing in the literature with immense potential of production at a large-scale, for medical and veterinary applications and as powerful tools for diagnostics or in therapy (Thomas et al. 2011). An antibody cocktail for the treatment of the Ebola virus infection (ZMapp™) was produced in *Nicotiana benthamiana* plants. ZMapp was authorized for emergency use during an outbreak of the Ebola virus infection and was shown to be safe in humans. Furthermore, this cocktail, showing some effect against Ebola highlighted the potential of PMPs on a global stage (Bishop 2015). The concept of vaccine creation in transgenic plants and edible leafy crops goes as back as far as 1992 and 1999, respectively. Indeed, the plants represent another emerging branch of molecular pharming that focuses on the development of safer and more potent vaccines and a large range of immunogenic antigens (Karg and Kallio 2009, Franconi et al. 2018). Some examples of plant-made vaccines, therapeutics or reagents, for immunization or treatment against different infectious diseases include influenza viruses (as the subtype of H1N1virus), hepatitis B virus, ebolaviruses, rabies virus, flaviviruses and other pathogens for global human pathologies (Chan and Daniell 2015, Rybicki 2017). Recently, the Canadian Company Medicago has developed an alternative high-yield system, called Proficia^{R}, from living plants as hosts to create effective vaccines against the emerging virus of infectious diseases (influenza, COVID-19) as well as antibodies and therapeutic proteins. The Proficia^{R} technology platform for producing vaccines uses the transient protein expression of virus-like particles (VLPs) in plant leaves by through vacuum infiltration. This transient protein expression process allows a performing genetic transformation after an incubation period of up to eleven days (ww.medicago.com/fr).

An important research direction is the production of edible vaccines using consumable plants without thermal processing (Shchelkunov and Shchelkunova 2010). Indeed, vaccine antigens have been expressed in vegetable crops (carrot and potato), fruits (banana and tomato) and green leafy crops (lettuce, spinach). Oral vaccines offer major benefits in terms of easy stable or transient transformation, scale-up production, absence of pathogen contaminations, elimination of a cold chain, long-term storage (seeds), bio encapsulation, Moreover, laborious extraction and purification mechanisms, capable of affecting the stability of the recombinant antigens, are not required. Nowadays, the plant-based edible vaccines, created from HBsAg (hepatitis B surface antigen) and CTB (cholera toxin subunit B) reached Phase I of clinical trial status (Criscuolo et al. 2019). Two challenges of plant-made edible vaccines are an improper glycosylation of the viral antigen and its production variability within transgenic plants. Another concern is a possible oral tolerance or an allergy to the transgenic plant once associated to the immunogen antigen. Moreover, one of the main hurdles for the delivery of vaccines or other therapeutic entities is the digestive system degrading or digesting the recombinant protein before it can perform its function (Thomas et al. 2011).

The use of plants as expression platforms for pharmaceuticals offers potential cost saving, both for manufacturing and for production in the large scale. PMPs also do not pose a risk for contamination with human pathogens as plants do not support their multiplication (Moustafa et al. 2016, Makhzoum et al. 2014a, Singhabahu et al. 2017, Makhzoum et al. 2013) as shown in (Table 2). Plants also allow for the production of more complex proteins and therefore more effective vaccines. One of the major concerns about PMPs is possible contamination of food with pharmaceutical proteins. This occurs through the contamination of non-transgenic plants that are consumed as food crops by people. However, there are still no foolproof methods devised that can avoid the possibility of such contamination. In theory, the rapid production, and lower costs inherent in plant-based manufacturing platforms are appealing for the development of biologics to combat HIV. However, low expression yields and appropriate post translational modifications in the system are important considerations. Proteins produced using plant systems are generally, structurally, and functionally comparable to their native counterparts. Differences in the host cellular machinery can impact post-translational modifications which influence biological activity. Plants can also be grown *en masse* without sophisticated equipment enabling rapid large-scale production. It is for these reasons that plant molecular pharming, as it is known, has been considered for the production of biologics for developing countries which lack appropriate infrastructure to manufacture low cost pharmaceuticals even though it is possible to build low cost manufacturing facilities and plants. In these areas where the rural poor have little access to modern medicine, plant-made pharmaceuticals could have the greatest potential. In the

Table 2: Comparison of expression systems used to produce recombinant HIV antigens.

	Bacteria	**Yeast**	**Mammalian Cell Culture**	**Transgenic Plants**	**Plant cell culture**	**Transient vectors**
Production cost	Moderate	Moderate	Expensive	Inexpensive	Inexpensive	Inexpensive
Timescale	Fast	Fast	Fast	Slower	Fast	Fast
Safety	Potential of contamination	Issues of contamination	Issues of contamination, human pathogens	Yes	Yes (highly contained)	Yes (highly contained)
Posttranslational modifications	No	Yes	Yes	Yes	Yes	Yes
Regulatory compliance	High	High	High	Complex	High	High
Yield	High	High	Moderate	High	High	High
Scalability	Moderate	Moderate	Moderate	Rapid	Rapid	Rapid
Protein stability	Yes	Yes	Yes	In seed, yes	Yes	Yes

case of HIV/AIDS, much effort has been made to generate plant-made vaccines and antiretroviral drugs which could be distributed easily and at a low cost to patients who need them most. Or still, microbicides that can prevent the HIV entry in the infected host cells. So, Hoelscher and collaborators have explored the possibility of using transplastomic plants as an inexpensive production platform for griffithsin (Hoelscher et al. 2018).

HIV Origin and its Types and Variants

The Human Immunodeficiency Viruses, HIV-1 and HIV-2, are zoonoses of the lentivirus subfamily (genus) of retroviruses (Cos et al. 2008). HIV-1 is found worldwide while HIV-2 remains largely confined to West Africa, because its geographical distribution is considerably less than for HIV-1 due to its low transmissibility (Nyamweya et al. 2013).

As detailed above, the sequence diversity of HIV-1 is a defining feature of the virus because of the high rate of errors incurred during replication and the absence of proofreading activity. The resulting sequence diversity is in the global diversity of the virus. The global population of HIV-1 is subdivided into at least 3 different clades, M, N and O, which likely represent at least three independent groups of HIV that humans can be infected with (Keele et al. 2006). M is the predominant group, with 11 sub-clades, A–K. It is recognized worldwide, and its sub-clades are differentiated by approximately 15% variability in Gag and 25% variability the HIV Envelope (Env). Subtype phylogenetic sub-structuring and inter-subtype recombinant forms are also recognized. In Africa, all subtypes and a diversity of recombinant types are recognized (Lihana et al. 2012). Subtype C is responsible for ~ 50% of all infections globally and is also the predominant circulating subtype in Africa. Subtypes A, B and C are the most common sub-types outside of Africa. The vast diversity of HIV-1 group M subtypes have arisen since the introduction of the virus into humans in approximately 1930. Considerable genetic variability of the virus has been described in infected individuals. The HIV-1 population in a patient is considered a quasi-species: a set of diverse viral sequences. That arises from a high frequency of mutation and recombination, with varying fitness over the course of the disease (Nowak et al. 1990, Zanini et al. 2015). The variability of the quasi-species is strongly influenced by the number of founder populations as individuals from high-risk populations may have been exposed multiple times with a concomitant increase in viral diversity and recombination between founders (Li et al. 2010). Modern HIV-1 vaccines have generally been designed to either elicit antibodies to protect from infection or cellular responses to control the virus following infections, but neither response has been capable of preventing viral escape thus far (Regoes et al. 2005, Dalod et al. 1998, Deng et al. 2015, Goulder and Watkins 2004, Mazzoli et al. 1997, Rinaldo et al. 1995, van Baalen et al. 1997). The greatest challenge to the development of effective vaccines and therapeutics for HIV-1 is the unprecedented viral diversity and the rapid mutation rate of the virus.

Structure of the HIV-1 Virus

HIV-1 is an enveloped virus of approximately 145 nm in diameter of the spherical-shape and surrounded by a lipoprotein layer as an envelope. The most outer lipoprotein layer of the viral envelope (Env) comes from the host cell's membrane when the virus is released in the extracellular spaces. Consequently, the Env of one mature virion carries some host glycoproteins which facilitate its ability to invade the next cell. The HIV-1 envelope protein complex is initially produced as a highly glycosylated precursor which is then proteolytically cleaved into two subunits (Chan and Kim 1998). Functional spikes of the envelope consist extensively of two types of glycosylated trimers, i.e., three glycoproteins gp120 and three glycoproteins gp41. These spikes arise through noncovalently linked heterodimers of trimers gp120 (Extracellular subunits) and trimers gp 41 (Transmembrane subunits) (Engelman and Cherepanov 2012, Pritchard et al. 2015). The two subunits of the viral envelope are derived from the polyprotein precursor P160 which undergoes proteolytic processing by furin proteases (Chan et al. 1997, Chan and Kim 1998). The structural characteristics of the HIV-1 Env spike are critical to its unusual ability to escape neutralizing antibodies and counter the host immune defenses through a diversity of functions (Lythgo 2004). Located under the lipoprotein envelope, the matrix (MA) is composed of p17 proteins with, in its center, a capsid (CA) made of p24 proteins. As for the p7 proteins, they form the nucleocapsid (NC) which protects the viral RNA. Each trimer interacts with a CD4 receptor and its subsequent rearrangement is essential for the infection of target cells and therefore is the sole target for broadly neutralizing antibodies (Pritchard et al. 2015). The identification of the Env Epitope-focused antigenic domains (EADs) represents a promising collection of possible targets in the rational design of HIV-1 vaccines (Wang et al. 2018).

The HIV genome consists of two copies of single-stranded viral RNA, which are packed within the core of the viral particles. Viral single-stranded RNA comprises of nine functional genes, including gag, pol, vif, vpr, tat, rev, vpu, env and nef genes that are arranged in three different reading frames (Ataie Kachoie et al. 2018). The genes encode structural/enzymatic proteins, regulatory and accessory proteins that are required in the HIV-1 replication (Ataie Kachoie et al. 2018), i.e., in total, 15 individual proteins (Watts et al. 2009, Frankel and Young 1998). The gag, pol and env genes code for the Gag, Gag-Pol and Env polyprotein precursors. Gag and P24 genes are very conserved in HIV-1 genome and are currently used for vaccine production (Ataie Kachoie et al. 2018). The three structural Pol proteins, which are protease p12 (PR), reverse transcriptase p51 (RT) and integrase p32 (IN), have essentially enzymatic functions and are encapsulated within the capsid along with the RNA genome (Frankel and Young 1998). Viral RNA is protected by the nucleocapsid.

Reverse transcriptase p66/p51 retro transcribe viral RNA into viral single stranded cDNA, then into double stranded DNA, which is integrated by the integrase p32 into the nuclear DNA of the host cell. All retroviruses, including HIV, are furnished with a reverse transcriptase that converts its single-stranded RNA genome into a double-stranded DNA counterpart that can then be integrated into the genome of infected cells (Korber et al. 2000). The HIV Gag and Pol genes express two large polyprotein precursors, Gag (p55) and Gag-Pol (p160) that are cleaved, at 12 protease cleavage sites (PCS), to lead to 13 individual proteins, at the end of the budding process and during the release of virions from infected cell (Abdurahman et al. 2009). The cleavage of Gag (p55) by HIV protease, takes place at PCS of 1 to 6, leading to the structural proteins: p17 (MA), p24 (CA), p2, (NC), p1, p6. While the cleavage of Gag-Pol (160p) (PCS of 7–12) results in the production of the enzymatic proteins: protease (PR), reverse transcriptase (RT), RT-RNase H- (IN) integrase, and Nef (Li et al. 2017).

Nef is a pathogenic factor expressed by primate lentivirus. HIV-1 virions, produced by cells that expressed Nef, and infect new target cells with higher efficiency and sensitivity decreased to neutralizing antibodies (Lai et al. 2011). The nucleocapsid protein (p7) interacts noncovalently with the viral genome, p17 (MA) anchors internal face of the viral envelope and p24 capsid protein encapsulates the HIV genome (Lythgo 2004). Other genes encode accessory viral proteins which subvert the host immune response and therefore enable efficient replication of the virus. For example: Vpu aids in virion assembly and Vif maintains infection efficiency (Emerman and Malim 1998).

Life Cycle of HIV

The first phase of the HIV-1 replication cycle begins with the virus adhesion to the host cells, in first, to CD4 + T lymphocytes then, macrophages, dendritic cells and astrocyte cells that have CD4 receptors and are susceptible to HIV. Following the discovery of HIV-1, studies demonstrated that CD4 is a receptor for HIV and, over a decade later, the chemokine receptors CCR5 and CXCR4 were identified as co-receptors for HIV-1 (Prakash 2010). Env glycoproteins (gp 120 and gp 41) were organized into trimeric complexes on the virion surface, mediate HIV-1 entry into target cells (Chan and Kim 1998, Kwong et al. 2000). The gp120 envelope glycoprotein, through the one of its two CD4-domain sites, binds the CD4 receptor on host cells, for triggering the conformational change in gp120 and making a possible attachment to its second site to the chemokine receptor 5 (CCR5) or 4 (CXCR4) (Kwong et al. 2000). Binding the CD4 receptor at the one chemokine receptor allows a signal to be transmitted to gp41 and to initiate the membrane fusion (Li et al. 2009). The close contact between the HIV and the target cell, as well as the insertion of a fusion peptide into the host membrane leads to a channel which ensures the fusion of the viral envelope to the cell membrane and the penetration of the virus content into the target cells (Wilen et al. 2012), by an endocytosis process. The fusion peptide can also promote the fusion of infected cells with uninfected neighboring cells (Chan and Kim 1998). After its entry into the host cell, the capsid is taken up by an endosome that induces the release of the capsid contents into the cytoplasm. Subsequently, the HIV-1 genomic RNA encapsulated in the conical capsid core is delivered inside the target cell cytosol (Li et al. 2009) as well as the enzymes essential RT for reverse transcription and IN for integration into the host genome. Using a cellular lysine tRNA molecule, as a primer, the viral genome is transcribed by HIV-1 reverse transcriptase (RT), first into single-stranded cDNA. Then, the cDNA is converted into a double-strand DNA by the DNA-dependent DNA polymerase activity of RT (Blood 2016).

HIV-1 genomic RNA encapsulated in the conical capsid core is delivered inside the target cell cytosol (Li et al. 2009) as well as the enzymes essential RT for reverse transcription and IN for integration into the host genome. Using a cellular lysine tRNA molecule, as a primer, the viral genome is transcribed by HIV-1 reverse transcriptase (RT), first into single-stranded cDNA. Then, cDNA is converted into double-strand DNA by the DNA-dependent DNA polymerase activity of RT (Blood 2016).

Assembling, budding and maturation of HIV-1 virus call for complex mechanisms to be brought into play which occurs during the last cycle step and results in the release and formation of the mature HIV-1 virus. Assembly into immature virus particles takes place at the plasma membrane of the infected host cell within specialized membrane micro-domains (Westley et al. 2012). The viral polyprotein Gag orchestrates this assembly process and, its expression is sufficient to lead to the formation of numerous virions

Table 3: Immunosuppressive attributes of HIV-1 proteins that have been produced in plants.

HIV-1 protein/ peptide	Immunosuppressive functions	Other Pathological functions	Notes
gp41	Immunosuppressive (ISU) domain, increased expression of immunosuppressive cytokines IL6 and IL-10, inhibits the activation of T- cell mitogen stimulated PBMCs, suppresses the humoral response of rats injected with wt gp41 compared to an ISU-inactivated form. Transmembrane (TM) domain, inhibits T cell proliferation by CD3-specific antibodies *in vitro*. Fusion Peptide, (FP), interacts with T cell receptor complexes to inhibit antigen-specific T cell activation and cytokine release, B and T cell hypo -responsiveness, failure of the antigen-induced PBMC-proliferative response, drastic lymphopenia in immunized chimpanzees. Immunosuppressive Loop-Associated Determinant, (ISLAD), inhibits antigen-specific lympho-proliferation by APC stimulation and IFN gamma pro-inflammatory cytokine release through their interaction with T cell receptor alpha transmembrane domains.	Neurotoxin	ISU and ISLAD are highly conserved retroviral domains, ISU-inactivated virions have not been found in HIV-1 infected individuals
gp120	Inhibits proliferation of polyclonal PBMC T cell populations and T cell clonal lineages after application of recall antigens.	Neurotoxin	
Tat	Inhibits proliferation of antigen-stimulated PBMC in a dose dependent fashion, Enhancement of suppressor Treg cells. Increases macrophage production of IFN alpha, a marker of disease progression. Interacts with tubulin dimers to incite apoptosis through microtubule stabilization, an effect similar to taxol. Incites apoptosis of uninfected CD4+ T cells through up-regulation of the Fas ligand, FasL, and the casp-8 apoptotic protease		Immunostimulatory as well as IS activity noted, Tat is found in the cytoplasm of infected cells or uninfected cells through extracellular uptake. Transgenic expression of Tat in mice resulted in diverse pathologies such as formation of Kaposi sarcoma-like lesions
Nef	Prevents the CTL killing of Nef-expressing cells through inhibition of MHC1, disruption of CD4+ T cell lipid metabolism. Suppresses immunoglobulin class-switch in B cells	Disruption of lipid metabolism associated with AIDS	Has extracellular as well as intracellular functions
p24	Evasion of innate immunity to viral DNA PAMP in the cytosol		IS function appears to be inconsequential to vaccination, although this subject has not been studied systematically, no known extracellular IS functions

(Ataie Kachoie et al. 2018). Assembly is driven by this main viral structural polyprotein Gag which consists of four domains, namely, matrix (MA), capsid (CA), nucleocapsid (NC), and p6, and of two short spacer peptides, SP1 and SP2 (Novikova et al. 2019). The HIV-Gag and Gag-Pro-Pol polyproteins itself mediates all the essential events in virion assembly including binding the plasma membrane and concentrating the viral Env protein and package the genome RNA. Virion assembly packages two copies of the genome viral RNA, cellular t-RNA Lys, and Gag/Pol polyprotein precursors. Then, the virion acquires its lipid Env and Env protein spikes as it buds from the plasma membrane (Westley et al. 2012). HIV-1 buds at the plasma membrane of infected cells and the viral membrane is therefore derived from the cellular plasma membrane. The budding events, that releases the virion from the plasma membrane, is mediated by the host ESCRT (Endosomal sorting complexes required for transport) machinery (Westley et al. 2012). Virion maturation takes place upon the release of the virus particle from the cell membrane via the action of a host protease. The mature virion is ready to infect the next cell, which is targeted by interactions between surface gp120 (SU) and cell surface HIV-1 receptor (CD4) and CC or CXC chemokine co-receptors (Frankel and Young 1998). The formation of the HIV-1 particle with a conical core structure is a prerequisite for the infectivity of the virus particle (Abdurahman et al. 2009).

The Immune Response to HIV Infection

HIV infection is characterized by immune dysregulation and the characteristic depletion of CD4+ T cells. First, HIV-1 penetrates the host's $CD4^+$ T lymphocytes, cells carrying CD4 receptors, and subsequently it invades the macrophages and dendritic cells provided with the same receptors. Regardless of the mode of contamination, the virus settles in a few hours in the lymphoid tissue. The destruction of the lymph nodes and related immunological organs also play a major role in causing immuno-suppression observed in people with HIV infection (Mishra et al. 2016). HIV primarily infects CD4+ T-cells and monocytes (Lythgo 2004). In the early phase of infection, HIV-1 principally infects T-cells and cells of monocyte/macrophage lineage, which express CD4 cell surface proteins (Engelman and Cherepanov 2012, Jaworowski and Crowe 1999). Subsequently, a phase of clinical latency occurs during which the CD4+ T-cell numbers inexorably decline (Prakash 2010). Characteristic lethal pathologies such as chronic diarrhea, pneumonia, multifocal leukoencephalopathy or cancers such as multifocal leukoencephalopathy, lymphoma and Kaposi's sarcoma result from the progressive depletion of CD4+ T-cells (Dalgleish et al. 1984, Prakash 2010).

Plant-derived HIV-1 Vaccines

The development of an HIV-1 vaccine is considered to be "one of the most difficult challenges that biomedical science is confronting" (Esparza 2013). The search for an HIV-1 vaccine started with the recognition of HIV-1 as the causal agent of AIDS in 1983 (Esparza 2013). Despite sustained efforts over thirty years, only 5 of 218 vaccines in Phase I trials have moved to phase IIb/III efficacy trials and no viable vaccine candidate has progressed successfully to licensing (Esparza 2013). The only successful HIV-1 vaccine clinical trial to date is the Phase III Thai Trial (RV144), as described above. The trial yielded 60% efficacy at 12 months and 31.2% efficacy at 3.5 years. No reduction in viral load was observed in volunteers who became infected during the course of the trial (Rerks-Ngarm et al. 2009, Robb et al. 2012). Given the difficulties of conferring sterilizing immunity, a realistic HIV-1 vaccine may reasonably be expected to confer partial protection and possibly also reduce the viral load following acquisition which in turn would delay disease progression (Regoes et al. 2005). Despite the challenge of creating an HIV-1 vaccine, the "Paris Statement" from the 9th International AIDS Society HIV Science Conference in 2017 cited a renewed commitment to new approaches in the development of prophylactic and therapeutic vaccines as one of their 5 main goals. Work thus far has shown that future efforts will likely require highly innovative approaches (Allen et al. 2005, Bar et al. 2010, Deng et al. 2015, Goulder and Watkins 2004, Keele et al. 2006, Korber et al. 2000, Li et al. 2010, Lihana et al. 2012, Sharp et al. 2001, Zanini et al. 2015).

HIV Epitopes in the Plant System: Safety and Containment Issues

Research on plants as production houses for HIV proteins continues to gain momentum (Table 4). Immunodeficiency resulting from infection by HIV is controlled by an intricate array of factors. Beyond the direct destruction of immune cells or fatiguing the immune system through highly variable immunodominant epitopes, HIV-1 suppresses the immune system via the secondary pathogenicity or tolerogenic functions of many viral proteins (Ashkenazi et al. 2013, Cohen et al. 1999a, Tikhonov et al. 2003). Immunosuppressive (IS) functions of HIV-1 proteins are theorized to protect conserved regions of the virus that cannot avoid the immune response via escape mutation (Quintana et al. 2005). IS proteins affect vaccine efficacy in at least two ways: IS vaccine antigens suppress an immune response towards themselves or other vaccine antigens in a cocktail during immunization and after immunization; viral IS proteins suppress the vaccine-primed response during infection (Viscidi et al. 1989, Lai et al. 2011, Cohen et al. 1999b).

Risks Associated with Plant-made HIV Vaccines

An appreciation of the highly evolved IS or other toxic functions of HIV-1 proteins is essential to vaccine development in plants for food safety, transgene containment and regulatory issues (Rasty et al. 1996). The use of transgenic plants to produce HIV-1 antigens is regularly cited to be "safe" although it must be acknowledged that special risks posed by this technology are not applicable to other platforms (Matoba et al. 2004). Inappropriate consumption of food plants expressing HIV-1 antigens is especially relevant as food scarcity in areas of the world where HIV-1 incidence is high may result in use of recombinant vaccine food plants as a food source. For this reason alone, the use of non-food plants for the development of recombinant HIV-1 vaccines is of paramount importance. While HIV-1 vaccine antigens have been repeatedly proven safe when administered in conventional vaccination schedules, long-term consumption of food plants expressing HIV-1 proteins has not been specifically studied and may be hazardous. As some HIV-1 antigens have extracellular pathogenicity functions, cell-penetrating ability or the ability to insert into cell membranes, it is conceivable that HIV-1 antigens

Table 4: Examples of HIV epitopes and proteins bioengineered in plants.

Plant Species	**Gene(s) Originating from HIV**	**Protein/ Epitope(s) Expressed**	**Yield in Plant Tissue**	**Details of Immunogenicity Trials**	**References**
Tobacco, lettuce	ENV	C4(V3)6 from gp120	240 ug/g freeze-dried leaf	Mice, Oral administration of 24 ug freeze-dried powder, 4X, T cell proliferation and humoral response	(Govea-Alonso et al. 2013a)
Tobacco chloroplasts	ENV	Multi-HIV (epitopes from gp120, gp41)	16 ug/g fresh leaf mass	Mice, oral administration, 50 or 100 ug multi-HIV, 4X, T helper, CTL response, broad humoral response, virus neutralization	(Rubio-Infante et al. 2015)
Moss *Physcomitrella Patens*	ENV	Poly-HIV Epitopes from gp120and gp41 (CD4 region and V3 loop)	3.7 ug/g fresh weight	Mice, oral administration, antibody response increased	(Orellana-Escobedo et al. 2015)
Tobacco *Nicotiana benthamiana*	ENV	Gp41 P1 region fused to CTB	0.14 +/–0.03% TSP	Mice, intra-nasal doses, 5X High IgG, IgA titers	(Matoba et al. 2008)
Tobacco, carrot, tomato, Arabidopsis Tobacco chloroplast	GAG	P24 Fused to human IgA heavy chain	139 ng/g fresh weight 4.5% TSP 1.4% TSP	Mice, subcutaneous prime and boost of 10 ug, also oral administration, Stimulated antibody titers	(Azizi et al. 2010, Lindh et al. 2014a, Obregon et al. 2006)
Tomato, potato, spinach	TAT	CBT-tat	300–500 ug/g in spinach	Mice, intra-peritoneal, intramuscular, orally, 3X Antibody responses	(Cueno et al. 2010b)
Tobacco protoplasts	NEF	Nef-p27	0.18–0.7% TSP, 40% TSP in tomato and tobacco leaves	None available	(Breuer et al. 2006)

could exert a negative influence on cells of the GI tract and adjacent tissues (Mangino et al. 2007).

Transgenic Tat-expressing mice may be a valid model for long-term oral Tat exposure considering the cell-penetrating capability of Tat. Various studies on transgenic Tat mice were reported to have Kaposi's Sarcoma-like lesions, other tumors of diverse etiology, cardiomyopathy and decreased synthesis of the antioxidant glutathione (GSH) possibly associated with increased drug toxicity and aberrant regulation of cytokines (Choi et al. 2000, Corallini et al. 1993, Garza et al. 1996, Kundu et al. 1999, Prakash et al. 1997, Raidel et al. 2002, Vogel et al. 1988). Also, long-term oral exposure to antigens is known to initiate antigen tolerance, an outcome that could increase HIV-1 susceptibility (Wu et al. 2013). Increased production of oral tolerance induction mediators, IL-10 and TGF-b, occur when there is continuous feeding (Wang et al. 2013). Escape of vaccine antigens with immunosuppressive properties from pharmaceuticals to food crops may result in serious immune dysfunction if the contaminated food plant is frequently consumed, including increased pathogen susceptibility and cancer incidence similar to those observed during the progression to AIDS (Vogel et al. 1988).

Methods of containing plant transgenes will render 'pharming' of vaccine antigens in plants safer for human health and food production. The adoption of non-food plants or chloroplast expression decreases the possible risk of entry of the vaccine into the food chain. Chloroplast transformation, which has been shown to result in the high expression of several HIV-1 antigens in plants, prevents the release of transgenes through pollen, as the plastid genome is maternally inherited (Ruf et al. 2007, Barta et al. 1986, Scotti et al. 2009, McCabe et al. 2008). Chloroplast expression of HIV-1 proteins are phytotoxic in several plant species and may not be suitable for high volume industrial production of vaccine antigens (Zhou et al. 2008, Scotti et al. 2009). By harvesting *Nicotiana tabacum*, a non-food plant that is commonly used for plant pharming, before flowering, the possibility of gene leakage into the environment through pollen2 and seed dispersion would be greatly reduced. The use of tobacco cultivars that flower only under short day conditions has been proposed as a further means to prevent flowering and the release of pollen or seeds to the environment (McCabe et al. 2008).

Env gp160-derived Subunit Vaccines

The Env glycoprotein is the primary target for HIV vaccine development, as it is targeted by neutralizing antibodies during infection (Meador et al. 2017, Meng et al. 2002). The success of prophylaxis in recombinant MAb studies using anti-Env antibodies gives credence to the possibility that if a humoral response of sufficient breadth, strength and longevity to Env could be elicited by a vaccine, it could protect from HIV-1 infection. However, it has not yet been possible to induce a broadly neutralizing humoral response to HIV-1 through vaccination that approaches the efficacy of broadly neutralizing monoclonal antibodies (bNMAbs) supplied intravenously or as a microbicide. The elicitation of potent and broadly neutralizing antibodies is a hallmark of a protective vaccine.

The humoral response to HIV ENV in seropositive individuals is comprised of a spectrum of neutralization breadth and potency that develops after infection. Anti-

ENV bNMAbs and the epitopes that they bind to have been studied to gain insight as to why the development of these antibodies is delayed until they are no longer useful (Zhou et al. 2010). ENV bNAb development is associated with several unusual features (Bonsignori et al. 2011, Liao et al. 2011). Anti-HIV bNMAbs display a biased selection of segments during V(D)J recombination, suggesting that only a small percentage of V(D)J recombinants have the potential to undergo further affinity maturation into Env-binding Abs. After V(D)J recombination, anti-Env antibodies experience an elevated rate and unique spectrum of somatic hypermutation (Pancera et al. 2010). Somatic hypermutation at immunoglobulin (Ig) loci is stimulated in naive and memory B cells by antigen exposure and vaccination during B cell proliferation. Affinity maturation occurs through the action of activation-induced cytidine deaminase (AID) to enhance the affinity of antibodies to a specific epitope. Subsequently, cells that produce the highest affinity to B-cell receptors are selected for further proliferation and class-switch recombination (Rajewsky 1996). Env bNAbs are characterized by an unusually high rate of somatic mutation in the Variable Heavy Chain region (VH) genes that ultimately contact the antigen. The heavy-chain third complementary-determining regions (HCDR3) of known EnvHIV bNAbs from humans is occasionally longer than antibodies isolated from other viral infections, an oddity that results from a high indel rate. The importance of the HCDR3 has also been observed during the humoral response to Env antigens in other species. Immunization of cows with the well-ordered BG505 HIV-1 ENV trimer resulted in antibodies with a broad and potent neutralization of > 90% of isolates from a 117 cross-clade virus panel after a relatively short period of time, < 1 year, and after only two boosts. All 10 ENV-binding mAbs isolated from cows had ultra-long HCDR3 sequences, demonstrating that extension of the HCDR3s region to contact structurally occluded residues of Env is a key feature in the formation of many anti-HIV-1 NAbs. Humans and some other vertebrates produce HCDR3s with an average length of 12–16 amino acids (aa) while the cow HCDR3 repertoire extends from an average length of 26 to over 70 aa, suggesting that they are more easily predisposed to elicit antibodies with this feature

Given that highly mutated antibodies HCDR3s often interact with both viral and self-antigens, human B cells that express them are susceptible to inactivation. Inactivation of B cell lineages because of auto or poly-reactivity such as the lipid-binding attribute of some anti-MPER antibodies reduces the population of B cells capable of responding to Env antigens (Haynes et al. 2005, Huang et al. 2012, Irimia et al. 2016, Nemazee and Weigert 2000, Wardemann and Nussenzweig 2007, Wardemann et al. 2003). Env-binding antibodies are also distinct outside of the antigen binding surface. Antibody Framework Regions of human bNAbs also demonstrates a higher-than-average mutation rate. Framework Regions provide structure to the antibody without directly contacting the epitope surface. A study of framework mutations in anti-HIV bNMAbs found that reversion of framework mutations to the un-mutated germinal state resulted in a pronounced loss of antigen binding and an elevated IC50 in a neutralization assay for 12 of 13 bNAbs (REHECK) investigated in the study (Klein et al. 2013).

Together, the array of changes incurred by bNAb-producing B cells suggests that HIV-1 has evolved mechanisms to evade the host antibody response necessitating radical structural changes of human antibodies to bind and neutralize the virus. Failure of some seropositive patients to produce ENV bNAbs may be either because a suite of rare and often unfavorable events must occur to produce them or because bNAb-negative individuals lack the genetic machinery to enact the unusual B cell affinity maturation or selection required to produce HIV-1 bNAbs (Alam et al. 2011). The inability to produce unusual antibodies or a failure to inactivate auto or poly-reactive B cells may be beneficial in an HIV seronegative state (Gelmez et al. 2014, Palacios et al. 2010, Sutton et al. 2013). If the immune capabilities of the general population are reflected in those of the HIV seropositive population and the development of bNAbs during infection is similar to their development during vaccination, then more than 75% of people would fail to respond to an Env-based HIV vaccine with bNAbs (McGuire et al. 2014). The production, testing, and optimization of Env antigens remains an unwavering goal in HIV vaccinology because it is hoped that Env antigens and vaccination regimes can be engineered to accelerate the rate of bNAb development.

Plant-derived Env Subunit Vaccines: Multi-Epitope Vaccines displaying the V3 Region of gp120

The V3 region of gp120 was considered to be a "Principal Neutralizing Determinant" (PND) of the envelope protein soon after the identification of the virus. The region forms a loop on the protein surface with a role in co-receptor specificity and cell-type tropism (Huang et al. 2012, Hung et al. 1999, Shioda et al. 1991, Trujillo et al. 1996). Antibodies directed to this region can neutralize HIV and preventing infection in chimpanzees with a live virus isolate. Several clinical trials featured V3 loop antigens (Emini et al. 1990). Problematically, the V3 region is highly variable; antibodies to this linear epitope are often not cross-neutralizing and the region is not exposed on many HIV-1 isolates as a result of the variable conformation and the extensive glycosylation of Env (Kwong et al. 2002, Palker et al. 1988, Steimer et al. 1991, Wyatt and Sodroski 1998, Zolla-Pazner 2004).

To enhance the breadth of neutralization conferred by V3, researchers created antigens with multiple fused copies of V3 and expressed the fusion product in plants (Govea-Alonso et al. 2013a). C4(V3)6, a protein consisting of epitopes of the Env variable V3 region of gp120 from 5 isolates tethered to the conserved, C4 region, a T-helper and CTL epitope and a region critical for CD4-binding and cell-entry, was successfully expressed as a nuclear gene in tobacco and lettuce without observable effects on plant growth (Govea-Alonso et al. 2013a, Patterson et al. 2001). Sera from lettuce immunized mice was shown to bind specifically to IIIB, CC and RF epitopes of V3, the C4 region and gp120, demonstrating that immunity toward several represented epitopes and a native gp120 molecule was initiated by the vaccine. When isolated spleen cells from lettuce C4(V3)6-vaccinated mice were stimulated with rc4(V3)6, enhanced proliferation of lymphocytes was detected in a FACS assay compared to controls, demonstrating that the oral vaccine activated CD4+ T cell proliferation in mice in addition to humoral factors (Govea-Alonso et al. 2013b).

Success with the expression and immunogenicity testing of the C4(V3)6 antigen motivated further efforts by the same research group to study other multi-epitope Env fusion proteins. "Multi-HIV" is a Kilo Dalton (KD) fusion protein of 8 Env conformational and linear epitopes, 5 from gp120 and 3 from gp41, that had been previously reported to elicit neutralizing humoral antibodies, cytotoxic T lymphocytes (CTLs) or T helper cells during vaccination (Rosales-Mendoza et al. 2014). To increase the neutralization breadth:6 linear MPER epitope variants of the ELDKWAbNMAb2F45-binding domain of gp41 and 8 epitope variants of the STSIRGKV domain, a linear epitope in V1/V2 region of gp120 that elicits potent neutralization of HIV strain IIIB were integrated into the antigen (Wu et al. 1995). Multi-HIV was expressed in tobacco chloroplasts as a means to increase expression of the recombinant antigen and to create a transgenic plant that was amenable to containment. Multi-HIV plants were phenotypically indistinguishable from controls which suggested that the recombinant protein did not act as a phytotoxin.

The immunogenicity of the complete spectrum of Multi-HIV epitopes has not been studied. However, relevant preliminary characterizations of the protein were made. First, the Multi-HIV protein sequence was subjected to several *in silico* analyses to assess the similarity of the recombinant fusion protein to wild-type antigens and to predict its antigenicity. The 3D structure of Multi-HIV was generated with the protein structure homology server SWISS-MODEL and compared to the known 3D crystal structures of those same epitopes on the complete gp120 and gp41 proteins. This analysis showed that the epitopes on the fusion protein retained their native conformation in some cases, for instance, the V3 loop array, but often adopted divergent structures, likely dependent on interactions with adjacent epitope features. Predicted accessibility of Multi-HIV epitopes to humoral antibodies was assessed by the Hoop and Woods hydrophilicity analysis, which predicts solvent-exposed regions of the fusion protein. Several regions of Multi-HIV such as the KQIINMWQEVGKAMYA domain of gp120 and the NWFDITNWLWKKKK domain of gp41 had strongly hydrophobic profiles in the analysis, making them unlikely contributors to antigenicity of the vaccine protein. Western blotting with sera from HIV seropositive patients recognized the expected the 32 kDa band from Multi-HIV plants but not from wildtype (WT) plants, suggesting that at least one of the vaccine epitopes was correctly structured. Structuring multi-epitope proteins to prevent unwanted inter-epitope interactions or enhance immunogenicity, for instance by the use of linkers or scaffold proteins, may improve presentation of the selected epitopes to the immune system (Nezafat et al. 2016, Wu et al. 2016, Makhzoum et al. 2014b).

An analysis of Multi-HIV immune responses was conducted over several studies (Rosales-Mendoza et al. 2014, Ruf et al. 2007). Sera from BALB/c mice fed four weekly doses of freeze-dried and ground Multi-HIV plant tissue produced antibodies that bound the C4(V3)6 protein on ELISA plates. In a subsequent paper, it was demonstrated that oral immunization of Multi-HIV stimulates the production of antigen-specific antibodies that bind several peptides including: the C4(V3)6 protein, the gp41ELDKWAp peptide, Multi-HIV produced in *E. coli* and a V3 synthetic peptide derived from the sequence of the HIV-1 isolate MN. Increased proliferation of CD4+ and CD8+ splenocytes derived from Multi-HIV immunized mice was also observed following stimulation with recombinant C4(V3)6, a synthetic gp41ELDKWAp peptide, or C4 peptides. These studies show that Multi-HIV might constitutes an immunogenic antigen that can induce both cellular and humoral responses.

Another attempt to produce an Env antigen in plants involved the expression of a chimeric, multi-epitope protein in the moss *Physcomitrella patens*. Moss has several features which renders it a desirable vehicle to produce a recombinant vaccine antigen. From a biotechnological perspective, it readily undergoes homologous recombination, enabling targeted insertion of the transgene possible. In addition, it is amenable to protein glyco-engineering, thereby allowing structural modifications of the antigen glycocalyx to improve antigenicity. From a pharmaceutical perspective, it can be produced in industrial-sized cultures that conform to Good Manufacturing Practice (GMP) requirements and it is non-toxic when administered parenterally.

The antigen the researchers expressed in moss was termed Poly-HIV. This antigen consists of the C4 region of gp120, a V3 loop sequence and 5 ELDKWA variants from gp41. The recombinant antigen was produced at the expected size of 35 kDa in moss at levels up to 3.7 μg g^{-1} fresh weight of moss protonema. The moss Poly-HIV protein reacted with immune sera from animals injected with the C4(V3)6 protein or a tandemly repeated ELDWKA peptide in the ELISA. However, reactivity of the moss Poly-HIV to immune sera from HIV-1 seropositive patients was not assayed, a benchmark test of structural similarity of a recombinant antigen to wild-type HIV-1. Sub-cutaneous injection of 34 ng of Poly-HIV (derived from 10 mg of moss tissue) weekly for 4 weeks to BALC/c mice resulted in significantly increased binding of immune sera towards an ELDKWA synthetic peptide in ELISA. These preliminary immunological findings may provide evidence that this moss species is a viable candidate for recombinant anti-HIV vaccine production (Orellana-Escobedo et al. 2015).

The P1/MPR Epitope of gp41

The gp41 protein of Env is responsible both for fusion of the viral membrane with the host cell and the movement of the virus across the mucosa in a process called transcytosis. Transcytosis is a non-fusion, microtubule-mediated process that transfers HIV-1 virions over a mucosal surface where they are captured by dendritic cells and delivered to susceptible CD4+ T cells. The P1 region of the gp41 protein, which spans the ELDKWA epitope, has been identified as a key region associated with this process because of its ability to interact with the glycosphingolipidcerebroside galactosyl-ceramide (GalCer) rafts on the surface of epithelial cells (Alfsen et al. 2001, Magérus-Chatinet et al. 2007, Yu et al. 2008). Binding of Abs to this region prevents infection by inhibition of mucosal transcytosis, a desirable vaccine outcome (Alfsen et al. 2001, Bomsel et al. 1998). Blocking this early phase in mucosal transmission by anti-ELDKWA antibodies such as 2F5 prevents mucosal transcytosis of HIV-1 *in vitro* by inhibition of GalCer binding (Alfsen et al. 2001, Parker et al. 2001). 2F5 antibodies were also capable of blocking vaginal transmission of HIV-1 in macaques exposed to HIV-1 (Mascola et al. 2000). Transcytosis-blocking activity was also observed by anti-ELDKWA S-IgA derived from the colostrum of seropositive mothers (Mascola et al. 2000). Anti-gp41 IgA antibodies that bind to the ELDKWA epitope and block HIV-1 transcytosis are produced in the cervicovaginal fluid of highly exposed seronegative individuals and may constitute a natural barrier to infection (Bélec et al. 2001, Tudor et al. 2009).

Tobacco plants have been used to transiently express the P1 region of gp41 bound to cholera toxin B (CTB)-referred to as either CTB-P1 or CTB–$MPR_{649–684}$ (Bélec et al. 2001, Matoba et al. 2009). CTB is a non-toxic subunit of Cholera toxin (CT) that binds to mucosal GM1 (monosialotetrahexosylganglioside) and enhances association with mucosal membranes as well as stimulating the development of mucosal immunity to co-administered antigens (Fukuyama et al. 2015, George-Chandy et al. 2001, Liljeqvist et al. 1997). The adjuvant action of CTB is also conferred to chemically tethered or genetically-fused antigens (Dertzbaugh et al. 1990). Antigens attached to CTB have a significantly increased IgA immune response when applied to a mucosal surface. The CTB-P1 fusion protein was first transiently expressed in *Nicotiana benthamiana* leaves. It was found to be capable of binding GM1 gangliosides in a ganglioside ELISA and reacted with anti-CTB Abs on a Western Blot (Matoba et al. 2004). *N. benthamiana* plants were later stably transformed with the CTB–MPR$_{649-684}$ construct. The fusion protein was detected at low levels, 0.14 ± 0.03% TSP, although this value could be under-estimated as a result of CTB oligomerization into pentamers (Daniell et al. 2001). CTB–MPR$_{649-684}$ was shown to be capable of oligomerization in *N. benthamiana* extracts, a state necessary for binding GM1-ganglioside and the adjuvant action of CTB (Holmgren et al. 1975, Lesieur et al. 2002). Further investigation of the strength of this CTB–MPR$_{649-684}$ binding to the GM1 ganglioside in a competitive ELISA showed that it inhibited the GM1 ganglioside binding of biotinylated CTB with an IC_{50} (41.5 nM) that notably higher than the equivalent *E. coli* produced protein (8.5 nM) suggesting that the interaction was weaker (Matoba et al. 2008). This could potentially be attributed to glycosylation of the plant Derived (PD) protein near the GM1 ganglioside binding site which may have obscured the interaction (Matoba et al. 2009). Strong binding of the anti-MPER antibodies 2F5 and 4E10 to the CTB–MPR$_{649-684}$ protein demonstrated that the HIV-1 epitope was exposed on the fusion protein and that the conformation of the epitope was retained.

Immunogenicity of the PD CTB–MPR$_{649-684}$ was tested in mice by quantification of the serum IgG and vaginal IgA response after vaccination. The vaccination schedule consisted of 5 weekly intra-nasal doses of the CTB–MPR$_{649-684}$ liposome-conjugated protein administered with the CT adjuvant, followed by an intraperitoneal immunization at week 9. A serum IgG response to the primes was only detected in 2 of 7 animals and was very weak in responders. The IgG response was significantly boosted at 9 weeks and was found to persist at up to 19 weeks. Consistent with the stimulation of mucosal immunity with CTB, the vaginal IgA response to the intra-nasal prime was considerably stronger than the IgG response, and the response was also strongly enhanced by the intraperitoneal boost. The IgA response, however, proved short-lived as it declined at 19 weeks. No information was reported to support the neutralization or transcytosis-blocking efficiency of the antibodies stimulated by the CTB–MPR$_{649-684}$ vaccination. As antibodies raised to the CTB–MPR$_{649-684}$ epitope produced in *E. coli* were shown to be transcytosis-blocking in the absence of neutralizing activity, it would have been informative to investigate these responses further (Matoba et al. 2008).

Gag-derivative Subunits and Virus-Like Particles (VLP) Production in Plants: p24

Gag is the target of cellular immunity in that Gag responses correlate with reduced viral load in nonhuman primates and human studies. But so far, no study has successfully reported the production of a VLP in plants that present the near full-length HIV Env on a Gag VLP, However, some subunits of Gag have been expressed. The Gag derivative p24 shows promise for a subunit vaccine. p24 is one of several antigens that are produced during the proteolytic cleavage of the Gag polyprotein Pr55 Gag in the host cytoplasm. Cleavage of Pr55 Gag produces the p24 central capsid protein, the p17 N-terminal matrix protein of the virus, the p7 nucleocapsid domain protein and the p6 C-terminal domain protein. The p24 sequence is conserved between viral subtypes and therefore may provide greater protection in areas with high HIV-1 diversity. Also, a high serum anti-p24 titer is strongly correlated with long-term non-progression in seropositive patients, suggesting a prominent role of p24 neutralization in disease control (Binley et al. 1997, Dyer et al. 2002, Kiepiela et al. 2007).

Gag-derivative Subunits and VLP Production in Plants: Production Issues

Expression of p24 has been studied in several plant species including tobacco, carrot and tomato (Zhang et al. 2002). Studies on the expression and yield of Gag and its proteolytic products have identified compartmentalization as a key factor in protein yield. When the complete p24 protein was expressed in *Arabidopsis* and carrot, with or without an SEKDEL tag, thereby targeting the translated protein to the ER or the cytosol, respectively. The SEKDEL sequence was found to increase the amount of p24 protein in Arabidopsis compared to the expression of p24 without the SEKDEL tag from 26 ng to 139 ng protein/g fresh weight. The recombinant SEKDEL-tagged p24 protein structure from both plant species was structurally similar to recombinant *E. coli* and WT HIV-1 p24 as it bound anti-*E. coli* p24 antibodies in the ELISA assay. This binding was neutralized by serum from an HIV-1 seropositive patient. Native or codon-optimized p24 expression in the tobacco chloroplast increased the accumulation by 2.5 and 4.5% total soluble protein (TSP), respectively, compared to nuclear transformation (0.35% TSP) or TMV-vectored transformation (~ 0.8% TSP) (Pérez-Filgueira et al. 2004, Zhang et al. 2002). The chloroplast-derived p24 was unsurprisingly devoid of any post-translational modifications (PTM), thus rendering it suitable for administration without further processing to remove potentially allergenic moieties. In contrast, p24 produced in tobacco had a molecular weight of approximately 32 kDa, suggestive of O-glycosylation at the Ser49 and Thr124 predicted sites (Zhang et al. 2002).

Increased expression of p24 in *N. tabacum* was also obtained by fusing it to the constant regions α2-α3 of a human IgA heavy chain (Nuttall et al. 2002). Since recombinant IgA antibodies are expressed at high levels in plants between 1–8% of total soluble proteins, researchers hypothesized that fusing an IgA sequence to an HIV-1 antigen might improve expression levels (De Neve et al. 1993, Hiatt et al. 1989, Ma et al. 1995). A p24-specific ELISA titration assay estimated the mean expression of the p24/α2-α3 fusion protein from the cauliflower mosaic virus (CMV) promoter. The highest yields recorded was 1.4% TSP, which was approximately 13-fold higher than the expression of p24 alone. This approach improved p24 accumulation to levels suitable for commercial production, and this is a notable improvement over yields observed in other nuclear transformation studies (McCabe et al. 2008). The role of the α2-α3 sequence in increased expression was not addressed in the study but several observations suggested possible explanations. Low secretion of p24/α2-α3 protein to the media suggested that the Ig sequences may have caused retention in a post-golgi compartment of the endomembrane system, as might be expected from previous observation of recombinant Ig proteins in plants (Frigerio et al. 2000). A concomitant exposure to folding factors, such as chaperones, in this compartment was theorized to have changed folding associated with aggregation and stability (Nuttall et al. 2002). In addition, the α2-α3 peptide allows association of proteins into higher order aggregates that may stabilize the fusion proteins. Supporting this theory, on Western blots, p24/α2-α3 protein was observed to exist as monomers, dimers and possibly as trimers and tetramers (Obregon et al. 2006).

Gag-derivative Subunits and VLP Production in Plants

It was suggested that the α2-α3 sequence may improve the immunogenicity of p24 by interaction with Fc alpha receptors, thereby facilitating the uptake and presentation of p24 during the cellular adaptive response. Subcutaneous prime and boosting at 3 and 8 weeks, respectively with 10 μg of purified *E. coli* p24-histidine tag (His) or PD-derived p24/α2-α3 demonstrated antigen-specific stimulation of splenocytes from immunized BALB/c mice in a dose-dependent fashion. The *E. coli*-derived antigen was a significantly more powerful stimulant when applied to splenocytes from either *E. coli* p24-his (His Tag) or PD-derived p24/α2-α3 immunized animals. The addition of the α2-α3 sequence did not enhance the cellular adaptive response as predicted, as the α2-α3 fragment was derived from human IgA and tested in BALB/c mice. It remains to be seen if the fusion antigen will be more immunoreactive in humans. P24/α2-α3 also boosted anti-p24 IgG and IgA titers in immunized mice, demonstrating that the PD protein is immunogenic.

Other studies have also demonstrated that PD p24 and other Gag derivatives elicit an immune response in vaccinated animals. Mice primed by oral administration of SEKDEL-tagged and untagged Arabidopsis p24 or SEKDEL-tagged carrot p24 with a subsequent intramuscular (IM) boost of non-PD p24 protein were capable of eliciting an antigen-specific-IgG response (Lindh et al. 2014b). When administered as the prime in an oral prime/intramuscular boost regime, Arabidopsis p24 immunogenicity was repeatedly found to be dose dependent. When mice were administered plant tissue with a low or high p24 expression either by tube feeding (20 ng vs. 460 ng p24) or self-directed feeding (200 ng vs 460 ng p24), the post-boost immune response was clearly present in animals that received the low p24 dose and was faint to undetectable level in animals that received the high dose. Unexpectedly, the administration of the low oral dose with full-length cholera toxin, a strong mucosal adjuvant, eliminated immunogenicity of this oral prime (Lycke and Holmgren 1986, Vajdy and Lycke 1992). As oral tolerance to antigens in humans and other mammals is avoided by a systemic prime, it is conceivable that oral tolerance, promoted by the oral primes, was initiated by high p24 doses or the enhancing effect of the CT adjuvant (Azizi et al. 2010). Immunogenic vaccine regimes in this and other oral vaccine studies with p24 were characterized by occasional animals that were completely unresponsive to the vaccine, a situation that is suboptimal for an HIV-1 vaccine (Lindh et al. 2008). Failure to respond to the oral vaccine regime and the unresponsive immunity at high doses may be another result of tolerance mechanisms associated with the GI route, suggesting that further optimization of oral vaccine components or schedules will be necessary to create a predictable product.

Optimizing Vaccine-Membrane Interactions to Enhance Immunogenicity: VLPs

A feature of the Gag protein that is of interest to vaccine development is the observation that the protein assembles into VLPs. VLPs are self-assembling particles either with or without a membrane component, referred to as enveloped or non-enveloped VLPs, respectively. Gag protein vaccines can assemble into enveloped or non-enveloped forms. Enveloped VLPs are tubular or spherical, ~ 100 nm, membrane-bound vesicles formed by budding from a host membrane. VLP formation has been observed upon the expression of Gag proteins in Baculovirus/insect cells, *E. coli*, mammalian cells, yeast and plants.

VLPs are immunogenic and have been observed to elicit a humoral and cellular immune response when administered without adjuvant (Doan et al. 2005). In addition to acting as Gag immunogens, enveloped VLPs can present membrane-associated viral proteins to the immune system in a more relevant natural state than a soluble subunit vaccine. HIV pseudovirions, VLPs created by Gag proteins that display Env proteins on the VLP membrane, have been developed for HIV vaccination. However the reported level of immunogenicity in animal models has been variable and no HIV pseudovirion has reached clinical trials (Buonaguro et al. 2005, Deml et al. 1997, Hammonds et al. 2005, Hammonds et al. 2007, Paliard et al. 2000). It is now acknowledged that the low density of the Env glycoprotein on these VLPs is not ideal for the induction of neutralizing antibodies. Furthermore, VLPs (like native virions) contain aberrantly folded Env species which misdirect the immune response from functional but less abundant well-ordered trimers. VLPs are licensed for the prevention of Hepatitis A virus, hepatitis B virus, human papillomavirus and as an influenza vaccine adjuvant (Herzog et al. 2009). A plant derived VLP vaccine for prevention of Avian H5N1 Influenza has reached the Phase I clinical trial stage and a Good Manufacture Practice (GMP) study for the production of a PD Norwalk virus VLP was reported to be in preparation for a Phase 1 clinical trial (Lai and Chen 2012, Landry et al. 2010).

The results with HIV-1 VLPs at a clinical trial have been unclear. Post-vaccination antibodies to Env for both p24 and p17 and a p24-specific lympho proliferative response were detected in some seronegative subjects during a phase 1 trial of a yeast-derived HIV-1 p17/p24:Ty-VLP. However later, no p24-specific antibodies or a gag-specific CTL response was detected in seronegative subjects enrolled in the Phase II trial. Tests of this vaccine as an HIV therapy failed to demonstrate effective suppression of CD4+ cell decline or a delay in the onset of AIDS in seropositive subjects (Lindenburg et al. 2002, Martin et al. 1993, Peters et al. 1997, Smith et al. 2001, Weber et al. 1995). These clinical trial results demonstrate the necessity of further exploration and optimization of HIV-1 VLPs.

Production Issues

Several challenges have been faced with respect to plant production of HIV-1vaccines. Failure to isolate stable transgenic regenerants expressing $Pr55^{gag}$ in the cytosol of *N. benthamiana* was considered a possible indicator of Gag toxicity in this subcellular compartment. Targeting to the endoplasmic reticulum (ER) or chloroplast in stable transformants resulted in regenerants but $Pr55^{gag}$ expression was low. Transiently transformed *N. benthamiana* leaves also produced low $Pr55^{gag}$ levels in all three compartments. In a subsequent study, Gag protein expression in the cytosol, apoplast, ER and mitochondria of transiently transformed *N. benthamiana* was non-detectable by Western blot using polyclonal anti-p24 antiserum. On the contrary, transient expression in the chloroplast resulted in pronounced bands at the expected size, 41 and 24 kDa. Failure to express $Pr55^{gag}$ in the cytosol was investigated further. Cytosol-targeted p24, p17 and deltap17 ($Pr55^{gag}$ without p17), transgenes were expressed at the mRNA level but only p24 and deltap17 proteins were produced, thus pointing to a role for p17 as the inhibitory factor. All three constructs, including p17 were expressed at both the mRNA and protein level in chloroplasts, indicating that this compartment is suitable for the expression of Gag. As myristoylation is important for the association of p17/p24 to the membrane, it was considered possible that non-membrane-bound p17/p24 could be cytotoxic, although other equally valid theories could explain the increased yields of myristoylated p17/p24 forms (Meyers et al. 2008). Loss of the myristoylation signal sequence reduced yields of protein targeted to the cytosol, ER and chloroplast, 9-fold, ~ 2 fold and 50 to 70-fold, respectively (Meyers et al. 2008).

While the expression of $Pr55^{gag}$ in the chloroplast potentiates significant yield increases, high expression levels in this compartment are phytotoxic and would preclude the association with co-expressed Env which needs to traffic through the secretory pathway for folding/processing/glycosylation. Autotrophic and heterotrophic *N. tabacum* plants with high $Pr55^{gag}$ expression had a bleached and stunted phenotype and a drastically (–50%) reduced TSP content, compared to the empty vector control and low yield transformants (Scotti et al. 2009). A seedling lethal phenotype was also observed but rare survivors were fostered under conditions of dimmed lighting, high nutrient supplementation and high humidity (Scotti et al. 2015). Some bleached plants were capable of flowering and self-fertilization to produce viable seed (Scotti et al. 2015). While reproductive competence in these bleached plants is experimentally convenient, it is likely that they would be stress-prone and difficult to foster successfully in a high production environment.

To improve transgenic production of the subunit vaccine, researchers sought to identify the cellular and molecular basis of $Pr55^{gag}$ phytotoxicity. Microscopically, leaves of the high expressing line, NS40, lacked a differentiated palisade cell layer and contained only mesophyll-like cells. Ultra-structural TEM micrographs of mesophyll cells revealed that NS40 thylakoid membranes were rudimentary, dispersed, and malformed compared to the dense and organized thylakoid stacks of control plants. Immunogold labeling with p17 antibodies showed that $Pr55^{gag}$ was associated with thylakoid membranes. Co-localization of $Pr55^{gag}$ with thylakoid membrane marker enzymes substantiated a thylakoid membrane localization. $Pr55^{gag}$ has an affinity to the phosphatidylinositol-(4, 5) bisphosphate moiety of cell membranes through a highly basic region and protein myristoylation. As myristoylation is not known to occur in chloroplasts, the membrane association of $Pr55^{gag}$ and disruption of thylakoid structure and function was theorized to occur through a non-specific lipophilicity of Gag towards the major non-phosphorous glycero-galactolipids thylakoid lipids, such as monogalactosyldiacylglycerol and digalactosyldiacylglycerol. Optimization of Gag expression in chloroplasts may require deletion of membrane-association residues if this alteration does not interfere with immunity to WT virus.

Stable transgenic expression in chloroplasts pointed to a strong dependence on transgene accessory sequences. The pNS40 construct had a 2-fold increase in mRNA and 20 to 30 fold increases in protein expression compared to a similar construct, pFA1. Both constructs were similar in that they were integrated into chloroplast DNA and were driven by the strong rrn promoter with the rbcL 3'UTR. The pNS40 construct had the rbc L5'UTR and a 6x His tag and a factor Xa site, whereas pFA1 had the T7g10 5'UTR and no tag. A pulse-chase experiment with 35S-labelled amino acids determined that the increased protein expression was a result of more efficient protein synthesis and not differences in protein stability.

p24/p17 VLPs

The potential of p24 and p17 to assemble into Virus-Like Particles and the increased number of epitopes compared to p24 was cited as an impetus to preferentially pursue the former in a subsequent study. The high density of CTL epitopes on p24 and the strong correlation between p24 neutralization and disease control provided the impetus to focus on the cellular immune response of mice to this antigen (Addo et al. 2003). The IFN-γ ELISPOT analysis was used to quantify interferon-gamma producing cells following vaccination. An intramuscular DNA Gag prime followed by an intra-muscular boost of 64 ng PD p24/p17 increased the number of spot forming units Gag CD8+ and Gag CD4+ T cells 2.3 and 4.7 fold, respectively compared to unboosted animals or those boost with a control (leaf tissue). The number of IFN-γ expressing CD8+ or CD4+ splenocytes in the p24/p17 boost treatment was not improved by a higher dose (646 ng), nor did it differ from a boost with the DNA Gag vaccine. An approximately equivalent effect of a DNA Gag prime and p17/p24 boost on Gag-specific total IgG and IgG subtypes IgG1, IgG2a and IgG2b titers in serum was also observed. Overall, the humoral stimulatory effect of the recombinant plant protein boost was similar to a DNA Gag boost and did not improve at a higher dose. Interestingly, intra-muscular administration of 646 ng or 64 PD p17/p24 did not induce a detectable humoral or cellular immune response in the absence of the priming vaccine. , Although the simplicity of a DNA Gag vaccine renders it a preferable boost with a DNA Gag prime when considered in isolation, the significant immunogenicity of the PD p24/p17 boost as well as its potential use as a VLP delivery system with other HIV-1 immunogens motivates further study of this recombinant antigen in plants.

Membrane Presentation of gp41-derived Antigens on VLPs

The potential to co-express the Gag polyprotein and a degenerate gp41 protein was investigated in *N. benthamiana*. The dgp41 is a chimeric antigen comprising of the MPER from aB-clade isolate and a transmembrane domain/cytoplasmic tailfrom a C-clade virus (Gong et al. 2014, Kessans et al. 2016). Despite its reduced size and chimeric-derivation, dgp41 was effectively bound by human anti-MPER 2F5 in a Western blot, an indication that the deconstructed protein preserves the integrity of this epitope (Kessans et al. 2013). As the gp41 trimer changes conformationally during fusion, it can elicit unique humoral factors, the presentation of gp41 in a manner most relevant to disease progression is paramount in a vaccine (Bélec et al. 2001, Broliden et al. 2001, Coëffier et al. 2000, Devito et al. 2000). An antigen spanning the gp41 MPER and part of the C-terminal heptad repeat that was presented on a liposome to mimic insertion of the peptide in a membrane was found to elicit systemic and mucosal antibodies in both humans and macaques and was protective against a vaginal SHIV viral challenge in macaques (Bomsel et al. 2011, Leroux-Roels et al. 2013). To elicit a relevant immune response to this weak antigen, dgp41 was co-expressed with Gag to form a VLP presenting the gp41 antigen in a state that mimics their natural association with the cell membrane.

Optimizing Vaccine-membrane Interactions to Enhance Immunogenicity: VLPs and CTB conjugates

The successful production of VLPs in *N. benthamiana* was followed by several mice immunogenicity studies with gag/dgp41 VLPs (Kessans et al. 2016). In the first study, the effects of vaccination route and use of VLPs as either the prime or the boost were tested. The studied antigens were inoculated either systemically via intra-peritoneal administration or mucosally via intra-nasal administration. In this study, immunological responses to VLPs were compared to those mounted against CTB-MPER, a fusion protein consisting of gp41 MPER tethered to the Cholera toxin B subunit (CTB) that was produced in *Escherichia coli*. It is unclear why a similar PD form of this protein was not investigated in this study. The gp41 MPER region, residues 649–684, contains the highly conserved Katinger sequence, ELDKWA, which binds mucosal cerebroside galactosyl-ceramide rafts during mucosal transcytosis and is bound by bNMabs 2F5 and 4E10 (Alfsen and Bomsel 2002, Bomsel 1997, Matoba et al. 2008, Meng et al. 2002, Muster et al. 1993). This fusion protein, mixed with the CT adjuvant, elicited serum IgG and mucosal IgA that specifically blocked transcytosis of a clade D isolate in mice and a clade B isolate in rabbit in *ex vivo* tissue studies (Matoba et al. 2008, Matoba et al. 2004). Therefore, the *E. coli*-derived CTB-MPER antigen used in the VLP study is predicted to bind to the nasal membranes during IN priming. Generally, the interaction of VLPs with host membranes is expected to be dependent on the presence of membrane-binding proteins on the VLP surface. Specific instances of membrane-binding have been characterized for Norovirus and Parvovirus VLPs (Bally et al. 2012, Bally et al. 2011, Nasir et al. 2015, Nasir et al. 2014). Although the dgp41 VLP component has the potential to interact with host membranes through dgp41 or host Mor Dendritic cells through size or surface characteristics of the VLP, the host membrane-binding of the Gag/dgp41 VLP has not been studied (Neutra and Kozlowski 2006). The two vaccine forms are expected to interact differently with the host mucosa and these interactions will likely influence the strength of the response.

The serum anti-Gag responses towards VLPs administered as IP or IN primes were quantified. Anti-gag serum IgG was effectively elicited in response to systemic VLP primes, demonstrating that the PD VLP is immunogenic in mice. Mucosal immunization generally resulted in a delayed gag immune response compared to systemic immunization. For instance, systemic VLP vaccination elicited detectable anti-gag serum by the fourth IP prime at 3 weeks, whereas anti-gag antibodies were only detected after 4 mucosal primes and 1 or 2 IP boosts between 8 and 10 weeks in the mucosal vaccination schedules. A systemic prime with VLP provoked stronger anti-Gag serum IgG than a mucosal VLP prime. The difference between administration routes for the immunogenicity of the VLP may be relevant to the protection for individuals at risk of blood-borne infections. For instance, individuals susceptible to exposure via injection drug use or occupational needle stick injury may accrue greater benefit from systemic VLP vaccination.

Serum anti-membrane proximal external region (MPER) responses to either the VLP or CTB-MPER antigen were weak despite the lengthy mucosal vaccination schedule, which consisted of 4 primes at weeks 0,1,2 and 3 and 2 boosts at weeks 7 and 8. Intranasal priming with either VLP or CTB-MPER was more effective than systemic priming with VLP. However, ELISA generally failed to detect anti-MPER IgG or IgA until the inception or completion of the boosts at week 7 or 8, confirming previous observations of the weak immunoreactivity of gp41-derived antigens. Dgp41 possesses the immunosuppressive TMD which may account for the poor immunogenicity in mice. CTB-MPER partially spans the C-heptad repeat region of gp41, which does not contain any known immunosuppressive domain. The weak immunogenicity of the MPER domain of CTB-MPER was also observed in a previous study and was thought to be the result of the weak immunogenicity of MPER as well as the immunodominant influence of CTB (Matoba et al. 2008). Neither format, whether conjugated to CTB or displayed on a VLP, strongly enhanced the immunogenicity of this difficult antigen.

As most HIV-1 infections occur across a mucosal membrane, the ability to elicit anti-Gag and anti-MPER IgA in the vagina and rectum is of considerable importance. Mucosal vaccination is expected to result in stronger mucosal immunity compared to systemic vaccination as mucosal homing receptors that target the migration of IgA antibody secreting cells to mucosal surfaces are not induced by systemically-activated B cells. Overall, CTB-MPER proved to be slightly more evocative of serum anti-MPER IgG than dgp41 when administered either systematically or mucosally. Fecal and vaginal anti-MPER IgA responses were weak and delayed in most treatments regardless of exposure route or antigen. A strong mucosal anti-MPER IgA response was observed on the vaginal mucosa after intranasal priming with CTB-MPER. However, after 10 weeks, endpoint titers were highly variable and undetectable in some animals. Other studies also found intranasally-administered CTB or CTB-conjugated antigens capable of eliciting mucosal IgA antibodies in the vagina (Bergquist et al. 1997, Johansson et al. 1998, Kozlowski et al. 1997, Kozlowski et al. 2002, Rudin et al. 1998). Because vaginal exposure constitutes a considerable risk, this data suggests that the intranasal vaccination route may be a viable option with stronger antigen candidates conjugated to CTB.

Gag is known to elicit a significant T cell response that suppresses viral load and increases the CD4+ T cell count (Jiao et al. 2006, Kiepiela et al. 2007, Koup et al. 1994, Novitsky et al. 2003). To test the ability of plant derived Gag on gag/dgp41 VLPs to stimulate proliferation of CD8 and CD4+ splenocytes, researchers vaccinated mice with VLPs, collected splenocyctes and tested them for a Gag-specific response in the IFN-γ ELISPOT assay. An intranasal or intraperitoneal prime and boost with VLP resulted in stimulation of Gag+ CD8 splenocytes. Intraperitoneal VLP alone stimulated Gag+ CD4 splenocytes. This finding demonstrates a quantitative and qualitative difference between the site of VLP administration and the Gag cellular immune response. While more work is necessary to develop a strong gp41 antigen for presentation on VLPs, especially for mucosal vaccination, this pioneering work did succeed in demonstrating that PD VLPs can incite humoral and cellular immunity to HIV-1 antigens to some degree.

An attenuated, replicating vaccinia poxvirus strain, NYVAC-KC, has been created expressing either Gag or dgp41 to be assessed for immunogenicity in a prime/boost regime with PD HIV-1 Gag/dgp41 VLPs (Meador et al. 2017). A replicating poxvirus was chosen as the prime in this study to build upon the successes of previous studies that used attenuated, replicating canarypox or cytomegalovirus as either the prime in the RV144 clinical trial or to enhance CD8+ T cell-derived immunity and clear an SIV infection in macaques, respectively (Hansen et al. 2013a, Hansen et al. 2013b, Jones and Peterlin 1994). CD8+ T cells require continuous antigenic stimulation to maintain their responsiveness and low-level replication of an attenuated virus may be an effective means to provide this in a vaccine. The virus (referred to as VV) were tested as a prime with either a VLP or VV/VLP boost in mice. Despite the use of the same Gag and dgp41 sequences in both the plant VLPs and vaccinia virus vectors, serum IgG and mucosal IgA stimulation over the course of the vaccination schedule resulted from the PD VLP boost alone. The failure of VVs to prime the humoral response was presumed to result from the failure of NYVAC-KC-Gag and NYVAC-KC-dgp41 virus delivered in separate virions to co-infect mouse cells and produce Gag/dgp41 VLPs derived from mouse cell membranes *in vivo*. The VV vaccine had a significant stimulatory effect on the CD8+ T cell response when delivered in conjunction with VLPs, as was expected from previous studies. Upon exposure of isolated mice splenocytes to 5 immunodominant ZM96 Gag epitopes, a VV prime followed by a VV/VLP boost was found to elicit a higher magnitude of Gag-specific IFN-γ+ CD8+ T cell responses than a VV prime/VLP boost, or either VV or VLP when administered alone. This mirrors the findings of the RV144 trial which showed protection after prime-boosting with the recombinant canarypox vaccine ALVAC-HIV and gp120 envelope glycoprotein AIDSVAX B/E vaccines, but not after vaccination with either component alone. It appears that, in mice, the viral and the VLP components also interact to stimulate cellular immunity to HIV-1. It would be interesting to see how co-delivery of Gag and dgp41 in a single viral vector, which is predicted to increase the production of virus-derived VLP in mouse cells, will affect humoral and cellular stimulation of immunity by the VV/VLP vaccine.

Tat

The Tat (trans-activator of transcription) protein of HIV-1 is a conserved, non-structural protein found in all primate lentiviruses (Jones and Peterlin 1994). It is an 86 to 104 amino-acids, 14 kDa polypeptide that acts as a nuclear trans-activator of transcription and processivity. It activates the transcription of the viral long terminal repeat (LTR) as well as host genes such as the viral co-receptors CCR5 and CXCR4, thereby potentiating viral replication and reinfection, respectively (Huang et al. 1998, Secchiero et al. 1999). Tat also has a role in the development of several AIDS pathologies such as Kaposi's sarcoma and AIDS-associated neurocognitive disorders, such as HIV Dementia. These pathologies have been directly ascribed to several discreet Tat domains that promote immunosuppression (see Table 1) and angiogenesis, including endothelial cell migration, invasion of the extracellularmatrix, proliferation and vascular differentiation (Barillari and Ensoli 2002). The Tat cysteine-rich domain, amino acids 22–39, possesses the CCF (Cys-Cys-Phe) sequence characteristic of several achemokines that allows it to interact with chemokine receptors CCR2 and CCR3 and attract uninfected monocytes and monocyte-derived dendritic cells towards infected cells (Albini et al. 1998). The Basic Domain, aa 49–57, mimics angiogenic growth factors to promote endothelial cell growth and induce the migration and invasion of endothelial and AIDS-associated Kaposi's Sarcoma cells. The basic domain also facilitates uptake of extracellular Tat into the cytoplasm (Vives et al. 1997). Movement of extracellular-derived Tat from the cytoplasm to the nucleus via the Tat nuclear localization sequence can result in the trans-activation of latent HIV-1 genomes (Frankel and Pabo 1988, Green and Loewenstein 1988). The Tat RGD domain, aa 73–86, mimics matrix molecules fibronectin or vitronectin to interact with cell adhesion receptors and incite endothelial cell chemotaxis and cell aggregation to promote re-infection (Brake et al. 1990). Tat has evolved an impressive diversity of secondary immunosuppressive and pathogenicity functions that have been overlaid on a protein with a primary functional role in the viral life cycle.

An anti-Tat vaccine will be key to HIV-1 suppression, if not sterilizing immunity (Goldstein et al. 2001). Given that Tat is one of the first genes expressed after infection, it has been suggested that a vaccine targeting the protein could prevent establishment of infected cells and release of progeny virions (Wu and Marsh 2001). Further, as Tat incites migration of T cells towards immature dendritic cells, the primary site of HIV-1 invasion of the mucosa, a Tat vaccine may block this critical, early step in HIV-1 infection (Izmailova et al. 2003). Because Tat can easily be taken up from the extracellular environment and transported to the nucleus, Tat antibodies prevent the activation of latent virus inside cells (Barillari and Ensoli 2002, Eguchi et al. 2001, Steinaa et al. 1994, Tyagi et al. 2001). A Tat vaccine may delay or prevent Kaposi's Sarcoma and AIDS-associated neurotoxicity and neurocognitive disorders, as both pathologies are tied directly to Tat through a diversity of factors including T-cell activation and cytokine release (Barillari and Ensoli 2002, Barillari et al. 1992, Sabatier et al. 1991, Nath et al. 1996, Conant et al. 1998, Wesselingh et al. 1993, Rappaport et al. 1999). Long-term non-progressors were found to have a high titer of Tat antibodies compared to progressors and immediate seroconverters, which had no humoral response to Tat. A similar association with high Tat-specific CTL frequency and LTNP was detected in seropositive individuals (van Baalen et al. 1997). Immunization of macaques with Tat or Tat toxoid, a chemically inactivated form of Tat, was not able to protect them from infection with simian-human immunodeficiency virus (SHIV) 89.6PD but had lower viral loads after eight weeks of exposure, higher CD4+ T cell counts and a lower expression of chemokine receptors, an indicator of disease progression (Pauza et al. 2000). A therapeutic Tat vaccine, Tat Oyi (European Clinical trial data base, ID: 2012-000374-36), reduced viral rebound, decreased HIV-1 DNA in peripheral blood and increased CD4+ T cell counts in HIV+ patients after interruption of antiretroviral therapy at five to seven months after the inception of vaccination (Loret et al. 2016). An effective Tat vaccine is a key goal in improving the health of infected individuals and the longevity of HIV-1 treatments (Cohen et al. 1999b, Re et al. 1996).

Tat production has been attempted in tomato, potato, and spinach plants. PD Tat reacts with monoclonal antibodies directed towards HIV-1Tat and can trans-activate transcription of an HIV-1 LTR-luciferase construct in HeLa cells, demonstrating that the recombinant protein is biologically active. A CTB-Tat fusion protein was successfully produced in potato and confirmed by Western blotting. The recombinant protein formed monomers and pentamers (Kim et al. 2004). Problematically, expression of native Tat can be phytotoxic in some species. Tat-producing tomato plants failed to form roots and seeds and appear stunted, and chlorotic with progressive necrosis (Cueno et al. 2010a, Cueno et al. 2010b). Control tomato plants growing 3 cm away from Tat-expressing plants on an agar surface also displayed symptoms of Tat phytotoxicity. Using a disc-blot immunoassay and a Tat-specific antibody, Tat was shown to diffuse into agar, evidence that Tat is secreted outside of cells and taken up by nearby plants. Elimination of the The tripeptide Arg-Gly-Asp (RGD) motif prevented the movement of Tat into the agar medium (Cueno et al. 2010a). Expressing RGD-mutated Tat may then be a means to prevent movement from the zone of infiltration during transient expression, although this experiment has not been performed. Elimination of the Arg-rich motif, but not the RGD motif, reduced the phytotoxicity of Tat but did not prevent its movement into the agar medium (Cueno et al. 2010a). It is of pharmaceutical interest that plants without the Arg-rich motif will allow the secretion of non-phytotoxic Tat to a growing medium from whence it may be more easily purified than from plant tissues. The immunological equivalence of an Arg-rich motif-mutated or an RGD-mutated Tat protein to native Tat has not been tested. In another study, the phytotoxic effects of native Tat were largely prevented by expressing Tat from the tomato E8 fruit-specific promoter (reference). There were no vegetative or reproductive differences between the WT and these native-Tat plants except that Tat-transformed plants failed to produce seeds. Spinach expressing Tat derived from a viral vector did not display symptoms of phytotoxicity, an indication that some species may be more prone to Tat phytotoxicity than others (Karasev et al. 2005).

PD Tat can initiate an anti-Tat humoral response although cellular responses to PD Tat are weak. Administration of tomato fruit-derived native Tat by intraperitoneal, intramuscular or orally to BALB/cmice at 0, 14 and 28 days resulted in increasing levels of anti-Tat antibodies similar to those observed after immunization with a synthesized Tat (Ramírez et al. 2007). Oral feeding of lyophilized Tat-tomato fruit extracts to mice resulted in a pronounced stimulation of anti-IgG1, anti-IgG2a, and anti-IgA antibodies in sera. Another study that administered recombinant Tat derived from tomato-fruit extracts intra-dermally to mice detected a heightened anti-Tat IgG antibody titer compared to control inoculation with untransformed tomato extracts (Cueno et al. 2010b). In contrast, 3 weekly oral feedings with one gm spinach expressing Tat from a viral vector did not produce a detectable anti-Tat humoral response in mice. Interestingly, animals subsequently administered with a Tat DNA vaccine had a stronger anti-Tat humoral response after the oral feedings of Tat-spinach compared to control animals, suggesting that, although Tat levels were high in transformed spinach, 300–500 μgTat/g leaf tissue, this treatment may require an adjuvant for greater efficacy or may be so high as to have resulted in antigen tolerance (Azizi et al. 2010, Karasev et al. 2005). Neutralization of extracellular Tat by antibodies may be particularly important for preventing immunosuppressive and neurotoxic effects of Tat. In this regard, sera from orally immunized mice could neutralize the trans-activation function of recombinant Tat at a 1:100 dilution when added to HeLa cells with a HIV-1 LTR-luciferase construct. Anti-Tat IgA antibodies were also detected by ELISA in mice vaginal washes and fecal pellets, confirming that mucosal stimulation of IgA by tomato-Tat was effective on mucosal surfaces relevant to HIV-1 infection. Importantly, mucosal immunity was stimulated with oral immunization of Tat-tomato extracts without the addition of adjuvants. A cellular immune response in PD-Tat immunized mice was detected by the presence of IFN-y secreting CD8+ spleen cells after exposure to a synthetic Tat peptide by the ELISPOT assay in a single study (Cueno et al. 2010b). Together, these results demonstrate the potential for a plant-based vaccine specific for Tat.

Nef

Nef (Negative Factor) is a small, 27 to 35 kDa, phosphorylated and myristoylated viral accessory protein of HIV-1, HIV-2 and SIV that is dispensable for infection and the virus life cycle but promotes viral replication and immune escape. The complexity of Nef functions, which cannot be covered in their entirety here, is suggested by comparison of the gene expression profiles of WT HIV-1 vs a Nef-deleted strain. This study revealed a set of 98 host genes exclusively deregulated as a result of Nef action, including genes involved in signaling, apoptosis, transcription and lipid metabolism (Shrivastava et al. 2016). Nef up-regulates transcription of viral genomes and enhances various down-stream aspects of viral replication as well (Shrivastava et al. 2016). By increasing the rate of endocytosis and lysosomal degradation of membrane-bound CD4 on infected cells, Nef prevents superinfection of nascent, budding virions through re-fusion with the host cell membrane instead of being released into the extracellular fluid (Aiken et al. 1994). Nef acts as an immunosuppressant by inhibiting antigen display and killing of infected cells by CTL (Collins et al. 1998). It does this by disrupting the trafficking of nascent MHC I HLA-A and HLA-B to the cell membrane while avoiding HLA-C and HLA-E, thus preventing recognition/destruction of the infected cell by NK surveillance for aberrant MCH I display (Cohen et al. 1999a, Collins et al. 1998, Mangasarian et al. 1999). As a pathogenicity factor, evidence from mice infected with an HIV-1 strain with deletions in all genes but Nef showed that Nef is responsible for the development of typical AIDS symptoms including diarrhea, edema, weakness, hypo-activity and wasting. Nef functions appear to be important for disease progression as humans and macaques infected with Nef-deleted strains of HIV-1 and SIV do not progress or progress slowly to AIDS with low viral loads and elevated CD4+ T cell counts (Calugi et al. 2006, Daniel et al. 1992, Kestier III et al. 1991, Kirchhoff et al. 1995, Learmont et al. 1999). Elite long-term non-progression in seropositive individuals infected with Nef-deleted strains raised the hope of using live attenuated virus as a vaccine. When it was revealed that several individuals infected with Nef-deleted strains had sustained significant immunological damage related to the HIV-1 infection, and that adult macaques exposed to Nef-deleted strains eventually progressed to AIDS. Nef is superfluous to disease progression in infant macaques, therefore, the idea of using a Nef-deleted strain as a live, attenuated vaccine was put to rest (Baba et al. 1995, Greenough et al. 1999, Hofmann-Lehmann et al. 2003, Learmont et al. 1999). The observation that the loss of Nef function could delay the progression to AIDS supported the development of Nefas a subunit vaccine target for therapeutic immunization (Asakura et al. 1996, Cosma et al. 2003, Harrer et al. 2005, Muthumani et al. 2002).

Studies describing the production of Nef antigens in plants is preliminary as all reports of PD Nef describe expression issues which precluded further analysis of the immunogenicity or antigenicity. The researchers expressed full length Nef with and without a myristoylation signal, p27 and p27 Nef, respectively, and a truncated p25 Nef variant derived from natural infections in tobacco protoplasts. Analysis of protein stability in a pulse-chase experiment revealed that these Nef variants were detectable and stable when targeted to the cytosol (Marusic et al. 2007). There was no evidence for the association of cytosolic Nef with the cell membrane of transgenic tobacco protoplasts. This result was contrary to expectations for a myristoylated protein as plant cells can recognize signals for this PTM and this was taken as evidence that Nef was not myristoylated in tobacco cells (Marusic et al. 2007, Podell and Gribskov 2004). Targeting the protein into the secretory pathway by the use of P1 signal peptide, which often leads to increased recombinant protein expression, did not increase the yield of any Nef form in the study. Nef protein expression and glycosylation could be detected at the pulse phase, but, after the 5-hour chase, the proteins were barely detectable. The structure of Nef contains a large, 70 residue, flexible region, which constitutes ~ 50% of the protein. It may be that this region is detected as misfolded and is degraded by the endoplasmic reticulum quality control system of the secretory pathway (Arold and Baur 2001, de Virgilio et al. 2008).

Subsequently, the researchers chose to create stable transformants of p27mut and p25 in tobacco, which both lack the myristoylation signal. The myristoylation signal confers immunosuppressive properties to Nef that might be undesirable in a vaccine but also may influence immunogenicity (Peng and Robert-Guroff 2001, Stoddart et al. 2003, Peng et al. 2006, Liang et al. 2002). Myristoylation affects the tertiary and quaternary structure of the Nef protein, which may explain the observed differences in the immunogenicity of modified and unmodified Nef (Breuer et al. 2006, Dennis et al. 2005). Stable tobacco transformants produced approximately 0.18% p27mut TSP and 0.7% p25 TSP, in the cytosol. Further study of the immune correlates of the PD-myristoylated and unmyristoylated forms will guide development of PD-Nef for pharmaceutical purposes.

Successful Nef expression but low yields laid the way for a further study focused on improving protein expression. Transformation of tobacco chloroplasts with Nef resulted in high Nef protein expression in some transformants; however, plants with the highest Nef protein yields were yellow or had yellow sectors, an indication of phytotoxicity. Re-engineering promoter elements of the transgene to prevent rearrangements associated with lowered expression resulted in high expression of a p24/Nef fusion protein to ~ 40% of TSP in tomato and tobacco leaves. In this case, transformed plants were also yellowish, a phenotype that may be tied to lowered chloroplast protein expression, as indicated by reduced levels of Rubisco in p24/Nef transformed plants. The high expression of the p24/Nef fusion protein was not maintained in tomato fruit, an unfortunate outcome for use of p24/Nef transformed tomato as an oral antigen. The low level of p24/Nef protein in mature fruit was considered either a result of low stability of the protein in mature structures, as was observed in leaves, or because plastids are not found at high levels in mature fruit (Zhou et al. 2008). Another fusion of the complete zeolin seed storage protein gene sequence with Nef increased the stability of Nef in transgenic tobacco leaves and resulted in the formation of insoluble protein bodies in the ER and expression of the fusion protein to 1.5% TSP (de Virgilio et al. 2008). The pharmacological relevance of the zeolin-Nef fusion protein awaits immunological and allergenicity testing considering seed storage proteins are common PD allergens (Breiteneder et al. 2000).

In *N. benthamiana*, researchers attempted to improve the yields of transiently expressed Nef by co-infiltration with Agrobacterium strains expressing the viral suppressors of gene silencing P25 of Potato Virus X and P19 of either Artichoke Mottled Crinkle virus (AMCV-P19) or Tomato Bushy Stunt virus (Nef/TBSV-P19) (Lombardi et al. 2009). Nef/TBSV-P19 and Nef/AMCV-P19 improved Nef expression over Nef alone by preventing PTGS, as indicated by reduced expression of Nef-specific siRNAs. The best candidate AMCV-P19 increased the expression of Nef4.4 fold, to 1.33% TSP, over plants that expressed Nef alone. While Nef yields are low in stably transformed plants, this study increased the diversity of techniques available to improve expression to industrially relevant levels.

Plant-derived HIV-1 Antibodies

The Env glycoproteins of HIV-1 are present on the HIV virion surface and present the only functionally relevant target for neutralizing antibodies. The functional glycoprotein responsible for mediating entry into cells comprises of a heavily glycosylated trimer that is proteolytically processed during its synthesis. The Env complex binds to CD4 receptor to initiate infection, and subsequently engages a chemokine co-receptor, usually CCR5 or CXCR4. This enables binding to the target cell and fusion of the viral and host membranes (Dalgleish et al. 1984). During infection the structural rearrangement of Env, briefly exposes regions of the glycoprotein which are otherwise inaccessible.

(Kwong et al. 2000). Large, N-linked glycan residues on viral spike proteins obscure most of the protein surface and given that they are derived from the host are poorly immunogenic. During the evolution of infection these glycans can be lost or can shift to yield a constantly changing structural ornament that stymies immune recognition (Kwong et al. 2000, Wyatt and Sodroski 1998).

Plant-derived Broadly Neutralizing Monoclonal Antibodies

The immune systems of seropositive patients can occasionally elicit a humoral response via the production of broadly neutralizing anti-Env antibodies. The evolution of HIV epitopes in an infected person incites the production of humoral factors that bind virions emitted from latently infected cells and prevent reinfection with a variable efficacy. Microbicidal bNMAbs are intended to prevent mucosal infection when administered vaginally, orally or rectally in situations where barriers to infection, such as condom use or abstinence, have a low accessibility and a poor ARV compliance (Ma et al. 2015).

2G12

Early studies of single PD bnMAb expression and function were instrumental in fostering the diversity of PD bnMAb products that have been described. Low glycosylation of the rice 2G12 did not prevent HIV epitope binding. Although, 2G12 produced in rice was less potent than 2G12 produced in Chinese hamster ovary (CHO) cells in the HIV syncytium inhibition assay, it was substantially more potent than 2G12 produced in *Nicotiana tobacco.* This highlights the impact of the expression host on the functionality of recombinant antibodies and is probably related to species-specific differences in glycosylation (Vamvaka et al. 2016).

Maximizing the yield of recombinant proteins and its cost is an important determinant of the feasibility of a biologic. A study examined the effect of Elastin-like polypeptide (ELP) fusions with 2G12 on yield and functionality in tobacco (Floss et al. 2009). Fusion of ELP repeats to a protein is a simple and inexpensive method to facilitate protein purification that could be performed in resource-poor countries (Floss et al. 2009, Floss et al. 2008). The light and heavy chains of 2G12 were fused to a 100x ELP sequence and stably transformed *N. tabacum* (Nt) plants were created in all combinations of single and double 2G12-ELP chains (Floss et al. 2009). Western blot analysis showed that ELP functioned to increase 2G12-ELP fusion protein expression in single and double ELP fusion lines except ^{Nt}2G12 heavy ELP. An array of solitary or defective chains, aggregates and degradation products were observed in ELP and non-ELP samples, a factor that could influence product performance or it impacts folding/assembly. Inverse transition cycling (ITC) precipitation facilitated protein purification from leaves and seeds and served to remove degradation products that lacked an ELP fusion. This purification of ELP fusion mAbs from degradation products may have been a factor in the increased antigen binding activity of the high temperature (50°C) Nt2G12-ELP samples. The leaf ^{Nt}2G12 and ^{Nt}2G12-ELP N-linked glycosylation profiles were highly similar: both contained predominantly oligomannose-type (OMT) (> 94%) glycans and a small proportion of complex-type (CT) glycans. This glycan profile conforms to expectations for an

ER-targeted protein. The seed N-linked glycosylation profiles of ^{Nt}2G12 and ^{Nt}2G12-ELP were highly dissimilar despite common KDEL targeting sequences. Seed ^{Nt}2G12 N-linked glycans were a mixture of single N-acetylglucosamine (GlcNAc) (45%), OMT (46%), and CT glycans 8% whereas Nt2G12-ELP had OMT (94%), CT (3%) and only 1% single GlcNAc residues, an outcome likely related to the observed differences in protein trafficking for ELP and non-ELP mAbs in seed cells (Floss et al. 2009). The divergence in N-linked glycosylation did not have pronounced functional consequences on kinetic rate and equilibrium constants of antigen binding, as ^{Nt}2G12-ELP, ^{Nt}2G12 and ^{CHO}2G12 samples were largely similar. The ELP fusion had a pronounced negative effect on IC_{50} in the HIV syncytium inhibition assay, however. While leaf Nt2G12 showed a stronger inhibition of HIV than ^{CHO}2G12, all leaf-derived ^{Nt}2G12-ELP fusion proteins had a significantly lower inhibition of HIV compared to either Nt2G12 or ^{CHO}2G12. The *in vitro* potency of Nt2G12-ELP may be improved by reducing degradation products, optimizing the number of ELP repeats or using self-cleaving ELP tags (Floss et al. 2009).

Investigations of recombinant 2G12 production in these plant species were a prelude to the EU regulatory acceptance for manufacture of a recombinant tobacco 2G12 destined for use in a vaginal microbiocide (Ma et al. 2015). Using the European Medicines Agency (EMA), 'Guideline on the quality of biological active substances produced by the stable transgene expression in higher plants' (EMEA/CHMP/BWP/48316/2006), researchers developed a compliant Good Manufacturing Process (GMP) and described the results of tests important for regulatory acceptance, product safety and quality control for 3 batches of the active pharmaceutical ingredient (Ma et al. 2015). A batch in the study was defined as a single harvest of leaf material from T6 plants grown and processed on the same day, usually ~ 250 kg of tobacco leaves. Parameters of special concern with a tobacco-derived mAb product were protein A and nicotine contamination, appropriate glycosylation and confirmation of biological activity (HIV neutralization) (Ma et al. 2015). The effects of administration of a single, intra-vaginal dose of 2G12 to New Zealand white rabbits were reported, including observations of negative outcomes such as irritation, tissue pathology, vaginal irritation, post-mortem organ weights and immunogenicity. Given that no negative toxicological or immunogenic outcomes were identified, the persistence of 2G12 in the rabbit vaginal tract was measured by ELISA. 2G12 was found to persist for at least 24 hours in all test subjects which was considered acceptable for an intra-vaginal microbicide. Permission for a clinical trial authorization was then obtained from the UK Medicines and Healthcare products Regulatory Agency (MHRA) in April 2011 (Eudract No. 2009-015609-38). The 2G12 product proved to have only a few mild adverse effects in clinical testing and was deemed safe, a best-case scenario for a Phase I clinical trial of a product derived from plant molecular ‘pharming’.

b12

b12 was stably expressed in *N. benthamiana* plants as the native human b12 IgG1 and as a b12-cyanovirin-N (b12-CY-N) fusion protein (Sexton et al. 2009). Cyanovirin-N binds the gp120 glycocalyx whereas b12 binds the gp120 peptide. The fusion mAb is expected to result in a molecule with 4 potential HIV binding sites in contrast to the 2 sites on the native b12 mAb (Sexton et al. 2009). Cyanovirin-N fused to the C-terminus of the heavy chain did not hinder correct mAb assembly. Recombinant protein of the expected size was accompanied by minor bands of lower molecular weight in both mAb b12 and b12-CY-N extracts, suggesting that both forms had cleavage sites susceptible to proteolysis. Both b12 and b12/CV-N bound the gp120 peptide and b12/CV-N also bound gp120 at the carbohydrate epitope.

2F5

PD 2F5 was originally produced in *Nicotiana tabacum* (Nt) L. cv. bright yellow 2 (BY-2) suspension cells (Sack et al. 2007). Although the yield in this system was low (1.8 mg/L suspension culture) the mAb assembled correctly and no inactive products were observed during binding of BY-2 2F5 to ARP7073 (a synthetic gp41 epitope) in an electrophoretic mobility shift assay. Affinity comparisons by ELISA or surface plasmon resonance found ^{Nt}2F5 had an equal or slightly lower affinity or activity compared with ^{CHO}2F5 (Sack et al. 2007). ^{Nt}2F5 had an 89% lower binding capacity (RU) to ARP7073 and a slightly higher dissociation constant with the effector FcγRI than ^{CHO}2F5, possibly because of binding instability conferred by the SEKDEL signal on the ^{Nt}2F5 light chain. Despite the slightly lower binding affinity and slightly less favorable binding kinetics, HIV inhibition by ^{Nt}2F5 in the syncytium assay was 3-fold lower than ^{CHO}2F5, an indication that this expression system is not suitable for pharmaceutical grade 2F5 expression.

Stability of PD bNMAbs *in vivo*

An attempt was made to increase the longevity of a PD bnMAbs for use in a microbiocide by creating recombinant secretory IgA (SIgA) by transgenesis in tobacco or transient expression in *N. benthamiana* (Paul et al. 2014). SIgAs are present at mucosal surfaces and have a higher stability in secretory environments where low pH or the presence of proteases may reduce the stability of other immunoglobulin types. The feasibility of producing functional secretory SIgA in plants had been previously established (Ma et al. 1995). The dimeric IgA, J chain and secretory component were expressed and properly assembled into ^{Nt}SIgA 2G12 in tobacco and *N. benthamiana* (Paul et al. 2014). ^{Nt}SIgA 2G12 demonstrated the strongest ability to aggregate HIV particles but had an equivalent or somewhat reduced ability to neutralize neutralization-resistant HIV isolates compared to ^{Nt}2G12 IgG. Native SIgA interacts with the cellular receptors FcαR (CD89) and DC-SIGN (CD209) to trigger anti-pathogen effector functions. However, ^{Nt}SIgA was only found to interact with DC-SIGN (CD209). As predicted, the persistence of ^{Nt}2G12 SIgA was approximately 10x longer than ^{Nt}2G12 IgG in cervical vaginal mucosal secretions (100 min vs 1000 min half-life). While ^{Nt}SIgA has a favorable functional profile *in vitro*, the commercial production of ^{Nt}SIgA 2G12 may be impeded by the expense of purification as this product did not bind to Protein A and had to be purified with an even more expensive alternative, Protein L (Paul et al. 2014).

Plant Virus Expression Vectors to Produce Vaccines and Microbicides Against HIV

Plant viruses have also been used to transiently express HIV vaccine antigens (Table 5). Plant viruses have a number of advantages; they can produce large quantities of proteins

Table 5: HIV antigens expressed by Plant Virus Expression Vectors.

Antigen	Virus Expression System	Expression host	%TSP	References
P24	Tomato bushy stunt virus	*N. benthamiana*	5%	(Zhang et al. 2000)
P24	TMV	*N. benthamiana*	10–15 times higher than previously	(Pérez-Filgueira et al. 2004)
P17/p24, Pr55Gag	TMV	*N. benthamiana*	0.46–2.0 ug/kg FW	(Meyers et al. 2008)
Gp41, GAG	TMV	*N. benthamiana*	9 mg/kg FW	(Kessans et al. 2013)
P24	BeYDV	*N. benthamiana*	3-7-fold higher orders of magnitude more	(Regnard et al. 2010)
gp41, aa731-752	CPMV	*N. benthamiana*	unclear	(McLain et al. 1995)
Mab 2G12	CPMV	*N. benthamiana*	100 mg/kg FW	(Sainsbury et al. 2010)
bnMab VRCO1	TMV	*N. benthamiana*	150 mg/kg	(Hamorsky et al. 2013)

in a short period of time (within a few days, depending on the virus/host plant system used), they can produce proteins which are toxic to plants (and are thus unable to be produced in transgenic plants) and they lack the same biocontainment issues and public concerns that are associated with genetically modified plants, or GMOs. On the other hand, vaccines produced by virus expression vectors are transient, so they cannot be stored as seed for the long term. Plant viruses which have been engineered to function as expression vectors for vaccine proteins include the positive-sense RNA viruses Cowpea mosaic virus (CPMV), Tobacco mosaic virus (TMV), Potato virus X (PVX), and the single-stranded DNA geminivirus Bean yellow mosaic virus (BeYDV).

Originally, full-length cDNA versions of plant virus genomes were used for vaccine production; more recently however, 'deconstructed' vector systems that have been specifically designed to lack the components necessary for virus encapsidation and spread have been engineered. These deconstructed versions are offered in the form of plasmids that harbor portions of the viral genome responsible for replication. These modules can be easily combined and added to plant cells via a technology known as agroinfiltration, to maximize levels of vaccine proteins produced. A common example of a deconstructed plant virus expression vector system is the MagnICON system, based on Tobacco mosaic virus and developed by the company ICON genetics (www.icongenetics.com/).

Plant virus expression systems have been used to express full length proteins of HIV-1 as vaccine candidates. Plant viruses have also been used as epitope presentation systems. Plant virus capsid proteins can self-assemble into virus-like particles (VLPs), structures that are highly ordered and repetitive on their surfaces, and as a result stimulate the immune system.

Plant Viruses which Express full HIV Proteins as Vaccines

p24 of HIV was expressed from Tomato bushy stunt virus (TBSV) coat protein in *N. benthamiana* to a level of 5% TSP (Zhang et al. 2000). p24 was also expressed from a Tobacco mosaic virus-based vector (Pérez-Filgueira et al. 2004). In this case, p24 was His-tagged and purified by immobilized metal affinity chromatography. The TMV vector has also been used to express plant codon-optimized versions of p17/p24 and Pr55Gag (VLPs composed of TMV-produced Pr55Gag were observed under SEM, however, plants infected with TMV expressing p24 did not produce any particles) (Meyers et al. 2008). TMV expressed HIV proteins were not immunogenic in mice as a homologous vaccine but were shown to significantly boost a humoral and T cell immune response primed by a Gag–based DNA vaccine. Similarly, deconstructed versions of gp41 (dgp41) as well as Gag were expressed using ICON Genetic's MagnICON TMV-based vector system (Kessans et al. 2013).

Plant Viruses Expressing HIV Epitopes

McLain and colleagues (McLain et al. 1996) used gp41 aa 731–752 fused to the CP of CPMV and expressed it in plants to produce chimeric virions. This stimulates the production of neutralizing antibodies in mice via subcutaneous injection, as well as IgA production when administered intranasally (Durrani et al. 1998). The 2F5 epitope of gp41 has been fused to the N-terminus of PVX CP. Mice immunized with chimeric virus particles raised a neutralizing antibody response. CPMV has also been used to express mAb 2G12 against HIV to give yields of 100 mg/kg FW (Sainsbury et al. 2010).

A CPMV HIV chimera expressing a 22 amino acid epitope of gp41 was inoculated into mice parenterally along with an alum adjuvant and was capable of generating a strong response against three different strains of HIV-1, as indicated by a high titer of neutralizing antibodies. The authors found that there was a neutralizing and non-neutralizing response, and that removing the portion of the epitope responsible for the non-neutralizing response actually increased the neutralizing response (Peyret and Lomonossoff 2015). It was found that the best adjuvant to elicit a strong immunogenic response in mice, in particular, a proliferative T cell response was Quil A, although an immunogenic response was observed in the absence of any adjuvant as well (McInerney et al. 1999).This suggests that CPMV particles themselves have adjuvant-like properties.

Antiretroviral Agents Expressed Using Plant Virus Expression Vector Technology

Broadly neutralizing monoclonal antibody (bNMAb) VRC01 has been produced in *Nicotiana benthamiana* plants using a tobamovirus replicon vector (Hamorsky et al. 2013). Broadly neutralizing antibodies can act as topical microbicides, and thus protect against exposure to HIV-1. VRC01 IgG1 accumulated 5 to 7 days post-inoculation and generated 150 mg bnMAb/kg of fresh leaf material. Plant-made VRC01 was purified by protein A affinity chromatography followed by hydrophobic-interaction chromatography.

The plant derived MAb strongly resembled the human derived antibody with respect to its neutralization properties toward HIV-1. When combined with other microbicide candidates, plant-derived VRC01 exhibited a synergistic effect with the antiviral lectin griffithsin, the CCR5 antagonist maraviroc, and the reverse transcriptase inhibitor tenofovir.

Conclusions

The moderate success of the RV144 HIV vaccine trial demonstrates the feasibility of vaccine-induced protection against HIV acquisition in humans for the first time. A significant burden of HIV/AIDS remains in Southeast Asia and sub-Saharan Africa, however, the novel technologies for producing potential vaccines and other anti-HIV biologics must address the challenges of cost, an under developed infrastructure related to the medical sphere and manufacturing vaccines , the challenge of maintaining a cold chain and poor accessibility in remote locations. Plant-derived vaccines, monoclonal antibodies and antiretroviral agents are a promising prospect with these challenges in mind. Indeed, there is an increasing number of plant-based manufacturing companies, such as Protalix which makes the therapeutic drug GCD for Gaucher's disease in carrot suspension cells and Plantform which makes antibodies in tobacco plants to combat Ebola virus and HIV/AIDS. However, none of these are established in developing countries as considerable capital investment is still needed. This chapter provides in detail the strides taken toward improving accessibility of HIV therapeutics to resource poor countries using plant expression platforms. The next steps towards actualizing plant-based medicines will require collaboration between multiple stakeholders, including governments, industries, and NGOs. By accessing current pipelines that have been prepared for the administration of medicines, the capital required to build a manufacturing plant and food to resource poor settings, it will be possible to provide plant-based biologics to prevent and or block HIV/AIDS to those in need.

References

Abdurahman, S., Á. Végvári, M. Levi, S. Höglund, M. Högberg, W. Tong, I. Romero, J. Balzarini, and A. Vahlne. 2009. Isolation and characterization of a small antiretroviral molecule affecting HIV-1 capsid morphology. Retrovirology 6(1): 1–10.

Addo, M.M., X.G. Yu, A. Rathod, D. Cohen, R.L. Eldridge, D. Strick, M.N. Johnston, C. Corcoran, A.G. Wurcel, C.A. Fitzpatrick, and M.E. Feeney. 2003. Comprehensive epitope analysis of human immunodeficiency virus type 1 (HIV-1)-specific T-cell responses directed against the entire expressed HIV-1 genome demonstrate broadly directed responses, but no correlation to viral load. Journal of Virology 77(3): 2081–2092.

Aiken, C., J. Konner, N.R. Landau, M.E. Lenburg, and D. Trono. 1994. Nef induces CD4 endocytosis: requirement for a critical dileucine motif in the membrane-proximal CD4 cytoplasmic domain. Cell 76: 853–864.

Alam, S.M., H.-X. Liao, S.M. Dennison, F. Jaeger, R. Parks, K. Anasti, A. Foulger, M. Donathan, J. Lucas, L. Verkoczy, and N. Nicely. 2011. Differential reactivity of germline allelic variants of a broadly neutralizing HIV-1 antibody to a gp41 fusion intermediate conformation. Journal of Virology 85(22): 11725–11731.

Albini, A., R. Benelli, D. Giunciuglio, T. Cai, G. Mariani, S. Ferrini, and D.M. Noonan. 1998. Identification of a novel domain of HIV tat involved in monocyte chemotaxis. Journal of Biological Chemistry 273(26): 15895–15900.

Alfsen, A., P. Iniguez, E. Bouguyon, and M. Bomsel. 2001. Secretory IgA specific for a conserved epitope on gp41 envelope glycoprotein inhibits epithelial transcytosis of HIV-1. The Journal of Immunology 166(10): 6257–6265.

Alfsen, A., and M. Bomsel. 2002. HIV-1 gp41 envelope residues 650-685 exposed on native virus act as a lectin to bind epithelial cell galactosyl ceramide. Journal of Biological Chemistry 277(28): 25649–25659.

Allen, T.M., M. Altfeld, S.C. Geer, E.T. Kalife, C. Moore, K.M. O'sullivan, I. DeSouza, M.E. Feeney, R.L. Eldridge, E.L. Maier, and D.E. Kaufmann. 2005. Selective escape from CD8+ T-cell responses represents a major driving force of human immunodeficiency virus type 1 (HIV-1) sequence diversity and reveals constraints on HIV-1 evolution. Journal of Virology 79(21): 13239–13249.

Arold, S.T., and A.S. Baur. 2001. Dynamic Nef and Nef dynamics: How structure could explain the complex activities of this small HIV protein. Trends in Biochemical Sciences 26(6): 356–363.

Asakura, Y., K. Hamajima, J. Fukushima, H. Mohri, T. Okubo, and K. Okuda. 1996. Induction of HIV-1 Nef-specific cytotoxic T lymphocytes by Nef-expressing DNA vaccine. American Journal of Hematology 53(2): 116–117.

Asbach, B., J.P. Meier, M. Pfeifer, J. Köstler, and R. Wagner. 2018. Computational design of epitope-enriched HIV-1 Gag antigens with preserved structure and function for induction of broad CD8+ T cell responses. Scientific Reports 8(1): 1–15.

Ashkenazi, A., O. Faingold, and Y. Shai. 2013. HIV-1 fusion protein exerts complex immunosuppressive effects. Trends in Biochemical Sciences 38(7): 345–9.

Ataie Kachoie, E., S.A.A. Behjatnia, and S. Kharazmi. 2018. Expression of Human Immunodeficiency Virus type 1 (HIV-1) coat protein genes in plants using cotton leaf curl Multan betasatellite-based Vector. Plos One, 13(1): e0190403.

Azizi, A., H. Ghunaim, F. Diaz-Mitoma, and J. Mestecky. 2010. Mucosal HIV vaccines: A holy grail or a dud? Vaccine 28(24): 4015–4026.

Baba, T.W., Y.S. Jeong, D. Pennick, R. Bronson, M.F. Greene, and R.M. Ruprecht. 1995. Pathogenicity of live, attenuated SIV after mucosal infection of neonatal macaques. Science 267(5205): 1820–1825.

Bally, M., A. Gunnarsson, L. Svensson, G. Larson, V.P. Zhdanov, and F. Höök. 2011. Interaction of single viruslike particles with vesicles containing glycosphingolipids. Physical Review Letters 107(18): 188103.

Bally, M., K. Dimitrievski, G. Larson, V.P. Zhdanov, and F. Höök. 2012. Interaction of virions with membrane glycolipids. Physical Biology 9(2): 026011.

Bar, K.J., H. Li, A. Chamberland, C. Tremblay, J.P. Routy, T. Grayson, C. Sun, S. Wang, G.H. Learn, C.J. Morgan, and J.E. Schumacher 2010. Wide variation in the multiplicity of HIV-1 infection among injection drug users. Journal of Virology 84(12): 6241–6247.

Barillari, G., L. Buonaguro, V. Fiorelli, J. Hoffman, F. Michaels, R.C. Gallo, and B. Ensoli. 1992. Effects of cytokines from activated immune cells on vascular cell growth and HIV-1 gene expression. Implications for AIDS-Kaposi's sarcoma pathogenesis. The Journal of Immunology 149(11): 3727–3734.

Barillari, G., and B. Ensoli. 2002. Angiogenic effects of extracellular human immunodeficiency virus type 1 Tat protein and its role in the pathogenesis of AIDS-associated Kaposi's sarcoma. Clinical Microbiology Reviews 15(2): 310–326.

Barta, A., K. Sommergruber, D. Thompson, K. Hartmuth, M.A. Matzke, and A.J. Matzke. 1986. The expression of a nopaline synthase—human growth hormone chimaeric gene in transformed tobacco and sunflower callus tissue. Plant Molecular Biology 6(5): 347–357.

Bélec, L., P.D. Ghys, H. Hocini, J.N. Nkengasong, J. Tranchot-Diallo, M.O. Diallo, V. Ettiegne-Traore, C. Maurice, P. Becquart, M. Matta, and A. Si-Mohamed. 2001. Cervicovaginal secretory antibodies to human immunodeficiency virus type 1 (HIV-1) that block viral transcytosis through tight epithelial barriers in highly exposed HIV-1–seronegative African women. The Journal of Infectious Diseases 184(11): 1412–1422.

Bergquist, C., E.L. Johansson, T.E.R.E.S.A. Lagergård, J. Holmgren, and A. Rudin. 1997. Intranasal vaccination of humans with recombinant cholera toxin B subunit induces systemic and local antibody responses in the upper respiratory tract and the vagina. Infection and Immunity 65(7): 2676–2684.

Binley, J.M., P.J. Klasse, Y. Cao, I. Jones, M. Markowitz, D.D. Ho, and J.P. Moore. 1997. Differential regulation of the antibody responses to Gag and Env proteins of human immunodeficiency virus type 1. Journal of Virology 71(4): 2799–2809.

Bishop, B.M. 2015. Potential and emerging treatment options for Ebola virus disease. Annals of Pharmacotherapy 49(2): 196–206.

Blood, G.A.C. 2016. Human immunodeficiency virus (HIV). Transfusion Medicine and Hemotherapy 43(3): 203.

Bomsel, M., M. Heyman, H. Hocini, S. Lagaye, L. Belec, C. Dupont, and C. Desgranges. 1998. Intracellular neutralization of HIV transcytosis across tight epithelial barriers by anti-HIV envelope protein dIgA or IgM. Immunity 9(2): 277–287.

Bomsel, M., D. Tudor, A.-S. Drillet, A. Alfsen, Y. Ganor, M.-G. Roger, N. Mouz, M. Amacker, A. Chalifour, L. Diomede, and G. Devillier. 2011. Immunization with HIV-1 gp41 subunit virosomes induces mucosal antibodies protecting nonhuman primates against vaginal SHIV challenges. Immunity 34(2): 269–280.

Bomsel, M. 1997. Transcytosis of infectious human immunodeficiency virus across a tight human Epithelial Cell Line Barrier. Nature Medicine 3(1): 42–47.

Bonsignori, M., K.-K. Hwang, X. Chen, C.-Y. Tsao, L. Morris, E. Gray, D.J. Marshall, J.A. Crump, S.H. Kapiga, N.E. Sam, and F. Sinangil. 2011. Analysis of a clonal lineage of HIV-1 envelope V2/V3 conformational epitope-specific broadly neutralizing antibodies and their inferred unmutated common ancestors. Journal of Virology 85(19): 9998–10009.

Brake, D.A., C. Debouck, and G. Biesecker. 1990. Identification of an Arg-Gly-Asp (RGD) cell adhesion site in human immunodeficiency virus type 1 transactivation protein, tat. The Journal of Cell Biology 111(3): 1275–1281.

Breiteneder, H., and C. Ebner. 2000. Molecular and biochemical classification of plant-derived food allergens. Journal of Allergy and Clinical Immunology 106(1): 27–36.

Breuer, S., H. Gerlach, B. Kolaric, C. Urbanke, N. Opitz, and M. Geyer. 2006. Biochemical indication for myristoylation-dependent conformational changes in HIV-1 Nef. Biochemistry 45(7): 2339–2349.

Broliden, K., J. Hinkula, C. Devito, P. Kiama, J. Kimani, D. Trabbatoni, J.J. Bwayo, M. Clerici, F. Plummer, and R. Kaul. 2001. Functional HIV-1 specific IgA antibodies in HIV-1 exposed, persistently IgG seronegative female sex workers. Immunology Letters 79(1-2): 29–36.

Buonaguro, L., M. Visciano, M. Tornesello, M. Tagliamonte, B. Biryahwaho, and F.M. Buonaguro. 2005. Induction of systemic and mucosal cross-clade neutralizing antibodies in BALB/c mice immunized with human immunodeficiency virus type 1 clade A virus-like particles administered by different routes of inoculation. Journal of Virology 79(11): 7059–7067.

Burton, D.R., R. Ahmed, D.H. Barouch, S.T. Butera, S. Crotty, A. Godzik, D.E. Kaufmann, M.J. Mcelrath, M.C. Nussenzweig, and B. Pulendran. 2012. A blueprint for HIV vaccine discovery. Cell Host & Microbe 12(4): 396–407.

Calugi, G., F. Montella, C. Favalli, and A. Benedetto. 2006. Entire genome of a strain of human immunodeficiency virus type 1 with a deletion of nef that was recovered 20 years after primary infection: large pool of proviruses with deletions of env. Journal of Virology 80(23): 11892–11896.

Caputo, A., R. Gavioli, S. Bellino, O. Longo, A. Tripiciano, V. Francavilla, C. Sgadari, G. Paniccia, F. Titti, A. Cafaro, F. Ferrantelli, P. Monini, F. Ensoli, and B. Ensoli. 2009. HIV-1 Tat-based vaccines: An overview and perspectives in the field of HIV/AIDS vaccine development. International Reviews of Immunology 28(5): 285–334.

Chan, D.C., D. Fass, J.M. Berger, and P.S. Kim. 1997. Core structure of gp41 from the HIV envelope glycoprotein. Cell 89(2): 263–273.

Chan, D.C., and P.S. Kim. 1998. HIV entry and its inhibition. Cell 93(5): 681–684.

Chan, H.T., and H. Daniell. 2015. Plant-made oral vaccines against human infectious diseases—are we there yet? Plant Biotechnology Journal 13(8): 1056–1070.

Choi, J., R.-M. Liu, R.K. Kundu, F. Sangiorgi, W. Wu, R. Maxson, and H.J. Forman. 2000. Molecular mechanism of decreased glutathione content in human immunodeficiency virus type 1 Tat-transgenic mice. Journal of Biological Chemistry 275(5): 3693–3698.

Coëffier, E., J.-M. Clément, V. Cussac, N. Khodaei-Boorane, M. Jehanno, M. Rojas, A. Dridi, M. Latour, R. El Habib, F. Barré-Sinoussi, and M. Hofnung. 2000. Antigenicity and immunogenicity of the

HIV-1 gp41 epitope ELDKWA inserted into permissive sites of the MalE protein. Vaccine 19(7-8): 684–693.

Cohen, G.B., R.T. Gandhi, D.M. Davis, O. Mandelboim, B.K. Chen, J.L. Strominger, and D. Baltimore. 1999a. The selective downregulation of class I major histocompatibility complex proteins by HIV-1 protects HIV-infected cells from NK cells. Immunity 10(6): 661–671.

Cohen, S.S., C. Li, L. Ding, Y. Cao, A.B. Pardee, E.M. Shevach, and D.I. Cohen. 1999b. Pronounced acute immunosuppression *in vivo* mediated by HIV Tat challenge. Proceedings of the National Academy of Sciences 96(19): 10842–10847.

Collins, K.L., B.K. Chen, S.A. Kalams, B.D. Walker, and D. Baltimore. 1998. HIV-1 Nef protein protects infected primary cells against killing by cytotoxic T lymphocytes. Nature 391(6665): 397–401.

Conant, K., A. Garzino-Demo, A. Nath, J.C. Mcarthur, W. Halliday, C. Power, R.C. Gallo, and E.O. Major. 1998. Induction of monocyte chemoattractant protein-1 in HIV-1 Tat-stimulated astrocytes and elevation in AIDS dementia. Proceedings of the National Academy of Sciences 95(6): 3117–3121.

Corallini, A., G. Altavilla, L. Pozzi, F. Bignozzi, M. Negrini, P. Rimessi, F. Gualandi, and G. Barbanti-Brodano. 1993. Systemic expression of HIV-1 tat gene in transgenic mice induces endothelial proliferation and tumors of different histotypes. Cancer Research 53(22): 5569–5575.

Cos, P., L. Maes, A. Vlietinck, and L. Pieters. 2008. Plant-derived leading compounds for chemotherapy of human immunodefiency virus (HIV) infection–an update (1998–2007). Planta Medica 74(11): 1323–1337.

Cosma, A., R. Nagaraj, S. Bühler, J. Hinkula, D.H. Busch, G. Sutter, F.D. Goebel, and V.J.V. Erfle. 2003. Therapeutic vaccination with MVA-HIV-1 nef elicits Nef-specific T-helper cell responses in chronically HIV-1 infected individuals. Vaccine 22(1): 21–29.

Criscuolo, E., V. Caputo, R.A. Diotti, G.A. Sautto, G.A. Kirchenbaum, and N. Clementi. 2019. Alternative methods of vaccine delivery: An overview of edible and intradermal vaccines. Journal Of Immunology Research.

Cueno, M.E., Y. Hibi, K. Imai, A.C. Laurena, and T. Okamoto. 2010a. Impaired plant growth and development caused by human immunodeficiency virus type 1 Tat. Transgenic Research 19(5): 903–913.

Cueno, M.E., Y. Hibi, K. Karamatsu, Y. Yasutomi, K. Imai, A.C. Laurena, and T. Okamoto. 2010b. Preferential expression and immunogenicity of HIV-1 Tat fusion protein expressed in tomato plant. Transgenic Research 19(5): 889–895.

Dalgleish, A.G., P.C. Beverley, P.R. Clapham, D.H. Crawford, M.F. Greaves, and R.A. Weiss. 1984. The CD4 (T4) antigen is an essential component of the receptor for the AIDS retrovirus. Nature 312(5996): 763–767.

Dalod, M., M. Harzic, I. Pellegrin, B. Dumon, B. Hoen, D. Sereni, J.-C. Deschemin, J.-P. Levy, A. Venet, and E. Gomard. 1998. Evolution of cytotoxic T lymphocyte responses to human immunodeficiency virus type 1 in patients with symptomatic primary infection receiving antiretroviral triple therapy. Journal of Infectious Diseases 178(1): 61–69.

Daniel, M.D., F. Kirchhoff, S.C. Czajak, P.K. Sehgal, and R.C. Desrosiers. 1992. Protective effects of a live attenuated SIV vaccine with a deletion in the nef gene. Science 258(5090): 1938–1941.

Daniell, H., S.-B. Lee, T. Panchal, and P.O. Wiebe. 2001. Expression of the native cholera toxin B subunit gene and assembly as functional oligomers in transgenic tobacco chloroplasts. Journal of Molecular Biology 311(5): 1001–1009.

De Neve, M., M. De Loose, A. Jacobs, H. Van Houdt, B. Kaluza, U. Weidle, M. Van Montagu, and A. Depicker. 1993. Assembly of an antibody and its derived antibody fragment in Nicotiana and Arabidopsis. Transgenic Research 2(4): 227–237.

De Virgilio, M., F. De Marchis, M. Bellucci, D. Mainieri, M. Rossi, E. Benvenuto, S. Arcioni, and A. Vitale. 2008. The human immunodeficiency virus antigen Nef forms protein bodies in leaves of transgenic tobacco when fused to zeolin. Journal of Experimental Botany 59(10): 2815–2829.

Deml, L., R. Schirmbeck, J. Reimann, H. Wolf, and R. Wagner. 1997. Recombinant Human Immunodeficiency Pr55gagVirus-like Particles Presenting Chimeric Envelope Glycoproteins Induce Cytotoxic T-Cells and Neutralizing Antibodies. Virology 235(1): 26–39.

Deng, K., M. Pertea, A. Rongvaux, L. Wang, C.M. Durand, G. Ghiaur, J. Lai, H.L. Mchugh, H. Hao, H. Zhang, and Margolick, J.B. 2015. Broad CTL response is required to clear latent HIV-1 due to dominance of escape mutations. Nature 517(7534): 381–385.

Dennis, C.A., A. Baron, J.G. Grossmann, S. Mazaleyrat, M. Harris, and J. Jaeger. 2005. Function, & Bioinformatics. Co-translational myristoylation alters the quaternary structure of HIV-1 Nef in solution. PROTEINS: Structure, Function, and Bioinformatics 60(4): 658–669.

Dertzbaugh, M.T., D.L. Peterson, and F.L. Macrina. 1990. Cholera toxin B-subunit gene fusion: structural and functional analysis of the chimeric protein. Infection and Immunity 58(1): 70–79.

Devito, C., K. Broliden, R. Kaul, L. Svensson, K. Johansen, P. Kiama, J. Kimani, L. Lopalco, S. Piconi, J.J. Bwayo, and F. Plummer. 2000. Mucosal and plasma IgA from HIV-1-exposed uninfected Individuals Inhibit HIV-1 Transcytosis Across Human Epithelial Cells. The Journal of Immunology 165(9): 5170–5176.

Doan, L.X., M. Li, C. Chen, and Q. Yao. 2005. Virus-like particles as HIV-1 vaccines. Reviews in Medical Virology 15(2): 75–88.

Dorgham, K., N. Pietrancosta, A. Affoune, O. Lucar, T. Bouceba, S. Chardonnet, C. Pionneau, C. Piesse, D. Sterlin, and P. Guardado-Calvo. 2019. Reverse immunology approach to define a new HIV-gp41-neutralizing epitope. Journal of Immunology Research.

Durrani, Z., T.L. Mcinerney, L. Mclain, T. Jones, T. Bellaby, F.R. Brennan, and N.J. Dimmock. 1998. Intranasal immunization with a plant virus expressing a peptide from HIV-1 gp41 stimulates better mucosal and systemic HIV-1-specific IgA and IgG than oral immunization. Journal of Immunological Methods 220(1-2): 93–103.

Dyer, W.B., H. Kuipers, M.W. Coolen, A.F. Geczy, J. Forrester, C. Workman, and J.S. Sullivan. 2002. Correlates of antiviral immune restoration in acute and chronic HIV type 1 infection: sustained viral suppression and normalization of T cell subsets. AIDS Research and Human Retroviruses 18(14): 999–1010.

Eguchi, A., T. Akuta, H. Okuyama, T. Senda, H. Yokoi, H. Inokuchi, S. Fujita, T. Hayakawa, K. Takeda, M. Hasegawa, and Nakanishi, M. 2001. Protein transduction domain of HIV-1 Tat protein promotes efficient delivery of DNA into mammalian cells. Journal of Biological Chemistry 276(28): 26204–26210.

Emerman, M., and M.H. Malim. 1998. HIV-1 regulatory/accessory genes: Keys to unraveling viral and host cell biology. Science 280(5371): 1880–1884.

Emini, E.A., P. Nara, W. Schleif, J. Lewis, J. Davide, D.R. Lee, J. Kessler, S. Conley, S. Matsushita, and S.D. Putney. 1990. Antibody-mediated *in vitro* neutralization of human immunodeficiency virus type 1 abolishes infectivity for chimpanzees. Journal of Virology 64(8): 3674–3678.

Engelman, A., and P. Cherepanov. 2012. The structural biology of HIV-1: Mechanistic and therapeutic insights. Nature Reviews Microbiology 10(4): 279.

Esparza, J.J.V. 2013. A brief history of the global effort to develop a preventive HIV vaccine. Vaccine 31(35): 3502–3518.

Floss, D.M., M. Sack, J. Stadlmann, T. Rademacher, J. Scheller, E. Stöger, R. Fischer, and U. Conrad. 2008. Biochemical and functional characterization of anti-HIV antibody–ELP fusion proteins from transgenic plants. Plant Biotechnology Journal 6(4): 379–391.

Floss, D.M., M. Sack, E. Arcalis, J. Stadlmann, H. Quendler, T. Rademacher, E. Stoger, J. Scheller, R. Fischer, and U. Conrad. 2009a. Influence of elastin-like peptide fusions on the quantity and quality of a tobacco-derived human immunodeficiency virus-neutralizing antibody. Plant Biotechnol Journal 7(9): 899–913.

Franconi, I., O. Theou, L. Wallace, A. Malagoli, C. Mussini, K. Rockwood, and G. Guaraldi. 2018. Construct validation of a Frailty Index, an HIV Index and a Protective Index from a clinical HIV database. PloS one 13(10): e0201394.

Frankel, A.D., and C.O. Pabo. 1988. Cellular uptake of the tat protein from human immunodeficiency virus. Cell 55(6): 1189–1193.

Frankel, A.D., and J.A. Young. 1998. HIV-1: Fifteen proteins and an RNA. Annual Review of Biochemistry, 67(1): 1–25.

Frigerio, L., N.D. Vine, E. Pedrazzini, M.B. Hein, F. Wang, J.K.-C. Ma, and A. Vitale. 2000. Assembly, secretion, and vacuolar delivery of a hybrid immunoglobulin in plants. Plant Physiology 123(4): 1483–1494.

Fukuyama, Y., K. Okada, M. Yamaguchi, H. Kiyono, K. Mori, and Y.J.P.O. Yuki. 2015. Nasal administration of cholera toxin as a mucosal adjuvant damages the olfactory system in mice. PLoS One 10(9): e0139368.

Garza, H., O. Prakash, and D.J. Carr. 1996. Aberrant regulation of cytokines in HIV-1 TAT72-transgenic mice. Journal of Immunology 156(10): 3631–3637.

Gelmez, M.Y., A.B. Teker, A.D. Aday, A.S. Yavuz, T. Soysal, G. Deniz, M.J.L. Aktan, and Lymphoma. 2014. Analysis of activation-induced cytidine deaminase mRNA levels in patients with chronic lymphocytic leukemia with different cytogenetic status. Leukemia & lymphoma 55(2): 326–330.

George-Chandy, A., K. Eriksson, M. Lebens, I. Nordström, E. Schön, and J.J.I. Holmgren. 2001. Immunity Cholera toxin B subunit as a carrier molecule promotes antigen presentation and increases CD40 and CD86 expression on antigen-presenting cells. Infection and Immunity 69(9): 5716–5725.

Goldstein, G., G. Tribbick, and K.J.V. Manson. 2001. Two B cell epitopes of HIV-1 Tat protein have limited antigenic polymorphism in geographically diverse HIV-1 strains. Vaccine 19(13-14): 1738–1746.

Gong, Z., S.A. Kessans, L. Song, K. Dörner, H.H. Lee, L.R. Meador, J. Labaer, B.G. Hogue, T.S. Mor, and P.J.P.S. Fromme. 2014. Recombinant expression, purification, and biophysical characterization of the transmembrane and membrane proximal domains of HIV-1 gp41. Protein Science 23(11): 1607–1618.

Goulder, P.J., and D.I.J.N.R.I. Watkins. 2004. HIV and SIV CTL escape: implications for vaccine design. Nature Reviews Immunology 4(8): 630–640.

Govea-Alonso, D.O., E.E. Gómez-Cardona, N. Rubio-Infante, A.L. García-Hernández, J.T. Varona-Santos, M. Salgado-Bustamante, S.S. Korban, L. Moreno-Fierros, S.J.P.C. Rosales-Mendoza, and O. Tissue. 2013a. Culture Production of an antigenic C4 (V3) 6 multiepitopic HIV protein in bacterial and plant systems. Plant Cell, Tissue and Organ Culture (PCTOC) 113(1): 73–79.

Govea-Alonso, D.O., N. Rubio-Infante, A.L. García-Hernández, J.T. Varona-Santos, S.S. Korban, L. Moreno-Fierros, and S.J.P. Rosales-Mendoza. 2013b. Immunogenic properties of a lettuce-derived C4 (V3) 6 multiepitopic HIV protein. Planta 238(4): 785–792.

Green, M., and P.M.J.C. Loewenstein. 1988. Autonomous functional domains of chemically synthesized human immunodeficiency virus tat trans-activator protein. Cell 55(6): 1179–1188.

Greenough, T.C., J.L. Sullivan, and R.C.J.N.E.J.O.M. Desrosiers. 1999. Declining CD4 T-cell counts in a person infected with nef-deleted HIV-1. New England Journal of Medicine 340(3): 236–237.

Haddox, H.K., A.S. Dingens, S.K. Hilton, J. Overbaugh, and J.D. Bloom. 2018. Mapping mutational effects along the evolutionary landscape of HIV envelope. Elife 7: e34420.

Hammonds, J., X. Chen, T. Fouts, A. Devico, D. Montefiori, and P.J.J.O.V. Spearman. 2005. Induction of neutralizing antibodies against human immunodeficiency virus type 1 primary isolates by Gag-Env pseudovirion immunization. Journal of Virology 79(23): 14804–14814.

Hammonds, J., X. Chen, X. Zhang, F. Lee, and P.J.V. Spearman. 2007. Advances in methods for the production, purification, and characterization of HIV-1 Gag–Env pseudovirion vaccines. Vaccine 25(47): 8036–8048.

Hamorsky, K.T., T.W. Grooms-Williams, A.S. Husk, L.J. Bennett, K.E. Palmer, and N. Matoba. 2013. Efficient single tobamoviral vector-based bioproduction of broadly neutralizing anti-HIV-1 monoclonal antibody VRC01 in Nicotiana benthamiana plants and utility of VRC01 in combination microbicides. Antimicrob Agents Chemother 57(5): 2076–86.

Hansen, S.G., M. Piatak, Jr., A.B. Ventura, C.M. Hughes, R.M. Gilbride, J.C. Ford, K. Oswald, R. Shoemaker, Y. Li, and M.S.J.N. Lewis. 2013a. Immune clearance of highly pathogenic SIV infection. Nature 502(7469): 100.

Hansen, S. G., J.B. Sacha, C.M. Hughes, J.C. Ford, B.J. Burwitz, I. Scholz, R.M. Gilbride, M.S. Lewis, A.N. Gilliam, and A.B.J.S. Ventura. 2013b. Cytomegalovirus vectors violate CD8+ T cell epitope recognition paradigms. Science 340(6135): 1237874.

Harrer, E., M. Bauerle, B. Ferstl, P. Chaplin, B. Petzold, L. Mateo, A. Handley, M. Tzatzaris, J. Vollmar, and S.J.A.T. Bergmann. 2005. Therapeutic vaccination of HIV-1-infected patients on HAART with a recombinant HIV-1 nef-expressing MVA: safety, immunogenicity and influence on viral load during treatment interruption. Antiviral Therapy 10(2): 285–300.

Haynes, B.F., J. Fleming, E.W.S. Clair, H. Katinger, G. Stiegler, R. Kunert, J. Robinson, R.M. Scearce, K. Plonk, and H.F.J.S. Staats. 2005. Cardiolipin polyspecific autoreactivity in two broadly neutralizing HIV-1 antibodies. Science 308(5730): 1906–1908.

Herzog, C., K. Hartmann, V. Künzi, O. Kürsteiner, R. Mischler, H. Lazar, and R.J.V. Glück. 2009. Eleven years of Inflexal® V—a virosomal adjuvanted influenza vaccine. Vaccine 27(33): 4381–4387.

Hiatt, A., R. Caffferkey, and K.J.N. Bowdish. 1989. Production of antibodies in transgenic plants. Nature 342(6245): 76–78.

Hoelscher, M., N. Tiller, A.Y.-H. Teh, G.-Z. Wu, J.K. Ma, and R. Bock. 2018. High-level expression of the HIV entry inhibitor griffithsin from the plastid genome and retention of biological activity in dried tobacco leaves. Plant Molecular Biology 97(4): 357–370.

Hofmann-Lehmann, R., J. Vlasak, A.L. Williams, A.-L. Chenine, H.M. Mcclure, D.C. Anderson, S. O'Neil, and R.M.J.A. Ruprecht. 2003. Live attenuated, nef-deleted SIV is pathogenic in most adult macaques after prolonged observation. Aids 17(2): 157–166.

Holmgren, J., I. Lönnroth, J. Månsson, and L.J.P.O.T.N.A.O.S. Svennerholm. 1975. Interaction of cholera toxin and membrane GM1 ganglioside of small intestine. Aids 17(2): 2520–2524.

Huang, L., I. Bosch, W. Hofmann, J. Sodroski, and A.B.J.J.O.V. Pardee.1998. Tat protein induces human immunodeficiency virus type 1 (HIV-1) coreceptors and promotes infection with both macrophage-tropic and T-lymphotropic HIV-1 strains. Journal of Virology 72(11): 8952–8960.

Huang, J., G. Ofek, L. Laub, M.K. Louder, N.A. Doria-Rose, N.S. Longo, H. Imamichi, R.T. Bailer, B. Chakrabarti, and S.K.J.N. Sharma. 2012. Broad and potent neutralization of HIV-1 by a gp41-specific human antibody. Nature 491(7424): 406–412.

Hung, C.-S., N. Vander Heyden, and L.J.J.O.V. Ratner.1999. Analysis of the critical domain in the V3 loop of human immunodeficiency virus type 1 gp120 involved in CCR5 utilization. Journal of Virology 73(10): 8216–8226.

Irimia, A., A. Sarkar, R.L. Stanfield, and I.A.J.I. Wilson. 2016. Crystallographic identification of lipid as an integral component of the epitope of HIV broadly neutralizing antibody 4E10. Immunity 44(1): 21–31.

Izmailova, E., F.M. Bertley, Q. Huang, N. Makori, C.J. Miller, R.A. Young, and A.J.N.M Aldovini. 2003. HIV-1 Tat reprograms immature dendritic cells to express chemoattractants for activated T cells and macrophages. Nature Medicine 9(2): 191.

Jaworowski, A., and S.M. Crowe. 1999. Does HIV cause depletion of CD4+ T cells in vivo by the induction of apoptosis? Immunology and Cell Biology 77(1): 90–98.

Jiao, Y., J. Xie, T. Li, Y. Han, Z. Qiu, L. Zuo, and A.J.J.J.O.A.I.D.S. Wang. 2006. Correlation between gag-specific CD8 T-cell responses, viral load, and CD4 count in HIV-1 infection is dependent on disease status. JAIDS Journal of Acquired Immune Deficiency Syndromes 42(3): 263–268.

Johansson, E.-L., C. Rask, M. Fredriksson, K. Eriksson, C. Czerkinsky, J.J.I. Holmgren, and Immunity. 1998. Antibodies and antibody-secreting cells in the female genital tract after vaginal or intranasal immunization with cholera toxin B subunit or conjugates. Infection and Immunity 66(2): 514–520.

Jones, K.A. and B.M.J.A.R.O.B. Peterlin. 1994. Control of RNA initiation and elongation at the HIV-1 promoter. Annual Review of Biochemistry 63(1): 717–743.

Karasev, A.V., S. Foulke, C. Wellens, A. Rich, K.J. Shon, I. Zwierzynski, D. Hone, H. Koprowski, and M.J.V. Reitz. 2005. Plant based HIV-1 vaccine candidate: Tat protein produced in spinach. Vaccine 23(15): 1875–1880.

Karg, S.R., and P.T.J.B.A. Kallio. 2009. The production of biopharmaceuticals in plant systems. Biotechnology Advances 27(6): 879–894.

Keele, B.F., F. Van Heuverswyn, Y. Li, E. Bailes, J. Takehisa, M.L. Santiago, F. Bibollet-Ruche, Y. Chen, L.V. Wain, and F.J.S. Liegeois. 2006. Chimpanzee reservoirs of pandemic and nonpandemic HIV-1. Science 313(5786): 523–526.

Kessans, S.A., M.D. Linhart, N. Matoba, and T. Mor. 2013a. Biological and biochemical characterization of HIV-1 Gag/dgp41 virus-like particles expressed in Nicotiana benthamiana. Plant Biotechnol. Journal 11(6): 681–90.

Kessans, S.A., M.D. Linhart, L.R. Meador, J. Kilbourne, B.G. Hogue, P. Fromme, N. Matoba, and T.S.J.P.O. Mor. 2016. Immunological characterization of plant-based HIV-1 Gag/Dgp41 virus-like particles. PLoS One 11(3): e0151842.

Kestier, III, H.W., D.J. Ringler, K. Mori, D.L. Panicali, P.K. Sehgal, M.D. Daniel, and R.C.J.C. Desrosiers. 1991. Importance of the nef gene for maintenance of high virus loads and for development of AIDS. Cell 65(4): 651–662.

Kiepiela, P., K. Ngumbela, C. Thobakgale, D. Ramduth, I. Honeyborne, E. Moodley, S. Reddy, C. De Pierres, Z. Mncube, and N.J.N.M. Mkhwanazi. 2007. CD8+ T-cell responses to different HIV proteins have discordant associations with viral load. Nature Medicine 13(1): 46.

Kim, T.-G., R. Ruprecht, and W.H.J.P.E. Langridge. 2004. Purification Synthesis and assembly of a cholera toxin B subunit SHIV 89.6 p Tat fusion protein in transgenic potato. Protein Expression and Purification 35(2): 313–319.

Kirchhoff, F., T.C. Greenough, D.B. Brettler, J.L. Sullivan, and R.C.J.N.E.J.O.M. Desrosiers. 1995. Absence of intact nef sequences in a long-term survivor with nonprogressive HIV-1 infection. New England Journal of Medicine 332(4): 228–232.

Klein, F., R. Diskin, J.F. Scheid, C. Gaebler, H. Mouquet, I.S. Georgiev, M. Pancera, T. Zhou, R.-B. Incesu, and B.Z.J.C. Fu. 2013. Somatic mutations of the immunoglobulin framework are generally required for broad and potent HIV-1 neutralization. Cell 153(1): 126–138.

Korber, B., M. Muldoon, J. Theiler, F. Gao, R. Gupta, A. Lapedes, B. Hahn, S. Wolinsky, and T.J.S. Bhattacharya. 2000. Timing the ancestor of the HIV-1 pandemic strains. Science 288(5472): 1789–1796.

Koup, R., J.T. Safrit, Y. Cao, C.A. Andrews, G. Mcleod, W. Borkowsky, C. Farthing, and D.D.J.J.O.V. Ho. 1994. Temporal association of cellular immune responses with the initial control of viremia in primary human immunodeficiency virus type 1 syndrome. Journal of Virology 68(7): 4650–4655.

Kozlowski, P.A., S. Cu-Uvin, M.R. Neutra, T.P.J.I. Flanigan, and Immunity.1997. Comparison of the oral, rectal, and vaginal immunization routes for induction of antibodies in rectal and genital tract Secretions Of Women. Infection and Immunity 65(4): 1387–1394.

Kozlowski, P.A., S.B. Williams, R.M. Lynch, T.P. Flanigan, R.R. Patterson, S. Cu-Uvin, and M.R.J.T.J.O.I. Neutra. 2002. Differential induction of mucosal and systemic antibody responses in women after nasal, rectal, or vaginal immunization: Influence of the menstrual cycle. The Journal of Immunology 169(1): 566–574.

Kundu, R.K., F. Sangiorgi, L.-Y. Wu, P.K. Pattengale, D.R. Hinton, P.S. Gill, and R.J.B. Maxson. 1999. Expression of the human immunodeficiency virus-Tat gene in lymphoid tissues of transgenic mice is associated with B-cell lymphoma. Blood, The Journal of the American Society of Hematology 94(1): 275–282.

Kwong, P.D., R. Wyatt, S. Majeed, J. Robinson, R.W. Sweet, J. Sodroski, and W.A. Hendrickson. 2000. Structures of HIV-1 gp120 envelope glycoproteins from laboratory-adapted and primary isolates. Structure Structure 8(12): 1329–1339.

Kwong, P.D., M.L. Doyle, D.J. Casper, C. Cicala, S.A. Leavitt, S. Majeed, T.D. Steenbeke, M. Venturi, I. Chaiken, and M.J.N. Fung. 2002. HIV-1 evades antibody-mediated neutralization through conformational masking of receptor-binding sites. Nature 420(6916): 678–682.

Lai, H. and Q.J.P.C.R. Chen. 2012. Bioprocessing of plant-derived virus-like particles of Norwalk virus capsid protein under current good manufacture practice regulations. Plant Cell Reports 31(3): 573–584.

Lai, R.P., J. Yan, J. Heeney, M.O. Mcclure, H. Göttlinger, J. Luban, and M.J.P.P. Pizzato. 2011. Nef decreases HIV-1 sensitivity to neutralizing antibodies that target the membrane-proximal external region of PLoS Pathogens 7(12): e1002442.

Landry, N., B.J. Ward, S. Trépanier, E. Montomoli, M. Dargis, G. Lapini, and L.-P.J.P.O. Vézina. 2010. Preclinical and clinical development of plant-made virus-like particle vaccine against avian H5N1 Influenza. PloS one 5(12): E15559.

Learmont, J.C., A.F. Geczy, J. Mills, L.J. Ashton, C.H. Raynes-Greenow, R.J. Garsia, W.B. Dyer, L. Mcintyre, R.B. Oelrichs, and D.I.J.N.E.J.O.M. Rhodes. 1999. Immunologic and virologic status after 14 to 18 years of infection with an attenuated strain of HIV-1—a report from the Sydney Blood Bank Cohort. New England Journal of Medicine 340(22): 1715–1722.

Leroux-Roels, G., C. Maes, F. Clement, F. Van Engelenburg, M. van den Dobbelsteen, M. Adler, M. Amacker, L. Lopalco, M. Bomsel, A. Chalifour, and S. Fleury. 2013. Randomized phase I: Safety, immunogenicity and mucosal antiviral activity in young healthy women vaccinated with HIV-1 Gp41 P1 peptide on virosomes. PloS one 8(2): e55438.

Lesieur, C., M.J. Cliff, R. Carter, R.F. James, A.R. Clarke, and T.R. Hirst. 2002. A kinetic model of intermediate formation during assembly of cholera toxin B-subunit pentamers. Journal of Biological Chemistry 277(19): 16697–16704

Li, H., K.J. Bar, S. Wang, J.M. Decker, Y. Chen, C. Sun, J.F. Salazar-Gonzalez, M.G. Salazar, G.H. Learn, C.J. Morgan, and J.E. Schumacher. 2010. High multiplicity infection by HIV-1 in men who have sex with men. PLoS Pathogens 6(5): e1000890.

Li, H., R.W. Omange, F.A. Plummer, and M. Luo. 2017. A novel HIV vaccine targeting the protease cleavage sites. AIDS Research and Therapy, 14(1): 1–5.

Li, Y., A.K. Kar, and J. Sodroski. 2009. Target cell type-dependent modulation of human immunodeficiency virus type 1 capsid disassembly by cyclophilin A. Journal of Virology 83(21): 10951–10962.

Liang, X., T.-M. Fu, H. Xie, E.A. Emini, and J.W.J.V. Shiver. 2002. Development of HIV-1 Nef vaccine components: Immunogenicity study of Nef mutants lacking myristoylation and dileucine Motif In Mice. Vaccine 20(27-28): 3413–3421.

Liao, H.-X., X. Chen, S. Munshaw, R. Zhang, D.J. Marshall, N. Vandergrift, J.F. Whitesides, X. Lu, J.-S. Yu, and K.-K.J.J.O.E.M. Hwang. 2011. Initial antibodies binding to HIV-1 gp41 in acutely infected subjects are polyreactive and highly mutated. Journal of Experimental Medicine 208(11): 2237–2249.

Lihana, R.W., D. Ssemwanga, A.L. Abimiku, and N.J.A.R. Ndembi. 2012. Update on HIV-1 diversity in Africa: a decade in review. Aids Rev. 14(2): 83–100.

Liljeqvist, S., S. Ståhl, C. Andreoni, H. Binz, M. Uhlen, and M.J.J.O.I.M. Murby. 1997. Fusions to the cholera toxin B subunit: Influence on pentamerization and GM1 binding. Journal of Immunological Methods 210(2): 125–135.

Lindenburg, C.E., I. Stolte, M.W. Langendam, F. Miedema, I.G. Williams, R. Colebunders, J.N. Weber, M. Fisher, and R.A.J.V. Coutinho. 2002. Long-term follow-up: no effect of therapeutic vaccination with HIV-1 p17/p24: Ty virus-like particles on HIV-1 disease progression. Vaccine 20(17-18): 2343–2347.

Lindh, I., I. Kalbina, S. Thulin, N. Scherbak, H. Sävenstrand, A. Bråve, J. Hinkula, A. Strid, and S.J.A. Andersson. 2008. Feeding of mice with Arabidopsis thaliana expressing the HIV-1 subtype C p24 antigen gives rise to systemic immune responses. Apmis 116(11): 985–994.

Lindh, I., A. Brave, D. Hallengard, R. Hadad, I. Kalbina, A. Strid, and S. Andersson. 2014a. Oral delivery of plant-derived HIV-1 p24 antigen in low doses shows a superior priming effect in mice compared to high doses. Vaccine 32(20): 2288–93.

Lindh, I., A. Bråve, D. Hallengärd, R. Hadad, I. Kalbina, A. Strid, and S.J.V. Andersson. 2014b. Oral delivery of plant-derived HIV-1 p24 antigen in low doses shows a superior priming effect in mice compared to high doses. Vaccine 32(20): 2288–2293.

Lombardi, R., P. Circelli, M.E. Villani, G. Buriani, L. Nardi, V. Coppola, L. Bianco, E. Benvenuto, M. Donini, and C.J.B.B. Marusic. 2009. High-level HIV-1 Nef transient expression in Nicotiana benthamiana using the P19 gene silencing suppressor protein of Artichoke Mottled Crinckle Virus. BMC Biotechnology 9: 96.

Loret, E.P., A. Darque, E. Jouve, E.A. Loret, C. Nicolino-Brunet, S. Morange, E. Castanier, J. Casanova, C. Caloustian, and C.J.R. Bornet. 2016. Intradermal injection of a Tat Oyi-based therapeutic HIV vaccine reduces of 1.5 log copies/mL the HIV RNA rebound median and no HIV DNA rebound following cART interruption in a phase I/II randomized controlled clinical trial. Retrovirology 13: 21.

Lycke, N., and J.J.I. Holmgren.1986. Strong adjuvant properties of cholera toxin on gut mucosal immune responses to orally presented antigens. Immunology 59(2): 301.

Lythgo, P.A. 2004. Molecular virology of HIV-1 and current antiviral strategies. Bio. Teach. J. 2: 82–88.

Ma, J.K.-C., A. Hiatt, M. Hein, N.D. Vine, F. Wang, P. Stabila, C. Van Dolleweerd, K. Mostov, and T.J.S. Lehner. 1995. Generation and assembly of secretory antibodies in plants. Science 268(5211): 716–719.

Ma, J.K., J. Drossard, D. Lewis, F. Altmann, J. Boyle, P. Christou, T. Cole, P. Dale, C.J. Van Dolleweerd, V. Isitt, D. Katinger, M. Lobedan, H. Mertens, M.J. Paul, T. Rademacher, M. Sack, P.A. Hundleby, G. Stiegler, E. Stoger, R.M. Twyman, B. Vcelar, and R. Fischer. 2015. Regulatory approval and a first-in-human phase I clinical trial of a monoclonal antibody produced in transgenic tobacco plants. Plant Biotechnology Journal 13(8): 1106–20.

Magérus-Chatinet, A., H. Yu, S. Garcia, E. Ducloux, B. Terris, and M.J.V. Bomsel. 2007. Galactosyl ceramide expressed on dendritic cells can mediate HIV-1 transfer from monocyte derived dendritic cells to autologous T cells. Virology 362(1): 67–74.

Makhzoum, A., R. Benyammi, K. Moustafa, and J. Trémouillaux-Guiller. 2014a. Recent advances on host plants and expression cassettes' structure and function in plant molecular pharming. BioDrugs 28(2): 145–159.

Makhzoum, A., S. Tahir, M.E.O. Locke, J. Trémouillaux-Guiller, and K. Hefferon. 2014b. An *in silico* overview on the usefulness of tags and linkers in plant molecular pharming. Plant Science Today 1(4): 201–212.

Makhzoum, A.B., P. Sharma, M.A. Bernards, and J. Trémouillaux-Guiller. 2013. Hairy roots: An ideal platform for transgenic plant production and other promising applications. Phytochemicals, Plant Growth, and the Environment. Springer. New York, NY, 95–142.

Mangasarian, A., V. Piguet, J.-K. Wang, Y.-L. Chen, and D.J.J.O.V. Trono. 1999. Nef-induced CD4 and major histocompatibility complex class I (MHC-I) down-regulation are governed by distinct determinants: N-terminal alpha helix and proline repeat of Nef selectively regulate MHC-I trafficking. Journal of Virology 73(3): 1964–1973.

Mangino, G., Z.A. Percario, G. Fiorucci, G. Vaccari, S. Manrique, G. Romeo, M. Federico, M. Geyer, and E.J.J.O.V. Affabris. 2007. *In vitro* treatment of human monocytes/macrophages with myristoylated recombinant Nef of human immunodeficiency virus type 1 leads to the activation of mitogen-activated protein kinases, IκB kinases, and interferon regulatory factor 3 and to the release of beta interferon. Journal of Virology 81(6): 2777–2791.

Martin, S.J., A. Vyakarnam, R. Cheingsong-Popov, D. Callow, K.L. Jones, J.M. Senior, S.E. Adams, A.J. Kingsman, P. Matear, and F.M.J.A. Gotch, 1993. Immunization of human HIV-seronegative volunteers with recombinant p17/p24: Ty virus-like particles elicits HIV-1 p24-specific cellular and humoral immune responses. AIDS (London, England) 7(10): 1315–1323.

Marusic, C., J. Nuttall, G. Buriani, C. Lico, R. Lombardi, S. Baschieri, E. Benvenuto, and L.J.B.B. Frigerio. 2007. Expression, intracellular targeting and purification of HIV Nef variants in tobacco cells. BMC Biotechnology 7(1): 12.

Mascola, J.R., G. Stiegler, T.C. Vancott, H. Katinger, C.B. Carpenter, C.E. Hanson, H. Beary, D. Hayes, S.S. Frankel, and D.L.J.N.M. Birx. 2000. Protection of macaques against vaginal transmission of a pathogenic HIV-1/SIV chimeric virus by passive infusion of neutralizing antibodies.Nature Medicine 6(2): 207.

Matoba, N., A. Magérus, B.C. Geyer, Y. Zhang, M. Muralidharan, A. Alfsen,C.J. Arntzen, M. Bomsel, and T.S.J.P.O.T.N.A.O.S. Mor. 2004. A mucosally targeted subunit vaccine candidate eliciting HIV-1 transcytosis-blocking Abs. Proceedings of the National Academy of Sciences 101(37): 13584–13589.

Matoba, N., T.A. Griffin, M. Mittman, J.D. Doran, A. Alfsen, D.C. Montefiori, C.V. Hanson, M. Bomsel, and T.S.J.C.H.R. Mor. 2008. Transcytosis-blocking abs elicited by an oligomeric immunogen based on the membrane proximal region of HIV-1 gp41 target non-neutralizing epitopes. Current HIV Research 6(3): 218–229.

Matoba, N., H. Kajiura, I. Cherni, J.D. Doran, M. Bomsel, K. Fujiyama, and T.S.J.P.B.J. Mor. 2009. Biochemical and immunological characterization of the plant-derived candidate human immunodeficiency virus type 1 mucosal vaccine CTB–MPR649–684. Plant Biotechnology Journal 7(2): 129–145.

Mazzoli, S., D. Trabaironi, S.L. Caputo, S. Piconi, C. Blé, F. Meacci, S. Ruzzante, A. Salvi, F. Semplici, and R.J.N.M. Longhi. 1997. HIV-specific mucosal and cellular immunity in HIV-seronegative partners of HIV-seropositive individuals. Nature Medicine 3(11): 1250.

Mccabe, M.S., M. Klaas, N. Gonzalez-Rabade, M. Poage, J.A. Badillo-Corona, F. Zhou, D. Karcher, R. Bock, J.C. Gray, and P.J.J.P.B.J. Dix. 2008. Plastid transformation of high-biomass tobacco variety Maryland Mammoth for production of human immunodeficiency virus type 1 (HIV-1) p24 antigen. Plant Biotechnology Journal 6(9): 914–929.

Mcguire, A.T., J.A. Glenn, A. Lippy, and L.J.J.O.V. Stamatatos. 2014. Diverse recombinant HIV-1 Envs fail to activate B cells expressing the germline B cell receptors of the broadly neutralizing anti-HIV-1 Antibodies PG9 And 447-52D. Journal of Virology 88(5): 2645–2657.

Mcinerney, T.L., F.R. Brennan, T.D. Jones, and N.J. Dimmock. 1999. Analysis of the ability of five adjuvants to enhance immune responses to a chimeric plant virus displaying an HIV-1 peptide. Vaccine Vaccine 17(11-12): 1359–1368.

Mclain, L., C. Porta, G.P. Lomonossoff, Z. Durrani, and N.J. Dimmock. 1995. Human immunodeficiency virus type 1-neutralizing antibodies raised to a glycoprotein 41 peptide expressed on the surface of a plant virus. AIDS Res. Hum. Retroviruses AIDS Research and Human Retroviruses 11(3): 327–334.

Mclain, L., Z. Durrani, L.A. Wisniewski, C. Porta, G.P. Lomonossoff, and N.J.J.V. Dimmock. 1996. Stimulation of neutralizing antibodies to human immunodeficiency virus type 1 in three strains of mice immunized with a 22 amino acid peptide of gp41 expressed on the surface of a plant virus. Vaccine 14(8): 799–810.

Meador, L.R., S.A. Kessans, J. Kilbourne, K.V. Kibler, G. Pantaleo, M.E. Roderiguez, J.N. Blattman, B.L. Jacobs, and T.S.J.V. Mor. 2017. A heterologous prime-boosting strategy with replicating Vaccinia virus vectors and plant-produced HIV-1 Gag/dgp41 virus-like particles. Virology 507: 242–256.

Meng, G., X. Wei, X. Wu, M.T. Sellers, J.M. Decker, Z. Moldoveanu, J.M. Orenstein, M.F. Graham, J.C. Kappes, and J.J.N.M. Mestecky. 2002. Primary intestinal epithelial cells selectively transfer R5 HIV-1 to CCR5+ cells. Nature medicine 8(2): 150.

Meyers, A., E. Chakauya, E. Shephard, F.L. Tanzer, J. Maclean, A. Lynch, A-L. Williamson, and E.P.J.B.B. Rybicki. 2008a. Expression of HIV-1 antigens in plants as potential subunit vaccines. BMC Biotechnology 8: 53.

Mishra, R.C., A. Singh, L.D. Tiwari, A.J.C.S. Grover, and Chaperones. 2016. Characterization of 5′ UTR of rice ClpB-C/Hsp100 gene: Evidence of its involvement in post-transcriptional regulation. Cell Stress and Chaperones 21(2): 271–283.

Moustafa, K., A. Makhzoum, and J. Trémouillaux-Guiller. 2016. Molecular farming on rescue of pharma industry for next generations. Critical Reviews in Biotechnology 36(5): 840–850.

Muster, T., F. Steindl, M. Purtscher, A. Trkola, A. Klima, G. Himmler, F. Rüker, and H.J.J.O.V. Katinger. 1993. A conserved neutralizing epitope on gp41 of human immunodeficiency virus type 1. Journal of Virology 67(11): 6642–6647.

Muthumani, K., M. Bagarazzi, D. Conway, D. Hwang, V. Ayyavoo, D. Zhang, K. Manson, J. Kim, J. Boyer, and D.J.J.O.M.P. Weiner. 2002. Inclusion of Vpr accessory gene in a plasmid vaccine cocktail markedly reduces Nef vaccine effectiveness *in vivo* resulting in CD4 cell loss and increased viral loads in rhesus macaques. Journal of Medical Primatology 31(4-5): 179–185.

Nasir, W., J. Nilsson, S. Olofsson, M. Bally, and G.E.J.V. Rydell. 2014. Parvovirus B19 VLP recognizes globoside in supported lipid bilayers. Virology 456: 364–369.

Nasir, W., M. Bally,, V.P. Zhdanov, G.R. Larson, and F.J.T.J.O.P.C.B. Höök. 2015. Interaction of Virus-Like Particles With Vesicles Containing Glycolipids: Kinetics Of Detachment. The Journal of Physical Chemistry B 119(35): 11466–11472.

Nath, A., K. Psooy, C. Martin, B. Knudsen, D. Magnuson, N. Haughey, and J.D.J.J.O.V. Geiger. 1996. Identification of a human immunodeficiency virus type 1 Tat epitope that is neuroexcitatory and neurotoxic. Journal of Virology 70(3): 1475–1480.

Nemazee, D., and M.J.J.O.E.M. Weigert. 2000. Revising B cell receptors. The Journal of experimental Medicine 191(11): 1813–1818.

Neutra, M.R., and P.A.J.N.R.I. Kozlowski. 2006. Mucosal vaccines: the promise and the challenge. Nature Reviews Immunology 6(2): 148–158.

Nezafat, N., Z. Karimi, M. Eslami, M. Mohkam, S. Zandian, Y.J.C.B. Ghasemi, and Chemistry. 2016. Designing an efficient multi-epitope peptide vaccine against Vibrio cholerae via combined immunoinformatics and protein interaction based approaches. Computational Biology and Chemistry 62: 82–95.

Novikova, M., Y. Zhang, E.O. Freed, and K. Peng. 2019. Multiple roles of HIV-1 capsid during the virus replication cycle. Virologica Sinica 34(2): 119–134.

Novitsky, V., P. Gilbert, T. Peter, M.F. Mclane, S. Gaolekwe, N. Rybak, I. Thior, T. Ndung'u, R. Marlink, and T.H.J.J.O.V. Lee. 2003. Association between virus-specific T-cell responses and plasma viral load in human immunodeficiency virus type 1 subtype C infection. Journal of Virology 77(2): 882–890.

Nowak, M.A., R.M. May, and R.M.J.A. Anderson. 1990. The evolutionary dynamics of HIV-1 quasispecies and the development of immunodeficiency disease. Aids 4(11): 1095–1103.

Nuttall, J., N. Vine, J.L. Hadlington, P. Drake, L. Frigerio, and J.K.C.J.E.J.O.B. Ma. 2002. ER-resident chaperone interactions with recombinant antibodies in transgenic plants. European Journal of Biochemistry 269(24): 6042–6051.

Nyamweya, S., A. Hegedus, A. Jaye, S. Rowland-Jones, K.L. Flanagan, and D.C. Macallan. 2013. Comparing HIV-1 and HIV-2 infection: Lessons for viral immunopathogenesis. Reviews in Medical Virology 23(4): 221–240.

Obregon, P., D. Chargelegue, P.M. Drake, A. Prada, J. Nuttall, L. Frigerio, and J.K. Ma. 2006a. HIV-1 p24-immunoglobulin fusion molecule: A new strategy for plant-based protein production. Plant Biotechnology Journal 4(2): 195–207.

Orellana-Escobedo, L., S. Rosales-Mendoza, A. Romero-Maldonado, J. Parsons, E.L. Decker, E. Monreal-Escalante, L. Moreno-Fierros, and R. Reski. 2015. An Env-derived multi-epitope HIV chimeric protein produced in the moss Physcomitrella patens is immunogenic in mice. Plant Cell Reports 34(3): 425–33.

Palacios, F., P. Moreno, P. Morande, C. Abreu, A. Correa, V. Porro, A.I. Landoni, R. Gabus, M. Giordano, and G.J.B. Dighiero. 2010. High expression of AID and active class switch recombination might accounts for a more aggressive disease in unmutated CLL patients: link with an activated microenvironment in CLL disease. Blood, The Journal of the American Society of Hematology 115(22): 4488–4496.

Paliard, X., Y. Liu, R. Wagner, H. Wolf, J. Baenziger, C.M.J.A.R. Walker, and H. Retroviruses. 2000. Priming of strong, broad, and long-lived HIV type 1 p55gag-specific CD8+ cytotoxic T cells after administration of a virus-like particle vaccine in rhesus macaques. AIDS Research and Human Retroviruses 16(3): 273–282.

Palker, T.J., M.E. Clark, A.J. Langlois, T.J. Matthews, K.J. Weinhold, R.R. Randall, D.P. Bolognesi, and B.F.J.P.O.T.N.A.O.S. Haynes. 1988. Type-specific neutralization of the human Immunodeficiency Virus With Antibodies To Env-Encoded Synthetic Peptides. Proceedings of the National Academy of Sciences 85(6): 1932–1936.

Pancera, M., J.S. Mclellan, X. Wu, J. Zhu, A. Changela, S.D. Schmidt, Y. Yang, T. Zhou, S. Phogat, and J.R.J.J.O.V. Mascola. 2010. Crystal structure of PG16 and chimeric dissection with somatically related PG9: Structure-function analysis of two quaternary-specific antibodies that effectively neutralize HIV-1. Journal of Virology 84(16): 8098–8110.

Pankrac, J., K. Klein, P.F. Mckay, D.F. King, K. Bain, J. Knapp, T. Biru, C.N. Wijewardhana, R. Pawa, and D.H. Canaday. 2018. A heterogeneous human immunodeficiency virus-like particle (VLP) formulation produced by a novel vector system. NPJ vaccines 3(1): 1–10.

Parker, C.E., L.J. Deterding, C. Hager-Braun, J.M. Binley, N. Schülke, H. Katinger, J.P. Moore, and K.B.J.J.O.V. Tomer. 2001. Fine definition of the epitope on the gp41 glycoprotein of human immunodeficiency virus type 1 for the neutralizing monoclonal antibody 2F5. Journal of Virology 75(22): 10906–10911.

Patterson, L.J., F. Robey, A. Muck, K. Van Remoortere, K. Aldrich, E. Richardson, W.G. Alvord, P.D. Markham, M. Cranage, M.J.A.R. Robert-Guroff, and H. Retroviruses. 2001. A conformational C4 peptide polymer vaccine coupled with live recombinant vector priming is immunogenic but does not protect against rectal SIV challenge. AIDS Research and Human Retroviruses 17(9): 837–849.

Paul, M., R. Reljic, K. Klein, P.M. Drake, C. Van Dolleweerd, M. Pabst, M. Windwarder, E. Arcalis, E. Stoger, F. Altmann, C. Cosgrove, A. Bartolf, S. Baden, and J.K. Ma. 2014. Characterization of a plant-produced recombinant human secretory IgA with broad neutralizing activity against HIV. mAbs 6: 1585–1597.

Pauza, C.D., P. Trivedi, M. Wallace, T.J. Ruckwardt, H. Le Buanec, W. Lu, B. Bizzini, A. Burny, D. Zagury, and R.C.J.P.O.T.N.A.O.S. Gallo. 2000. Vaccination with tat toxoid attenuates disease in simian/HIV-challenged macaques. Proceedings of the National Academy of Sciences 97(7): 3515–3519.

Peng, B., and M.J.I.L. Robert-Guroff. 2001. Deletion of N-terminal myristoylation site of HIV Nef abrogates both MHC-1 and CD4 down-regulation. Immunology Letters 78(3): 195–200.

Peng, B., R. Voltan, A.D. Cristillo, W.G. Alvord, A. Davis-Warren, Q. Zhou, K.K. Murthy, and M.J.A. Robert-Guroff. 2006. Replicating Ad-recombinants encoding non-myristoylated rather than wild-type HIV Nef elicit enhanced cellular immunity. Aids 20(17): 2149–2157.

Pérez-Filgueira, D., B. Brayfield, S. Phiri, M. Borca, C. Wood, and T.J.J.J.O.V.M. Morris. 2004. Preserved antigenicity of HIV-1 p24 produced and purified in high yields from plants inoculated with a tobacco mosaic virus (TMV)-derived vector. Journal of Virological Methods 121(2): 201–208.

Peters, B., R. Cheingsong-Popov, D. Callow, R. Foxall, G. Patou, K. Hodgkin, and J.J.J.O.I. Weber. 1997. A pilot phase II study of the safety and immunogenicity of HIV p17/p24: VLP (p24-VLP) in asymptomatic HIV seropositive subjects. Journal of Infection 35(3): 231–235.

Peyret, H., and G.P. Lomonossoff. 2015. When plant virology met Agrobacterium: The rise of the deconstructed clones. Plant Biotechnology Journal 13(8): 1121–1135.

Podell, S., and M.J.B.G. Gribskov. 2004. Predicting N-terminal myristoylation sites in plant proteins. BMC genomics 5: 37.

Prakash, D. 2010. Phytochemical molecules. International Journal of Pharma and Bio Sciences 1: 2.

Prakash, O., S. Teng, M. Ali, X. Zhu, R. Coleman, R.A. Dabdoub, R. Chambers, T.Y. Aw, S.C. Flores, and B.H. Joshi. 1997. The human immunodeficiency virus type 1 Tat protein potentiates zidovudine-induced cellular toxicity in transgenic mice. Archives of Biochemistry and Biophysics 343(2): 173–180.

Pritchard, L.K., D.J. Harvey, C. Bonomelli, M. Crispin, and K.J. Doores. 2015. Cell-and protein-directed glycosylation of native cleaved HIV-1 envelope. Journal of Virology 89(17): 8932–8944.

Quintana, F.J., D. Gerber, S.C. Kent, I.R. Cohen, and Y.J.T.J.O.C.I. Shai. 2005. HIV-1 fusion peptide targets the TCR and inhibits antigen-specific T cell activation. The Journal of clinical investigation 115(8): 2149–2158.

Raidel, S.M., C. Haase, N.R. Jansen, R.B. Russ, R.L. Sutliff, L.W. Velsor, B.J. Day, B.D. Hoit, A.M. Samarel, W.J.A.J.O.P.-H. Lewis, and C. Physiology. 2002. Targeted myocardial transgenic expression of HIV Tat causes cardiomyopathy and mitochondrial damage. American Journal of Physiology-Heart and Circulatory Physiology 282(5): H1672–H1678.

Rajewsky, K.J.N. 1996. Clonal Selection And Learning In The Antibody System. Nature 381(6585): 751–758.

Ramírez, Y.J.P., E. Tasciotti, A. Gutierrez-Ortega, A.J.D. Torres, M.T.O. Flores, M. Giacca, M.Á.G.J.C. Lim, and V. Immunology. 2007. Fruit-specific expression of the human immunodeficiency virus type 1 tat gene in tomato plants and its immunogenic potential in mice. Clinical and Vaccine Immunology 14(6): 685–692.

Rappaport, J., J. Joseph, S. Croul, G. Alexander, L.D. Valle, S. Amini, and K.J.J.O.L.B. Khalili. 1999. Molecular pathway involved in HIV-1-induced CNS pathology: Role of viral regulatory protein, Tat. Journal of Leukocyte Biology 65(4): 458–465.

Rasty, S., P. Thatikunta, J. Gordon, K. Khalili, S. Amini, and J.C. Glorioso. 1996. Human immunodeficiency virus tat gene transfer to the murine central nervous system using a replication-defective herpes simplex virus vector stimulates transforming growth factor beta 1 gene expression. Proceedings of the National Academy of Sciences 93(12): 6073–8.

Rathore, U., M. Purwar, V.S. Vignesh, R. Das, A.A. Kumar, S. Bhattacharyya, H. Arendt, J. Destefano, A. Wilson, and C. Parks. 2018. Bacterially expressed HIV-1 gp120 outer-domain fragment immunogens with improved stability and affinity for CD4-binding site neutralizing antibodies. Journal of Biological Chemistry 293(39): 15002–15020.

Re, M.C., G. Furlini, M. Vignoli, E. Ramazzotti, G. Zauli, M.J.C. La Placa, and D.L. Immunology. 1996. Antibody against human immunodeficiency virus type 1 (HIV-1) Tat protein may have influenced the progression of AIDS in HIV-1-infected hemophiliac patients. Clinical Diagnostic Laboratory Immunology 3(2): 230–232.

Regnard, G.L., R.P. Halley-Stott, F.L. Tanzer, Ii. Hitzeroth, and E.P. Rybicki. 2010. High level protein expression in plants through the use of a novel autonomously replicating geminivirus shuttle vector. Plant Biotechnology Journal 8(1): 38–46.

Regoes, R.R., I.M. Longini, Jr., M.B. Feinberg, and S.I.J.P.M. Staprans. 2005. Preclinical assessment of HIV vaccines and microbicides by repeated low-dose virus challenges. PLoS Medicine 2(8): e249.

Rerks-Ngarm, S., P. Pitisuttithum, S. Nitayaphan, J. Kaewkungwal, J. Chiu, R. Paris, N. Premsri, C. Namwat, M. De Souza, and E.J.N.E.J.O.M. Adams. 2009. Vaccination with ALVAC and

AIDSVAX to prevent HIV-1 infection in Thailand. New England Journal of Medicine 361(23): 2209–2220.

Rinaldo, C., X-L. Huang, Z. Fan, M. Ding, L. Beltz, A. Logar, D. Panicali, G. Mazzara, J. Liebmann, and M.J.J.O.V. Cottrill. 1995. High levels of anti-human immunodeficiency virus type 1 (HIV-1) memory cytotoxic T-lymphocyte activity and low viral load are associated with lack of disease in HIV-1-infected long-term nonprogressors. Journal of Virology 69(9): 5838–5842.

Robb, M.L., S. Rerks-Ngarm, S. Nitayaphan, P. Pitisuttithum, J. Kaewkungwal, P. Kunasol, C. Khamboonruang, P. Thongcharoen, P. Morgan, and M.J.T.L.I.D. Benenson. 2012. Risk behaviour and time as covariates for efficacy of the HIV vaccine regimen ALVAC-HIV (vCP1521) and AIDSVAX B/E: A post-hoc analysis of the Thai phase 3 efficacy trial RV The Lancet Infectious Diseases 12(7): 531–537.

Rosales-Mendoza, S., N. Rubio-Infante, E. Monreal-Escalante, D.O. Govea-Alonso, A.L. García-Hernández, J.A. Salazar-González, O. González-Ortega, L.M. Paz-Maldonado and L. Moreno-Fierros. 2014. Chloroplast expression of an HIV envelop-derived multiepitope protein: towards a multivalent plant-based vaccine. Plant Cell, Tissue and Organ Culture (PCTOC) 116(1): 111–123.

Rubio-Infante, N., D.O. Govea-Alonso, A. Romero-Maldonado, A.L. García-Hernández, D. Ilhuicatzi-Alvarado, J.A. Salazar-González, S.S. Korban, S. Rosales-Mendoza, and L.J.M.B. Moreno-Fierros. 2015. A plant-derived multi-HIV antigen induces broad immune responses in orally immunized mice. Molecular Biotechnology 57(7): 662–674.

Rudin, A., E.-L. Johansson, C. Bergquist, and J. Holmgren. 1998. Differential kinetics and distribution of antibodies in serum and nasal and vaginal secretions after nasal and oral vaccination of humans. Infection and Immunity 66(7): 3390–3396.

Rudometov, A.P., A.N. Chikaev, N.B. Rudometova, D.V. Antonets, A.A. Lomzov, O.N. Kaplina, A.A. Ilyichev, and L.I. Karpenko. 2019. Artificial anti-HIV-1 immunogen comprising epitopes of broadly neutralizing antibodies 2F5, 10E8, and a peptide mimic of VRC01 discontinuous epitope. Vaccines 7(3): 83.

Ruf, S., D. Karcher, and R.J.P.O.T.N.A.O.S. Bock. 2007. Determining the transgene containment level provided by chloroplast transformation. Proceedings of the National Academy of Sciences 104(17): 6998–7002.

Rybicki, E. P. 2017. Plant-made vaccines and reagents for the One Health initiative. Human Vaccines & Immunotherapeutics 13(12): 2912–2917.

Sabatier, J., E. Vives, K. Mabrouk, A. Benjouad, H. Rochat, A. Duval, B. Hue, and E.J.J.O.V. Bahraoui. 1991. Evidence for neurotoxic activity of tat from human immunodeficiency virus type 1. Journal of Virology 65(2): 961–967.

Sack, M., A. Paetz, R. Kunert, M. Bomble, F. Hesse, G. Stiegler, R. Fischer, H. Katinger, E. Stoeger, and T. Rademacher. 2007. Functional analysis of the broadly neutralizing human anti-HIV-1 antibody 2F5 produced in transgenic BY-2 suspension cultures. The FASEB Journal 21(8): 1655–1664.

Sainsbury, F., M. Sack, J. Stadlmann, H. Quendler, R. Fischer, and G.P.J.P.O. Lomonossoff. 2010. Rapid transient production in plants by replicating and non-replicating vectors yields high quality functional anti-HIV antibody. PLoS One 5(11): e13976.

Scotti, N., F. Alagna, E. Ferraiolo, G. Formisano, L. Sannino, L. Buonaguro, A. De Stradis, A. Vitale, L. Monti, and S.J.P. Grillo. 2009. High-level expression of the HIV-1 Pr55 gag polyprotein in transgenic tobacco chloroplasts. Planta 229(5): 1109–1122.

Scotti, N., L. Sannino, A. Idoine, P. Hamman, A. De Stradis, P. Giorio, L. Maréchal-Drouard, R. Bock, and T.J.T.R. Cardi. 2015. The HIV-1 Pr55 gag polyprotein binds to plastidial membranes and leads to severe impairment of chloroplast biogenesis and seedling lethality in transplastomic tobacco plants. Transgenic Research 24(2): 319–331.

Secchiero, P., D. Zella, S. Capitani, R.C. Gallo, and G.J.T.J.O.I. Zauli. 1999. Extracellular HIV-1 tat protein up-regulates the expression of surface CXC-chemokine receptor 4 in resting CD4+ T cells. The Journal of Immunology 162(4): 2427–2431.

Sexton, A., S. Harman, R.J. Shattock, and J.K.-C.J.T.F.J. Ma. 2009a. Design, expression, and characterization of a multivalent, combination HIV microbicide. The FASEB Journal 23(10): 3590–3600.

Sharp, P.M., E. Bailes, R.R. Chaudhuri, C.M. Rodenburg, M.O. Santiago, and B.H. Hahn. 2001. The origins of acquired immune deficiency syndrome viruses: Where and when? Philosophical Transactions of the Royal Society of London. Series B: Biological Sciences 356(1410): 867–876.

Shchelkunov, S.N., and G.A. Shchelkunova. 2010. Plant-based vaccines against human hepatitis B virus. Expert Review of Vaccines 9(8): 947–955.

Shih, S.M.H., and P.M. Doran. 2009. Foreign protein production using plant cell and organ cultures: advantages and limitations. Biotechnology Advances 27(6): 1036–1042.

Shioda, T., J.A. Levy, and C. Cheng-Mayer. 1991. Macrophage and T cell-line tropisms of HIV-1 are determined by specific regions of the envelope gp! 20 gene. Nature 349(6305): 167–169.

Shrivastava, S., J. Trivedi, and D. Mitra. 2016. Gene expression profiling reveals Nef induced deregulation of lipid metabolism in HIV-1 infected T cells. Biochemical and Biophysical Research Communications 472(1): 169–174.

Sijmons, P.C., B.M. Dekker, B. Schrammeijer, T.C. Verwoerd, P.J. Van Den Elzen, and A. Hoekema. 1990. Production of correctly processed human serum albumin in transgenic plants. Bio/technology 8(3): 217–221.

Singhabahu, S., K. Hefferon, and A. Makhzoum. 2017. Transgenesis and plant molecular pharming. Transgenesis and Secondary Metabolism, Springer 571.

Smith, D., I. Gow, R. Colebunders, I. Weller, S. Tchamouroff, J. Weber, F. Boag, G. Hales, S. Adams, G. Patou, and D.A. Cooper. 2001. Therapeutic vaccination (p24-VLP) of patients with advanced HIV-1 infection in the pre-HAART era does not alter CD4 cell decline. HIV Medicine 2(4): 272–275.

Steimer, K.S., C.J. Scandella, P.V. Skiles, and N.L. Haigwood. 1991. Neutralization of divergent HIV-1 isolates by conformation-dependent human antibodies to Gp120. Science 254(5028): 105–108.

Steinaa, L., A.M. Sørensen, J.O. Nielsen, and J.E.S. Hansen. 1994. Antibody to HIV-1 Tat protein inhibits the replication of virus in culture. Archives of Virology 139(3): 263–271.

Stephenson, K.E., G.H. Neubauer, U. Reimer, N. Pawlowski, T. Knaute, J. Zerweck, B.T. Korber, and D.H. Barouch. 2015. Quantification of the epitope diversity of HIV-1-specific binding antibodies by peptide microarrays for global HIV-1 vaccine development. Journal of Immunological Methods 416: 105–123.

Stoddart, C.A., R. Geleziunas, S. Ferrell, V. Linquist-Stepps, M.E. Moreno, C. Bare, W. Xu, W. Yonemoto, P.A. Bresnahan, J.M. McCune, and W.C. Greene. 2003. Human immunodeficiency virus type 1 Nef-mediated downregulation of CD4 correlates with Nef enhancement of viral pathogenesis. Journal of Virology 77(3): 2124–2133.

Sutton, L.-A., A. Agathangelidis, C. Belessi, N. Darzentas, F. Davi, P. Ghia, R. Rosenquist, and K. Stamatopoulos. 2013. Antigen selection in B-cell lymphomas—tracing the evidence. Seminars in cancer biology. Academic Press 23: 399–409.

Thomas, D.R., C.A. Penney, A. Majumder, and A.M. Walmsley. 2011. Evolution of plant-made pharmaceuticals. International Journal of Molecular Sciences 12(5): 3220–3236.

Tikhonov, I., T.J. Ruckwardt, G.S. Hatfield, and C.D. Pauza. 2003. Tat-neutralizing antibodies in vaccinated macaques. Journal of Virology 77(5): 3157–3166.

Trujillo, J.R., W.-K. Wang, T.-H. Lee, and M. Essex. 1996. Identification of the envelope V3 loop as a determinant of a CD4-negative neuronal cell tropism for HIV-1. Virology 217(2): 613–617.

Tudor, D., M. Derrien, L. Diomede, A. Drillet, M. Houimel, C. Moog, J.-M. Reynes, L. Lopalco, and M.J.M.I. Bomsel. 2009. HIV-1 gp41-specific monoclonal mucosal IgAs derived from highly exposed but IgG-seronegative individuals block HIV-1 epithelial transcytosis and neutralize CD4+ cell infection: An IgA gene and functional analysis. Mucosal Immunology 2(5): 412–426.

Tyagi, M., M. Rusnati, M. Presta, and M. Giacca. 2001. Internalization of HIV-1 tat requires cell surface heparan sulfate proteoglycans. Journal of Biological Chemistry 276(5): 3254–3261.

Vajdy, M., and N.Y. Lycke. 1992. Cholera toxin adjuvant promotes long-term immunological memory in the gut mucosa to unrelated immunogens after oral immunization. Immunology 75(3): 488.

Vamvaka, E., R.M. Twyman, A.M. Murad, S. Melnik, A.Y.H. Teh, E. Arcalis, F. Altmann, E. Stoger, E. Rech, J.K. Ma, and P. Christou. 2016. Rice endosperm produces an underglycosylated and potent form of the HIV-neutralizing monoclonal antibody 2G12. Plant biotechnology Journal 14(1): 97–108.

van Baalen, C.A., O. Pontesilli, R.C. Huisman, A.M. Geretti, M.R. Klein, F. de Wolf, F. Miedema, R.A. Gruters, and A.D. Osterhaus. 1997. Human immunodeficiency virus type 1 Rev-and Tat-specific cytotoxic T lymphocyte frequencies inversely correlate with rapid progression to AIDS. Journal of General Virology 78(8): 1913–1918.

Van Schooten, J., and M.J. Van Gils. 2018. HIV-1 immunogens and strategies to drive antibody responses towards neutralization breadth. Retrovirology 15(1): 1–11.

Viscidi, R.P., K. Mayur, H.M. Lederman, and A.D. Frankel. 1989. Inhibition of antigen-induced lymphocyte proliferation by Tat protein from HIV-1. Science 246(4937): 1606–1608.

Vives, E., P. Brodin, and B. Lebleu. 1997. A truncated HIV-1 Tat protein basic domain rapidly translocates through the plasma membrane and accumulates in the cell nucleus. Journal of Biological Chemistry 272(25): 16010–16017.

Vogel, J., S.H. Hinrichs, R.K. Reynolds, P.A. Luciw, and G. Jay. 1988. The HIV tat gene induces dermal lesions resembling Kaposi's sarcoma in transgenic mice. Nature, 335(6191): 606–611.

Wang, H., X. Chen, D. Wang, C. Yao, Q. Wang, J. Xie, X. Shi, Y. Xiang, W. Liu, and L. Zhang. 2018. Epitope-focused immunogens against the CD4-binding site of HIV-1 envelope protein induce neutralizing antibodies against auto-and heterologous viruses. Journal of Biological Chemistry 293(3): 830–846.

Wang, X., A. Sherman, G. Liao, K.W. Leong, H. Daniell, C. Terhorst, and R.W. Herzog. 2013. Mechanism of oral tolerance induction to therapeutic proteins. Advanced Drug Delivery Reviews 65(6): 759–773.

Wardemann, H., S. Yurasov, A. Schaefer, J.W. Young, E. Meffre, and M.C. Nussenzweig. 2003. Predominant autoantibody production by early human B cell precursors. Science 301(5638): 1374–1377.

Wardemann, H., and M.C. Nussenzweig. 2007. B-cell self-tolerance in humans. Advances in Immunology 95: 83–110.

Watts, J.M., K.K. Dang, R.J. Gorelick, C.W. Leonard, J.W. Bess Jr., R. Swanstrom, C.L. Burch, and K.M. Weeks. 2009. Architecture and secondary structure of an entire HIV-1 RNA genome. Nature 460(7256): 711–716.

Weber, J., R. Cheinsong-Popov, D. Callow, S. Adams, G. Patou, K. Hodgkin, S. Martin, F. Gotch, and A. Kingsman. 1995. Immunogenicity of the yeast recombinant p17p24: Ty virus-like particles (p24-VLP) in healthy volunteers. Vaccine 13(9): 831–834.

Wesselingh, S.L., C. Power, J.D. Glass, W.R. Tyor, J.C. McArthur, J.M. Farber, J.W. Griffin, and D.E. Griffin. 1993. Intracerebral cytokine messenger RNA expression in acquired immunodeficiency syndrome dememtia. Annals of Neurology: Official Journal of the American Neurological Association and the Child Neurology Society 33(6): 576–582.

Westley, B.P., A.K. DeLong, C.S. Tray, D. Sophearin, E.M. Dufort, E. Nerrienet, L. Schreier, J.I. Harwell, and R. Kantor. 2012. Prediction of treatment failure using 2010 World Health Organization Guidelines is associated with high misclassification rates and drug resistance among HIV-infected Cambodian children. Clinical Infectious Diseases 55(3): 432–440.

Wilen, C.B., J.C. Tilton, and R.W. Doms. 2012. HIV: Cell binding and entry. Cold Spring Harbor perspectives in medicine, 2(8): a006866.

Wu, M., M. Li, Y. Yue, and W. Xu. 2016. DNA vaccine with discontinuous T-cell epitope insertions into HSP65 scaffold as a potential means to improve immunogenicity of multi-epitope Mycobacterium tuberculosis vaccine. Microbiology and Immunology 60(9): 634–645.

Wu, Y., and J.W. Marsh. 2001. Selective transcription and modulation of resting T cell activity by preintegrated HIV DNA. Science 293(5534): 1503–1506..

Wu, Y., D. Liang, Y. Wang, M. Bai, W. Tang, S. Bao, Z. Yan, D. Li, and J. Li. 2013. Correction of a genetic disease in mouse via use of CRISPR-Cas9. Cell Stem Cell 13(6): 659–662.

Wu, Z., S.C. Kayman, W. Honnen, K. Revesz, H. Chen, S. Vijh-Warrier, S.A. Tilley, J. McKeating, C. Shotton, and A. Pinter. 1995. Characterization of neutralization epitopes in the V2 region of human immunodeficiency virus type 1 gp120: role of glycosylation in the correct folding of the V1/V2 domain. Journal of Virology 69(4): 2271–2278.

Wyatt, R., and J. Sodroski. 1998. The HIV-1 envelope glycoproteins: fusogens, antigens, and immunogens. Science 280(5371): 1884–1888.

Yu, H., A. Alfsen, D. Tudor, and M. Bomsel. 2008. The binding of HIV-1 gp41 membrane proximal domain to its mucosal receptor, galactosyl ceramide, is structure-dependent. Cell Calcium 43(1): 73–82.

Zanini, F., J. Brodin, L. Thebo, C. Lanz, G. Bratt, J. Albert, and R.A. Neher. 2015. Population genomics of intrapatient HIV-1 evolution. Elife 4: e11282.

Zhang, G., C. Leung, L. Murdin, B. Rovinski, and K.A. White. 2000. In planta expression of HIV-1 p24 protein using and RNA plant virus-based expression vector. Molecular Biotechnology 14(2): 99–107.

Zhang, G.G., L. Rodrigues, B. Rovinski, and K.A. White. 2002. Production of HIV-1 p24 protein in transgenic tobacco plants. Molecular Biotechnology 20(2): 131–136.

Zhou, F., J.A. Badillo-Corona, D. Karcher, N. Gonzalez-Rabade, K. Piepenburg, A.M.I. Borchers, A.P. Maloney, T.A. Kavanagh, J.C. Gray, and R. Bock. 2008. High-level expression of human immunodeficiency virus antigens from the tobacco and tomato plastid genomes. Plant Biotechnology Journal 6(9): 897–913.

Zhou, T., I. Georgiev, X. Wu, Z.Y. Yang, K. Dai, A. Finzi, Y. Do Kwon, J.F. Scheid, W. Shi, L. Xu, and Y. Yang. 2010. Structural basis for broad and potent neutralization of HIV-1 by antibody VRC01. Science 329(5993): 811–817.

Zolla-Pazner, S. 2004. Identifying epitopes of HIV-1 that induce protective antibodies. Nature Reviews Immunology 4(3): 199–210.

Index

Q

R

S

T

U

V

W

Y

Z

About the Editors

Professor Abdullah Makhzoum received his Ph.D. from the University of Tours in France and completed his postdoctoral fellowship as well as a position as visitor Scientist at Western University in Ontario Canada. Abdullah has published multiple research papers, chapters, and reviews, and He is now an Associate professor in the Department of Biological Sciences and Biotechnology at BIUST University Botswana. His research interests include most aspects of plant Biotechnology and its applications such as plant secondary metabolism, plant molecular pharming, phytoremediation, and Plant Bioinformatics.

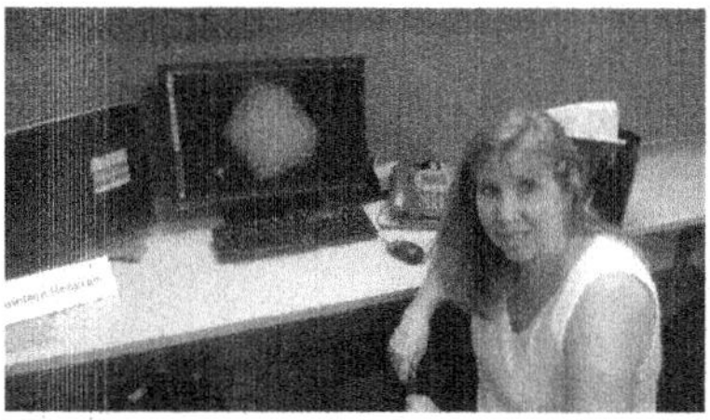

Professor Kathleen Hefferon received her Ph.D. from the Department of Medical Biophysics, University of Toronto, and completed her postdoctoral fellowship at Cornell University. Kathleen has published multiple research papers, chapters, and reviews, and has written three books. Kathleen is the Fulbright Canada Research Chair of Global Food Security and has been a visiting professor at the University of Toronto over the past year. Her research interests include virus expression vectors, food security agricultural biotechnology, and global health. Kathleen lives in New York with her husband and two children